Edible Electronics for Smart Technology Solutions

Shilpa Mehta
Auckland University of Technology, New Zealand

Fadi Al-Turjman
AI and Robotics Institute, Near East University, Nicosia, Turkey & Faculty of Engineering, University of Kyrenia, Turkey

Published in the United States of America by
IGI Global
701 E. Chocolate Avenue
Hershey PA, USA 17033
Tel: 717-533-8845
Fax: 717-533-8661
E-mail: cust@igi-global.com
Web site: https://www.igi-global.com

Copyright © 2025 by IGI Global. All rights reserved. No part of this publication may be reproduced, stored or distributed in any form or by any means, electronic or mechanical, including photocopying, without written permission from the publisher.
Product or company names used in this set are for identification purposes only. Inclusion of the names of the products or companies does not indicate a claim of ownership by IGI Global of the trademark or registered trademark.

Library of Congress Cataloging-in-Publication Data

CIP Data Pending
ISBN: 979-8-3693-5573-2
eISBN: 979-8-3693-5575-6

Vice President of Editorial: Melissa Wagner
Managing Editor of Acquisitions: Mikaela Felty
Managing Editor of Book Development: Jocelynn Hessler
Production Manager: Mike Brehm
Cover Design: Phillip Shickler

British Cataloguing in Publication Data
A Cataloguing in Publication record for this book is available from the British Library.

All work contributed to this book is new, previously-unpublished material.
The views expressed in this book are those of the authors, but not necessarily of the publisher.

Table of Contents

Preface xxii

Chapter 1
Edible Electronics for Smart Technology Solutions: Edible Electronics for
Food Packaging 1
    *Babitha Hemanth, Sahyadri College of Engineering and Management,
        India*
    *Meghashree Meghashree, Sahyadri College of Engineering and
        Management, India*
    , Sahyadri College of Engineering and Management, India
    *Rutesh R. Kalungade, Sahyadri College of Engineering and
        Management, India*

Chapter 2
Edible Electronics for Food Quality Monitoring 23
    Ushaa Eswaran, Anna University, Chennai, India
    Vivek Eswaran, Medallia, India
    Keerthna Murali, Dell, India
    Vishal Eswaran, CVS Health, India
    *E. Kannan, Vel Tech Rangarajan Dr. Sagunthala R&D Institute of
        Science and Technology, India*

Chapter 3
Enhancing Food Quality Monitoring Processes Through Innovative
Microcontroller-Based Systems 49
    Saurabh Chandra, Bennett University, Greater Noida, India
    *Shreenidhi K. S., Department of Biotechnology, Rajalakshmi
        Engineering College, Chennai, India*
    *Harishchander Anandaram, Department of Artificial Intelligence,
        Amrita Vishwa Vidyapeetham, Coimbatore, India*
    *Ketki P. Kshirsagar, Department of Electronics and Telecommunication,
        Vishwakarma Institute of Technology, Pune, India*
    *N. R. Rajagopalan, Department of Chemistry, St. Joseph's College of
        Engineering, OMR, Chennai, India*

Chapter 4
Smart Food Quality Monitoring by Integrating IoT and Deep Learning for Enhanced Safety and Freshness 79

 Kavitha Kumari K. S., *Department of Electrical and Electronic Engineering, Aarupadai Veedu Institute of Technology, Vinayaka Mission's Research Foundation (Deemed), India*

 J. Samson Isaac, *School of Engineering and Technology, Karunya Institute of technology and Sciences, Coimbatore, India*

 V. G. Pratheep, *Department of Mechatronics Engineering, Kongu Engineering College, Perundurai, India*

 M. Jasmin, *Department of Electronics and Communication Engineering, New Prince Shri Bhavani College of Engineering and Technology, Chennai, India*

 A. Kistan, *Department of Chemistry, Panimalar Engineering College, Chennai, India*

 Sampath Boopathi, *Department of Mechanical Engineering, Muthayammal Engineering College, Namakkal, India*

Chapter 5
Edible Energy Harvesting: Powering the Future of Smart Food Technology ... 111

 Dipan Kumar Das, *Centurion University of Technology and Management, India*

 Padmaja Patnaik, *Centurion University of Technology and Management, India*

 Sudip Kumar Das, *Dr. C.V. Raman University, India*

 Mandakini Barala, *Centurion University of Technology and Management, India*

 Nibedita Nayak, *Centurion University of Technology and Management, India*

Chapter 6
Edible Electronics for Smart Detection of Food Spoilage 149

 S. N. Kumar, *Amal Jyothi College of Engineering, India*

 Nikki John Kannampilly, *Amal Jyothi College of Engineering, India*

 Jomin Joy, *Amal Jyothi College of Engineering, India*

Chapter 7
Advancements in Edible Electronics and Robotics for Smart Food Packaging
Processes 167
    Sudhakara Rao J., Department of Life Science (Microbiology and
        Food Science & Technology), GITAM University (Deemed),
        Visakhapatnam, India
    Gitanjali Behera, Department of Life Science (Microbiology and
        Food Science & Technology), GITAM University (Deemed),
        Visakhapatnam, India
    Shreenidhi K. S., Department of Biotechnology, Rajalakshmi
        Engineering College, Chennai, India
    Harishchander Anandaram, Department of Artificial Intelligence,
        Amrita Vishwa Vidyapeetham, Coimbatore, India
    G. Durai Muthu Mani, Department of Biochemistry, SRM Arts and
        Science College, Kattankulathur, India

Chapter 8
From Ingestible to Edible: Quantifying and Analyzing Edible Electronics'
Role in Environment Monitoring and Future Advancement............................. 197
    Snehasis Dey, College of Engineering Bhubaneswar, BPUT University,
        India

Chapter 9
Edible Electronics' Role in Healthcare: Application, Challenges, and Future
Research 217
    Snehasis Dey, College of Engineering Bhubaneswar, India
    Kadambini Himanshu, College of Engineering Bhubaneswar, India

Chapter 10
Integrating Internet of Things (IoT) With Edible Electronics 237
    Himadri Sekhar Das, Haldia Institute of Technology, India
    Subir Maity, KIIT University, India

Chapter 11
Recurrent Neural Network Technology in Smart Hydroponics Control
Systems 265
    Shanthalakshmi Revathy J., Velammal College of Engineering and
        Technology, India
    J. Mangaiyarkkarasi, NMSS Vellaichamy Nadar College, India
    Shilpa Mehta, Auckland University of Technology, New Zealand

Chapter 12
The Internet of Things, Automation in the Future, and Security Protocols for the Smart Grid Framework 281
    Ramiz Salama, Near East University, Turkey
    Fadi Al-Turjman, Near East University, Turkey

Chapter 13
Smart Grid Environment, Data Security in the Internet of Things, and Supply Chain Ecosystem Transformation 305
    Ramiz Salama, Near East University, Turkey
    Fadi Al-Turjman, Near East University, Turkey

Chapter 14
Mobile Cloud Computing and the Internet of Things Security and Privacy 333
    Ramiz Salama, Near East University, Turkey
    Fadi Al-Turjman, Near East University, Turkey

Chapter 15
AI-Controlled Robotics in Smart Agricultural Systems: Enhancing Precision, Sustainability, and Productivity 351
    Dayana D. S., Department of Networking and Communications, School of Computing, SRM Institute of Science and Technology, India
    T. Venkatamuni, Department of Mechanical Engineering, VSB Engineering College, Karur, India
    A. Bhagyalakshmi, Department of Computer Science and Engineering, Vel Tech Rangarajan Dr. Sagunthala R&D Institute of Science and Technology, Chennai, India
    TYJ Naga Malleswari, Department of Networking and Communications, SRM Institute of Science and Technology, Kattankulathur, India
    S. Ushasukhanya, Department of Networking and Communications, School of Computing, SRM Institute of Science and Technology, Kattankulathur, India

Chapter 16
AI-Driven Disaster Forecasting by Integrating Smart Technology 383
> Periasamy J. K., Department of Computer Science and Engineering, Sri Sai Ram Engineering College, Chennai, India
> Kunduru Srinivasulu Reddy, Department of Mechanical Engineering, Sreenidhi Institute of Science and Technology, Hyderabad, India
> Prachi Rajendra Salve, Department of Computer Engineering, D.Y. Patil School of Engineering Academy, Pune, India
> S. Ushasukhanya, School of Computing, SRM Institute of Science and Technology, Kattankulathur, India
> T. Y. J. Naga Malleswari, Networking and Communications, SRM Institute of Science and Technology, Kattankulathur, India

Chapter 17
Deep Learning for Predictive Analytics in Environmental and Social Sciences 415
> Senthilkumar Thangavel, Cloud Solutions & Machine Learning Expert, USA

Chapter 18
Machine Vision System: An Emerging Technique for Quality Estimation of Tea.. 445
> Debangana Das, Silicon Institute of Technology, Bhubaneswar, India
> Shreya Nag, University of Engineering and Management, Kolkata, India
> Runu Banerjee Roy, Jadavpur University, India

Compilation of References 477

About the Contributors 519

Index 525

Detailed Table of Contents

Preface xxii

Chapter 1
Edible Electronics for Smart Technology Solutions: Edible Electronics for
Food Packaging 1
    *Babitha Hemanth, Sahyadri College of Engineering and Management,
        India*
    *Meghashree Meghashree, Sahyadri College of Engineering and
        Management, India*
    , Sahyadri College of Engineering and Management, India
    *Rutesh R. Kalungade, Sahyadri College of Engineering and
        Management, India*

The advent of edible electronics marks a groundbreaking advancement in the field of food packaging, ushering in a new era of sustainability, functionality, and consumer engagement. This abstract explores the transformative potential of edible electronics and its implications for the food industry. Edible electronics represent a paradigm shift where traditional packaging materials are augmented with electronic components, rendering them not only biodegradable but also functional beyond containment. By seamlessly integrating sensors, actuators, and other electronic elements into edible substrates, such as films or coatings, these innovative packaging solutions offer unparalleled benefits. One of the primary advantages of edible electronics is their ability to extend the shelf life of perishable goods through real-time monitoring of environmental conditions. This ensures optimal storage conditions and reduces food waste, addressing a critical challenge in the global food supply chain.

Chapter 2
Edible Electronics for Food Quality Monitoring 23
    Ushaa Eswaran, Anna University, Chennai, India
    Vivek Eswaran, Medallia, India
    Keerthna Murali, Dell, India
    Vishal Eswaran, CVS Health, India
    E. Kannan, Vel Tech Rangarajan Dr. Sagunthala R&D Institute of
        Science and Technology, India

This chapter explores the innovative applications of edible electronics for food quality monitoring, focusing on the development of edible tags and sensors embedded in food packaging materials. Edible electronics offer unique opportunities to enhance food safety and quality assurance by providing real-time monitoring of various parameters such as temperature, humidity, freshness, and contamination. The chapter discusses the design principles, fabrication techniques, and deployment strategies for edible electronic tags and sensors. It explores the integration of Internet of Things (IoT) technologies with edible electronics to enable seamless data transmission and analysis. Furthermore, the chapter examines the potential impact of artificial intelligence on enhancing the capabilities of edible electronics for food quality monitoring. Case studies and practical examples demonstrate the feasibility and effectiveness of using edible electronics in ensuring food safety and quality throughout the supply chain.

Chapter 3
Enhancing Food Quality Monitoring Processes Through Innovative
Microcontroller-Based Systems 49
    Saurabh Chandra, Bennett University, Greater Noida, India
    Shreenidhi K. S., Department of Biotechnology, Rajalakshmi
        Engineering College, Chennai, India
    Harishchander Anandaram, Department of Artificial Intelligence,
        Amrita Vishwa Vidyapeetham, Coimbatore, India
    Ketki P. Kshirsagar, Department of Electronics and Telecommunication,
        Vishwakarma Institute of Technology, Pune, India
    N. R. Rajagopalan, Department of Chemistry, St. Joseph's College of
        Engineering, OMR, Chennai, India

Advanced microcontroller technology is making food quality monitoring processes more accurate and efficient. The chapter describes the state of-the-art microcontrollers for enhancing the safety and quality of food by enabling real-time monitoring and data analysis. The study investigates sensors built into microcontrollers for the measurement of critical parameters in maintaining the quality of food, namely temperature, humidity, and pH. The chapter has highlighted a number of applications of advanced microcontrollers for tracking and controlling food storage facilities' conditions, as well as detecting spoilage and safety standards. It provided case studies on successful implementation and advantages of such technologies: higher accuracy, lower risk of human error, and cost-effectiveness. This chapter takes readers through state-of-the-art advances, providing useful insights to researchers and practitioners who aim at leveraging microcontroller innovations in the enablement of superior food quality monitoring processes with safer food supply chains.

Chapter 4
Smart Food Quality Monitoring by Integrating IoT and Deep Learning for
Enhanced Safety and Freshness 79
> Kavitha Kumari K. S., Department of Electrical and Electronic
>   Engineering, Aarupadai Veedu Institute of Technology, Vinayaka
>   Mission's Research Foundation (Deemed), India
> J. Samson Isaac, School of Engineering and Technology, Karunya
>   Institute of technology and Sciences, Coimbatore, India
> V. G. Pratheep, Department of Mechatronics Engineering, Kongu
>   Engineering College, Perundurai, India
> M. Jasmin, Department of Electronics and Communication Engineering,
>   New Prince Shri Bhavani College of Engineering and Technology,
>   Chennai, India
> A. Kistan, Department of Chemistry, Panimalar Engineering College,
>   Chennai, India
> Sampath Boopathi, Department of Mechanical Engineering,
>   Muthayammal Engineering College, Namakkal, India

This chapter investigates the integration aspects between the Internet of Things and deep learning technologies in efforts directed toward advancing food quality monitoring, thus enhancing issues of safety and food freshness in the supply chain. IoT sensors capture real-time data with regard to environmental conditions including temperature, humidity, and gas composition all through the food production and supply process. Such data is analyzed by deep learning algorithms to detect any kind of anomalies and predict the potential hazards to enable proactive actions toward timely intervention. The chapter deals some IoT devices used for monitoring, namely smart sensors and wearable technologies, while comparing the application of deep learning models for predictive and pattern recognition analytics. Case studies underscore this integrated approach to reducing spoilage, increasing shelf life, and meeting the requirements put forth by today's safety standards.

Chapter 5
Edible Energy Harvesting: Powering the Future of Smart Food Technology ... 111
    Dipan Kumar Das, Centurion University of Technology and
        Management, India
    Padmaja Patnaik, Centurion University of Technology and Management,
        India
    Sudip Kumar Das, Dr. C.V. Raman University, India
    Mandakini Barala, Centurion University of Technology and
        Management, India
    Nibedita Nayak, Centurion University of Technology and Management,
        India

Edible energy harvesting merges food science, biotechnology, and engineering to convert chemical energy from safe, ingestible substances into electrical power. This chapter explores the evolution of energy harvesting technologies, focusing on advancements in materials science and biocompatible electronics. It discusses the significance of edible energy harvesting in medical devices, sustainable technologies, and IoT applications. Safety and ethical considerations, along with recent innovations in fabrication techniques and regulatory frameworks, are analyzed. The chapter concludes with prospects for integrating energy harvesting components into food and addressing challenges for broader adoption in healthcare and environmental sustainability. Ethical and environmental considerations, including safety and biocompatibility, are crucial in advancing these technologies. The chapter also discusses challenges such as regulatory compliance and environmental impact mitigation, highlighting the pathway for future developments in edible energy harvesting.

Chapter 6
Edible Electronics for Smart Detection of Food Spoilage 149
    S. N. Kumar, Amal Jyothi College of Engineering, India
    Nikki John Kannampilly, Amal Jyothi College of Engineering, India
    Jomin Joy, Amal Jyothi College of Engineering, India

This chapter describes an innovative approaches which employs edible electronics sensors to monitor food in real time to prevent spoilage. In today scenario, food spoilage detection plays a vital role, since many packaged food items are there in the market. Edible sensors role is inevitable in food industry for spoilage detection of liqud and solid foods. This chapter discuss about various edible electronics sensors for food spoilage detection. Biosensors are widely used in many applications, the sensor values are processed by an IoT based system comprising of a microcontroller and internet connectivity enables the transfer of data through cloud network. Food spoilage detection enables the buyers to predict the shelf life and minimizes the wastage of food items.

Chapter 7
Advancements in Edible Electronics and Robotics for Smart Food Packaging Processes 167
    Sudhakara Rao J., Department of Life Science (Microbiology and Food Science & Technology), GITAM University (Deemed), Visakhapatnam, India
    Gitanjali Behera, Department of Life Science (Microbiology and Food Science & Technology), GITAM University (Deemed), Visakhapatnam, India
    Shreenidhi K. S., Department of Biotechnology, Rajalakshmi Engineering College, Chennai, India
    Harishchander Anandaram, Department of Artificial Intelligence, Amrita Vishwa Vidyapeetham, Coimbatore, India
    G. Durai Muthu Mani, Department of Biochemistry, SRM Arts and Science College, Kattankulathur, India

This chapter explores recent advancements in edible electronics and robotics for smart food packaging, focusing on sustainability, food safety, and consumer interaction. Edible electronics, designed for human consumption, offer new applications in food quality monitoring, including real-time detection of spoilage, monitoring nutritional content, and product authentication, thereby enhancing food safety Robotics and other technologies will enable innovative functionalities like dynamic packaging responses, automated freshness indicators, and active interventions to extend shelf life. The chapter delves into the intricacies of edible sensors and robotic systems, examining their materials, technologies, design, regulatory, safety, and ethical implications. The study explores the potential of IoT in packaging for supply chain management, personalized nutrition, and waste reduction, highlighting its integration into digital ecosystems, presenting a shift in food distribution, safety, and sustainability.

Chapter 8
From Ingestible to Edible: Quantifying and Analyzing Edible Electronics'
Role in Environment Monitoring and Future Advancement............................. 197
    Snehasis Dey, College of Engineering Bhubaneswar, BPUT University, India

Edible electronics symbolize a departure from the traditional concept of ingestible electronics, characterized by devices that are not only tailored for ingestion but also completely disintegrate in the body and can be safely released into the environment without the need for retrieval. In recent years, there has been a swift progression in the field of Green Electronics, marked by a significant rise in the number of research groups showcasing remarkable achievements in the realm of edible and biodegradable electronics. On this context this chapter caters to the study of edible electronics, its physical characterization and their role in environment monitoring. Thus, researchers are putting hard time out to discover the development of electronic components from food waste which controls the wastage and creation of proper sustainable approach towards edible electronics which also monitors the environment. This chapter dives into the edible electronics role in environment monitoring as well as its future advancement and current research.

Chapter 9
Edible Electronics' Role in Healthcare: Application, Challenges, and Future
Research 217
    Snehasis Dey, College of Engineering Bhubaneswar, India
    Kadambini Himanshu, College of Engineering Bhubaneswar, India

Edible electronics have emerged as a prominent technology within the healthcare industry over the past decade. The advancement of edible electronics signifies a notable progression in the realm of technology due to their composition of natural or synthetic food-based materials. These devices can be easily digested, absorbed, and processed by the human body to carry out their intended functions. Within edible electronics are a variety of components including resistors, conductors, transistors, capacitors, batteries, antennas, sensors, inductors, conductive binders' substrates, and dielectrics. The fundamental structure serves as the foundation of electronic devices, encompassing functionally active and passive elements, circuits, sensors, power supplies, and communication mechanisms. In many instances, edible materials showcase inherent electronic or ionic conductivity, or both.

Chapter 10
Integrating Internet of Things (IoT) With Edible Electronics 237
    Himadri Sekhar Das, Haldia Institute of Technology, India
    Subir Maity, KIIT University, India

The unification of Internet of Things (IoT) accompanying edible electronic devices marks a meaningful advancement in two together science and healthcare. This synergy integrates the relatedness and data processing wherewithal of IoT accompanying the innovative potential of edible electronics, offering unprecedented opportunities for monitoring and improving human health. By implanting IoT sensors and actuators within succulent substrates, in the way that biocompatible materials or ingestible sensors, a smooth and non-obtrusive monitoring whole is worked out. This abstract explores the arising flows, challenges, and potential applications of integrating IoT with edible electronics. From embodied cure to real-occasion well-being monitoring, this unification holds promise for transforming healthcare delivery and improving the feature of life. However, righteous concerns concerning privacy and dossier protection must be addressed to guarantee the responsible deployment concerning this transformational technology.

Chapter 11
Recurrent Neural Network Technology in Smart Hydroponics Control
Systems 265
    Shanthalakshmi Revathy J., Velammal College of Engineering and
        Technology, India
    J. Mangaiyarkkarasi, NMSS Vellaichamy Nadar College, India
    Shilpa Mehta, Auckland University of Technology, New Zealand

Agriculture is vital for global economies, providing food and raw materials. Efficient water use and land productivity are essential. Deep learning supports hydroponics, growing plants without soil. Long Short-Term Memory (LSTM) networks in an online system predict and optimize plant growth by analyzing humidity, temperature, and pH for irrigation. This tech-driven indoor farming reduces costs and labor while being independent of sunlight. Using IoT, a microcontroller kit with wireless sensors monitors and allows remote tracking of plant growth. This integration enhances sustainability and productivity, conserving resources and reducing manual labor. It meets the demand for higher-quality food while supporting economic development. This study presents a novel approach to predicting hydroponic system quality using LSTM networks enhanced with attention mechanisms. The model achieved an F1 Score of 94.86%, accuracy of 94.94%, precision of 97.72%, and specificity of 92.52%. These results highlight the model's effectiveness in precision agriculture.

Chapter 12
The Internet of Things, Automation in the Future, and Security Protocols for
the Smart Grid Framework 281
    Ramiz Salama, Near East University, Turkey
    Fadi Al-Turjman, Near East University, Turkey

The smart grid's mix of automation, the Internet of Things (IoT), and robust security measures is reshaping the landscape of energy management. This abstract provides a quick summary of the key components and implications of this innovative approach. Automation, monitoring, and management of energy resources in real time are made possible by the smart grid framework, which increases overall efficiency and system responsiveness. A network of linked sensors and equipment can transmit and analyze data more readily by using the Internet of Things, which makes it possible to provide predictive maintenance, demand forecasting, and a strong energy infrastructure. However, the interconnection of the smart grid necessitates stringent security measures to prevent assaults.

Chapter 13
Smart Grid Environment, Data Security in the Internet of Things, and Supply
Chain Ecosystem Transformation 305
    Ramiz Salama, Near East University, Turkey
    Fadi Al-Turjman, Near East University, Turkey

The supply chain ecosystem is undergoing a significant shift in today's business environment, mostly due to the adoption of smart grids and the integration of cutting-edge technologies like the Internet of Things (IoT). Unprecedented prospects for improved efficiency, real-time monitoring, and data-driven decision-making are presented by this paradigm change. But even in the middle of these developments, data security becomes an increasingly important issue. This study examines the various facets of the change of the supply chain ecosystem in the context of the Internet of Things and the smart grid, with a particular emphasis on the difficulties and solutions pertaining to data security. The smart grid and supply chain operations' integration of IoT devices create several weak points that are open to cyberattacks, underscoring the necessity of putting strong security measures in place as soon as possible.

Chapter 14
Mobile Cloud Computing and the Internet of Things Security and Privacy 333
    Ramiz Salama, Near East University, Turkey
    Fadi Al-Turjman, Near East University, Turkey

Security and privacy are critical factors in the rapidly emerging fields of mobile cloud computing and the Internet of Things (IoT). Because mobile devices are so widely used and IoT devices are becoming ingrained in many aspects of our life, it is increasingly imperative to protect sensitive data and respect user privacy. This abstract looks at the challenges, solutions, and technology surrounding security and privacy in the context of mobile cloud computing and the Internet of Things. Data protection at all stages of life, from processing and transmission to storage and retrieval, is one of the core concerns in this field. Data encryption techniques are crucial for shielding confidential information from unauthorized access or interception. Access control systems control user permissions and stop unauthorized access to resources, while robust identity management approaches guarantee the identities of people and devices.

Chapter 15
AI-Controlled Robotics in Smart Agricultural Systems: Enhancing Precision, Sustainability, and Productivity 351
    Dayana D. S., Department of Networking and Communications, School of Computing, SRM Institute of Science and Technology, India
    T. Venkatamuni, Department of Mechanical Engineering, VSB Engineering College, Karur, India
    A. Bhagyalakshmi, Department of Computer Science and Engineering, Vel Tech Rangarajan Dr. Sagunthala R&D Institute of Science and Technology, Chennai, India
    TYJ Naga Malleswari, Department of Networking and Communications, SRM Institute of Science and Technology, Kattankulathur, India
    S. Ushasukhanya, Department of Networking and Communications, School of Computing, SRM Institute of Science and Technology, Kattankulathur, India

AI-controlled robotics in smart agriculture systems have revolutionized farming practices by improving precision, sustainability, and productivity, marking a significant milestone in modern farming. AI allows real-time monitoring and decision-making through advanced machine learning algorithms, sensors, and autonomous systems to optimize resources like water, fertilizers, and pesticides. AI-based technologies are revolutionizing precision agriculture, reducing waste and environmental degradation while increasing yield and quality. Robotics is automating labor-intensive tasks like planting, harvesting, and weeding not only for efficiency but also to reduce human intervention. AI enables predictive analytics in disease detection and weather forecasting, providing farmers with actionable inputs at their doorstep. The chapter delves into the potential of AI-controlled robots in agriculture, highlighting their potential to improve food security, mitigate environmental harm, and foster sustainable farming practices.

Chapter 16
AI-Driven Disaster Forecasting by Integrating Smart Technology 383
    Periasamy J. K., Department of Computer Science and Engineering, Sri
        Sai Ram Engineering College, Chennai, India
    Kunduru Srinivasulu Reddy, Department of Mechanical Engineering,
        Sreenidhi Institute of Science and Technology, Hyderabad, India
    Prachi Rajendra Salve, Department of Computer Engineering, D.Y.
        Patil School of Engineering Academy, Pune, India
    S. Ushasukhanya, School of Computing, SRM Institute of Science and
        Technology, Kattankulathur, India
    T. Y. J. Naga Malleswari, Networking and Communications, SRM
        Institute of Science and Technology, Kattankulathur, India

This chapter explores how AI and smart technologies could altogether be integrated to bring revolutionary change in the fields of disaster forecasting and management. It will try to analyze through advanced algorithms and IoT sensors how these technologies can potentially advance a disaster-related prediction along with accuracy and timeliness. Important applications of real-time data collection, predictive modeling, and automated alerts collectively enhance response strategies as well as resource allocation. This chapter's discussion of the promise of AI merged with smart technologies—improved predictiveness, faster response times, and better risk assessment—perhaps weighs the potential liabilities and limitations of such applications, including data privacy issues and infrastructures sturdy enough to host such a system. This chapter draws on case studies and continuing research into the use of AI-driven systems in disasters to present insights about how they are changing practices in disaster management and outline future directions for the emerging field.

Chapter 17
Deep Learning for Predictive Analytics in Environmental and Social
Sciences 415
*Senthilkumar Thangavel, Cloud Solutions & Machine Learning Expert,
   USA*

Deep learning is currently one of the transformative technologies for predictive analytics, with deep learning providing robust methodologies to analyse complex data in both the environmental and social sciences. This chapter is concerned with looking into the use of some deep learning techniques, such as CNNs and RNNs, in modelling and predicting environmental phenomena and social trends. Deep learning techniques fit into the environmental sciences through the massive processing of remote sensing and observational data in climate modelling, disaster prediction, and ecologic monitoring. In the social sciences, it helps with sentiment analysis, behavioural predictions, and trend forecasting from large-scale social media and survey data. Key challenges include issues related to data quality and model interpretability; Highlighted case studies provide evidence of the effectiveness of deep learning in these domains.

Chapter 18
Machine Vision System: An Emerging Technique for Quality Estimation of
Tea.. 445

Debangana Das, Silicon Institute of Technology, Bhubaneswar, India
Shreya Nag, University of Engineering and Management, Kolkata, India
Runu Banerjee Roy, Jadavpur University, India

The electronic sensing platforms measure the aroma, the taste, and color profiles, respectively. Color of tea samples is measured by using colorimeter and the process is called colorimetry. Flavanols, which are the most important compounds present in tea are responsible for the dark color of tea. The higher the flavavols like catechin, epigallocatechin-gallate, theaflavin, thearubigin, etc, the better is the quality of tea. It is feasible to monitor the image data which has been captured by a camera using digital detection and analysis with an electronic eye instrument. In essence, the use of electronic eye in the field of quality estimation of tea is an emerging area and is of utmost interest to the tea-traders. Integration of the three sensory platforms, i.e., electronic nose, electronic tongue and electronic eye, has been used for quality estimation of the beverage.Combining a feature-level fusion method with pattern recognition correlations between sensory qualities and metabolic profiles can greatly increase the effectiveness and accuracy of prediction models.

Compilation of References 477

About the Contributors 519

Index 525

Preface

In recent years, the field of edible electronics (EE) has rapidly evolved, bringing together an intriguing intersection of technology and food sciences. With this exciting development, researchers have begun embedding electronic components in edible materials, opening up a plethora of possibilities for advancements in diverse industries. This emerging field draws upon various disciplines, including biomedicine, electronics, and neurology, to create innovative and nutritive electronics. Our goal in compiling *Edible Electronics for Smart Technology Solutions* is to highlight the profound impact that EE can have on creating an intelligent, interconnected world, offering solutions in health monitoring, food safety, sustainable energy, and beyond.

The potential benefits of edible electronics extend far beyond mere novelty. These technologies promise improved dining experiences, more sustainable approaches to energy, and enhanced food handling practices, all of which contribute to a broader vision of sustainability and technological progression. By fostering interconnectivity between industries and sectors, EE has the potential to address some of the most pressing challenges of our time.

Despite the exciting advancements, there are still significant challenges to overcome in this field. From developing suitable materials to refining components and integrating them into functional systems, much work remains to be done. This book delves into these challenges and provides a comprehensive exploration of the materials, devices, and technologies that are being developed to meet the growing demands of this industry.

One of the primary objectives of this book is to identify and promote emerging materials and innovative solutions for energy harvesting within the realm of EE. The latest solutions in energy harvesting promise to be highly efficient, cost-effective, and compact—essential qualities that meet the needs of modern industries. We believe that by offering an in-depth exploration of these developments, this book will serve as a valuable resource for expanding theoretical knowledge and practical understanding in this field.

Moreover, the methods and technologies discussed in this book are intended to be adaptable for researchers and professionals alike. We encourage readers to apply the concepts and methods presented here to their own research and professional endeavors, contributing to the continued growth and innovation in the field of edible electronics.

Edible Electronics for Smart Technology Solutions is designed as a comprehensive guide to this interdisciplinary field, covering materials, technologies, and solutions that can drive progress across multiple sectors. Whether you are a researcher, student, academician, or professional, we hope this book serves as an essential resource to further your understanding and involvement in the exciting world of edible electronics.

ORGANIZATION OF THE BOOK

Chapter 1: Edible Electronics for Smart Technology Solutions: Edible Electronics for Food Packaging

This chapter delves into the transformative role of edible electronics in food packaging, showcasing a new frontier in sustainability and consumer engagement. By incorporating biodegradable electronic components like sensors and actuators into food packaging materials, edible electronics can monitor real-time conditions, extending the shelf life of perishable goods and reducing food waste. The potential impact on the global food supply chain is immense, as these advanced packaging solutions ensure optimal storage conditions, making them a significant innovation for future food safety and sustainability.

Chapter 2: Edible Electronics for Food Quality Monitoring

Chapter 2 explores the innovative application of edible electronics in food quality monitoring, specifically focusing on the integration of sensors and tags within packaging materials. The chapter discusses the development of these edible devices to track vital food safety parameters, such as temperature, humidity, and contamination levels. Additionally, the potential of IoT technologies and artificial intelligence to enhance data analysis and ensure food quality throughout the supply chain is thoroughly examined, with case studies illustrating their real-world applications and efficacy.

Chapter 3: Innovative Microcontrollers for Enhancing Food Quality Monitoring Processes

This chapter highlights the critical role of advanced microcontrollers in food quality monitoring processes. By leveraging microcontrollers with built-in sensors, food safety and quality can be more accurately and efficiently controlled. Real-time monitoring of key parameters like temperature, humidity, and pH is enabled, enhancing food storage and reducing spoilage risks. The chapter provides case studies demonstrating the successful implementation of these technologies, emphasizing their advantages in cost-effectiveness and precision in food safety management.

Chapter 4: Smart Food Quality Monitoring by Integrating IoT and Deep Learning for Enhanced Safety and Freshness

Chapter 4 investigates the intersection of IoT and deep learning technologies in food quality monitoring, aimed at enhancing food safety and freshness across the supply chain. IoT sensors capture critical data on environmental conditions, which is then analyzed by deep learning algorithms to identify anomalies and predict potential risks. The chapter highlights various IoT devices used for monitoring, the predictive power of deep learning models, and case studies demonstrating their effectiveness in reducing spoilage and extending shelf life.

Chapter 5: Edible Energy Harvesting: Powering the Future of Smart Food Technology

This chapter explores the emerging field of edible energy harvesting, which combines food science and engineering to convert ingestible substances into electrical power. The chapter discusses its relevance in powering medical devices, sustainable technologies, and IoT applications. Key advancements in material science, biocompatibility, and energy harvesting techniques are covered, along with challenges related to safety, regulatory frameworks, and environmental sustainability. The future potential of this technology, particularly in healthcare and environmental monitoring, is also considered.

Chapter 6: Edible Electronics for Smart Detection of Food Spoilage

Edible electronics are revolutionizing the detection of food spoilage, offering a real-time solution to monitor both liquid and solid foods. This chapter focuses on the development of edible sensors that work in tandem with IoT systems to transmit data about food freshness and spoilage. By enabling real-time monitoring, these sensors predict shelf life and reduce food waste. The chapter explores the role of biosensors and the integration of IoT networks, providing insights into how this technology can enhance consumer safety and sustainability.

Chapter 7: Integrating Edible Electronics and Robotics in Smart Food Packaging Solutions

This chapter reviews cutting-edge developments in edible electronics and robotics, focusing on their applications in smart food packaging. It highlights how these technologies can monitor food freshness, detect spoilage, and even authenticate products in real time. The integration of robotics facilitates automated interventions like dynamic packaging responses and automated freshness indicators. The chapter addresses the technical, regulatory, and ethical considerations essential for the broader adoption of these innovative packaging solutions.

Chapter 8: From Ingestible to Edible: Quantifying and Analyzing Edible Electronics' Role in Environment Monitoring and Future Advancement

Chapter 8 shifts the focus to edible electronics as an environmentally sustainable technology. It examines the evolution from ingestible to biodegradable electronics that disintegrate after serving their purpose. The chapter discusses their applications in environmental monitoring, emphasizing the role of food waste-derived electronic components in reducing environmental impact. Through this lens, it also examines ongoing research and future advancements that could position edible electronics as a sustainable solution for environmental challenges.

Chapter 9: Edible Electronics in Healthcare: Applications, Challenges, and Future Research

This chapter explores the growing role of edible electronics in healthcare. These devices, composed of natural or synthetic food-based materials, are designed to be digested and absorbed by the human body while performing critical functions like monitoring health metrics. The chapter outlines the various components of edible electronics, including sensors, batteries, and communication mechanisms, while also addressing challenges related to safety, biocompatibility, and future research directions in this rapidly advancing field.

Chapter 10: Integrating the Internet of Things (IoT) with Edible Electronics

Chapter 10 presents the integration of IoT with edible electronics, a technological breakthrough that combines data connectivity with innovative edible devices. By embedding IoT sensors in edible substrates, a seamless and non-intrusive monitoring system is achieved, opening new possibilities for real-time health monitoring. The chapter explores emerging trends, challenges, and potential applications in healthcare, discussing the ethical concerns around data privacy and security, which are critical for the responsible deployment of this technology.

Chapter 11: Recurrent Neural Network Technology in Smart Hydroponics Control Systems

This chapter examines the application of recurrent neural networks (RNN) in optimizing hydroponic farming systems. It explores how Long Short-Term Memory (LSTM) networks analyze environmental data, such as humidity, temperature, and pH, to optimize plant growth and resource usage. IoT-enabled systems with wireless sensors and predictive models enhance sustainability and productivity in agriculture by automating irrigation and monitoring processes. The chapter showcases a model with high accuracy and precision, highlighting the potential for revolutionizing hydroponic farming.

Chapter 12: The Internet of Things, Automation in the Future, and Security Protocols for the Smart Grid Framework

Chapter 12 discusses how the integration of IoT and automation within the smart grid framework improves energy management. IoT-enabled sensors provide real-time data, enabling predictive maintenance and enhancing energy infrastructure efficiency. However, the increased interconnectivity also raises security concerns, necessitating stringent protocols to protect against cyberattacks. This chapter addresses the challenges of balancing automation with the need for robust security measures to ensure a safe and resilient smart grid.

Chapter 13: Smart Grid Environment, Data Security in the Internet of Things, and Supply Chain Ecosystem Transformation

The transformative impact of smart grid technology on the supply chain ecosystem is explored in this chapter. By integrating IoT devices into the grid, real-time data monitoring and decision-making capabilities are enhanced, leading to greater efficiency in supply chain operations. However, this increased connectivity also presents vulnerabilities in data security. The chapter analyzes the key challenges and solutions for securing IoT data in the smart grid and supply chain environments, emphasizing the need for proactive security measures.

Chapter 14: Mobile Cloud Computing and the Internet of Things: Security and Privacy Challenges

Chapter 14 focuses on the security and privacy challenges posed by the growing intersection of mobile cloud computing and IoT. As these technologies become ubiquitous, the need for robust data protection measures is critical. The chapter discusses encryption techniques, access control systems, and identity management protocols to safeguard sensitive information across various stages of data processing, transmission, and storage. It also highlights best practices for ensuring user privacy in this increasingly connected ecosystem.

Chapter 15: AI-Controlled Robotics in Smart Agricultural Systems: Enhancing Precision, Sustainability, and Productivity

This chapter examines how AI-controlled robotics are transforming agricultural practices by enhancing precision, sustainability, and productivity. By using machine learning algorithms and autonomous systems, AI optimizes resource use—such as water and fertilizers—while reducing environmental impact. Robotics further automate tasks like planting, harvesting, and weeding, improving efficiency and reducing labor costs. The chapter highlights the potential of AI-driven robotics to advance food security and foster more sustainable farming methods.

Chapter 16: AI-Driven Disaster Forecasting by Integrating Smart Technology

The final chapter explores the integration of AI and smart technology in disaster forecasting and management. By combining AI algorithms with real-time data from smart devices, it is possible to predict natural disasters with greater accuracy and take proactive measures to mitigate their impact. The chapter discusses the potential for AI-driven systems to revolutionize disaster management by enabling faster, more precise forecasting and response, offering new hope for reducing the human and economic toll of disasters.

Chapter 17: Deep Learning for Predictive Analytics in Environmental and Social Sciences

Deep learning has become a pivotal tool in predictive analytics, particularly within the environmental and social sciences. This chapter investigates the use of deep learning models, such as Convolutional Neural Networks (CNNs) and Recurrent Neural Networks (RNNs), for predictive tasks across these fields. In environmental science, deep learning facilitates climate modeling, disaster prediction, and ecological monitoring by processing large volumes of remote sensing and observational data. In the social sciences, it enables sentiment analysis, behavioral predictions, and trend forecasting by analyzing social media and survey data. The chapter highlights challenges such as data quality and model interpretability, offering case studies to demonstrate the effectiveness of deep learning in both domains.

Chapter 18: Machine Vision System: An Emerging Technique for Quality Estimation of Tea

This chapter discusses the innovative use of machine vision systems for assessing tea quality. It explains how electronic sensing platforms, including colorimeters, electronic noses, and electronic tongues, work together to measure the aroma, taste, and color profiles of tea. The quality of tea is often linked to flavanols like catechin and epigallocatechin-gallate, which influence the beverage's color. Machine vision systems, such as electronic eyes, capture image data to monitor these qualities digitally. The chapter explores how integrating these sensory platforms using feature-level fusion and pattern recognition can improve the accuracy and effectiveness of tea quality assessments, making this technique highly beneficial for tea traders.

In Conclusion

As we conclude *Edible Electronics for Smart Technology Solutions*, it is clear that we are standing at the forefront of a technological revolution that merges the digital world with the physical realm of food. Edible electronics, once a novel concept, is now a rapidly expanding field with applications that stretch across numerous industries. This book has aimed to provide a comprehensive understanding of the interdisciplinary nature of EE, from the materials and technologies involved to the real-world challenges and potential solutions.

The innovations explored within these chapters demonstrate the transformative potential of EE in areas such as health monitoring, food safety, and sustainable energy. However, the journey of discovery and development is far from complete. As researchers, students, and professionals engage with these new technologies, we are confident that the methods and strategies presented will spark further exploration, innovation, and practical applications in this exciting field.

We envision a future where edible electronics will not only enhance our daily lives but also contribute significantly to global sustainability efforts. The advancement of these smart technologies will allow for more efficient, cost-effective, and environmentally friendly solutions that address the current demands of industries worldwide.

In closing, we hope this book serves as both an inspiration and a practical guide for all those venturing into the world of edible electronics. The work ahead is challenging, but with continued dedication and innovation, the possibilities are boundless.

Chapter 1
Edible Electronics for Smart Technology Solutions:
Edible Electronics for Food Packaging

Babitha Hemanth
https://orcid.org/0009-0006-0685-3962
Sahyadri College of Engineering and Management, India

Meghashree Meghashree
Sahyadri College of Engineering and Management, India

Sahyadri College of Engineering and Management, India

Rutesh R. Kalungade
https://orcid.org/0009-0001-9860-6275
Sahyadri College of Engineering and Management, India

ABSTRACT

The advent of edible electronics marks a groundbreaking advancement in the field of food packaging, ushering in a new era of sustainability, functionality, and consumer engagement. This abstract explores the transformative potential of edible electronics and its implications for the food industry. Edible electronics represent a paradigm shift where traditional packaging materials are augmented with electronic components, rendering them not only biodegradable but also functional beyond containment. By seamlessly integrating sensors, actuators, and other electronic elements into edible

DOI: 10.4018/979-8-3693-5573-2.ch001

substrates, such as films or coatings, these innovative packaging solutions offer unparalleled benefits. One of the primary advantages of edible electronics is their ability to extend the shelf life of perishable goods through real-time monitoring of environmental conditions. This ensures optimal storage conditions and reduces food waste, addressing a critical challenge in the global food supply chain.

1. INTRODUCTION

The convergence of technology and sustainability has birthed a remarkable innovation in the realm of food packaging: edible electronics. Imagine a world where the very materials that wrap our food not only safeguard its freshness but also provide invaluable insights into its quality and safety. Edible electronics represent a paradigm shift in packaging design, where electronic components are seamlessly integrated into materials that are not just biodegradable but also consumable. This introduction delves into the essence of edible electronics, exploring what they are and the materials that drive their functionality.

Edible electronics, at their core, are electronic devices or components that are safe for consumption. Unlike conventional electronics, which rely on non-biodegradable materials like plastics and metals, edible electronics utilize organic or edible materials as substrates. These materials not only ensure the biocompatibility of the electronics but also enable them to be consumed without posing any harm to human health.

In essence, edible electronics represent a fusion of advanced technology and sustainable design principles, offering a novel approach to food packaging that is both functional and eco-friendly. By harnessing the potential of edible materials, these electronics pave the way for a future where packaging not only preserves food but also enriches the consumer experience while minimizing environmental impact.

Moreover, edible electronics offer a sustainable alternative to conventional electronic devices, which often contribute to environmental pollution and resource depletion. By leveraging renewable and biodegradable materials, such as cellulose, chitosan, and pectin, edible electronics not only reduce reliance on non-renewable resources but also mitigate the environmental impact of electronic waste.

One of the key advantages of edible electronics is their compatibility with biological systems, enabling seamless integration with the human body for applications such as medical diagnostics, targeted drug delivery, and monitoring of physiological parameters. For instance, ingestible sensors can provide real-time data on digestive processes, while edible drug delivery systems offer a non-invasive means of administering medication. Edible electronics hold significant potential for environmental monitoring and food safety, allowing for the detection of contaminants and pathogens in food and water sources. By harnessing the natural biodegradability of organic materials, these devices offer a sustainable solution for monitoring and improving public health.

In addition to their environmental and health benefits, edible electronics present exciting opportunities for personalized nutrition, interactive consumer experiences, and innovative culinary creations. From edible sensors that detect allergens or pathogens in food to interactive packaging that provides nutritional information or entertainment, the integration of electronics into edible materials transcends traditional boundaries, reshaping the future of food technology.

Material used in Edible Electronics

The materials used in edible electronics encompass a diverse array of organic compounds and food-grade substances. Edible polymers, such as gelatine, starches, and cellulose derivatives, form the foundation of these materials, providing both structural integrity and biocompatibility. These polymers are often combined with edible conductive materials, such as edible metals or conductive polymers, to facilitate the flow of electricity within the device.

Furthermore, edible electronics may incorporate encapsulation materials to protect the electronic components from moisture, enzymes, and other environmental factors encountered during ingestion. These encapsulates are typically derived from edible sources, such as proteins or polysaccharides, and are engineered to degrade harmlessly within the digestive system. Introduction:

The utilization of recycled food waste (Milan R. Radovanovic, Goran M. Stojanovic, Mitar Simic, Dragana Suvara, Lazar Milic, Sanja Kojic, and Biljana D. Skrbic, 2024) for the fabrication of edible electronic components represents a pioneering approach towards sustainable technology. Repurposing food waste for electronic applications. Edible electronics, characterized by their ability to be consumed safely, offer a promising avenue for reducing both electronic waste and food waste simultaneously. By ingeniously repurposing food by-products and waste streams, such as fruit peels, vegetable scraps, and expired products, into edible substrates and conductive materials, this research endeavours to not only mitigate environmental impact but also create value from otherwise discarded resources. This introduction sets the stage for a ground-breaking investigation into the fusion of sustainability and technology in the realm of edible electronics. These materials are processed and transformed into edible substrates and conductive elements suitable for electronic applications. Additionally, the paper may discuss the incorporation of encapsulation materials derived from food-grade sources to protect the electronic components from environmental factors. Overall, the utilization of recycled food waste as a raw material for edible electronics not only offers a sustainable solution for reducing food waste but also contributes to the development of eco-friendly technologies with potential applications in food packaging, sensing, and healthcare.

"Food-Based Edible and Nutritive Electronics" (Wenwen Xu, Haokai Yang, Wei Zeng, Todd Houghton, Xu Wang, Raghavendra Murthy, Hoejin Kim, Yirong Lin, Marc Mignolet, Huigao Duan, Hongbin Yu, Marvin Slepian, and Hanqing Jiang, 2017) a groundbreaking approach to electronic materials derived from natural foods and foodstuffs with minimal inorganic components. These materials are designed to be ingested and assimilated as metabolized nutrients, enabling the development of electronic devices that can function within the gastrointestinal tract. By creating edible electronic components, the research aims to revolutionize biomedical technologies by introducing electronics that can be consumed and utilized within the body. This innovative concept expands the possibilities of electronic devices beyond traditional non-degradable systems to edible and nutritive systems, opening up new avenues for healthcare monitoring and therapeutic applications.

It includes a variety of food-derived substances that are safe for consumption. These may encompass natural polymers such as gelatine, alginate, chitosan, or pectin, which are commonly found in foods and can be processed into biocompatible substrates for electronic device

2. Applications of Edible Electronics in Food Packaging

Edible electronics in food packaging present a transformative solution for enhancing food safety, quality assurance, and consumer engagement. These innovative technologies offer real-time monitoring capabilities, extending beyond traditional packaging to actively assess and preserve food freshness and integrity.

One application lies in the integration of edible sensors within packaging materials to monitor key parameters such as temperature, humidity, and gas composition. By providing continuous data on environmental conditions, these sensors enable early detection of spoilage or contamination, thereby reducing food waste and enhancing product safety.

Furthermore, edible electronics facilitate the implementation of smart packaging systems capable of interactive communication with consumers. QR codes or NFC tags embedded in edible packaging can provide access to detailed information about product origin, ingredients, and nutritional content, empowering consumers to make informed choices.

Another promising application is the incorporation of antimicrobial agents or preservatives into edible packaging materials, offering a natural and sustainable means of extending shelf life and maintaining food quality. These active packaging solutions can help mitigate microbial growth and oxidation, preserving the freshness and flavour of perishable goods.

Traditional packaging (Luk Jun Lam Iversen, Kobun Rovina, Joseph Merillyn Vonnie, Patricia Matanjun, Kana Husna Erna, Nasir Md Nur Aqilah, Wen Xia Ling Felicia, and Andree Alexander Funk, 2022) has been and continues to be widely used across numerous goods and industries, especially in the food industry. Packaging has a significant impact on consumers, particularly when purchasing food. When designing packaging, color, shape, material, and size are prioritized over other product characteristics, influencing how consumers perceive a product. The conventional packaging materials are paperboards, polyethylene films, glass jars, and metal cans. In contrast, eco-friendly and sustainable packaging has largely replaced conventional packaging over the past decade. Consumers are becoming more aware of the side effects of conventional packaging. The original intent of food packaging was to protect the contents from outside elements and lengthen the life of perishable goods. Traditional methods for determining whether food is still fresh, such as tasting and smelling it, have become obsolete due to packaging innovation. Overall, edible electronics revolutionize food packaging by offering advanced functionalities such as real-time monitoring, interactive communication, and active preservation. By enhancing food safety, quality, and transparency, these technologies contribute to a more sustainable and consumer-centric approach to food packaging and distribution.

The diverse applications of edible polymers (Monireh Kouhi, Molamma P. Prabhakaran, Seeram Ramakrishna, 2020) in various industries, including food, biomedical, cosmetics, energy, and water treatments. In the food industry, edible polymers in the form of particles and films are utilized for food production, preservation, and packaging, offering protection against spoilage and pathogenic microorganisms. In biomedical fields, edible polymers are employed in smart drug delivery, wound dressing, tissue engineering, and medical devices. Additionally, edible polymers find applications in cosmetics, energy production, and water/wastewater treatments. The figure showcases the versatility and wide-ranging potential of edible polymers in addressing different needs across multiple sectors, highlighting their importance in creating sustainable and innovative solutions.

Real-Time Monitoring of Food Freshness

Edible electronics facilitate real-time monitoring of food freshness, revolutionizing food safety and quality assurance. These electronic devices, often made from biocompatible materials, integrate seamlessly into food packaging, or can even be ingested safely. Equipped with sensors, they detect key indicators of freshness such as temperature, humidity, pH levels, and gas concentrations. For instance, they can identify spoilage by detecting gases emitted as food decays. The data collected by these devices is transmitted wirelessly to Smartphone's or other devices, providing consumers and producers with instant feedback on the food's condition. By enabling timely interventions, such as adjusting storage conditions or removing spoiled items from shelves, edible electronics help reduce food waste and mitigate health risks. This innovation holds promise for enhancing food supply chains, ensuring consumers receive fresher, safer products while also promoting sustainability.

Anti-Counterfeiting Measures

Edible electronics offer innovative anti-counterfeiting measures for food packaging, bolstering consumer safety and brand integrity. These electronics incorporate unique identifiers, such as RFID tags or printed electronic codes, directly onto packaging materials or within edible coatings. These identifiers are difficult to replicate, serving as digital fingerprints for authentication.

Furthermore, edible electronics can include tamper-evident features, like sensors that detect breaches in packaging integrity or changes in temperature indicative of tampering. This ensures consumers receive products in their original, unhampered state, instilling trust in the brand.

Additionally, these electronics enable real-time tracking throughout the supply chain, allowing manufacturers and regulators to monitor product movements and detect any irregularities. Data collected from these tracking systems can be securely stored and accessed, facilitating traceability and recall procedures if necessary.

Moreover, edible electronics can interact with block chain technology, creating an immutable record of each product's journey from production to consumption. This transparent ledger enhances transparency and accountability, making it exceedingly difficult for counterfeiters to infiltrate the supply chain undetected.

By combining these measures, edible electronics provide comprehensive anti-counterfeiting solutions, safeguarding both consumers and brands in the complex landscape of food packaging and distribution.

Interactive Packaging

Interactive packaging, leveraging technologies like QR codes and NFC (Near Field Communication), transforms passive product packaging into engaging experiences for consumers. QR codes, when scanned with a smartphone, provide instant access to product information, promotions, recipes, or even interactive games. NFC-enabled packaging allows users to tap their Smartphone's against the packaging to access similar content or perform tasks like reordering products or participating in loyalty programs.

These technologies enhance consumer engagement by providing valuable and personalized experiences, fostering brand loyalty and driving sales. Additionally, they offer opportunities for brands to gather valuable data on consumer behavior and preferences, enabling targeted marketing strategies and product improvements. Interactive packaging not only enriches the consumer experience but also opens up new avenues for brands to connect with their audience in innovative ways, ultimately strengthening brand-consumer relationships and driving business growth.

3. Benefits of Edible Electronics in Food Packaging

The world of food packaging is seeing a full change. Traditional methods, which rely largely on non-biodegradable plastics and are becoming increasingly unsustainable, are being replaced by creative solutions that encourage sustainability and improve the whole food experience. Edible electronics, a breakthrough technology, is at the forefront of this transition.

This section looks at the benefits of adding edible electronics into food packaging. The solution goes beyond simply eliminating waste, providing a complete revolution of food packaging. Consider intelligent packaging that actively monitors food quality, increasing shelf life and reducing spoiling. The edible layer can be used at interfaces as an interior heterogeneous component to segregate different meal portions (Harnkarnsujarit, 2017).

Consumer convenience is reaching new heights, with features such as time-temperature signals and self-dissolving packaging for easier preparation. Perhaps most crucially, edible electronics have the potential to transform food safety by including biosensors that detect foodborne viruses, thereby protecting consumer health.

The technological features of edible electronics, as well as their practical consequences for each benefit, will be investigated. According to Danyxa Piñeros-Hernandez et al. (2017), including agents into food packaging can increase shelf life, safety, and quality. We'll break down the environmental benefits, demonstrating how edible electronics can reduce waste and promote a more sustainable food system. Finally, we'll look at the possibilities for enhanced customer involvement, specifically how this technology could turn food packaging into a source of important information and even personalized experiences.

Improved Food Safety

One of the most significant benefits of incorporating edible electronics into food packaging is the enhancement of food safety measures. Traditional methods of ensuring food safety rely on periodic inspections, expiration dates, and visual inspections of packaging integrity. However, these methods can sometimes fall short in detecting potential hazards or ensuring the freshness of perishable products. Edible electronics offer a proactive and real-time solution to address these challenges.

Incorporating edible electronics into food packaging is a proactive step towards enhancing food safety. Edible sensors continuously monitor temperature, humidity, and gas composition, allowing for early detection of potential spoilage or contamination. This enables timelier measures to reduce risks. Smart labels or tags embedded with edible sensors give clear traceability across the supply chain by documenting the origin, production, and handling of foods. This strategy maintains the freshness and quality of perishable goods while enhancing consumer confidence. Compliance with regulatory standards has improved, leading to a safer food supply chain. Edible electronics also allow for real-time monitoring of certain contaminants or pathogens, which can aid in the prevention of foodborne illnesses. Freshness is maintained by utilizing sensor data to ensure optimal storage conditions.

Overall, edible electronics revolutionize food safety measures with their proactive monitoring capabilities. They enable swift responses to potential hazards, minimizing risks to public health. This technology contributes to reducing food waste by extending shelf life and ensuring optimal storage conditions. By preventing exposure to adverse environmental conditions, edible electronics reduce the risk of spoilage and contamination. With the advancement of technology, edible electronics hold immense promise for enhancing food safety standards globally.

Reduced Environmental Impact

Incorporating edible electronics into food packaging offers a sustainable solution to reduce environmental impact across the food supply chain. These biodegradable electronics, crafted from safe materials, break down harmlessly over time, mitigating the challenges associated with electronic waste disposal. By integrating electronic functionalities directly into food packaging, the need for separate components and additional materials is minimized, contributing to overall waste reduction efforts.

Moreover, edible electronics often utilize low-power components, optimizing energy consumption and embracing renewable energy sources for operation. This approach aligns with principles of the circular economy, promoting resource efficiency and minimizing carbon footprint. Through the adoption of edible electronics, companies can inspire sustainable practices within the food industry, driving consumer awareness and demand for eco-friendly packaging solutions. The reduced environmental impact of edible electronics extends throughout the product lifecycle, from production to disposal, offering a holistic approach to sustainability. By prioritizing these innovative technologies, the food industry can not only enhance safety and functionality but also foster a more environmentally conscious approach to packaging practices, contributing to a healthier planet for future generations.

Enhanced Consumer Engagement

The incorporation of edible electronics into food packaging transforms passive packaging into interactive platforms, significantly increasing customer involvement. These revolutionary technologies, which include sensors, displays, and interactive elements, provide personalized experiences based on individual tastes and dietary requirements. Smart labels embedded with edible sensors provide real-time information on freshness, quality, and nutritional content, allowing customers to make more educated decisions. By scanning a smart label with a smartphone or other device, users can receive personalized content such as recipe suggestions, nutritional recommendations, or allergies alerts.

Furthermore, gamification elements and entertainment features enhance packaging, making it more engaging and enjoyable. For example, interactive packaging may include QR codes or augmented reality experiences that unlock games, puzzles, or promotional offers. These interactive elements not only entertain consumers but also create memorable experiences that differentiate the brand in a competitive market. Edible electronics also serve as educational tools, highlighting the brand's commitment to quality, safety, and sustainability. Novel packaging designs and interactive features create excitement, driving consumer engagement and brand differentiation. Additionally, social sharing and connectivity are facilitated, allowing consumers to share their experiences on social media platforms, amplifying brand reach and fostering community engagement. Overall, the integration of edible electronics enhances consumer interaction with food packaging, creating memorable experiences and strengthening brand loyalty in a competitive market.

4. Challenges and Limitations

The future of food packaging could be better than we imagined. Edible electronics woven directly into the packaging offer a glimpse into a future in which sensors monitor freshness, indicators change color in response to temperature differences, and even minuscule circuits communicate with our telephones. This futuristic vision promises to improve food safety, reduce waste, and provide consumers with a more engaging experience. However, before edible electronics can become a reality, a number of crucial obstacles must be resolved.

This section discusses the three primary issues that are now impeding this technical approach. Understanding these limitations is critical for successfully incorporating edible electronics into food packaging. These challenges are not intended to interfere with progress, but rather to need imaginative solutions and a deep understanding of the complexities of material science, regulatory frameworks, and cost-efficiency. Let's take a look at three major sectors that require new approaches.

Safety Concerns and Regulatory Hurdles

The integration of edible electronics into food packaging introduces a host of safety concerns and regulatory challenges that must be carefully addressed to ensure consumer protection and regulatory compliance. One primary concern revolves around the safety of the materials used in edible electronics. While these materials are intended to be safe for consumption, ensuring their biocompatibility and non-toxicity is paramount to prevent any adverse health effects when ingested inadvertently.

The complex food safety rules complicate matters even more. Obtaining permission from competent agencies demands the development of certain standards, especially for edible electronics. Demonstrating the overall safety profile of these novel materials is crucial. Regulatory bodies must be convinced that these devices pose no health risks and will not jeopardize existing food safety safeguards. This requires a thorough data package that details the materials' biocompatibility, potential interactions with various food types, and the absence of any harmful byproducts from the manufacturing process.

Another key problem is to ensure the dependability and durability of edible electronics in food packaging. These devices must survive the rigors of food preparation, storage, transit, and handling while maintaining performance and safety. Furthermore, concerns about the migration of electronic components or pollutants from the packaging into the food matrix must be thoroughly investigated to avoid any potential health problems. To thoroughly investigate the safety and efficacy of edible electronics, rigorous testing methodologies, risk assessments, and regulatory frameworks must be devised. Furthermore, open communication with consumers regarding the safety and benefits of these technologies is critical for fostering confidence and adoption in the marketplace.

Shelf Life of Edible Electronics

One of the most difficult aspects of incorporating edible electronics into food packaging is determining how long these smart devices will last. Unlike standard electronics, which are meant to run for long periods of time, edible electronics must deal with the perishability of food products as well as the harsh environmental conditions faced throughout the supply chain. One important aspect is the durability of the materials used in edible electronics. These materials must be able to tolerate moisture, temperature variations, and other external variables while maintaining functioning and safety. However, maintaining the stability of edible electronics over a lengthy period poses substantial engineering hurdles, particularly when considering the biodegradability of these devices.

Additionally, the shelf life of edible electronics is closely tied to the shelf life of the food products they accompany. As perishable goods, food items are subject to spoilage and degradation over time, which can impact the performance and reliability of embedded electronic components. Ensuring that edible electronics remain functional throughout the anticipated shelf life of the packaged food requires careful consideration of storage conditions, packaging materials, and device design.

Another factor to consider is the long-term deterioration of edible electronics. While these gadgets are intended to decay naturally, ensuring their performance beyond their stated shelf life is a big task. To achieve maximum performance, careful engineering and testing must be conducted to balance the rate of degradation with the intended shelf life of packaged food goods.

Cost Considerations

Despite their enormous potential to revolutionize food packaging, edible electronics provide a substantial economic barrier that must be carefully addressed. While the concept of intelligent packaging that actively monitors freshness and communicates with customers is attractive, the technology must be commercially feasible before it can be widely adopted. It is critical to strike a balance between material costs, manufacturing complexity, and the potential benefits for consumers and food producers.

One primary cost consideration is the development and production of edible electronic materials. Unlike traditional packaging materials, which are typically inexpensive and widely available, edible electronics require specialized materials that are safe for consumption and biodegradable. Research and development efforts to identify and optimize these materials can be costly, particularly when considering factors such as biocompatibility, stability, and performance.

Furthermore, complex loading processes, incorporation into food packaging, and assembly processes are all part of food electronics manufacturing, increasing manufacturing costs is increased to produce industrially expensive food electronics, the manufacturing process must be scalable, accurate and reliable.

Solving cost problems requires a multi-pronged approach that balances innovation with economies of scale. Additionally, knowledge sharing can be facilitated for companies by fostering collaboration between stakeholders, research institutes and regulatory agencies. Simplification of manufacturing processes, economies of scale, and selection of the best materials can help reduce manufacturing costs, thereby feeding electronics manufacturers and their consumption can drop significantly.

CASE STUDIES

Case Study 1:

Background:

1. Innovative Approach: Tufts University researchers are pioneering silk-based edible electronics for food packaging, offering a fresh perspective on enhancing food safety and sustainability.

2. Real-time Monitoring: These novel materials enable real-time monitoring of food quality and safety parameters like freshness, temperature, and microbial contamination, promising significant advancements in food packaging technology.
    Objective:

    1. Novel Food Packaging Approach: The research project focused on developing an innovative method for food packaging by integrating electronic components into edible materials. This approach aimed to overcome limitations of traditional packaging methods and enhance food safety and freshness.
    2. Edible Sensors for Spoilage Detection: The primary goal was to create edible sensors capable of detecting key indicators of food spoilage, such as gases emitted during the decay process. By integrating these sensors into food packaging, the team aimed to provide real-time monitoring to ensure the freshness and safety of packaged food products.

    Approach:

1. Silk's Unique Properties: Led by Professor Fiorenzo Omenetto, the Tufts University team utilized silk due to its biocompatibility, flexibility, and biodegradability, making it an excellent substrate for embedding electronic components in food packaging.
2. Advanced Fabrication Techniques: The team employed sophisticated fabrication methods to create thin, flexible silk protein films. These films formed the basis for embedding various electronic components, such as sensors and microchips, with the goal of developing edible sensors capable of detecting specific gases indicative of food spoilage.

Results:

1. **Successful Demonstration**: Through extensive experimentation, the Tufts University team showcased the viability of silk-based edible electronics for food packaging.
2. **Prototype Sensors**: They developed sensors capable of detecting spoilage indicators like ammonia and hydrogen peroxide. Integrated into edible silk films, these sensors enable real-time monitoring of food freshness and safety.

Impact:

The integration of electronic components into edible materials, such as silk-based edible electronics, marks a groundbreaking advancement in food packaging technology, promising to revolutionize packaging, storage, and monitoring while enhancing food safety and reducing waste.

Future Directions:

Further research is required to enhance the performance, durability, safety, and scalability of silk-based edible sensors for potential widespread adoption in next-generation food packaging.

Conclusion:

1. Interdisciplinary Innovation: Professor Fiorenzo Omenetto's team at Tufts University has pioneered the convergence of materials science, electronics, and biotechnology to develop silk-based edible electronics for food packaging. This interdisciplinary approach has enabled the creation of edible

sensors capable of improving safety and freshness monitoring in packaged food products.
2. Promise for the Future: The research represents a significant step forward in food packaging technology, offering potential benefits such as reduced food waste, enhanced sustainability, and improved consumer safety. By harnessing the capabilities of silk-based edible electronics, the technology holds promise for revolutionizing the way food products are packaged and monitored in the future.

Case Study 2:

Background:

Led by Dr. Sarah Zhang, researchers at the University of California, Berkeley, are developing biodegradable edible sensors for smart food packaging, aiming to integrate electronic components into edible materials to provide real-time monitoring of food quality and safety, meeting the growing demand for safe and sustainable packaging solutions.

Objective:

1. **Designing Biodegradable Edible Sensors**: The research project focused on developing sensors capable of detecting crucial indicators of food spoilage, including pH levels and microbial activity, with the aim of enhancing food safety.
2. **Integration into Edible Films**: The team aimed to embed these sensors into biodegradable edible films, enabling real-time monitoring of packaged food products' freshness and safety, thereby providing valuable information to both consumers and retailers.

Approach:

UC Berkeley's research team utilized a multi-disciplinary approach, drawing inspiration from natural materials and sustainable manufacturing processes, to develop biodegradable edible sensors for smart food packaging, focusing on plant-based biopolymers like cellulose and starch as substrates for embedding electronic components.

Results:

1. Feasibility Demonstration: Through systematic experimentation, the UC Berkeley research team validated the feasibility of biodegradable edible sensors for smart food packaging, showcasing their potential for enhancing food safety and quality monitoring.
2. Prototype Sensor Development: The team successfully developed prototype sensors capable of accurately detecting changes in pH levels and microbial activity, offering real-time insights into the quality and safety of packaged food products, thus addressing crucial aspects of food spoilage detection.

Impact:

The integration of biodegradable edible sensors, utilizing electronic components within materials sourced from renewable sources, signifies a substantial leap forward in sustainable food packaging technology, addressing critical issues of plastic pollution and food waste.

Future Directions:

Further research and development are crucial to optimize the performance and scalability of biodegradable edible sensors for commercial use, addressing challenges such as sensor stability, biocompatibility, cost-effectiveness, regulatory compliance, and industry adoption.

Conclusion:

1. Advancing Sustainable Food Packaging: Dr. Sarah Zhang's research at UC Berkeley marks a significant advancement in developing biodegradable edible sensors, leveraging sustainable materials and advanced manufacturing techniques to enhance safety and freshness monitoring in food packaging.
2. Addressing Key Challenges: The team's innovative approach shows promise in tackling critical issues like plastic pollution, food waste, and consumer safety concerns, offering potential solutions through biodegradable edible sensors for smart food packaging.

Case Study 3:

Background:

SmartSnack™, a fictional company dedicated to revolutionizing the snack food industry through innovative packaging solutions, embarked on a groundbreaking project to develop smart packaging using edible electronics, with a focus on sustainability and consumer safety.

Objective:

The main goal of the project was to develop innovative snack packaging featuring edible sensors and electronic components. These cutting-edge packages aimed to offer instant updates on snack freshness, quality, and safety, ultimately bolstering consumer trust and curbing food wastage.

Approach:

Leveraging expertise in materials science, nanotechnology, and food engineering, SmartSnack™ spearheaded a collaboration with top researchers to engineer state-of-the-art edible sensors. These sensors were meticulously designed to detect crucial indicators of snack quality, including moisture levels, oxygen exposure, and temperature fluctuations.

Results:

1. Through rigorous research and development efforts, SmartSnack™ achieved a significant milestone by creating prototype smart packaging embedded with advanced edible sensors. These sensors were adept at continuously monitoring critical factors such as moisture levels, oxygen exposure, and temperature fluctuations in real-time.
2. SmartSnack™ revolutionized the snacking industry by seamlessly integrating cutting-edge technology into their packaging. With the ability to interact with the packaging via a smartphone app, consumers gained immediate access to valuable insights regarding snack freshness and quality, empowering them to make informed consumption decisions.

Impact:

SmartSnack™'s introduction of smart packaging marked a watershed moment in the snack food industry, reshaping norms around transparency and consumer involvement. With its ability to offer instant updates on snack freshness and safety, these innovative packages empowered consumers to make educated choices while shopping, fostering a culture of responsibility and minimizing food wastage.

Future Directions:

SmartSnack™ built upon their initial success by further innovating and refining their smart packaging technology. They explored new sensor capabilities, including the detection of allergens and contaminants, to enhance consumer safety and convenience.

Conclusion:

SmartSnack™ revolutionizes food packaging with edible electronics, blending tech innovation, sustainability, and safety. Their smart packaging boosts transparency, cuts food waste, and elevates the consumer journey. Leading in edible electronics, SmartSnack™ shapes the future of snack packaging.

Table 1. Comparison of case studies on edible electronics in food packaging

Criteria	Case Study 1: Tufts University	Case Study 2: UC Berkeley	Case Study 3: SmartSnack™
Research Leader	Professor Fiorenzo Omenetto	Dr. Sarah Zhang	SmartSnack™ in collaboration with top researchers
Objective	Develop silk-based edible electronics for food packaging	Develop biodegradable edible sensors for smart food packaging	Develop smart snack packaging with edible sensors and electronic components
Innovative Approach	Utilizing silk for biocompatible, flexible, and biodegradable sensors	Using plant-based biopolymers like cellulose and starch	Integrating materials science, nanotechnology, and food engineering
Key Features	Real-time monitoring of food quality, freshness, temperature, and microbial contamination	Real-time monitoring of pH levels and microbial activity in packaged food	Real-time monitoring of moisture levels, oxygen exposure, and temperature fluctuations
Materials Used	Silk protein films	Plant-based biopolymers (cellulose and starch)	Advanced edible sensors embedded in snack packaging

continued on following page

Table 1. Continued

Criteria	Case Study 1: Tufts University	Case Study 2: UC Berkeley	Case Study 3: SmartSnack™
Fabrication Techniques	Advanced fabrication methods to create thin, flexible silk films	Sustainable manufacturing processes	Collaboration with researchers to design state-of-the-art edible sensors
Prototype Development	Sensors capable of detecting spoilage indicators like ammonia and hydrogen peroxide	Prototype sensors for detecting changes in pH levels and microbial activity	Prototype smart packaging with sensors and smartphone app integration
Impact	Enhanced food safety, reduced waste, improved sustainability	Addressing plastic pollution, reducing food waste, improving safety	Transparency in snack freshness, consumer empowerment, reducing food waste
Future Directions	Enhance performance, durability, safety, scalability	Optimize performance, scalability, stability, cost-effectiveness, and regulatory compliance	Explore detection of allergens and contaminants for enhanced consumer safety
Conclusion	Interdisciplinary innovation, promise for reduced food waste and improved sustainability	Advancing sustainable packaging, addressing plastic pollution and food waste	Revolutionizing snack packaging, blending tech innovation, sustainability, and safety

Comparison Table of Case Studies on Edible Electronics in Food Packaging

Analysis

Objective Alignment: All three case studies aim to enhance food packaging technology through the integration of edible sensors and electronics. They share a common goal of improving food safety and reducing waste but approach it using different materials and techniques.

Materials and Techniques: Tufts University focuses on silk's unique properties, UC Berkeley utilizes plant-based biopolymers, and SmartSnack™ combines various scientific disciplines to innovate snack packaging. The choice of materials and fabrication methods varies based on the specific goals and challenges of each project.

Real-time Monitoring: Each case study emphasizes real-time monitoring of food quality indicators such as spoilage gases, pH levels, microbial activity, moisture, and temperature. This real-time aspect is crucial for ensuring food safety and freshness.

Impact and Future Directions: All projects have demonstrated successful prototypes with significant potential impacts on food safety, sustainability, and consumer empowerment. Future research directions focus on enhancing sensor performance, scalability, and expanding sensor capabilities.

This comparison table [Table 1] highlights the diverse approaches and innovations in the field of edible electronics for food packaging, showcasing the potential benefits and challenges associated with each project.

Future Outlook

1. Potential Advancements in Edible Electronics Technology:
    - Miniaturization and Enhanced Functionality: Edible electronics are likely to become even smaller and more sophisticated, allowing for seamless integration into various food packaging formats without compromising taste or safety.
    - Biodegadability and Sustainability: Continued research and development efforts may lead to the creation of edible electronic components that are fully biodegradable, further aligning with eco-friendly packaging initiatives.
    - Advanced Sensing Capabilities: Advancements in sensor technology may enable edible electronics to detect a wider range of contaminants, allergens, and nutritional information, providing consumers with even greater transparency and safety assurances.
    - Integration with Smart Devices: There is potential for edible electronics to seamlessly integrate with smart devices and platforms, allowing consumers to access real-time information about their food products and even interact with packaging for personalized experiences.
2. Market Trends and Growth Opportunities:
    - Increased Demand for Food Safety and Transparency: Growing consumer awareness and concerns regarding food safety and transparency are expected to drive the adoption of smart packaging solutions, including edible electronics, particularly in the wake of foodborne illness outbreaks and recalls.
    - Rise of Personalized Nutrition: With an increasing focus on personalized nutrition and dietary preferences, there is an opportunity for edible electronics to provide tailored information and recommendations to consumers, catering to individual needs and preferences.
    - Expanding Applications beyond Food: While the primary focus of edible electronics is currently on food packaging, there is potential for expansion into other industries such as pharmaceuticals, where ingestible sensors could revolutionize drug delivery and patient monitoring.
    - Emerging Markets and Global Expansion: As awareness of the benefits of smart packaging grows, particularly in emerging markets, there are significant opportunities for companies like SmartSnack™ to expand

their presence and capitalize on the growing demand for innovative packaging solutions worldwide.

Predictions for the Future of Edible Electronics in Food Packaging

1. Mainstream Adoption: Edible electronics are poised to become a standard feature in food packaging, with widespread adoption by major food manufacturers and retailers seeking to enhance product safety, traceability, and consumer engagement.
2. Customization and Personalization: The future of edible electronics will likely see an emphasis on customization and personalization, allowing food companies to tailor packaging solutions to individual consumer preferences, dietary restrictions, and nutritional needs.
3. Integration with Blockchain Technology: As blockchain technology gains traction in the food industry for its ability to provide transparent and immutable supply chain data, edible electronics may be integrated with blockchain platforms to provide real-time tracking and verification of food products from farm to fork.
4. Augmented Reality Experiences: With advancements in augmented reality (AR) technology, edible electronics could enable interactive packaging experiences, where consumers can use their smartphones or AR glasses to access additional product information, recipes, or promotional content directly from the packaging.
5. Health Monitoring and Feedback Systems: Edible electronics have the potential to evolve beyond passive sensors to actively monitor and provide feedback on consumer health metrics, such as hydration levels, nutrient intake, and even food sensitivities, offering personalized recommendations for healthier eating habits.
6. Regulatory Considerations and Standards: As edible electronics technology matures, there will be a need for clear regulatory guidelines and standards to ensure the safety, reliability, and efficacy of these innovations, as well as to address concerns regarding privacy and data security.
7. Collaboration and Cross-Industry Partnerships: Collaboration between food manufacturers, technology companies, healthcare providers, and regulatory agencies will be essential for driving innovation and establishing best practices in the development and implementation of edible electronics in food packaging.

CONCLUSION

Edible electronics offer a promising avenue for enhancing food packaging technology, promising improved safety, freshness monitoring, and sustainability. Integrating electronic components into edible materials presents opportunities to provide real-time information on food quality and safety, potentially reducing waste and enhancing consumer confidence. However, challenges such as safety, biocompatibility, regulatory compliance, scalability, cost, and sustainability need addressing for widespread adoption. Collaborative efforts between academia, industry, and regulators are crucial for advancing this technology. Despite hurdles, ongoing research and development efforts continue to drive progress. Edible electronics have the potential to revolutionize food packaging, ushering in a future where packaging not only protects but also actively monitors and enhances the safety and freshness of food products, contributing to a more sustainable and efficient food supply chain.

REFERENCES

Harnkarnsujarit, Nathdanai. "Glass-Transition and Non-Equilibrium States of Edible Films and Barriers." Elsevier eBooks, January 1, 2017. .DOI: 10.1016/B978-0-08-100309-1.00019-5

Kouhi, M., Prabhakaran, M. P., & Ramakrishna, S. (2020). Edible polymers: An insight into its application in food, biomedicine and cosmetics. *Trends in Food Science & Technology*, 103, 248–263. DOI: 10.1016/j.tifs.2020.05.025

Luk, Iversen Jun Lam, Kobun Rovina, Joseph Merillyn Vonnie, Patricia Matanjun, Kana Husna Erna, Nur'Aqilah Nasir Md, Felicia Xia Ling Wen, and Andree Alexander Funk. "The emergence of edible and food-application coatings for food packaging: a review." (2022): 20220435220.

Piñeros-Hernandez, D., Medina-Jaramillo, C., López-Córdoba, A., & Goyanes, S. (2017, February 1). Edible Cassava Starch Films Carrying Rosemary Antioxidant Extracts for Potential Use as Active Food Packaging. *Food Hydrocolloids*, 63, 488–495. Advance online publication. DOI: 10.1016/j.foodhyd.2016.09.034

Radovanović, Milan R., Goran M. Stojanović, Mitar Simić, Dragana Suvara, Lazar Milić, Sanja Kojić, and Biljana D. Škrbić. "Edible Electronic Components Made from Recycled Food Waste." Advanced Electronic Materials: 2300905.

Xu, W., Yang, H., Zeng, W., Houghton, T., Wang, X., Murthy, R., Kim, H., Lin, Y., Mignolet, M., Duan, H., Yu, H., Slepian, M., & Jiang, H. (2017). Food-Based Edible and Nutritive Electronics. *Advanced Materials Technologies*, 2(11), 1700181. DOI: 10.1002/admt.201700181

Chapter 2
Edible Electronics for Food Quality Monitoring

Ushaa Eswaran
Anna University, Chennai, India

Vivek Eswaran
https://orcid.org/0009-0002-7475-2398
Medallia, India

Keerthna Murali
https://orcid.org/0009-0009-1419-4268
Dell, India

Vishal Eswaran
https://orcid.org/0009-0000-2187-3108
CVS Health, India

E. Kannan
Vel Tech Rangarajan Dr. Sagunthala R&D Institute of Science and Technology, India

ABSTRACT

This chapter explores the innovative applications of edible electronics for food quality monitoring, focusing on the development of edible tags and sensors embedded in food packaging materials. Edible electronics offer unique opportunities to enhance food safety and quality assurance by providing real-time monitoring of various parameters such as temperature, humidity, freshness, and contamination. The chapter discusses the design principles, fabrication techniques, and deployment strategies for edible electronic tags and sensors. It explores the integration of Internet of Things (IoT) technologies with edible electronics to enable seamless

DOI: 10.4018/979-8-3693-5573-2.ch002

Copyright © 2025, IGI Global. Copying or distributing in print or electronic forms without written permission of IGI Global is prohibited.

data transmission and analysis. Furthermore, the chapter examines the potential impact of artificial intelligence on enhancing the capabilities of edible electronics for food quality monitoring. Case studies and practical examples demonstrate the feasibility and effectiveness of using edible electronics in ensuring food safety and quality throughout the supply chain.

1. INTRODUCTION TO EDIBLE ELECTRONICS FOR FOOD QUALITY MONITORING

1.1. Overview of Food Safety and Quality Monitoring

Food safety and quality assurance have become paramount concerns in today's globalized food industry. With increasing consumer awareness and stringent regulatory requirements, there is a growing demand for innovative technologies that can monitor and ensure the safety and quality of food products throughout the supply chain. Ensuring that food remains safe and of high quality from production to consumption is essential for protecting public health, maintaining consumer trust, and complying with regulations.

1.2. Specific Challenges in Food Quality Monitoring

Traditional methods of monitoring food quality, such as visual inspections and laboratory testing, can be time-consuming and insufficient in providing real-time data. These conventional approaches often fail to detect issues promptly, leading to potential health risks, food wastage, and economic losses. There is a pressing need for more advanced, efficient, and real-time solutions that can continuously monitor various parameters such as temperature, humidity, freshness, and contamination to ensure food safety and quality.

1.3. Research Gap and Novelty of the Study

Despite advances in food monitoring technologies, a significant gap remains in the ability to seamlessly integrate real-time monitoring solutions into food packaging and the food itself without causing harm. Edible electronics, an emerging field at the intersection of electronics and food science, offers a unique and promising solution to address these challenges. By integrating electronic components and circuits into edible materials, edible electronics provide a novel approach to real-time food quality monitoring. These components are designed to be safe for human consumption and can be ingested or discarded without causing any harm, thereby addressing the limitations of current methods.

1.4. Objectives and Contributions

This chapter delves into the innovative applications of edible electronics for food quality monitoring, focusing on the development of edible tags and sensors embedded in food packaging materials. The objectives of this work are to:

1. Explore the design principles and fabrication techniques for creating effective edible electronic tags and sensors.
2. Investigate the integration of Internet of Things (IoT) technologies with edible electronics to enable seamless data transmission and analysis.
3. Examine the potential impact of artificial intelligence (AI) on enhancing the capabilities of edible electronics for food quality monitoring.
4. Demonstrate the feasibility and effectiveness of using edible electronics through case studies and practical examples.

2. LITERATURE REVIEW

The concept of edible electronics has gained significant attention in recent years, with researchers exploring its potential applications in various fields, including food quality monitoring. This literature review aims to provide an overview of the existing research and studies related to edible electronics for food quality monitoring, highlighting the current state of knowledge and identifying gaps or areas that require further exploration.

2.1. Introduction to Edible Electronics

The concept of edible electronics dates back to the early 2000s when researchers began exploring the possibility of integrating electronic components into edible materials. One of the pioneering works in this field was conducted by researchers at Princeton University, who developed edible circuits using food-grade materials like gelatin and gold nanoparticles (Sharova, Alina S., et al., 2021). This groundbreaking study paved the way for further research into edible electronics and its potential applications. Despite this early progress, the field remains in its nascent stage, with many foundational challenges yet to be addressed.

2.2. Materials and Fabrication Techniques

A crucial aspect of edible electronics research focuses on the selection and development of suitable materials and fabrication techniques. Researchers have explored various conductive materials for edible electronics, including carbon-based inks, metallic nanoparticle-based inks, and electrolytic solutions (Baranov, Denis G., et al., 2017). These materials are typically deposited or printed onto edible substrates such as paper, biopolymers, or edible films using techniques like screen printing, inkjet printing, or aerosol jet printing.

However, the scalability and consistency of these techniques remain a concern. While initial studies demonstrate the feasibility of using these materials and methods, further work is needed to ensure reproducibility and cost-effectiveness in large-scale production. Additionally, the long-term biocompatibility and safety of some materials, particularly those involving nanoparticles, require more comprehensive evaluation.

2.3. Applications in Food Quality Monitoring

Edible electronics hold promise for monitoring various parameters related to food quality and safety. Below are key applications explored in existing literature:

- Temperature Monitoring: Temperature is critical for maintaining the quality and shelf life of many food products. Zou et al. (2024) developed edible temperature sensors integrated into food packaging. However, the response time and accuracy of these sensors in real-world conditions require further validation.
- Humidity and Moisture Monitoring: Excessive humidity can lead to spoilage. Lu et al. (2024) proposed an edible humidity sensor based on gelatin and carbon nanotubes. Despite promising results, issues such as sensor stability and long-term reliability under varying environmental conditions need to be addressed.
- Freshness and Spoilage Detection: Edible sensors detecting chemical markers of spoilage, such as volatile organic compounds, show potential. Sun et al. (2022) developed sensors using silk and gold nanoparticles. Yet, their sensitivity and selectivity in detecting specific spoilage markers remain inconsistent across different food types.
- Contamination and Allergen Detection: Yin et al. (2019) created sensors for detecting peanut allergens. Although effective in controlled settings, their real-world application in diverse and complex food matrices presents significant challenges.

- Gas Monitoring: Monitoring gases like carbon dioxide and ethylene, which affect ripening and spoilage, has been explored by Tripathi et al. (2024). While these sensors offer a novel approach to maintaining food quality, their integration into existing packaging systems without affecting the food product is complex.

2.4. Integration with IoT and Communication Technologies

To realize the full potential of edible electronics for food quality monitoring, researchers have explored integrating these technologies with IoT and wireless communication systems. Various protocols such as RFID, BLE, and LPWAN have been investigated (Anand et al., 2024). However, the power consumption, data transmission reliability, and cost of these systems in large-scale deployment remain significant hurdles. Additionally, the integration of these systems into current food supply chains poses logistical challenges.

2.5. Artificial Intelligence for Data Analysis and Decision-Making

AI techniques can enhance the capabilities of edible electronics for food quality monitoring. Machine learning and deep learning algorithms have been used for tasks such as predictive modeling and anomaly detection (Eni et al., 2024). While these techniques show promise, the quality and quantity of data from edible sensors often limit their effectiveness. Moreover, the complexity of AI models necessitates robust preprocessing and feature engineering, which are not trivial in the context of food monitoring.

2.6. Challenges and Future Directions

Despite significant progress, several challenges remain in the field of edible electronics for food quality monitoring:

- Scalability and Cost-effectiveness: Achieving scalable and cost-effective production of edible electronics is critical. Kapoor et al. (2024) explored optimizing material sourcing and manufacturing processes, but more work is needed to bring costs down while maintaining quality.
- Standardization and Regulatory Frameworks: Establishing standardized protocols and regulatory frameworks is essential for ensuring consistent quality and safety. Rout et al. (2022) highlighted the lack of comprehensive regulatory guidelines, which hampers widespread adoption.

- Data Security and Privacy: Edible electronics generate sensitive data that must be protected. Abid et al. (2024) discussed the need for robust security measures, but practical implementations are still lacking.
- Environmental Sustainability: The environmental impact of edible electronics, from material sourcing to disposal, must be considered. Shen et al. (2023) emphasized the need for sustainable practices, but current solutions are not yet fully eco-friendly.

Table 1. Overview of edible electronics applications in food quality monitoring

	Aspect	Summary
1	Introduction to Edible Electronics	The concept of integrating electronic components into edible materials originated in the early 2000s, with pioneering research conducted at Princeton University.
2	Materials and Fabrication Techniques	Research focuses on selecting and developing conductive materials and fabrication techniques suitable for edible electronics. Various materials and printing methods are explored.
3	Applications in Food Quality Monitoring	Edible electronics are utilized for monitoring parameters such as temperature, humidity, freshness, contamination, and gas levels in food products. Sensor development and applications are discussed.
4	Integration with IoT and Communication Technologies	Researchers investigate integrating edible electronics with IoT and wireless communication systems for data transmission to cloud-based platforms. Various wireless protocols are explored.
5	Artificial Intelligence for Data Analysis and Decision-Making	Integration of AI techniques with edible electronics enhances data analysis capabilities for tasks such as predictive modeling and supply chain optimization. Machine learning algorithms are applied.
6	Challenges and Future Directions	Challenges include scalability, cost-effectiveness, standardization, regulatory frameworks, data security, privacy, and environmental sustainability. Future research directions are outlined.

This literature review highlights the significant progress made in the field of edible electronics for food quality monitoring, covering various aspects such as materials, fabrication techniques, applications, integration with IoT and AI, and challenges. However, it also underscores the need for further research and development to address the remaining challenges and unlock the full potential of this innovative technology in ensuring food safety and quality throughout the supply chain.

3. METHODOLOGY

The development and implementation of edible electronic tags and sensors for food quality monitoring involve a multidisciplinary approach, combining principles from electronics, materials science, food science, and engineering. This section outlines the methodology employed in the research and development of these innovative devices, including materials selection, fabrication techniques, sensor design, IoT integration, and artificial intelligence implementation.

3.1. Materials Selection and Characterization

The selection of suitable materials is crucial for the fabrication of edible electronic devices. The materials must not only exhibit the desired electrical and mechanical properties but also meet the criteria for edibility and biocompatibility. The following steps are typically involved in materials selection and characterization:

- Screening and evaluation of potential edible materials: Researchers evaluate various edible materials, such as conductive inks, edible substrates, and electrolytes, based on their electrical conductivity, mechanical strength, and biocompatibility.
- Characterization techniques: The selected materials undergo rigorous characterization processes to assess their properties. This may include electrical characterization (e.g., conductivity measurements, impedance spectroscopy), chemical analysis (e.g., FTIR, XPS, EDX), and physical analysis (e.g., SEM, AFM, mechanical testing).
- Optimization and modification: Based on the characterization results, researchers may optimize or modify the materials to enhance their performance or compatibility with the fabrication processes.

3.2. Fabrication Techniques

The fabrication of edible electronic tags and sensors involves various techniques, depending on the materials and device architecture. Common fabrication methods include:

- Printing and deposition techniques: Conductive inks and materials can be deposited or printed onto edible substrates using techniques such as screen printing, inkjet printing, or aerosol jet printing. These techniques allow for precise patterning and deposition of materials.
- Encapsulation and packaging: To protect the electronic components and ensure their stability, edible encapsulation materials or packaging films are employed. This may involve techniques like dip-coating, spray-coating, or lamination processes.
- Integration of electronic components: Additional electronic components, such as sensors, antennas, or energy storage devices, may be integrated into the edible electronic devices using specialized assembly techniques.

3.3. Sensor Design and Development

Edible electronic sensors are designed to monitor various parameters related to food quality and safety, such as temperature, humidity, freshness, gas levels, and contamination. The sensor development process typically involves the following steps:

- Sensor type selection: Researchers select the appropriate sensor type based on the target parameter to be monitored, considering factors like sensitivity, selectivity, and response time.
- Transduction mechanism and working principle: The sensor's transduction mechanism and working principle are established, which governs how the target parameter is converted into a measurable electrical signal.
- Sensor fabrication and optimization: The sensor is fabricated using the selected materials and techniques, followed by optimization processes to enhance its performance, such as sensitivity tuning, calibration, and environmental testing.
- Integration and packaging: The sensor is integrated into the overall edible electronic device and packaged with appropriate encapsulation materials to protect it from environmental factors.

3.4. Integration with IoT and Communication Systems

To enable real-time monitoring and data transmission, edible electronic devices are integrated with Internet of Things (IoT) and wireless communication systems. The methodology involves:

- Selection of wireless communication protocol: Researchers evaluate and select the most suitable wireless communication protocol (e.g., RFID, BLE, LPWAN) based on factors like data rate requirements, power consumption, range, and scalability.
- Development of IoT gateways and edge devices: IoT gateways or edge devices are developed to interface with the edible electronic sensors, collect and preprocess data, and transmit it to cloud-based platforms or local servers.
- Integration with cloud platforms and data management systems: The collected data is transmitted to cloud-based platforms or local servers for secure storage, analysis, and visualization. These platforms facilitate data sharing and collaboration among stakeholders in the supply chain.

3.5. Artificial Intelligence Integration

Artificial intelligence (AI) techniques, particularly machine learning and deep learning algorithms, can enhance the capabilities of edible electronics for food quality monitoring. The methodology for AI integration involves:

- Data preprocessing and feature engineering: The collected sensor data undergoes preprocessing steps, such as cleaning, normalization, and feature extraction, to prepare it for AI model training.
- Model selection and training: Researchers select appropriate machine learning or deep learning models based on the task at hand (e.g., predictive modeling, anomaly detection, classification). These models are trained using the preprocessed sensor data.
- Model validation and performance evaluation: The trained AI models are validated using appropriate evaluation metrics and techniques, such as cross-validation, to assess their performance and generalization capabilities.
- Datasets and algorithms: Specific datasets relevant to food quality parameters (e.g., temperature, humidity, freshness) are utilized for model training and validation. Algorithms such as decision trees, support vector machines, neural networks, and ensemble methods are considered for different applications.
- Evaluation metrics: Common evaluation metrics such as accuracy, precision, recall, F1 score, and ROC-AUC are used to assess the performance of the AI models.

3.6. Testing and Evaluation

The developed edible electronic devices undergo rigorous testing and evaluation processes to assess their performance, reliability, and real-world applicability. The methodology includes:

- Laboratory-based testing and simulations: Initial testing is conducted in controlled laboratory environments, simulating various conditions and scenarios to evaluate the device's functionality, sensitivity, and robustness.
- Real-world pilot studies and field trials: Pilot studies and field trials are conducted in actual supply chain environments, such as transportation, storage, and retail settings, to assess the device's performance under real-world conditions.
- Data collection and analysis: During testing and field trials, data is collected from the edible electronic sensors and analyzed to validate the device's performance, identify potential issues, and make necessary improvements.

3.7. Ethical and Regulatory Considerations

The development and deployment of edible electronics for food quality monitoring must adhere to ethical principles and regulatory guidelines to ensure consumer safety and public trust. The methodology addresses:

- Food safety and consumer protection measures: Rigorous testing and evaluation are conducted to ensure that the edible electronic components and materials pose no health risks to consumers. Compliance with relevant food safety regulations and standards is paramount.
- Data privacy and security protocols: Robust data privacy and security protocols are implemented to protect the sensitive data collected by the edible electronic sensors, ensuring compliance with data protection regulations and preventing unauthorized access or misuse.
- Environmental impact assessments: The environmental impact of the edible electronic devices is assessed throughout their lifecycle, from material sourcing to disposal or biodegradation. Efforts are made to promote sustainable practices and minimize electronic waste.

4. DESIGN PRINCIPLES AND FABRICATION TECHNIQUES OF EDIBLE ELECTRONIC TAGS AND SENSORS

The design and fabrication of edible electronic tags and sensors require a careful consideration of various factors, including materials selection, biocompatibility, and manufacturing processes. The choice of materials is crucial, as they must not only be edible but also exhibit suitable electrical and mechanical properties for the intended application.

One of the most common materials used in edible electronics is edible inks, which are typically composed of food-grade conductive materials, such as carbon-based inks or metallic nanoparticle-based inks. These inks can be printed or deposited onto edible substrates, such as paper, biopolymers, or edible films, using techniques like screen printing, inkjet printing, or aerosol jet printing.

Another popular material for edible electronics is electrolytic solutions, which can be encapsulated within edible hydrogels or biopolymer matrices. These solutions can serve as electrolytes in electrochemical sensors or as conductive pathways in electronic circuits. Examples of edible electrolytes include sodium chloride (salt) solutions, fruit juices, and vinegar.

In addition to conductive materials, edible electronics may incorporate various functional components, such as sensors, antennas, and energy storage devices. These components can be fabricated using edible materials or coated with edible coatings to ensure their safety for human consumption.

The fabrication processes for edible electronic tags and sensors must be carefully designed to maintain the integrity and functionality of the electronic components while ensuring their edibility. This may involve the use of low-temperature processing techniques, such as room-temperature printing or solution-based deposition methods, to prevent degradation or contamination of the edible materials.

Furthermore, the encapsulation and packaging of edible electronic tags and sensors play a crucial role in their protection and shelf life. Edible coatings or edible packaging materials can be used to shield the electronic components from environmental factors, such as moisture, oxygen, and contaminants, while maintaining their edibility and safety.

5. REAL-TIME MONITORING OF FOOD PARAMETERS WITH EDIBLE ELECTRONICS

Edible electronic tags and sensors can be designed to monitor various parameters related to food quality and safety, providing real-time data and insights throughout the supply chain. Some of the key parameters that can be monitored using edible electronics include:

Temperature Monitoring:

Temperature monitoring is crucial for ensuring the safety and quality of various food products. Edible temperature sensors offer a non-invasive and convenient solution for monitoring temperature fluctuations in real-time. These sensors can be integrated into food packaging materials or directly embedded into food items. By continuously monitoring temperature variations during transportation, storage, and display, edible temperature sensors help prevent temperature abuse, which can lead to food spoilage and the growth of harmful bacteria. Moreover, real-time temperature data provided by these sensors enables timely interventions, such as adjusting storage conditions or implementing corrective measures, to maintain the quality and safety of food products.

Humidity Monitoring:

Humidity control is essential for preserving the quality and freshness of many food items, particularly those prone to moisture-related deterioration. Edible humidity sensors offer a practical solution for monitoring relative humidity levels within food packaging. These sensors can detect excess moisture, which can lead to mold growth, texture changes, and degradation of product quality. By providing real-time humidity data, edible sensors help ensure optimal moisture conditions for food preservation and extend the shelf life of perishable products. Additionally, early detection of high humidity levels allows for timely interventions, such as adjusting packaging materials or storage environments, to mitigate moisture-related issues and maintain product integrity.

Freshness Monitoring:

Maintaining the freshness of food products is critical for ensuring consumer satisfaction and reducing food waste.(Raak, N., Symmank, C., Zahn, S., Aschemann-Witzel, J., & Rohm, H. (2017)) Edible sensors designed to monitor freshness indicators offer a valuable tool for assessing the quality and shelf life of perishable items. These sensors can detect biochemical changes associated with food spoilage, such as pH variations, volatile organic compound (VOC) emissions, or enzymatic activity. By continuously monitoring these freshness indicators in real-time, edible sensors provide valuable insights into the condition of food products throughout the supply chain. Early detection of freshness degradation allows for prompt actions, such as adjusting storage conditions, optimizing inventory management, or implementing targeted marketing strategies to minimize waste and maximize product quality.

Contamination Detection:

Ensuring food safety is paramount in the food industry, and early detection of contaminants is crucial for preventing foodborne illnesses and protecting consumer health. Edible sensors equipped with specific receptors or probes can detect the presence of various contaminants, including pathogens, toxins, and allergens, in food products. These sensors can be designed to respond selectively to target analytes, providing rapid and sensitive detection capabilities. By continuously monitoring for contamination in real-time, edible sensors enable proactive measures to be taken, such as implementing corrective actions, initiating recalls, or conducting targeted investigations to identify and mitigate potential sources of contamination.

Gas Monitoring:

Gas monitoring is essential for controlling the ripening and spoilage of fresh produce and extending its shelf life. Edible gas sensors offer a versatile solution for monitoring gas levels within food packaging and storage environments. These sensors can detect gases such as carbon dioxide (CO_2), ethylene (C_2H_4), and oxygen (O_2), which play critical roles in the respiration and decay processes of fruits and vegetables. By continuously monitoring gas concentrations in real-time, edible sensors help optimize storage conditions, minimize respiratory losses, and prolong the freshness of fresh produce. Additionally, early detection of gas accumulation allows for timely interventions, such as adjusting storage atmospheres or implementing modified atmosphere packaging (MAP) techniques, to preserve product quality and reduce food waste.

The data collected from these edible electronic sensors can be transmitted wirelessly to IoT platforms or cloud-based systems for further analysis and decision-making. Real-time monitoring empowers stakeholders throughout the supply chain to take proactive measures, such as adjusting storage conditions, rerouting shipments, or issuing recalls, to maintain food quality and safety.

6. INTEGRATION OF IOT TECHNOLOGIES WITH EDIBLE ELECTRONICS

The integration of Internet of Things (IoT) technologies with edible electronics enables seamless data transmission and communication, unlocking the full potential of real-time food quality monitoring. IoT technologies facilitate the connectivity of edible electronic sensors and tags with central data repositories, allowing for efficient data collection, analysis, and decision-making.(Kamal, T., & Rahman, S. M. A. (2024)

One of the key components in this integration is the use of wireless communication protocols, such as Bluetooth Low Energy (BLE), Radio Frequency Identification (RFID), or Low-Power Wide-Area Networks (LPWAN). These protocols enable edible electronic tags and sensors to transmit data wirelessly to IoT gateways or edge devices, which can then relay the information to cloud-based platforms or local servers for further processing.

The choice of communication protocol depends on factors such as data rate requirements, power consumption, range, and scalability. For example, RFID technology can be suitable for short-range communication in warehouses or retail settings, while LPWAN protocols like LoRaWAN or NB-IoT are better suited for long-range communication across the supply chain.

Additionally, the integration of IoT technologies with edible electronics may involve the development of specialized IoT gateways or edge devices. These devices can be designed to interface with edible electronic sensors, collect and preprocess data, and transmit it to the cloud or central servers. They may also incorporate edge computing capabilities, allowing for local data processing and decision-making, reducing latency and improving response times.

Furthermore, cloud-based platforms and data management systems play a crucial role in the IoT integration with edible electronics. These platforms can securely store and manage the data collected from edible electronic sensors, enabling advanced analytics, visualization, and decision support tools. They can also facilitate data sharing and collaboration among various stakeholders in the supply chain, promoting transparency and traceability.

The mind map shown in the Figure 1 provides a visual representation of the intricate network of components and interconnections involved in the seamless integration of IoT technologies with edible electronics. Each node represents a key aspect, from materials selection to ethical considerations, while the edges delineate the flow of the methodology. This visualization offers a comprehensive overview of the multidisciplinary approach required for the development and implementation of edible electronic tags and sensors for food quality monitoring.

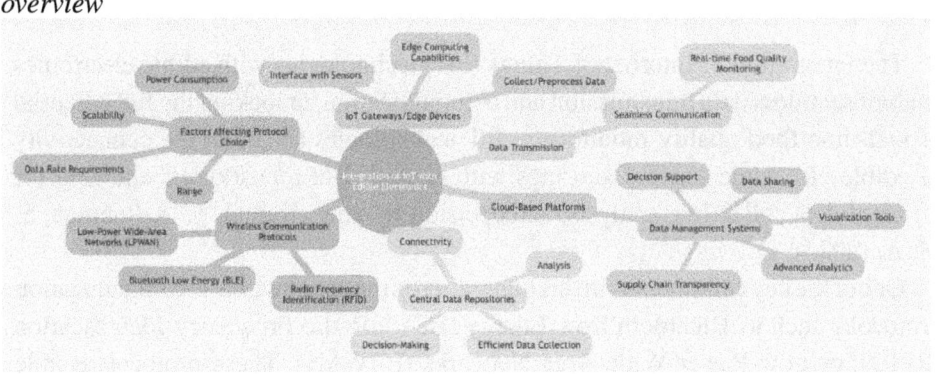

Figure 1. Integration of IoT technologies with edible electronics: a methodological overview

7. ARTIFICIAL INTELLIGENCE IN ENHANCING EDIBLE ELECTRONICS FOR FOOD QUALITY MONITORING

Artificial Intelligence (AI) techniques, particularly machine learning and deep learning algorithms, can significantly enhance the capabilities of edible electronics for food quality monitoring. By leveraging the vast amounts of data collected from edible electronic sensors, AI models can be trained to recognize patterns, make predictions, and optimize decision-making processes.

Predictive modeling and anomaly detection: Machine learning models can be trained on historical data from edible electronic sensors to identify patterns and correlations associated with food quality and safety issues. These models can then be used to predict potential quality issues or anomalies before they occur, enabling proactive measures to be taken.

Quality grading and classification: Deep learning algorithms, particularly convolutional neural networks (CNNs), can be applied to analyze sensor data and classify food products based on their quality or freshness levels. This can aid in automated grading and sorting processes, ensuring that only high-quality products reach consumers.

Optimized supply chain management: AI techniques can be employed to optimize supply chain operations by analyzing data from edible electronic sensors and factoring in various constraints and variables, such as transportation routes, storage conditions, and demand forecasts. This can lead to reduced waste, improved efficiency, and better preservation of food quality throughout the supply chain.

Predictive maintenance: By analyzing sensor data patterns, AI models can predict potential failures or degradation of edible electronic sensors or components. This information can be used to schedule preventive maintenance or replacements, ensuring the reliability and longevity of the monitoring systems.

Personalized recommendations: AI algorithms can be trained to analyze consumer preferences, dietary requirements, and other relevant data to provide personalized recommendations for food products based on their quality, freshness, and nutritional value.

The line graph shown in Figure 2 illustrates the performance improvement or accuracy gained by incorporating AI techniques in edible electronics for food quality monitoring over time. The x-axis represents the years from 2015 to 2022, while the y-axis indicates the accuracy percentage achieved by the edible electronics systems.

The graph shows a steady increase in accuracy over the years, indicating the progressive enhancement of edible electronics with AI integration. At the beginning of the timeline in 2015.

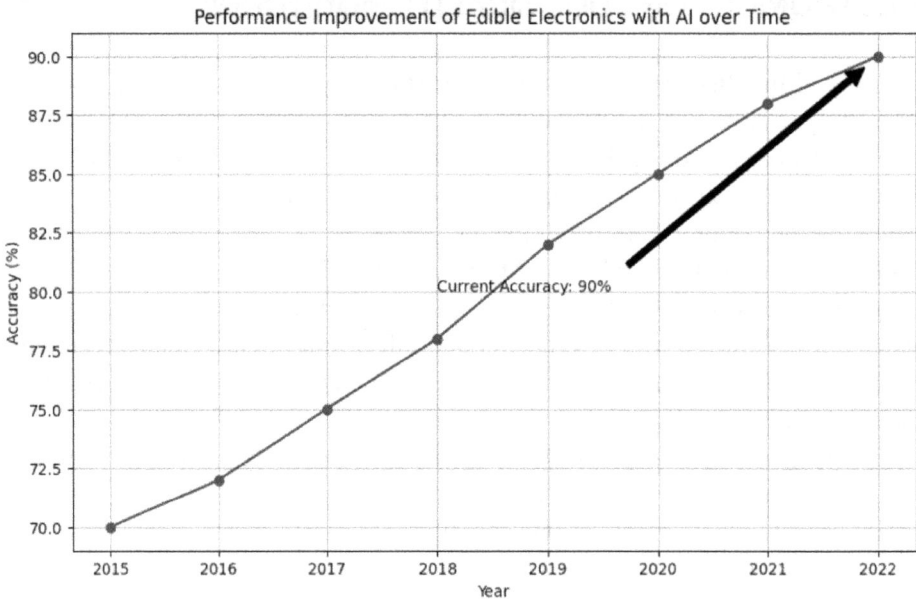

Figure 2. Performance improvement of edible electronics with AI over time

The integration of AI with edible electronics requires a robust data infrastructure, including secure data storage, processing capabilities, and advanced analytics tools. Additionally, the development and deployment of AI models necessitate careful consideration of factors such as data quality, model interpretability, and ethical implications to ensure responsible and trustworthy AI applications in the food industry.

8. CASE STUDIES: APPLICATIONS OF EDIBLE ELECTRONICS IN ENSURING FOOD SAFETY AND QUALITY

Edible electronics have been successfully applied in various real-world scenarios to enhance food safety and quality monitoring. Here are some case studies that illustrate the practical applications of this innovative technology:

Fresh produce monitoring: Researchers at Tufts University developed an edible sensor made from silk and gold nanoparticles that can be embedded into the stems of fruits and vegetables to monitor their freshness. The sensor detects the presence of plant-based enzymes associated with decay and spoilage, providing real-time freshness information to consumers and retailers.

Meat quality monitoring: A team of researchers from the University of Bologna, Italy, created an edible sensor made from gelatin and carbon nanotubes that can be integrated into meat packaging. The sensor can detect the presence of biogenic amines, which are indicators of spoilage, allowing for accurate monitoring of meat quality throughout the supply chain.

Intelligent food packaging: Companies like Insignia Technologies and Securetrace have developed intelligent food packaging solutions that incorporate edible electronic tags and sensors. These tags can monitor temperature, humidity, and other environmental conditions, providing traceability and quality assurance for various food products, such as dairy, meat, and fresh produce.

Smart labels for allergen detection: Researchers from Harvard University and the Massachusetts Institute of Technology (MIT) developed an edible sensor that can detect the presence of specific allergens, such as peanuts or gluten, in food products. These sensors can be integrated into smart labels or packaging, providing valuable information to consumers with food allergies or intolerances.

Cold chain monitoring: Companies like Tempronics and Emerson have developed edible temperature monitoring solutions for cold chain applications. These solutions incorporate edible electronic tags or labels that can track and record temperature data during transportation and storage, ensuring the proper handling of temperature-sensitive food products.

The bar chart shown in Figure 3 illustrates the distribution of applications of edible electronics across various sectors within the food industry, including fresh produce, meat, dairy, and packaged foods. Each bar represents the number of applications observed in a specific sector, providing insight into the prevalence of edible electronics technology in different areas of food production, processing, and packaging.

In the context of the provided content, the bar chart serves to visually convey the diversification of edible electronics applications, showcasing how this innovative technology is being utilized across multiple segments of the food supply chain. For instance, it highlights the significant presence of edible electronics in monitoring fresh produce and packaged foods, while also demonstrating its relevance in sectors such as meat and dairy products.

The distribution of applications of edible electronics across various sectors of the food industry is depicted in the accompanying bar chart. It showcases the prevalence of this technology in monitoring different segments of the food supply chain, with notable concentrations observed in fresh produce and packaged foods. The chart provides valuable insight into the diverse applications of edible electronics, illustrating its significance in ensuring food safety and quality across multiple sectors.

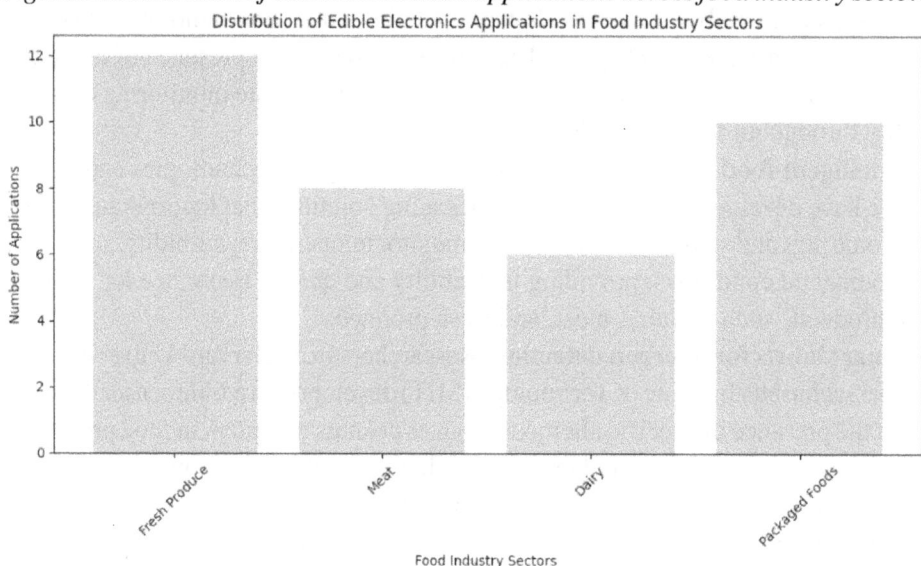

Figure 3. Distribution of edible electronics applications across food industry sectors

These case studies demonstrate the diverse applications of edible electronics in various sectors of the food industry, from fresh produce to meat, dairy, and packaged foods. As the technology continues to evolve and gain broader adoption, it holds the potential to revolutionize food quality monitoring and enhance consumer confidence in the safety and quality of the products they consume.

9. CHALLENGES AND FUTURE DIRECTIONS IN EDIBLE ELECTRONICS FOR FOOD QUALITY MONITORING

While edible electronics offer promising solutions for food quality monitoring, there are several challenges and future directions that need to be addressed to fully realize their potential:

Scalability and cost-effectiveness: To achieve widespread adoption, edible electronic tags and sensors need to be produced at scale in a cost-effective manner. This may require the development of new manufacturing techniques and the optimization of material sourcing and production processes.

Standardization and regulatory compliance: Establishing standardized protocols and regulatory frameworks for the development, testing, and deployment of edible electronics is crucial. This will ensure consistent quality, safety, and performance across different products and applications.

Integration with existing supply chain systems: Seamless integration of edible electronics with existing supply chain management systems, including enterprise resource planning (ERP) and warehouse management systems (WMS), is essential for efficient data sharing and decision-making.

Power and energy sources: While edible electronics can be designed to be low-power, the development of efficient and safe energy sources or energy harvesting techniques is crucial for their long-term operation and sustainability.

Data security and privacy: As edible electronics generate and transmit sensitive data related to food quality and supply chain operations, robust data security measures and privacy protocols need to be implemented to protect against potential cyber threats or data breaches.

Environmental impact and sustainability: The development and deployment of edible electronics should consider their environmental impact and strive for sustainability throughout their lifecycle, from material sourcing to disposal or biodegradation.

Multidisciplinary collaboration: Advancements in edible electronics require collaboration across multiple disciplines, including electronics, food science, materials science, chemistry, and computer science. Fostering interdisciplinary research and knowledge sharing will drive innovation and accelerate progress in this field.

Consumer acceptance and education: Educating consumers about the benefits and safety of edible electronics is essential for gaining public trust and acceptance. Effective communication strategies and transparency regarding the technology and its applications will be crucial for widespread adoption.

The challenges and potential solutions or future directions in edible electronics for food quality monitoring are summarized in Table 2. These challenges encompass scalability, cost-effectiveness, standardization, regulatory compliance, integration with existing supply chain systems, energy sources, data security, environmental impact, interdisciplinary collaboration, and consumer acceptance. Each challenge is accompanied by potential solutions or future directions aimed at addressing them. By proactively tackling these challenges and exploring innovative solutions, the field of edible electronics holds immense promise for revolutionizing food quality monitoring, ensuring safety, sustainability, and consumer trust throughout the food supply chain.

Table 2. Challenges and future directions in edible electronics for food quality monitoring

Challenges	Potential Solutions/Future Directions
Scalability and cost-effectiveness	- Development of new manufacturing techniques
	- Optimization of material sourcing and production processes
	- Economies of scale through mass production
Standardization and regulatory compliance	- Establishment of standardized protocols and regulatory frameworks
	- Compliance with food safety regulations and standards
Integration with existing supply chain systems	- Seamless integration with ERP and WMS systems
	- Interoperability with existing technologies and platforms
Power and energy sources	- Development of efficient and safe energy sources
	- Exploration of energy harvesting techniques
Data security and privacy	- Implementation of robust data security measures and privacy protocols
	- Encryption of sensitive data during transmission and storage
Environmental impact and sustainability	- Consideration of environmental impact throughout the lifecycle
	- Promotion of sustainable practices, such as biodegradable materials
Multidisciplinary collaboration	- Fostering collaboration across electronics, food science, and more
	- Knowledge sharing and interdisciplinary research initiatives
Consumer acceptance and education	- Education and communication strategies to inform consumers
	- Transparency regarding technology benefits and safety

By addressing these challenges and continuously exploring new frontiers, edible electronics have the potential to revolutionize food quality monitoring, enhancing food safety, reducing waste, and promoting sustainable practices throughout the global food supply chain.

10. ETHICAL AND REGULATORY CONSIDERATIONS IN DEPLOYING EDIBLE ELECTRONICS IN THE FOOD INDUSTRY

As with any emerging technology in the food industry, the deployment of edible electronics raises ethical and regulatory considerations that must be carefully addressed. Ensuring the safety and well-being of consumers, as well as maintaining public trust, is of paramount importance.(Gbashi, S., & Njobeh, P. B. (2024))

Food safety and consumer protection: Edible electronic components and materials must undergo rigorous testing and evaluation to ensure they pose no health risks to consumers. Comprehensive food safety regulations and standards should be established to govern the development, manufacturing, and use of edible electronics in food products.

Transparency and labeling: Consumers have the right to be informed about the presence of edible electronic components in food products. Clear and transparent labeling practices should be implemented to disclose the use of edible electronics, their purpose, and any potential risks or considerations.

Data privacy and security: The data collected by edible electronic sensors may contain sensitive information about food products, supply chain operations, and potentially consumer preferences or behaviors. Robust data privacy and security measures must be in place to protect against unauthorized access, misuse, or breaches of this data.

Environmental impact and sustainability: The production, use, and disposal of edible electronic components should be evaluated for their potential environmental impact. Efforts should be made to promote sustainable practices, such as the use of biodegradable or compostable materials, and minimizing electronic waste.

Ethical considerations in AI and decision-making: As artificial intelligence is integrated with edible electronics for data analysis and decision-making, ethical considerations related to algorithmic bias, transparency, and accountability must be addressed. AI systems should be designed and deployed with fairness, privacy, and human oversight in mind.

Regulatory frameworks and standardization: Regulatory bodies and industry organizations should collaborate to establish clear guidelines, standards, and certification processes for the development and deployment of edible electronics in the food industry. This will ensure consistency, quality control, and consumer protection across different regions and jurisdictions.

Public engagement and education: Engaging with the public, addressing concerns, and providing educational resources about the benefits, safety, and ethical considerations of edible electronics is crucial for building trust and acceptance of this technology.

By proactively addressing these ethical and regulatory considerations, the food industry can foster responsible innovation and ensure that the deployment of edible electronics aligns with societal values, consumer protection, and sustainable practices.

11. CONCLUSION: POTENTIAL IMPACT AND OPPORTUNITIES OF EDIBLE ELECTRONICS IN ENHANCING FOOD SAFETY AND QUALITY

Edible electronics represent a groundbreaking innovation with the potential to revolutionize food quality monitoring and enhance food safety and quality throughout the supply chain. By integrating electronic components and sensors into edible materials and food packaging, edible electronics enable real-time monitoring of various parameters, such as temperature, humidity, freshness, and contamination.

The integration of edible electronics with Internet of Things (IoT) technologies and artificial intelligence (AI) techniques unlocks new opportunities for seamless data transmission, advanced analytics, and optimized decision-making processes. IoT connectivity enables efficient data collection and sharing among stakeholders, while AI algorithms can process and analyze the vast amounts of data generated by edible electronic sensors, enabling predictive modeling, automated decision-making, and supply chain optimization.

The applications of edible electronics span various sectors of the food industry, from fresh produce and meat to dairy and packaged foods. Case studies have demonstrated the successful implementation of edible electronic tags and sensors for monitoring freshness, detecting allergens, tracking temperature fluctuations, and ensuring the quality and safety of food products throughout the supply chain.

However, realizing the full potential of edible electronics requires addressing challenges related to scalability, cost-effectiveness, standardization, data security, and environmental sustainability. Multidisciplinary collaboration, along with robust regulatory frameworks and consumer education, will be crucial for the responsible development and deployment of this technology.

As the world grapples with the challenges of ensuring food security, reducing waste, and promoting sustainable practices, edible electronics offer a promising solution. By enhancing food quality monitoring and enabling proactive measures throughout the supply chain, edible electronics can contribute to minimizing food losses, extending shelf life, and improving overall food safety and quality.

Moreover, edible electronics have the potential to empower consumers by providing them with transparent and accurate information about the products they consume. Smart labels and packaging integrated with edible electronic sensors can convey real-time data on freshness, nutritional content, and potential allergens, allowing consumers to make informed choices aligned with their dietary preferences and health requirements.

Beyond the food industry, the advancements in edible electronics may pave the way for innovative applications in other sectors, such as healthcare and environmental monitoring. For instance, edible sensors could be developed to monitor biological markers or deliver targeted drug release, enabling personalized medicine and improved patient outcomes. Similarly, edible sensors could be employed to detect and monitor environmental pollutants or contaminants in water sources, contributing to sustainable environmental management.

As with any disruptive technology, the journey toward widespread adoption of edible electronics will require collaborative efforts from various stakeholders, including researchers, industry players, regulatory bodies, and consumers. By fostering interdisciplinary collaborations, establishing robust regulatory frameworks, and promoting public trust and acceptance, the food industry can harness the full potential of edible electronics to enhance food safety, reduce waste, and contribute to a more sustainable and resilient global food system.

The future of edible electronics is promising, and its impact on food quality monitoring has the potential to reshape the way we approach food safety, quality assurance, and supply chain management. As this innovative technology continues to evolve, it will open new avenues for innovation, driving positive change and contributing to a safer, more transparent, and more sustainable food industry for generations to come.

REFERENCES

Abid, H. M. R., Khan, N., Hussain, A., Anis, Z. B., Nadeem, M., & Khalid, N. (2024). Quantitative and qualitative approach for accessing and predicting food safety using various web-based tools. *Food Control*, 162, 110471. DOI: 10.1016/j.foodcont.2024.110471

Anand, R., Pandey, D., Gupta, D. N., Dharani, M. K., Sindhwani, N., & Ramesh, J. V. N. (2024). Wireless Sensor-based IoT System with Distributed Optimization for Healthcare. In Anand, R., Juneja, A., Pandey, D., Juneja, S., & Sindhwani, N. (Eds.), *Meta Heuristic Algorithms for Advanced Distributed Systems.*, DOI: 10.1002/9781394188093.ch16

Baranov, D. G., Zuev, D. A., Lepeshov, S. I., Kotov, O. V., Krasnok, A. E., Evlyukhin, A. B., & Chichkov, B. N. (2017). All-dielectric nanophotonics: The quest for better materials and fabrication techniques. *Optica*, 4(7), 814–825. DOI: 10.1364/OPTICA.4.000814

Eni, L. N., Groenewald, E. S., Hamidi, I. A., & Garg, A. (2024). Optimizing Supply Chain Processes through Deep learning Algorithms: A Managerial Approach. *Journal of Informatics Education and Research*, 4(1). Advance online publication. DOI: 10.52783/jier.v4i1.567

Gbashi, S., & Njobeh, P. B. (2024). Enhancing food integrity through artificial intelligence and machine learning: A comprehensive review. *Applied Sciences (Basel, Switzerland)*, 14(8), 3421. DOI: 10.3390/app14083421

Kamal, T., & Rahman, S. M. A. (2024). *Productivity optimization in the electronics industry using simulation-based modeling approach*. International Journal of Research in Industrial Engineering., DOI: 10.22105/riej.2024.445760.1423

Kapoor, A., Ramamoorthy, S., Sundaramurthy, A., Vaishampayan, V., Sridhar, A., Balasubramanian, S., & Ponnuchamy, M. (2024). Paper-based lab-on-a-chip devices for detection of agri-food contamination. *Trends in Food Science & Technology*, 147, 104476. DOI: 10.1016/j.tifs.2024.104476

Landi, G., Pagano, S., Granata, V., Avallone, G., La Notte, L., Palma, A. L., Sdringola, P., Puglisi, G., & Barone, C. (2024). Regeneration and Long-Term Stability of a Low-Power Eco-Friendly Temperature Sensor Based on a Hydrogel Nanocomposite. *Nanomaterials (Basel, Switzerland)*, 14(3), 283. DOI: 10.3390/nano14030283 PMID: 38334553

Lu, P., Guo, X., Liao, X., Liu, Y., Cai, C., Meng, X., Wei, Z., Du, G., Shao, Y., Nie, S., & Wang, Z. (2024). Advanced application of triboelectric nanogenerators in gas sensing. Nano Energy, 126, 109672. ISSN 2211-2855. https://doi.org/DOI: 10.1016/j.nanoen.2024.109672

Raak, N., Symmank, C., Zahn, S., Aschemann-Witzel, J., & Rohm, H. (2017). Processing- and product-related causes for food waste and implications for the food supply chain. *Waste Management (New York, N.Y.)*, 61, 461–472. DOI: 10.1016/j.wasman.2016.12.027 PMID: 28038904

Rout, S., Tambe, S., Deshmukh, R. K., Mali, S., Cruz, J., Srivastav, P. P., Amin, P. D., Gaikwad, K. K., Andrade, E. H. de A., & de Oliveira, M. S. (2022). Recent trends in the application of essential oils: The next generation of food preservation and food packaging. *Trends in Food Science & Technology*, 129, 421–439. DOI: 10.1016/j.tifs.2022.10.012

Sharova, Alina S., et al. "Edible electronics: The vision and the challenge." *Advanced Materials Technologies* 6.2 (2021): 2000757.

Shen, S. C., Khare, E., Lee, N. A., Saad, M. K., Kaplan, D. L., & Buehler, M. J. (2023). Computational Design and Manufacturing of Sustainable Materials through First-Principles and Materiomics. *Chemical Reviews*, 123(5), 2242–2275. DOI: 10.1021/acs.chemrev.2c00479 PMID: 36603542

Sun, H., Cao, Y., Kim, D., & Marelli, B. (2022). Biomaterials technology for Agro-Food resilience. *Advanced Functional Materials*, 32(30), 2270173. Advance online publication. DOI: 10.1002/adfm.202270173

Tripathi, S., Hakim, L., Gupta, D., Deshmukh, R. K., & Gaikwad, K. K. (2024). Recent Trends in Films and Gases for Modified Atmosphere Packaging of Fresh Produce. In Novel Packaging Systems for Fruits and Vegetables (1st ed., pp. 27). Apple Academic Press. eBook ISBN: 9781003415701 DOI: 10.1201/9781003415701-3

Yin, H.-Y., Fang, T. J., Li, Y.-T., Fung, Y.-F., Tsai, W.-C., Dai, H.-Y., & Wen, H.-W. (2019). Rapidly detecting major peanut allergen-Ara h2 in edible oils using a new immunomagnetic nanoparticle-based lateral flow assay. *Food Chemistry*, 271, 505–515. DOI: 10.1016/j.foodchem.2018.07.064 PMID: 30236709

Zou, Y., Ren, Z., Xiang, Y., Guo, H., Yang, G.-Z., & Tao, G. (2024). Flexible fiberbotic laser scalpels: Material and fabrication challenges. *Matter*, 7(3), 758–771. DOI: 10.1016/j.matt.2024.01.007

Chapter 3
Enhancing Food Quality Monitoring Processes Through Innovative Microcontroller-Based Systems

Saurabh Chandra
Bennett University, Greater Noida, India

Shreenidhi K. S.
https://orcid.org/0000-0002-6844-0454
Department of Biotechnology, Rajalakshmi Engineering College, Chennai, India

Harishchander Anandaram
https://orcid.org/0000-0003-2993-5304
Department of Artificial Intelligence, Amrita Vishwa Vidyapeetham, Coimbatore, India

Ketki P. Kshirsagar
Department of Electronics and Telecommunication, Vishwakarma Institute of Technology, Pune, India

N. R. Rajagopalan
Department of Chemistry, St. Joseph's College of Engineering, OMR, Chennai, India

ABSTRACT

Advanced microcontroller technology is making food quality monitoring processes

DOI: 10.4018/979-8-3693-5573-2.ch003

more accurate and efficient. The chapter describes the state of-the-art microcontrollers for enhancing the safety and quality of food by enabling real-time monitoring and data analysis. The study investigates sensors built into microcontrollers for the measurement of critical parameters in maintaining the quality of food, namely temperature, humidity, and pH. The chapter has highlighted a number of applications of advanced microcontrollers for tracking and controlling food storage facilities' conditions, as well as detecting spoilage and safety standards. It provided case studies on successful implementation and advantages of such technologies: higher accuracy, lower risk of human error, and cost-effectiveness. This chapter takes readers through state-of-the-art advances, providing useful insights to researchers and practitioners who aim at leveraging microcontroller innovations in the enablement of superior food quality monitoring processes with safer food supply chains.

INTRODUCTION

In times of awareness about the safety and quality of food, technology has a very significant role in the integrities associated with the quality checking of our food. As global food systems have been increasingly complex, the need for sophisticated tools required to monitor and manage food quality along the supply chains has grown greatly. Among the most revolutionary changes within the line of this business development are an extensive application of advanced microcontrollers. These small but powerful devices stand on the very frontline of food quality monitoring process improvement, which they take up to such a level of precision, efficiency, and reliability(Popa et al., 2019).

Microcontrollers are integrated circuits that act like the brains of various electronic devices, endowing them with the capability to perform specific tasks by processing inputs from sensors and running programmed instructions. Because they are versatile and adoptive in nature, they can be used within a broad spectrum of areas, including food quality monitoring. Incorporating these tiny controllers into monitoring systems enables food producers, distributors, and retailers to monitor, in real time, critical parameters that impact the quality of food, such as temperature, humidity, and pH (Grazioli et al., 2020).

The most significant advantages of microcontrollers in food quality monitoring involve their easy interfacing with sensors of environmental conditions. For example, temperature sensors placed in storage facilities or transportation units may provide regular information on the conditions that surround the stored food products. Real-time monitoring is, therefore, very crucial in keeping the foods within an optimum range of temperature levels to prevent spoilage and maintain quality. The sensor for humidity will also monitor the levels of moisture, which, for instance, is very important in products such as grains and dried fruits that are highly sensitive to changes in humidity (Dias et al., 2020).

Besides environmental monitoring, microcontrollers can also be applied in the monitoring and controlling of chemical and biological factors affecting food quality. For instance, pH sensors can measure acidity or alkalinity in a food product, important for products like dairy, meat, and beverages. Through the incorporation of such sensors with microcontrollers, producers can ascertain that their products are within the desired pH range to prevent spoilage and ensure safety (Kaushik & Singh, 2013).

The application of microcontrollers is not limited to monitoring; they also have an important role in controlling and automating the processes involved in food processing and storage. Automated systems can be programmed to immediately alter environments based on sensor data. In this case, when a temperature sensor detects that a storage unit is approaching a critical temperature threshold, the microcontroller will consequently trigger cooling systems on in an attempt to bring it back into the desired range. Such degrees of automation dismiss human errors and maintain consistent quality control (Kaushik & Singh, 2013).

With microcontrollers, it also becomes possible to collect and process large volumes of data emanating from monitoring systems. Advanced micro-controllers are equipped with connectivity capabilities, such as Wi-Fi or Bluetooth, which can directly send the information collected to centralized databases or even to cloud platforms. Further analysis of this data for trending and forecasting potential issues will provide actionable insights toward improving food quality and food safety. Analytics on such data can reveal trends in the supply chain, such as recurring temperature fluctuations in transit, for example, where appropriate mitigation action may be taken (Venkatesh et al., 2017).

Case studies demonstrate the business benefit of integrating microcontrollers into systems for monitoring food quality. A major food retailer implemented a network of temperature sensors using microcontrollers throughout its cold storage and logistic facilities. Real-time data from this monitoring device enabled the company to rapidly detect and act upon temperature deviations, hence reduced spoilage rates and ensured product safety. Another good example is that of a dairy producer who uses pH sensors integrated with microcontrollers that track acidity through processing. This system enables milk to stay in an optimal pH range, improving product consistency and extending the shelf life of the product (Garcia-Breijo et al., 2013).

There are but a few complications or hurdles in the way of applying microcontroller-based systems to food quality monitoring, despite apparent benefits. The basic cost of advanced monitoring systems can be somewhat high, and there is a potential learning curve in incorporating new technologies into current processes. Moreover, reliability and accuracy in sensors and microcontrollers are key, as inaccuracies can further lead to false readings and possible safety issues with food (Jedermann et al., 2014).

The challenges associated with the adaptation of such micro-controller technology are, therefore, negligible compared to the accrued long-term benefits. Key benefits include increased accuracy in monitoring, reduced spoilage, increased safety standard compliance, and operational efficiency. With the continuous advancement in technology, future developments in micro-controllers are likely to introduce even more advanced features, including improved processing power, lower power consumption, and more efficient data analytics (Bandal & Thirugnanam, 2016).

Microcontrollers are redefining the concept of food quality monitoring by availing data in real time with accuracy and in an actionable way to improve the safety and quality of food products. Integration with monitoring systems will enable proactive management of environmental and chemical factors, automation of control processes, and making sophisticated data analysis possible. While the food industry will see rapid changes, microcontrollers will remain in the middle of maintaining standards regarding food safety and quality (Li et al., 2006). Innovation going on in microcontrollers will promise further enhancement in the capability for providing exciting opportunities for the improvement of food quality monitoring processes and for a better food safety chain in times to come.

Scope

This chapter examines the integration of advanced microcontroller technologies into food quality monitoring systems, focusing on the continuous, real-time measurement of temperature, humidity, and pH critical parameters. It details how embedded sensor technologies within the microcontrollers enhance food safety and quality through real-time, accuracy-driven monitoring and analysis of data. Chapters discussed topics such as: food storage applications, spoilage detection, and safety standard compliance, supplementary case studies that demonstrate the benefits of the technology: better accuracy, lesser human error, and cost-effectiveness. This chapter is addressed to researchers and professionals who intend to tap into these innovations for safer, more efficient food supply chains.

Objectives

The aims of this chapter are to discuss and present the role of advance microcontrollers as enhancements for processes to monitor food quality. The analysis will show how the real-time data from in-situ sensors increases the accuracy and reliability of food safety interventions. The main focus will be laid on such critical factors as temperature, humidity, and pH, respectively. In addition, the chapter will demonstrate microcontrollers' practical applications with respect to preventing food spoilage and meeting safety standards and optimal storage conditions. Case studies are utilized to address the cost effectiveness, reduced human error, and the operational efficiency gained from microcontroller technology in food monitoring systems.

BACKGROUND: MICRO-CONTROLLERS

Definition and Functionality

A microcontroller is a small integrated circuit designed to control an activity in an electronic device. Basically, it's a CPU for embedded systems, having a processor, memory, and I/O peripherals on one chip. Microcontrollers range from simple operation circuitries to complex mechanical operations, as they can be programmed for a wide range of applications and respond to several input signals.

The tasks that a microcontroller usually performs include decoding and processing input signals, executing calculations, and performing the commands of control. In food quality monitoring systems, for example, a microcontroller gets data from sensors that measure parameters such as temperature and humidity, then processes it using pre-identified algorithms, changing the control mechanisms or sending warnings if the conditions go out of threshold values. In this way, it ensures that monitoring and automation are done in real time and efficiency and accuracy are maintained in all operations.

Types of Microcontrollers

Microcontrollers are available in a number of types, each suited to different applications based on their processing power, memory, and peripheral support (Li et al., 2006). The most common types of microcontrollers are:

a) 8-bit Microcontrollers: These are the least complex and most affordable kinds of microcontrollers, which can be utilized even for very simple applications. They normally have limited processing power and memory but may be used for applications requiring very little computation, such as simple sensor data collection and simple control operations. Examples include the Intel 8051 and Microchip PIC series.

b) 16-bit Microcontrollers: With increased processing power and memory compared to 8-bit microcontrollers, their 16-bit brethren find uses in more complex applications. They offer higher performance and, therefore, are found in systems that require moderate computation and higher resolution. Well-known members

in this category include the Texas Instruments MSP430 series and Microchip PIC24 series.
c) 32-bit Microcontrollers: High-performance microcontrollers, normally used in those applications that require enormous processing power and memory. The on-chip features in these microcontrollers are advanced, enabling them to run and support advanced applications like real-time processing of data, communication, and multimedia applications. Examples include the ARM Cortex-M series and STMicroelectronics STM32 series.
d) Specialized Microcontrollers: Other than the general types, there are also specialized microcontrollers for special uses. An example would be that some microcontrollers come with integrated wireless communication capabilities, like Wi-Fi or Bluetooth, intended to be used in IoT devices. Others could include A/D converters and digital signal processors, DSPs; making them more functional and specialized for particular fields.

Main Capabilities and Features

They come with various features and abilities that make them helpful in a wide range of applications (Vergara et al., 2007). Here are just a few:

a) Power: The power of a microcontroller, decided by clock speed and architecture, is different. Higher clock speeds tell how fast a device executes instructions; on the other hand, newer architectures will run programs much more efficiently and provide better performance. For example, ARM Cortex yields far better performance and efficiency. The processing power comes in handy while dealing with higher calculations and real-time processing.
b) Memory: Most microcontrollers have different kinds of memory, including ROM, which holds the firmware; RAM, which stores temporary data; and EEPROM, used to store configuration data. The quantity and type of memory will affect the capability of a microcontroller in handling and storing data.
c) I/O Ports: I/O ports provide the way for a microcontroller to interface with external devices and sensors. It normally includes digital and analog inputs/outputs that provide functions like reading sensor data and driving actuators. The number and type of I/O ports determine the adaptability of a micro-controller to different applications.
d) Communication Interfaces: Most microcontrollers have multiple communication interfaces like SPI, I2C, UART, and USB. These interfaces allow a microcontroller to communicate with other devices, sensors, and network systems.

e) Timers and Counters: The functions of timers and counters are to measure time intervals and drive events or operations dependent on time. These two devices are the very bases for event scheduling, PWM, and time-based application management.
f) Analog-to-Digital Converters: These are interfaces used in the conversion of analog signals from sensors into digital data for processing by a microcontroller. This capability is very important in applications related to the monitoring of analog parameters, such as temperature and pH.
g) Power Management: Since most gadgets operate on batteries, therefore, efficient power management becomes a necessity. Most microcontrollers feature low-power operation modes and techniques to reduce power consumption in order to increase battery life.
h) Software and Programming: Most microcontrollers are programmed in various development environments and languages, including, but not limited to C, C++, and about the Assembly language. By flexibility in programming, customization and optimization can be performed so as to suit particular needs that an application may have.

It is for this reason that microcontrollers remain at the core of modern electronic systems, given their diversified blend of processing power, memory, and peripheral capabilities that allow for precise control and monitoring. Knowledge of the types, features, and capabilities is important to unleash the potential of a microcontroller in food quality monitoring by enhancing accuracy, efficiency, and automation. As technology evolves, so does the evolution of microcontroller technology, promising further expansion of their applications and capabilities well into the future, thereby driving innovation across a wide range of fields.

INTEGRATION WITH SENSORS

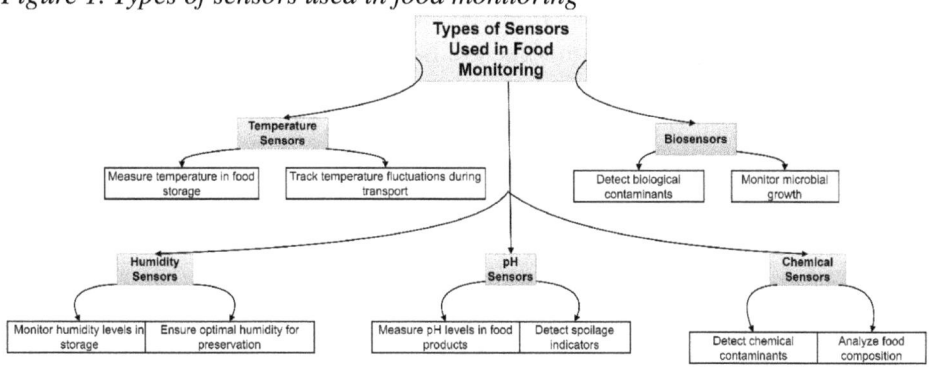

Figure 1. Types of sensors used in food monitoring

Types of Sensors Used in Food Monitoring:

- **Temperature Sensors**
  - Measure temperature in food storage
  - Track temperature fluctuations during transport
- **Biosensors**
  - Detect biological contaminants
  - Monitor microbial growth
- **Humidity Sensors**
  - Monitor humidity levels in storage
  - Ensure optimal humidity for preservation
- **pH Sensors**
  - Measure pH levels in food products
  - Detect spoilage indicators
- **Chemical Sensors**
  - Detect chemical contaminants
  - Analyze food composition

Types of Sensors Used in Food Monitoring

In food quality monitoring, sensors play an indispensable role in measuring various environmental and product parameters that affect safety and quality. The integration of sensors with microcontrollers allows for real-time data acquisition and precision control in maintaining the right conditions to keep the food products safe. The section discusses different sensor types utilized for monitoring foods, including temperature, humidity, pH, chemical, and biological sensors. Figure 1 clearly and structuredly presents the various sensors utilized in food quality monitoring, showcasing their specific applications and relevance (Nath, 2024; Wang et al., 2018).

Temperature Sensors Temperature is one of the most critical parameters in maintaining food quality because improper temperature conditions lead to food spoilage, bacterial multiplication, and loss of nutritional value. Temperature sensors contribute to monitoring and controlling products' temperature along the entire supply chain: from the storage of raw materials to processing, transportation, and preparation of food.

Thermistors: These are temperature sensors that exhibit a large change in electrical resistance with temperature. They have a reputation for high accuracy and stability over a wide temperature range. Thermistors find widespread application in applications where accurate temperature measurement is required, such as monitoring units of refrigeration and storage conditions. The RTD usually works on the principle of change in electrical resistance of metals with temperature and normally uses platinum. They are highly accurate and stable; hence, they best suit applications that require highly accurate temperature control, like in food processing and quality.

Thermocouples: These are a couple of sensors that develop voltage proportional to the difference in the temperature value in two different metals. Because of the wide temperature range and robustness, thermocouples find many applications in industries. In food monitoring, normally thermocouples are employed in oven-freezer environments.

Humidity Sensors: One of the most important environmental parameters in maintaining food quality, humidity levels are closely monitored, especially in products which readily absorb moisture like grains, dried fruits, and bakery products. Humidity sensors regulate and monitor the moisture level of storage environments against spoilage and deterioration in quality.

Capacitive Humidity Sensors: Capacitive humidity sensors work by measuring the relative humidity through change in capacitance of some hygroscopic dielectric material. These sensors offer a very high degree of accuracy and stability, thus finding applications in warehouse condition monitoring and drying process control. Resistive humidity sensors consist of a hygroscopic material whose electrical resistance changes with humidity. Such sensors are generally used in applications requiring moderate accuracy and economy, like household appliances or simple food storage systems. These sensors work on the principle of using optical light to measure the moisture content in a sample.

pH Sensors: pH sensors are needed to control acidity or alkalinity in food products, which can influence the taste and texture of a particular product, but it will ultimately endanger the safety of the product. Control at proper pH ranges needs to be observed to avoid microbial growth and ensure quality in a product.

Glass Electrode pH Sensors: These are some of the most common devices used in pH measurements for liquid food products. The glass membrane directly interfaces with the sample solution, eliciting a voltage proportional to the solution's pH. These sensors ensure a high degree of accuracy and see wide applications in food industries, for example, in the processing of dairy products, and in the beverage industry(Pachiappan et al., 2024; Upadhyaya et al., 2024).

pH Sensors using Ion-Selective Field-Effect Transistors (ISFETs): ISFET pH sensors use an ion-selective field-effect transistor for measurement purposes. Their response times are rather fast; thus, they are used in applications that need very fast and precise pH measurements, such as in food testing and quality control.

Combined Electrodes: Combined pH electrodes are sensing electrodes that incorporate a glass electrode and a reference electrode in one body. They find their application in the measurement of pH in food products and offer an easy way of making measurements both in and out of the laboratory.

Chemical and Biological Sensors

Various chemical and biological sensors, utilized in different food industries, usually detect the presence of specific compounds or microorganisms that could affect the safety and quality of foods. It gives enlightened insight into the chemical constitution and microbial load of food products(Rivas-Sánchez et al., 2019).

Gas Sensors: target gases that are usually indicative of food ripeness, such as CO_2 and ethylene. Other emerging applications of gas sensors are in controlled atmosphere packaging and monitoring systems as an extension of shelf life and quality maintenance of products.

Spectroscopic sensors: It uses light to analyze the chemical composition of food products. The normally used techniques in these sensors include near infrared NIR and Raman, which determine various attributes like moisture content, fat level, and adulteration. These sensors undertake non-destructive analysis and are applied in the quality control and assurance processes.

Biosensors: These biosensors are fabricated for the detection of a biological agent such as bacteria and pathogens by incorporating some kind of biological recognition elements: antibodies or enzymes. They are very important in ensuring that food is safe by detecting microorganisms present in food products. Examples include biosensors applied in rapid testing methodologies and monitoring food safety systems.

The integration of sensors with microcontrollers within the application area of food quality monitoring increases the accuracy and efficiency of monitoring of critical parameters. Temperature, humidity, and pH and chemical and biological sensors are also critical because they deliver real-time data that is so important to maintain conditions at an optimal value within the food chain. It is here that the food producers and processors can use these sensors to their best advantage to obtain better control, reduce spoilage, and provide the highest quality standards within the food supply chain. As development in technologies goes on and on, more advanced sensors will be developed, hence enhancing further the capabilities and effectiveness of food quality monitoring systems.

REAL-TIME MONITORING AND DATA ACQUISITION

Figure 2. States and transitions involved in real-time monitoring and data acquisition processes

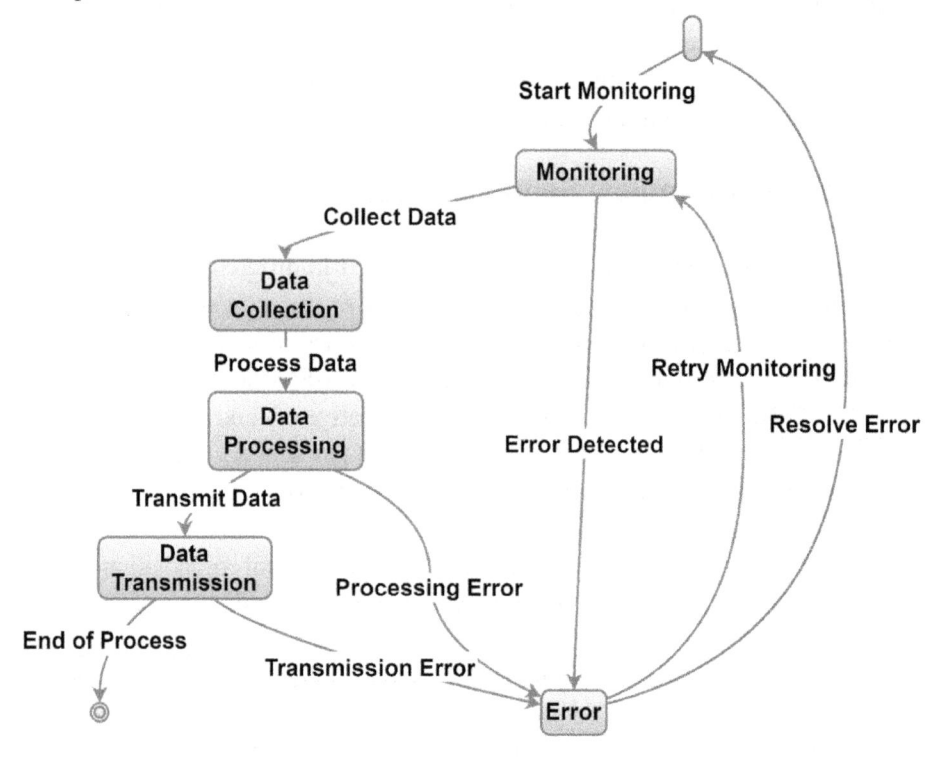

Data Collection Methods

Real-time monitoring in food quality management can be successfully done only when the means of data collection are appropriate. It involves several ways of collecting data from sensors and other sources to present timely and correct information to decision-making and control (Alam et al., 2021; Nath, 2024). The figure 2 depicts the states and transitions involved in real-time monitoring and data acquisition processes.

- Direct Measurement: These sensors involve direct measurement, and they continuously collect data from the environment or the product. Such examples include but are not limited to temperature sensors that give real-time quantitative readings of storage environment temperature and pH sensors that measure the acidity of food products. These are usually converted to digital signals which the microcontrollers process.
- Sampling: This technique involves taking a block of data from within a set or batch at equal intervals. This method is normally used where continuous monitoring is not feasible due to an expensive process or because of much complexity. For example, in large-scale production of food products, samples could be collected periodically from each batch for checking pH or contaminants regarding the quality.
- Automated Data Collection: Automatic systems record data through robotics, automatic machinery, software, and electronic hardware. Human interference is eliminated by such a setup. For example, in a food processing plant, automated systems can be designed where conveyor belts transport samples to sensors that will analyze them. This guarantees uniformity and effectiveness during the collection of information.
- Manual Data Collection: Operators measure data by hand using handheld instruments or devices. It is less common in high-tech systems; however, generally, this system is being used for conducting spot checks or when automated systems are not available.

Real-Time Monitoring Technologies

The real-time monitoring technologies enable the continuous observation of the parameters of food quality and immediate feedback, which enables immediate action if there is some sort of problem (Bandal & Thirugnanam, 2016).

- Internet of Things: IoT principally connects sensors and devices to the internet for remote monitoring and control. In the context of food quality monitor-

ing, IoT-enabled sensors may transmit data to cloud-based platforms, which can then be analyzed and accessed from anywhere, enabling real-time updates and alerts that allow for active management in food quality.
- Wireless Sensor Networks: WSN represents different sensors spread over a network with the capability to observe many parameters. All sensor nodes send their information back to a central gateway or coordinator that consolidates the information and sends it back to the monitoring system. WSN has been quite useful in large-scale food storage facilities where several parameters have to be monitored at different locations.
- Real-time data analytics: It involves the processing of data at the time of collection itself. Hence, it provided with immediate insights and actionable information. Techniques such as streaming data analysis and edge computing are done to gain real-time insights from sensor data to find trends, anomalies, and problems a long period of time before they grow out of proportion.
- Embedded Systems: These systems integrate sensors, micro-controllers, and software into a single unit for the purpose of monitoring and controlling the parameters of food quality. These systems operate uninterruptedly and provide real-time feedback related to temperature, humidity, and pH levels.

Data Transmission and Communication Protocols

The accurate representation of data transmission from sensors to monitoring systems and stakeholders has to be ensured with effective communication (Jedermann et al., 2014).

- Serial Communications: The embedded systems use serial communication, usually in the form of Universal Asynchronous Receiver/ Transmitter or UART. UARTs are relatively simple and inexpensive but are normally used to enable communication within embedded systems over short distances.
- I2C: It is a multi-master, multi-slave communication protocol used in the connection of sensors and microcontrollers over short distances. Several devices are capable of communicating on two wires-data and clock-which can be fitted for those applications that involve many sensors and devices.
- SPI stands for Serial Peripheral Interface: This is the synchronous serial communication between sensors and microcontrollers, used for high-speed data transfer. SPI utilizes lines for data, clock, and control signals separately; hence, it is quite fast in communication and reliable, especially in those applications that need high data rates.
- Wireless Communication: The protocols of wireless communication are used for sending data from sensors to the remote monitoring systems, including

Wi-Fi, Bluetooth, and Zigbee. Such protocols allow a wide range of data rates and areas, thus providing flexible and scalable communication solutions in food quality monitoring.
- Cellular Networks: Cellular communication protocols like GSM, 3G, 4G, and 5G can achieve wide area data transmission. Cellular networks are applied in applications for remote monitoring where internet connectivity is availed, enabling real-time data access and control from any location.
- Low-Power Wide Area Networks: LPWAN technologies are LoRaWAN and Sigfox, which allow for low-power, long-distance wireless communication. These protocols fit applications that demand less frequently transmitted data over long distances, such as monitoring remote or rural food storage facilities.
- Data Protocols and Standards: MQTT, or Message Queuing Telemetry Transport, and HTTP, or Hypertext Transfer Protocol, are the protocols and standards used to format and structure the data for transmission and interpretation. These protocols ensure compatibility between different systems and allow the smooth exchange of information.

In general, real-time monitoring and acquisition of data form a crucial part of effective food quality management. Precise and continuous observation of the key parameters can be enabled through various methods of data collection, along with advanced monitoring technologies. At the same time, protocols for transmission and communication at all levels are of great importance in ensuring accurate and timely data transfer for establishing decisions and responding to possible problems. Furthermore, with the future advance of technology, improvements in real-time monitoring and data acquisition will continue to increase this capability to uphold high standards in food quality and safety.

AUTOMATION AND CONTROL SYSTEMS

Control of Temperature and Humidity Automatically

Automated temperature and humidity control systems automatically ensure that the conditions of food storage, processing, and transportation meet the optimum requirements. These are advanced technological systems designed to protect specified environmental parameters for the purpose of maintaining the quality and safety of food. These automated temperature and humidity systems, integrated with sensors, microcontrollers, and control algorithms, provide a very accurate and reliable way of managing the levels of temperature and humidity (Dias et al., 2020; Prajwal et al., 2020). Figure 3 illustrates automatic processes for controlling temperature and humidity in response to changing environmental conditions.

Figure 3. Processes for controlling temperature and humidity automatically with changing environmental conditions

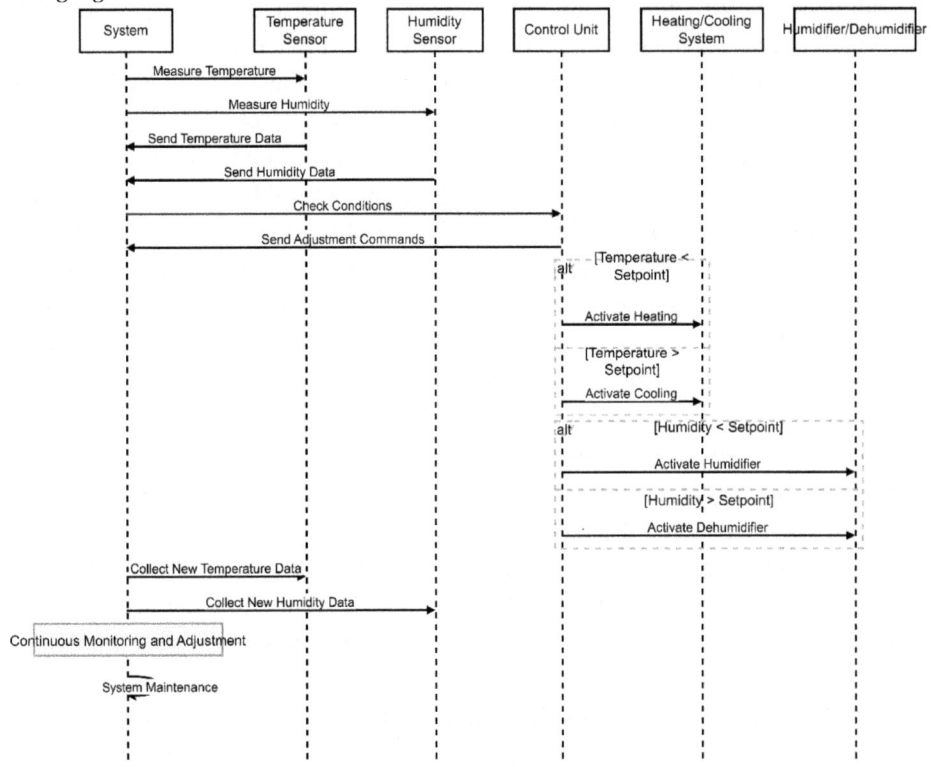

Sensors and Measurement: Automation of all control systems involves the use of sensors that may continuously give a reading of the levels of temperature and humidity. Temperature sensors include thermocouples or RTDs that give quick, accurate, and true readings of environmental air temperatures. Humidity sensors can also involve capacitive or resistive types that calculate moisture levels in the air. The data provided from these sensors is transmitted to a central control unit or a micro-controller.

Microcontrollers and Data Processing: The microcontroller will be serving as the brain of the control system by receiving data from sensors and using that to execute control algorithms to regulate the variables of environmental conditions. Instantly, it will process and compare real-time data against some predefined setpoints to identify if any change is called for. For instance, if the temperature increases beyond a certain predetermined threshold, it activates cooling systems that bring it within a set level.

Actuators and Control Mechanisms: These are the mechanisms responsible for implementing the adjustments instructed by the micro-controller. In temperature control systems, this can include heating elements or even cooling fans. In a system designed to regulate humidity, the actuators may involve humidifiers or dehumidifiers. It is at this point that the control system makes these actuators work, based on data provided by the sensors, to keep the environment within the set parameters.

Feedback and Adjustment: Feedback loops are really the heart of automated control systems. Sensors provide continuous data about the present state of the environment so that adjustments may be made by the micro-controller in real time. For example, if the humidity gets too low, the system starts turning on a humidifier until the optimum level is reached. In the same way, if the temperature falls outside the accepted range, the system automatically adjusts heating or cooling systems to bring the temperature back into the accepted range.

Integration with Other Systems: Generally, the temperature and humidity control system is integrated into an overall facility-wide environmental monitoring or control system. An inventory management system at a food processing plant, for instance, would integrate with the temperature and humidity control system along with quality control systems to ensure that all food quality aspects are controlled.

Changing Environmental Conditions

For both food safety and high-quality product realization, exact regulation of environmental conditions is very important(Kavitha et al., 2023; Maguluri et al., 2023). The following advantages are realized in automated systems for efficient adjustment of these conditions:

Real-time Adjustment: Automated systems ensure that there is continuous sensor feedback and real-time adjustment. This dynamic response further assures the temperatures and humidity levels will remain within narrow tolerances, which is important to inhibit spoiling and maintain product quality.

Energy Efficiency: Automation can boost energy efficiency greatly by ensuring the heating and cooling systems run optimally. For example, rather than running the cooling systems at full capacity, an automated system could adjust cooling powers based on current actual temperature values, resulting in lower energy use and operating costs.

Consistency and Reliability: Computer-controlled systems ensure not only consistency but also reliability in environmental control. Unlike manual adjustments, which could be prone to human error or variability, automation ensures that conditions remain within the desired envelope with minimal risk of quality issues or spoilage.

Most of the modern automated control systems include remote monitoring and control. This allows operators to view current data and implement changes from a distance to give them more flexibility in controlling the environmental conditions. Remote capabilities also enable quick responses to any issues that may arise.

This may include automated data logging of environmental conditions over time that can be evaluated for trends, examination of control system performance, and informed decisions for operational improvements. Data logging also provides valuable documentation for regulatory compliance and quality assurance.

Integration with IoT: The integration with the Internet of Things makes the automated control systems advanced. Integration of IoT into the automated systems makes them act in communication with other devices and systems for predictive maintenance over the systems where the defect that may arise is detected and sorted out before affecting the operation.

This calls for an automated temperature and humidity system, very essential in environmental control for food quality and safety. These systems would typically couple appropriate sensors and micro-controllers with effective actuators so as to run the environmental variables precisely and effectively. Real-time setting options, the path toward energy efficiency, and easy interfacing with other systems ensure consistent and reliable control. With advancements in technology, automated control systems would increasingly contribute significantly toward food safety and quality management.

DATA ANALYTICS AND INTERPRETATION

Data Analytics Techniques

Data analysis techniques are necessary in actionable insights from a huge amount of data that is collected using food quality monitoring systems. These techniques allow the estimation to be made regarding the current status of food quality, identification of possible issues, and making informed decisions accordingly(Alam et al., 2021; Li et al., 2006; Nath, 2024).

Descriptive analytics summarizes historic data in order to understand performance. Techniques include the use of mean, median, mode, and standard deviation to describe characteristics of the data in order to show a concise picture of what happened. For example, descriptive analytics can show temperature and humidity averages in a food storage facility over time, allowing general judgment as to whether the conditions generally meet or don't meet the required standards.

EDA: In EDA, it is expected to find out the hidden pattern, anomaly, and relationship from visual and statistical data. It uses different techniques like histograms, scatter plots, and box plots that can find insight into the path to build further analysis. EDA summarizes the distribution of data, outliers, and the correlation between variables, for instance, the relationship between humidity levels and spoilage rates.

Statistical Analysis: The use of mathematical techniques for the testing of hypotheses and inferences about a population based on sample data. Regression analysis, hypothesis testing, and ANOVA will be employed to establish the importance of the relationship or pattern observed. For instance, regression analysis will describe in detail how variations in temperature can affect the life span of a product.

Multivariate Analysis: Multivariate analysis is the technique where the relationship of more than one variable is studied at a time. Techniques of data reduction, such as principal component analysis and factor analysis, identify latent factors responsible for food quality. This will also be helpful in understanding such complex interactions of various factors together, such as temperature, humidity, and pH.

Predictive Analytics in Food Quality

Predictive analytics uses historical data and statistical algorithms to forecast future events and trends. To this end, predictive analytics allows food quality management to anticipate problems that are most likely to occur way in advance. This is instrumental in taking proactive measures aimed at ensuring the quality and safety of the products.

Forecasting: The goal of forecasting is to estimate future values of key parameters based on historical data. Time series analysis and machine learning models, such as ARIMA or LSTM, use past patterns to foresee future temperature and humidity levels. In this way, planning and adjustments could be made to avoid conditions that may result in spoilage or quality degradation.

Predictive modeling uses algorithms to build models that can predict certain outcomes from input data. Techniques, such as logistic regression, decision trees, and random forests, are included in modeling probability events of spoilage or contamination due to temperature fluctuations or humidity levels. This model could now be used in implementing any control strategy that improves quality management processes.

Anomaly detection helps in identifying patterns out of the ordinary and different from expected standards. This involves various techniques ranging from clustering to machine learning-based anomaly detection techniques using methods such as isolation forests, which identify outliers in sensor data to point out impending problems about malfunctioning equipment or breaches in standards concerning food safety.

Anomaly detection does, indeed, provide timely opportunities for intervention with corrective measures.

Risk Profiling: This is the determination of the likelihood occurring and the impact that potential risks may have on the quality of food. Predictive analytics can qualify the risks based on patterns and trends identified from the historical data, therefore helping in prioritizing interventions and resources. For example, predictive models can determine the risk of contamination with regard to environmental conditions and any history of contamination incidents.

Trend Analysis and Reporting

Trend analyses and reporting are useful for monitoring long-term changes in food quality, while decisions are supported with data insights(Sankar et al., 2023; Subha et al., 2023).

Trend Analysis: This is an analysis done over time to identify patterns and changes. Techniques include moving averages, seasonal decomposition, and trend lines, assisting in the explanation of long-term trends and cycles in food quality parameters. Trends in temperature and humidity data may be analyzed for their seasonal variation in order to help in adjusting storage conditions accordingly.

Data Visualization: Data visualization techniquesCharts, graphs, and dashboards-are utilized to communicate insightful data in a comprehensible and useable format. Techniques for visualization make better use of complex data more readily and intuitively present findings to stakeholders-be it via line charts in trend analysis or heatmaps in the identification of hotspots of anomalies.

Reporting: Through scheduled reporting, information about a set of key metrics and findings reached through the analysis of the data is given. Reports could be customized to point out specified aspects of food quality, such as deviations of set-points, trends of spoilage rates, or efficiency of control measures. Proper reporting will ensure that relevant stakeholders-quality managers and regulatory bodies-are kept posted about food quality status and the action required for it.

Benchmarking refers to performance measurements made against a predefined standard or best practice in the industry. Organizations may evaluate the food quality management processes and identify specific areas for improvement with this trend and performance measures analysis against benchmark performance.

Data analytics and interpretation are crucial for the betterment of the quality of food management. It thus helps organizations learn about various food quality parameters through different techniques of data analytics, predictive analytics, and trend analysis, using which valuable insights are obtained to anticipate all sorts of problems and arrive at quick decisions. Proper reporting and visualization of data further assist decision-making and make the quality management processes continuously optimum. With continuous technological advancement, the integration of sophisticated data analytics tools and techniques will continue to drive improvement in food quality and safety.

Figure 4. Challenges and considerations when using innovative microcontrollers in food quality monitoring

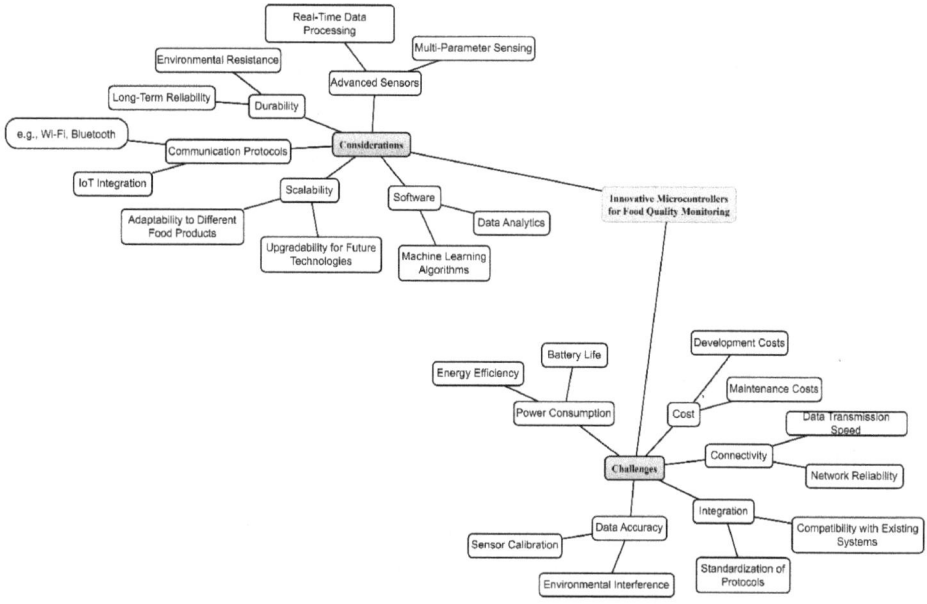

CHALLENGES AND CONSIDERATIONS

Figure 4 highlights the challenges and considerations associated with the use of innovative microcontrollers in food quality monitoring. The report emphasizes the importance of data accuracy, power consumption, connectivity, integration, cost, advanced sensors, communication protocols, software, durability, and scalability in the system (Dias et al., 2020).

Challenges in Cost and Implementation

Advanced food quality monitoring systems are liable to pose certain challenges on various grounds of cost and implementation. These challenges may affect the effectiveness or efficiency level of the overall system and hence, the management and planning need to be done with due care.

a) Initial Investment: The high initial investment in the installation of an integrated monitoring system, including sensors, microcontrollers, data acquisition systems, and software, is considered one of the main drawbacks. The sensors and advanced micro-controllers can also be rather costly, thus making the overall price an obstruction for most SMEs. Besides, there is an increased cost concerning installation and calibration.

b) Operational Costs: In addition to the initial investment in a system, the operational costs include continuous maintenance, calibration, and upgrades that may be needed for the monitoring system. Maintenance is a regular process to ensure that the system sustains its accuracy and reliability, while calibration is done periodically to maintain sensor accuracy. These will obviously add up over time and affect the overall budget meant for food quality management.

c) Skill Development and Training: Advanced systems require very specialized knowledge and skills in their implementation. Training of staff for operating the system effectively and maintaining it is crucial, though it is expensive and time-consuming. Thus, organizations must invest in the training programs and hire or develop such skilled personnel at the same time who have the capability to manage and analyze the data produced by the monitoring system.

d) Integration Complexity: The process of integration usually turns out to be complex because this involves the incorporation of new monitoring systems with existing infrastructure. This, in turn, requires modification of the old processes, addition of more hardware, or software adjustments. Compatibility at all levels needs to be ensured in order to avoid disruption of any operations and sustain operational efficiency.

Sensor Precision and Reliability

The food quality monitoring systems greatly require high accuracy and reliability of sensors. A sensor that is not accurate or reliable will result in incorrect data, hence decisions made based on such information will be affected and may further lead to an impact on food safety(Pachiappan et al., 2024).

a) Calibration: Sensors need periodic calibration to arrive at correct measures. Calibration consists of setting the sensor to a known standard and is called for to offset drift or any other modification in sensor performance over time. Inability to calibrate sensors will result in the derivation of incorrect data with a fall in the reliability of the system performance.
b) Environmental Conditions: These include temperature, humidity, even electromagnetic interference. For example, extreme levels of heat or excess humidity may influence the performance of the temperature and humidity sensors. The selection of sensors should, therefore, be based on efficiency in prevailing conditions at food processing or storage.
c) Sensor Degradation: Sensors are prone to deterioration due to friction and corroding conditions or contamination. Sensors need periodic replacement to continuously provide their output accurately and reliably; therefore, periodic maintenance is necessary. Use strategies to potentially identify issues so that predictive maintenance does not impact system performance.
d) Calibration and Testing: Sensors need periodic testing and recalibration to maintain accuracy. Establishing a proper calibration schedule and procedures ensures that sensors give reliable data and any discrepancies are promptly corrected.

Integration with Existing Systems

There are a number of challenges that are associated with the integration of new food quality monitoring systems with existing infrastructure. These need to be resolved in order to ensure that such installations are smooth and effective.

- System Compatibility: The new monitoring systems shall also be compatible with the existing hardware and software. It shall include format compatibility of data, communication protocols, and interfaces. Incompatibility can lead to data integration problems that may need further adjustments or customized solutions.
- Data integration means that in implementing data from new sensors, it has to be suitably planned towards being integrated with the existing management of data. Aggregation, processing, and analysis of data from different sources shall have to be done in concert. It is important that the new system interfaces the use of existing data platforms or databases for coherent data analysis and reporting.
- Operational Disruptions: The adoption of new monitoring systems actually causes disruptions in the current operation process. This is because installation and integration will cause disturbances in normal processes for a

considerable period of time. Proper planning and scheduling are needed to minimize disruptions to ensure a smooth transition of operations.
- System Integration Costs: The integration cost of any new systems to the already existing infrastructure is very high. Software customization, hardware modification, and system testing come under such costs. Organizations will be needed to make a budget for these integration costs and prepare resources for them.

The issues related to the implementation of a food quality monitoring system are basically related to the three foregoing aspects: cost, sensor accuracy, and system integration. The adverse factors include high capital and operational costs, which could be minimized if only they are diligently planned and managed. Sensor accuracy and reliability depend on calibration and routine maintenance. Their seamless integration into the existing systems would need consideration at the level of compatibility, data integration, and operation impact. Addressing these challenges in advance enables organizations to make their monitoring systems for food quality even more effective and efficient, thus improving safety and quality management in the food sector.

Figure 5. Future trends and innovations in innovative microcontrollers for enhancing food quality monitoring processes

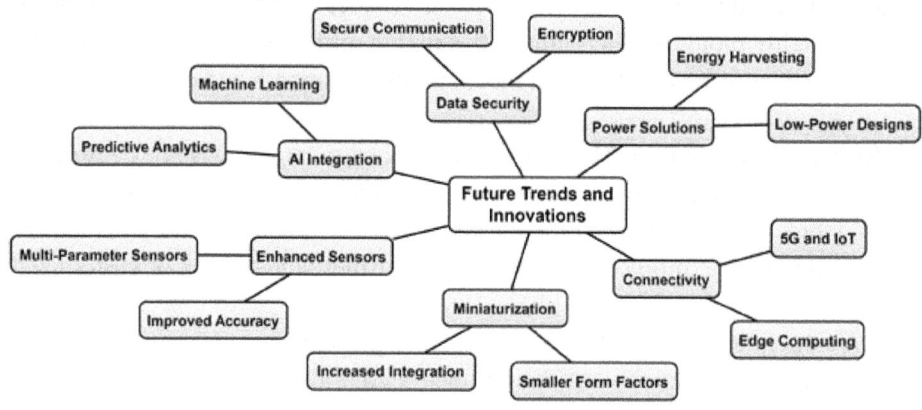

FUTURE TRENDS AND INNOVATIONS

Figure 5 outlines future trends and innovations in innovative microcontrollers for improving food quality monitoring processes. The future of food quality monitoring is going to be further characterized by rapid developments in micro-controller technology, emerging technologies, and innovative developments. These advances in microprocessors will see the application of more powerful, connected, secure, and smarter devices, hence driving the development of more sophisticated and efficient monitoring systems. Further, advanced sensors, AI, blockchain, and AR/VR will go a long way in making food quality monitoring more accurate, transparent, and effective. As these technologies continue to evolve, they will continue to drive the future landscape of food quality management towards more proactive, more efficient, and more reliable systems(Nath, 2024; Susanti et al., 2022; Ya'acob et al., 2021).

Advances in Microcontroller Technology

- Higher Processing Power: The future microcontrollers will have higher processing power, hence would be capable of performing complex functions that enable data-intensive applications. This would, in fact, enable more sophisticated data analysis to allow real-time decisions and effective control of food quality monitoring systems. Such improvements could be possible with the incorporation of multi-core processors and advanced architectures to enable quick processing of information for faster response times.
- Better Connectivity: New microcontrollers are most likely to be built with new connectivities such as 5G, Wi-Fi 6, and Bluetooth 5.0. These newer technologies will be able to achieve higher speeds and therefore faster data transfer, hence allowing for real-time monitoring and remote management of food-quality systems. Upcoming connectivity will also enable microcontrollers to integrate with IoT networks, hence enhancing system interoperability and the sharing of data.
- Low Power Consumption: Future microcontrollers will focus on higher performance without compromise on low power consumption. Power-efficient designs using low-power sleep modes and advanced power management techniques will extend battery life in wireless sensor nodes and reduce operational costs. This is very important for applications in remote or battery-operated monitoring systems.

Higher security features represent development, which increases in cybersecurity threats, making sophisticated future microcontrollers more secure. These assure protection against data and system integrity violations. Advanced features that add to the security of food quality monitoring systems against possible cyber-attacks include hardware-based encryption, secure booting, and tamper detection.

Emerging Technologies in Food Quality Monitoring

- Next-Generation Sensors: The new generation of sensors will dramatically raise the level of monitoring quality in food. Some of the new developing technologies, nanosensors, and biosensors are more sensitive, selective, and capable of real-time detection of contaminants, spoilage indicators, and other quality parameters. Such advanced sensors make the monitoring systems more effective with more accurate and reliable data.
- Artificial Intelligence and Machine Learning: AI and ML are about to change the way the process of monitoring food quality is done. Applications include predictive analytics, detection of anomalies, and automated decisions. Machine learning technologies will analyze voluminous data for finding patterns that predict potential defects that might occur before they happen. AI-driven systems amplify the accuracy of monitoring and enable being more proactive in the food quality management process.
- Food Quality Control: Food quality monitoring enters a new phase when blockchain technology is applied, as it will introduce total transparency and tamper-proof recordkeeping into the food supply chain. Recording each step of the supply chain on the blockchain enables full transparency in tracing the origin, handling, and quality of food products for all participants concerned. This technology will enhance traceability, accountability, and confidence in the safety of food.

AR and VR will find applications in the monitoring of food quality: immersive training, simulation, and visualization tools. Over and above this, options can further overlay real-time data and alerts on the physical environment through AR to better visualize and manage the parameters of food quality(Sekhar et al., 2024). VR can be used for training and also for scenario simulations with a view to improving the skill level of personnel concerned with quality management.

Potential Developments and Innovations

- Integration of Edge Computing: The integration of edge computing into systems that monitor food quality will enable data in real-time to be processed even at the sensor level. Since edge computing processes data either on the device or on another local device, it reduces latency and bandwidth, hence

enabling faster responses and efficient use of network resources. This will further improve the capability and performance of the system in real time.
- Smart Packaging: The intelligent labels and sensors imbedded in packaging materials themselves will open up new ways of monitoring the quality of food throughout its life cycle. It can provide information such as temperature, humidity, and freshness directly on the package in real time. Smart packaging will also improve traceability, reduce waste, and increase consumer confidence in the products.
- Integration of IoT with food quality monitoring systems: this will enable devices to talk with and share information with other devices seamlessly. IoT-enabled systems allow remote monitoring, automated alerts, and centralized management of parameters regarding the quality of food. The IoT, being a network, further smooths the operation of monitoring systems in real time with insights and control.
- Advanced Data Analytics Platforms: Future development in data analytics platforms will increasingly equip users with even more advanced, user-friendly tools to analyze data on food quality. The enabling technologies would include cloud-based analytics, advanced visualization, and integrated AI for deeper insights and more actionable information. These platforms support decision-making and continuous improvement in food quality management.

CONCLUSION

In this chapter, we looked at how microcontrollers and their associated technologies have transformed the way food quality is monitored. The core role of microcontrollers in devising sophisticated monitoring systems for data collection, analysis, and control in real time has been discussed. Advances in these systems-percentage processing power, increased connectivity, and energy efficiency-are opening ways to higher functionality and effective solutions in the management of food quality.

The study emphasizes the importance of sensors in monitoring systems, focusing on their functions of measuring temperature, humidity, pH, and chemical composition. Accuracy and reliability are crucial for system effectiveness. However, challenges in sensor performance and integration must be addressed for improved results.

The chapter discusses data analytics, the derivation of actionable insight, and interpretation from data gathered to drive meaningful insights. Predictive analytics are very much part of forecasting potential quality issues through trend analysis and real-time monitoring necessary for making the right informed decisions. These enhance a good ability to maintain high standards of food quality and address problems before they escalate.

Emerging trends and innovations in food quality monitoring are predicted to lead to revolutionary developments. Micro-controller technology advancements will enhance processing capability and connectivity, enabling more sophisticated systems. AI, blockchain, and smart packaging will contribute to increased accuracy, transparency, and efficiency in food quality monitoring.

As food quality monitoring systems continue to develop, the adoption of such innovations will be critical in furthering food safety and reducing waste to ensure that industry standards are met. With such advantages, organizations will have more control over the quality of food, which contributes to a much safer and more efficient food supply chain. The continuing integration of technology into monitoring systems and its development will no doubt shape the future with regard to food quality management, therefore bringing about change and innovation within the industry at all times.

REFERENCES

Alam, A. U., Rathi, P., Beshai, H., Sarabha, G. K., & Deen, M. J. (2021). Fruit quality monitoring with smart packaging. *Sensors (Basel)*, 21(4), 1509. DOI: 10.3390/s21041509 PMID: 33671571

Bandal, A., & Thirugnanam, M. (2016). Quality measurements of fruits and vegetables using sensor network. *Proceedings of the 3rd International Symposium on Big Data and Cloud Computing Challenges (ISBCC–16)*, 121–130. DOI: 10.1007/978-3-319-30348-2_11

Dias, R. M., Marques, G., & Bhoi, A. K. (2020). Internet of things for enhanced food safety and quality assurance: A literature review. *International Conference on Emerging Trends and Advances in Electrical Engineering and Renewable Energy*, 653–663.

Garcia-Breijo, E., Garrigues, J., Sanchez, L. G., & Laguarda-Miro, N. (2013). An embedded simplified fuzzy ARTMAP implemented on a microcontroller for food classification. *Sensors (Basel)*, 13(8), 10418–10429. DOI: 10.3390/s130810418 PMID: 23945736

Grazioli, C., Faura, G., Dossi, N., Toniolo, R., Abate, M., Terzi, F., & Bontempelli, G. (2020). 3D printed portable instruments based on affordable electronics, smartphones and open-source microcontrollers suitable for monitoring food quality. *Microchemical Journal*, 159, 105584. DOI: 10.1016/j.microc.2020.105584

Jedermann, R., Pötsch, T., & Lloyd, C. (2014). Communication techniques and challenges for wireless food quality monitoring. *Philosophical Transactions of the Royal Society A: Mathematical, Physical and Engineering Sciences, 372*(2017), 20130304.

Kaushik, S., & Singh, C. (2013). Monitoring and controlling in food storage system using wireless sensor networks based on zigbee & bluetooth modules. *International Journal of Multidisciplinary in Cryptology and Information Security*, 2(3).

Kavitha, C., Varalatchoumy, M., Mithuna, H., Bharathi, K., Geethalakshmi, N., & Boopathi, S. (2023). Energy Monitoring and Control in the Smart Grid: Integrated Intelligent IoT and ANFIS. In *Applications of Synthetic Biology in Health, Energy, and Environment* (pp. 290–316). IGI Global.

Li, Z., Wang, N., Raghavan, G., & Cheng, W. (2006). A microcontroller-based, feedback power control system for microwave drying processes. *Applied Engineering in Agriculture*, 22(2), 309–314. DOI: 10.13031/2013.20277

Maguluri, L. P., Ananth, J., Hariram, S., Geetha, C., Bhaskar, A., & Boopathi, S. (2023). Smart Vehicle-Emissions Monitoring System Using Internet of Things (IoT). In *Handbook of Research on Safe Disposal Methods of Municipal Solid Wastes for a Sustainable Environment* (pp. 191–211). IGI Global.

Nath, S. (2024). *Advancements in food quality monitoring: Integrating biosensors for precision detection*. Sustainable Food Technology.

Pachiappan, K., Anitha, K., Pitchai, R., Sangeetha, S., Satyanarayana, T., & Boopathi, S. (2024). Intelligent Machines, IoT, and AI in Revolutionizing Agriculture for Water Processing. In *Handbook of Research on AI and ML for Intelligent Machines and Systems* (pp. 374–399). IGI Global.

Popa, A., Hnatiuc, M., Paun, M., Geman, O., Hemanth, D. J., Dorcea, D., Son, L. H., & Ghita, S. (2019). An intelligent IoT-based food quality monitoring approach using low-cost sensors. *Symmetry*, 11(3), 374. DOI: 10.3390/sym11030374

Prajwal, A., Vaishali, P., & Sumit, D. (2020). Food quality detection and monitoring system. *2020 IEEE International Students' Conference on Electrical, Electronics and Computer Science (SCEECS)*, 1–4.

Rivas-Sánchez, Y. A., Moreno-Pérez, M. F., & Roldán-Cañas, J. (2019). Environment control with low-cost microcontrollers and microprocessors: Application for green walls. *Sustainability (Basel)*, 11(3), 782. DOI: 10.3390/su11030782

Sankar, K. M., Booba, B., & Boopathi, S. (2023). Smart Agriculture Irrigation Monitoring System Using Internet of Things. In *Contemporary Developments in Agricultural Cyber-Physical Systems* (pp. 105–121). IGI Global. DOI: 10.4018/978-1-6684-7879-0.ch006

Sekhar, K. Ch., Ingle, R. B., Banu, E. A., Rinawa, M. L., Prasad, M. M., & Boopathi, S. (2024). Integrating VR and AR for Enhanced Production Systems: Immersive Technologies in Smart Manufacturing. In *Advances in Computational Intelligence and Robotics* (pp. 90–112). IGI Global. DOI: 10.4018/979-8-3693-6806-0.ch005

Subha, S., Inbamalar, T., Komala, C., Suresh, L. R., Boopathi, S., & Alaskar, K. (2023). A Remote Health Care Monitoring system using internet of medical things (IoMT). *IEEE Explore*, 1–6.

Susanti, N. D., Sagita, D., Apriyanto, I. F., Anggara, C. E. W., Darmajana, D. A., & Rahayuningtyas, A. (2022). Design and implementation of water quality monitoring system (temperature, ph, tds) in aquaculture using iot at low cost. *6th International Conference of Food, Agriculture, and Natural Resource (IC-FANRES 2021)*, 7–11. DOI: 10.2991/absr.k.220101.002

Upadhyaya, A. N., Saqib, A., Devi, J. V., Rallapalli, S., Sudha, S., & Boopathi, S. (2024). Implementation of the Internet of Things (IoT) in Remote Healthcare. In *Advances in Medical Technologies and Clinical Practice* (pp. 104–124). IGI Global. DOI: 10.4018/979-8-3693-1934-5.ch006

Venkatesh, A., Saravanakumar, T., Vairamsrinivasan, S., Vigneshwar, A., & Kumar, M. S. (2017). A food monitoring system based on bluetooth low energy and Internet of Things. *International Journal of Engineering Research and Applications*, 7(3), 30–34. DOI: 10.9790/9622-0703063034

Vergara, A., Llobet, E., Ramírez, J., Ivanov, P., Fonseca, L., Zampolli, S., Scorzoni, A., Becker, T., Marco, S., & Wöllenstein, J. (2007). An RFID reader with onboard sensing capability for monitoring fruit quality. *Sensors and Actuators. B, Chemical*, 127(1), 143–149. DOI: 10.1016/j.snb.2007.07.107

Wang, X., Fu, D., Fruk, G., Chen, E., & Zhang, X. (2018). Improving quality control and transparency in honey peach export chain by a multi-sensors-managed traceability system. *Food Control*, 88, 169–180. DOI: 10.1016/j.foodcont.2018.01.008

Ya'acob, N., Dzulkefli, N., Yusof, A., Kassim, M., Naim, N., & Aris, S. (2021). Water quality monitoring system for fisheries using internet of things (iot). *IOP Conference Series. Materials Science and Engineering*, 1176(1), 012016. DOI: 10.1088/1757-899X/1176/1/012016

Chapter 4
Smart Food Quality Monitoring by Integrating IoT and Deep Learning for Enhanced Safety and Freshness

Kavitha Kumari K. S.
Department of Electrical and Electronic Engineering, Aarupadai Veedu Institute of Technology, Vinayaka Mission's Research Foundation (Deemed), India

J. Samson Isaac
School of Engineering and Technology, Karunya Institute of technology and Sciences, Coimbatore, India

V. G. Pratheep
https://orcid.org/0000-0002-2659-5474
Department of Mechatronics Engineering, Kongu Engineering College, Perundurai, India

M. Jasmin
https://orcid.org/0009-0007-9268-0937
Department of Electronics and Communication Engineering, New Prince Shri Bhavani College of Engineering and Technology, Chennai, India

A. Kistan
https://orcid.org/0000-0003-1334-4331
Department of Chemistry, Panimalar Engineering College, Chennai, India

Sampath Boopathi
https://orcid.org/0000-0002-2065-6539
Department of Mechanical Engineering, Muthayammal Engineering College, Namakkal, India

DOI: 10.4018/979-8-3693-5573-2.ch004

ABSTRACT

This chapter investigates the integration aspects between the Internet of Things and deep learning technologies in efforts directed toward advancing food quality monitoring, thus enhancing issues of safety and food freshness in the supply chain. IoT sensors capture real-time data with regard to environmental conditions including temperature, humidity, and gas composition all through the food production and supply process. Such data is analyzed by deep learning algorithms to detect any kind of anomalies and predict the potential hazards to enable proactive actions toward timely intervention. The chapter deals some IoT devices used for monitoring, namely smart sensors and wearable technologies, while comparing the application of deep learning models for predictive and pattern recognition analytics. Case studies underscore this integrated approach to reducing spoilage, increasing shelf life, and meeting the requirements put forth by today's safety standards.

INTRODUCTION

Food quality monitoring has become an important part of the present-day food supply chain. There is an increasing demand for safer, fresher, and more sustainable food products. Traditional methods for the control of food quality often mean periodic inspection and manual assessment. These are increasingly being supplemented, and in some cases superseded, by advanced technologies. Among them, the integration of IoT and deep learning stands out to become one of these transformative approaches that could enhance the effectiveness and efficiency of food quality monitoring (Tutul et al., 2023).

The IoT stands for a networked fabric of devices that gather, interact, and act upon multi-sourced data. In a food-quality context, IoT sensors are set up along food processing and distribution lines to keep track of environmental conditions related to temperature, humidity, gas composition, among others. Those sensors provide real-time data, which becomes critical in the maintenance of optimal conditions necessary for fresh food preservation and safety. For instance, temperature sensors track the condition within refrigeration units, whereas humidity sensors record the storage environment. Real-time data provides for immediate reaction to deviations outside of the set parameters, reducing the risks of spoliation and contamination (Khan et al., 2020).

The IoT environment nonetheless presents the challenge in the sheer volume and complexity of the production of data. That is where deep learning comes in—a subset of machine learning that uses neural networks with multiple layers. Deep learning algorithms are designed to analyze vast amounts of data for patterns or anomalies that may not be apparent to human observers. These algorithms can be applied to data from IoT sensors to truly understand the issues in food quality and forecasting any possible complications before their occurrence. Deep learning models, in particular, become very effective in the case of complex data processing and interpretation. For example, CNNs can efficiently analyze images and be used to detect visual signs of spoilage or quality degradation of food products. On the other hand, RNNs fit very well to analyze time-series data, like temperature or humidity changes over some time (Nayak et al., 2020). All these models could be integrated into one and handled a comprehensive system that would monitor the current situation and would also look forward to assess future trends and potential risks.

Some of the key advantages of the integration of IoT and deep learning in food quality monitoring include: First, it provides real-time monitoring. Traditional methods have gaps between collecting data and analyzing it. These often miss out on the opportunity for intervention at an early stage. IoT devices continuously collect data streams, and the same can be analyzed in real time with deep learning algorithms to detect any issues and resolve them on the spot. For example, if the temperature sensor senses an increase in temperature inside the refrigeration unit, deep learning model analysis on historical data will determine, in a very short time, whether this deviation might become a threat to food quality(Nerkar et al., 2023). In case of necessity, correction action, like adjusting the temperature or alerting staff for equipment inspection, would automatically be triggered.

Better predictive maintenance due to IoT and deep learning integration. Deep learning models predict when maintenance needs to be performed, prior to an occurring problem, through the analysis of historical data in association with pattern recognition related to equipment failures or quality issues. That approach is proactive in reducing downtime and minimizing the potential quality issue impact on the food supply chain. For instance, if a deep learning model recognizes a recurring pattern of temperature fluctuations in a refrigeration unit, it can use that information to predict when the unit is likely to fail and recommend the implementation of appropriate preventive maintenance measures. In addition, decision-making gets more efficient with this integrated approach that ensures actionable insights (Abass et al., 2024). Deep learning models can create detailed reports and visualization, those pinpoint trends, anomalies, and risks. It enables food producers and distributors to make effective decisions related to quality control, inventory management, and process improvement. For instance, if a deep learning model picks out that some specific batch of food products are continuously being exposed to undesirable conditions, it can recommend adjustments to the storage or transportation processes to mitigate the risk of spoilage.

Integration of IoT and Deep Learning in Food Quality Monitoring: Challenges. Despite the numerous benefits, the integration of IoT and deep learning in food quality monitoring also comes with some challenges. The principal one is ensuring that the data obtained from the IoT sensors is accurate and reliable. In fact, quality data is influenced by three factors: calibration of the sensor, environmental factors, and issues in the transmission of data. Besides, deep learning requires large volumes of quality data to be effectively trained (Manisha & Jagadeeshwar, 2023). Inadequate or biased data may lead to inaccurate predictions and less reliable insights. This is the reason why robust data validation and preprocessing techniques are of importance to ensure the reliability of a monitoring system.

Another challenge lies in the level of data privacy and security concerns involved. Constant collection and transmission of the data raise issues related to data protection and compliance. Most importantly, ensuring that data is securely stored and transmitted, with due consideration for privacy concerns, will help in the retention of trust and achievement of regulatory compliance (Sumathi et al., 2021).

Deep learning combined with IoT represents a step forward in monitoring the quality of food by bringing improved safety and freshness to your table through real-time monitoring, predictive maintenance, and actionable insights. These developments will only keep getting more sophisticated with advances in technology, bringing in efficiency and effectiveness in food quality management. Mastering the challenges in these technologies is the only way to achieve higher standards in quality, safety, and sustainability for food industry operations (Bhardwaj et al., 2022).

Background: Traditionally, food quality control has been performed through periodic checking and manual inspection. This is inefficient and full of errors. With the use of IoT technology today, there stands the potential for continuous environmental-condition monitoring in real-time due to factors of food quality such as temperature and humidity. At the same time, deep learning has taken over as a very powerful tool in the analysis of complex data patterns and predicting forthcoming problems. The integration of these technologies is, therefore, needed for a more precise and proactive way of managing food safety and freshness, bridging the limitations of conventional ways of improving general efficiency and effectiveness in the food supply chain .

Scope: IoT integrated with deep learning approaches in monitoring food quality is done in real-time data acquisition and analysis from the various phases of the food supply chain, such as production, storage, and distribution. This is from implementing IoT sensors that monitor environmental parameters such as temperature and humidity to the application of deep learning models in processing and interpreting the data. This also includes predictive analytics in order to forecast some possible quality issues, automated responses to mitigate the risks, enhanced food safety, extended shelf life, and improved quality management in general—with challenges of data accuracy and security.

Objectives: Principal objectives of integration of IoT and deep learning in food quality monitoring would be an improvement of real-time tracking on environmental conditions, predictive maintenance towards preventing spoliation, and actionable for decision-making. In essence, the aim is to allow both safety and freshness through the supply chain using IoT sensors for data collection and deep learning algorithms for advanced data analysis. It is also expected to reduce the costs associated with manual inspection, minimize quality issues, and support proactive measures in ensuring high standards of food quality and compliance.

INTERNET OF THINGS (IOT) IN FOOD QUALITY MONITORING

Figure 1. Internet of Things (IoT) in food quality monitoring

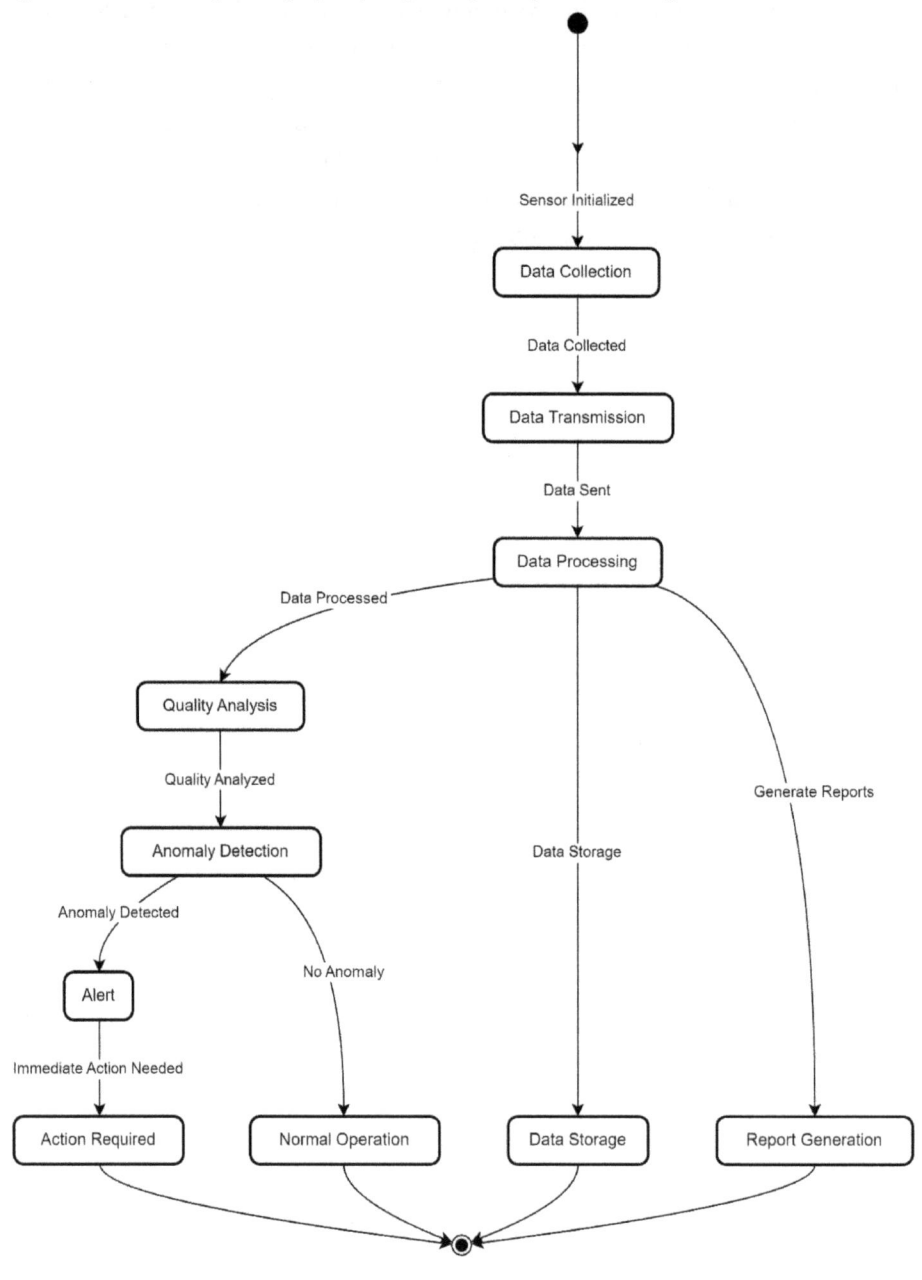

Figure 1 illustrates the stages of IoT-based food quality monitoring, including idle, data collection, data transmission, data processing, quality analysis, anomaly detection, alerting, action required, normal operation, data storage, and report generation. It provides a comprehensive overview of these states and their transitions(Atitallah et al., 2020; Manikandan et al., 2023).

IoT Technologies and Devices

The Internet of Things (IoT) encompasses a wide range of technologies and devices designed to collect, transmit, and act on data from interconnected sensors and systems. In food quality monitoring, IoT technologies play a crucial role by providing real-time insights into various environmental and operational parameters.

Sensor Technologies: The IoT is an all-encompassing term that describes technologies and devices developed to capture, transmit, and act upon data that, in turn, is generated by interconnected sensors and systems. For food quality monitoring, technologies of great importance in the development of real-time insight into environmental and operational parameters are related to the IoT. The core aspects of these techniques are sensor technology, mainly measuring some key parameters affecting food safety and freshness, including:

- Temperature Sensors: These play a vital role in monitoring storage and transportation conditions. The food products need to be in required temperature ranges to avoid any kind of spoilage. Some of the examples are thermocouples, resistance temperature detectors, and infrared sensors.
- Humidity Sensors: These sensors detect the moisture content in the storage environment. Higher than usual humidity promotes mold growth, which in turn promotes spoiling. Typical humidity sensors utilized to achieve this include capacitive and resistive sensors.
- Gas Sensors: They detect the gas concentration of CO_2 and ethylene, which shows ripening or deteriorating stages. The sensors utilized to detect the gases are electrochemical sensors and metal oxide sensors.
- pH Sensors: This category of sensor measures the acid or alkali phases of food products at a particular stage of the product line. The pH sensors offer appropriate readings of the product's acidic and basic phases of the product with electro-chemical methods.
- Optical Sensors: These sensors determine absorption and reflection of light. Here, it estimates changes in color, and therefore, it determines spoilage. These sensors are mostly incorporated with image processing.

Data Collection and Transmission:

Internet of things devices collect data through the different types of sensors. This gathered information is transmitted to one central system for analysis. The following are the critical elements noticed in data gathering and transmission(Bhardwaj et al., 2022):

- Acquisition of data: The collection of data on environmental conditions, status of the product, and key operation metrics from multiple sensors is relayed. The accrued data is usually continuously recorded and aggregated from several sources.
- Data Transmission: The accrued data is then transmitted through any of the available wireless communication protocols, such as Wi-Fi, Bluetooth, Zigbee, or cellular networks. The wireless links ensure that data is sent efficiently and securely to some central server or to a cloud-based platform.
- Data Storage: Information is stored in databases or cloud storage systems. This allows for scalable and secure ways to handle large chunks of information. Cloud platforms bring in several advantages like remote access, high availability, and integration with analytics tools.
- Data Integration: Most IoT systems would integrate with existing enterprise resource planning and supply chain management systems to drive seamless data flow and improved line of sight across a food supply chain.

Examples of IoT Applications in the Food Industry

- Smart Warehousing and Inventory Management: IoT technologies make a difference in warehousing and inventory management with the real-time monitoring of storage conditions. Smart warehouse systems, through temperature and humidity sensors, can maintain perfect conditions for perishable goods. It sends automated alerts in case of any deviation from acceptable ranges to take corrective actions in time. Further, IoT-enabled inventory management systems track stock levels and expiration dates to assure that products move out in time to reduce waste and ensure timely rotation(Ali et al., 2024; Boopathi, 2024c).
- Cold Chain Monitoring: Cold chain is an integral part of the food supply chain, focusing on the required temperature conditions of the perishable goods within this chain. The installations of IoT devices take place in cold storage facilities, refrigerated transportation, and delivery trucks. The continuous temperature conditions are tracked through temperature sensors and the information sent back through a central system. Should there be any chang-

es in temperature, it triggers automated alerts to the operators, who shall promptly act to prevent deterioration.
- Quality Control within Food Processing: IoT sensors are deployed in food processing units that track various parameters, such as temperature, pH, and gas concentrations. In real time, these data permit the checking of processing conditions against quality standards. Take, for instance, fermentation, one of the many processes that dairy products undergo. The temperature conditions and pH are checked through sensors to ensure the best possible conditions prevail to guarantee product quality. The use of sensor data by automated systems in adjusting processing parameters improves consistency and reduces the propensity for faults.
- Farm-to-Fork Traceability: IoT technologies make the food supply chain traceable right from farms to consumers. It furnishes information about where food products commence their journey. Sensors on farms track soil condition, crop health, and environmental factors. IoT devices at the time of transportation monitor parameters like temperature and humidity. This information is logged against product labels so that it tracks back right down to the end consumer about the origin and handling conditions of food products.
- Equipment Predictive Maintenance: IoT sensors track the condition of equipment involved in food production and processing. Now, analyzing data from vibration, temperature, and operation metrics, predictive maintenance systems can anticipate equipment failures before they take place. This proactive approach minimizes downtimes, reducing maintenance costs and ensuring uninterrupted production—both critical to the delivery of consistent product quality.
- Consumer Engagement and Smart Packaging: Another critical application of IoT technologies is in smart packaging solutions, which deliver information to consumers about food products in real-time. For instance, through the scanning of QR codes or NFC tags embedded in the packaging, consumers can have details on the freshness, handling, and safety of the products. This type of transparency will build up trust between the consumer and the product and further help in better decision-making at the point of sale.

Integration of IoT technologies into food quality monitoring enables extensive improvements in managing and maintaining the safety and freshness of foods. This improves the efficiency and effectiveness of the quality monitoring process at each stage of the supply chain through the application of sensor technologies, real-time data transmission, and numerous applications in this field. In the near future, these evolving technologies will have the potential to further improve safety, reduce waste, and increase transparency regarding foods.

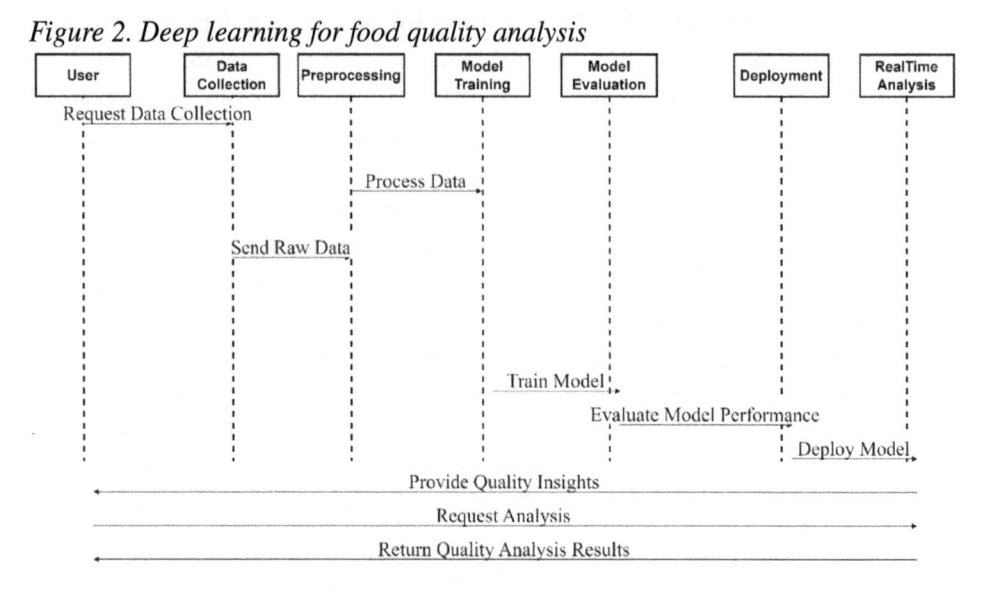

DEEP LEARNING FOR FOOD QUALITY ANALYSIS

Figure 2. Deep learning for food quality analysis

```
User  Data        Preprocessing  Model      Model        Deployment  RealTime
      Collection                 Training   Evaluation               Analysis

Request Data Collection
                        Process Data
           Send Raw Data
                                  Train Model
                                             Evaluate Model Performance
                                                          Deploy Model
                        Provide Quality Insights
                        Request Analysis
                        Return Quality Analysis Results
```

The process of applying deep learning for food quality analysis involves user request, data collection, pre-processing, training, evaluation, deployment, and real-time analysis as shown in Figure 2. This diagram illustrates the sequence of steps from data collection to real-time analysis.

Introduction to Deep Learning

Deep learning is the subset of machine learning that includes artificial neural networks with multiple layers—hence the term "deep." These networks have been designed in a manner such that they are capable of automatically learning and extracting features from raw data, thus taking over complex tasks like image and speech recognition, natural language processing, and predictive analytics. Deep learning algorithms analyze the food quality by processing huge data reams generated from IoT sensors and other sources to seek out patterns relevant to predicting food safety and freshness(Kumar et al., 2023; Ugandar et al., 2023).

Deep learning models have been quite efficient in treating high-dimensional data, such as images and time-series data, very common in food quality monitoring. These models may learn intricate patterns and relations in the data that are not so easily apparent through traditional analytical methods. Hence, deep learning becomes a really strong tool in helping to address all sorts of challenges posed by food quality management, with the ability for feature extraction and model training to be automated.

Important Algorithms and Models

Many deep learning algorithms and models are relevant to the field of food quality analysis. Some of them include the following(Boopathi, 2024a; KAV et al., 2023; Sangeetha et al., 2023):

- Convolutional Neural Networks (CNNs): Among most other neural networks, CNNs are explicitly designed to work with grid data, such as images. CNNs are mainly made up of some convolutional layers that convolve the input with filters in order to extract features, followed by a number of pooling layers for downsampling the data and then some fully connected layers for classification or regression. CNNs can be applied in food quality analysis by analyzing visual data from cameras and sensors for decay, contamination, or defect identification within a food product. For example, CNNs identify discoloration or texture change in fruits and vegetables, hence defining the quality status of produce.
- Recurrent Neural Networks: RNNs were specifically designed to handle sequential data and time-series analysis. They have intrinsic feedback loops, making them capable of maintaining some kind of memory about previous inputs. An especial form of these RNNs—the Long Short-Term Memory networks—turns out to be very good at learning long-term dependencies in time-series data. RNNs in food quality monitoring can study data produced by sensors as time series and understand trends or anomalies in temperature, humidity, or gas concentrations that may impact food quality.
- Autoencoders: These are neural networks used for unsupervised learning. An autoencoder is basically composed of an encoder that transforms the source data into a lower-dimensional representational space and a decoder that reconstructs the data from this representation. This makes them helpful in anomaly detection for food quality analysis, as they learn where the compact representation of normal data lies and thus detect deviations that might indicate spoilage or other quality issues.
- Generative Adversarial Networks (GANs): A generator and a discriminator constitute the most simple form of GAN, which is trained jointly for the two. The generator generates synthetic data, while the discriminator estimates the authenticity of the data. In cases where actual data is scarce, GANs can be used to generate synthetic examples to train deep learning models. This technique will benefit the domain of food quality analysis by generating simulat-

ed data for model training on rare spoilage conditions or new types of food products.

DRL combines deep learning with reinforcement learning in pursuit of the optimization of decision-making processes. It is applied in scenarios whereby a model would learn from many interactions with its environment. In food quality management, DRL can be used for operational decisions on storage conditions or maintenance scheduling based on real-time data and predicted outcomes.

Applications in Anomaly Detection and Predictive Analytics

Deep learning models offer huge advantages in the domain of both anomaly detection and predictive analytics related to food quality (Kumar et al., 2023; Sampath et al., 2022; Sangeetha et al., 2023).

Anomaly Detection: Deep learning excels at the identification of deviations from normal patterns. Models like autoencoders and CNNs can be trained on historical data to learn what high-quality food products normally look like, and then analyze new data in real time to flag anomalies that could mean the food has gone bad or has become contaminated. For example, a CNN trained on images of fruits and vegetables can detect tiny changes in color or texture, thus signaling a decline in quality. Similarly, auto-encoders are helpful in detecting anomalous patterns in time-series data, like spikes in temperature that have potential implications for food safety.

Predictive Analytics Deep learning models are also good at making predictions from past trends and patterns that have been detected. If one were speaking of food quality, predictive analytics can help give early warning before such problems actually arise. For instance, RNNs use time-series data from temperature sensors that predict future changes in temperatures and their potential effect on food quality. One could also use the CNN or LSTMs to predict rates of spoilage based on environmental conditions and historical trends. It allows food manufacturers and distributors to create ideal storage conditions, plan the right inventories, and ensure precautions to deliver high-quality products to their customers. Deep learning provides a very powerful method of analysis for food quality data with advanced capabilities in terms of anomaly detection and predictive analytics(Boopathi, 2023; Hema et al., 2023).

In other words, it is easier for the food industry to make more sense out of quality issues and come up with better ways of maintaining safety and freshness when guided by algorithms like CNNs, RNNs, auto-encoders, GANs, and DRL. The integration of deep learning with IoT and other data sources furthers food quality monitoring for improvements in efficiency, accuracy, and overall product quality.

INTEGRATION OF IOT AND DEEP LEARNING

System Architecture

In the case of deep learning and IoT integration, two strong technologies are combined and built on to provide an enhanced framework for monitoring food quality. The overall architecture for such system integration normally comprises the following key components (Bikash Chandra Saha, 2022; Hema et al., 2023; Syamala et al., 2023):

Figure 3. Integration of IOT and deep learning for food quality monitoring

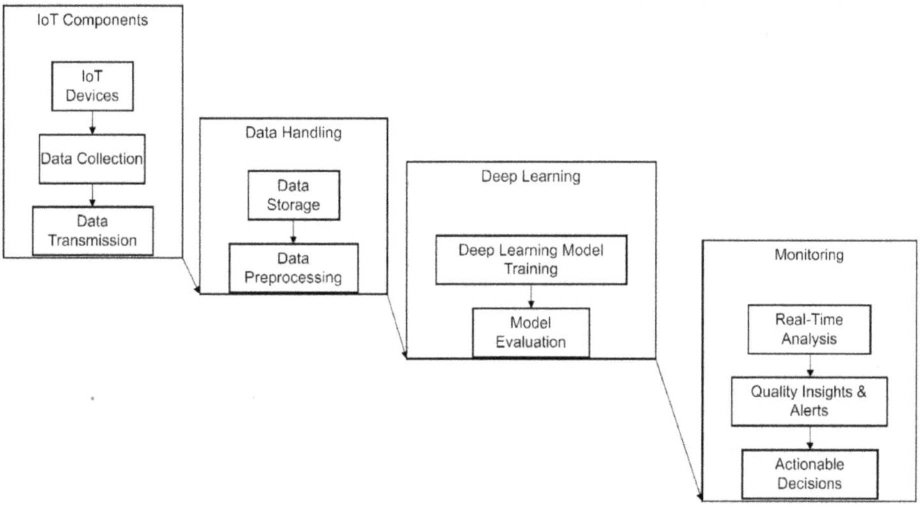

IoT devices and sensors are the most basic system constituents, intended to gather data from the physical environment. This includes temperature, humidity, gas, and pH sensors, as well as cameras for visual control. Such sensors could be located at a number of points along the food supply chain: in the facilities for production, storage, and transportation.

Data Collection and Transmission: In an uninterrupted cycle, an IoT sensor collects data and subsequently transmits it to a central data processing unit. This kind of transmission is aided by wireless communication protocols that define the nature of connectivity, which could be through Wi-Fi, Bluetooth, Zigbee, or cellular networks. The data is then relayed for processing at edge computing devices or directly to any cloud-based platform, contingent upon the design and requirements of the system(Hema et al., 2023; Syamala et al., 2023).

Edge Computing: Edge computing is the processing of data closer to where it is collected. This is particularly valuable in real-time applications requiring immediate analysis and action. Edge devices can preprocess data from IoT sensors, perform the first level of filtering and aggregation, and even execute lightweight deep models for the identification of immediate anomalies or triggering alerts.

Cloud-Based Data Processing: For more complex analysis, data is sent to a cloud platform where robust models of deep learning are deployed. The cloud infrastructure provides all the computational power to process large amounts of data and run sophisticated models. This accommodates a great variety of sources of data, which are then stored for further longitudinal analysis.

Deep models deployed either in the cloud or at edges depend on individual complexity and computational requirements form models like Convolutional Neural Networks, Recurrent Neural Networks, Autoencoders. These models learn the data for pattern recognition, anomaly detection, and trend recognition. Such outputs from these models further go to make insights and, in turn, recommendations.

User Interfaces and Dashboards: This is the final component, used to show users the results of analyses. Dashboards and user interfaces portray data, trends, and alerts to the customer in a form that they can understand. They allow operators and decision makers to monitor the quality of food in real-time and take the correct action accordingly.

Data Fusion and Processing

Data fusion and processing are very critical in the integration of IoT and deep learning. It is all about the integration of data streams from different sources and the use of advanced analytics techniques to get meaningful information(BOOPATHI et al., 2024; Upadhyaya et al., 2024; Venkateswaran et al., 2024).

Data Fusion: IoT systems generate heterogeneous data types, including numerical readings from sensors, images from cameras, and time-series data. Data fusion combines these different sources of data into one dataset. Techniques used in fusing data coming from different sensors and sources include feature extraction, normalization, and alignment. This fusion thus enables a deep learning model to scrutinize the food quality landscape from a more comprehensive perspective(Agrawal, Magulur, et al., 2023; Rahamathunnisa et al., 2023).

Data preprocessing: Before feeding data into deep learning models, a lot of preprocessing steps are needed to assure the quality and relevance of the data. This step involves cleaning data for noise and errors, normalization of values to some standard scales, handling missing or incomplete data, and other related tasks. For image data, it might involve resizing or cropping an image or even augmenting the content to increase the performance of the model.

Feature Extraction and Engineering: Deep learning models enable the training of relevant features that can bring out important aspects of the data. Feature extraction would be the identification of features in the raw data. In images, it could be edges or textures. For time series data, trends or periodic patterns are examples. A better performance for a model is given through feature engineering by creating raw data with informative inputs.

Model Training and Validation: Deep learning models learn through examples by using historical information. Training will adjust model parameters to minimize the prediction error. In the stage of validation, a model is checked for good generalization ability with respect to new, unseen data. Cross-validation and hyperparameter tuning are some of the techniques used in improving model accuracy and robustness.

Real-Time Processing: Real-time applications require quick processing so that the responses are immediate. Edge computing thus plays an important role in real-time data processing at the edges by doing initial analysis to cut down latency in the system. Even though complex models normally run in the cloud, edge devices can quite easily take up the preprocessing task and preliminary anomaly detection to provide instantaneous feedback.

Example Real-Time Integrated Solutions

Cold chain monitoring: In a cold chain monitoring system, IoT sensors track temperature and humidity conditions throughout a supply chain. The deep learning models analyze data from these sensors for anomaly detection, for example, deviations from optimal temperature ranges. For instance, a CNN can be used to analyze images of goods under refrigeration to identify major signs of spoilage. Real-time alerts will be raised if the conditions are outside acceptable limits, and this will enable corrective actions to be taken in time, for example, refrigeration settings being adjusted or shipments rerouted.

IoT sensors track temperature, pH, gas levels, and others in a food processing facility. Deep learning models fuse and process this data to ensure that the conditions of processing are within the specifications of the range. An autoencoder detects deviations from the normal pattern, and an RNN analyzes time-series data to anticipate issues. Finally, real-time dashboards provide operators with insights and recommendations so adjustments can be made to ensure the quality of the product.

Farm-to-Fork Traceability: IoT devices on farms track the condition of the soil, crop health, and environmental factors. These devices are connected to transportation sensors and deep learning models that track the journey from farm to consumer for food products. History and real-time data can be used to run predictive analytics, which can estimate the probability of an issue occurring with any product during its lifetime. Consumers are given fine-grained details about the origin and handling of their food through smart packaging.

The integration of IoT and deep learning presents major improvements to food quality monitoring, bridging real-time data collection with sophisticated analytics capabilities. Embedding IoT devices, edge computing, cloud-based processing, and deep learning models in the architecture enables end-to-end complete monitoring and proactive management of the safety and freshness of food products. This approach integrates data from all sources and does advanced analytics to support anomaly detection, issue prediction, and quality standards at each stage of the food supply chain.

REAL-TIME MONITORING AND PREDICTIVE MAINTENANCE

Figure 4. Real-time monitoring and predictive maintenance

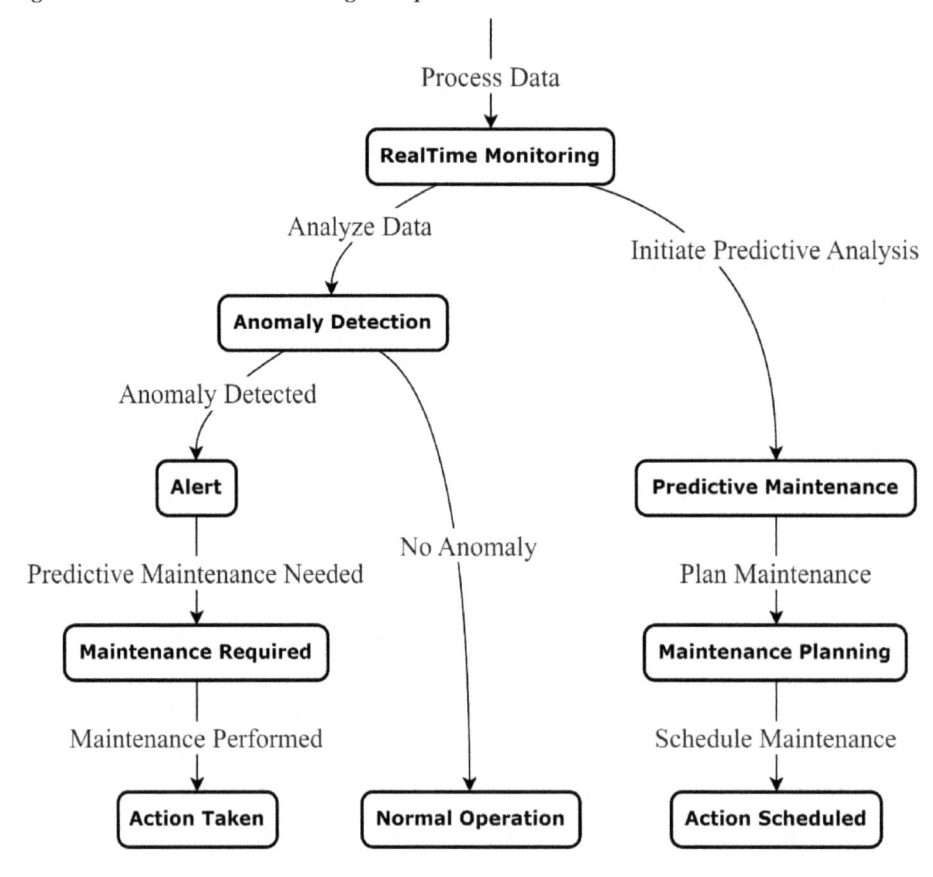

Figure 4 illustrates the process of real-time monitoring and predictive maintenance, from idle to anomaly detection, and from action taken to scheduled actions. It provides a high-level view of the states and transitions involved in these processes, enabling effective planning and scheduling of maintenance actions.

Real-Time Data Analysis

Real-time data analytics as far as modern food quality management systems are concerned is a description of real-time identification and reaction to conditions that may influence the safety of the product—food—and its freshness. This means continuous collection of data from various sensors and devices, its fast processing, and generating actionable insights(Dhanya et al., 2023; Tirlangi et al., 2024).

Data Acquisition: The IoT sensors, installed in production facilities, storage areas, and transportation vehicles, acquire continuous data on critical parameters such as temperature, humidity, gas concentrations, and visual cues. Continuous data streams from these sensing devices provide information on the current state of the environment and the condition of food products.

Data Transmission: The collected data is transmitted to the central processing units through wireless protocols like Wi-Fi, Bluetooth, Zigbee, or cellular networks that ensure real-time availability of data for analysis without huge delays.

Edge Computing: Process data closer to the source to reduce latency and permit real-time responses. This preliminary analysis on the edge devices is carried out in such a fashion that much of the noise gets filtered, and immediate anomalies get detected. For instance, in the event that a temperature sensor picks up a sudden spike, an alarm will be triggered or cooling systems brought down in real time by an edge device.

Real-Time Analytics: Advanced analytics platforms process data in real time to find patterns and anomalies. Machine learning algorithms—not least deep learning models—analyze the data for deviations from the normal condition. For example, a CNN can analyze images of food products produced to detect probable signs of spoilage or defects in packaging. Real-time dashboards give operators visualizations and alerts so they can respond effectively and quickly.

Alert Systems: Real-time monitoring systems provide with automated alert mechanisms that notify operators or managers in case there exist any deviations from accepted ranges. Such alerts may appear via email, SMS, or through integrated dashboard notifications. It is through these alerts that immediate corrective action will ensue, mitigating potential problems so as to maintain food quality.

Predictive Maintenance Strategies

This is predictive maintenance, which means using data-driven insights into the prediction and fixing of equipment before it fails. It includes avoiding unplanned equipment downtime, minimizing maintenance costs, and ensuring the continuous operation of food quality monitoring systems.

Data Collection and Integration: Predictive maintenance relies on data gathered from multiple sensors, which facilitates monitoring of the equipment's performance on parameters such as temperature, vibration, pressure, and operation metrics. This information is then integrated with historical maintenance records and operational data to build a full picture of equipment health.

Condition Monitoring: This becomes possible through the continuous monitoring of equipment conditions whereby the equipment can be identified at the first signs of wear and tear or malfunction. For example, vibration sensors detect any abnormal vibration in the machinery; this probably means an imbalance or misalignment. Temperature sensors monitor heating elements in an oven or a refrigeration unit to detect impending failures.

Machine Learning Models: Predictive maintenance utilizes machine learning algorithms in analyzing data from history and real-time observations. Regression analysis, classification algorithms, and deep learning networks are models for determining equipment failures through patterns in the data. For instance, looking at the time-series data from the equipment's sensors, an RNN may project potential malfunction based on historical trends(Rebecca et al., 2024; Saravanan et al., 2024).

Failure Prediction and Risk Assessment: Machine learning models in maintenance generate predictions of equipment failure or maintenance needs. This is determined by data patterns, historical performance, and real-time conditions. The risk assessment tools rank-order the jobs by the probability of failure and potential impact on operations.

Maintenance Scheduling: Scheduled maintenance activities, based on predictive insight, are executed proactively rather than reactively. This approach minimizes disruptions to operations and makes sure that maintenance happens before the issues lead to equipment breakdowns. Scheduled maintenance may be planned during off-peak hours or during production downtimes, not impacting food quality and production schedules.

Impact on Shelf Life and Spoilage Reduction

Real-time monitoring combined with predictive maintenance will affect extended shelf life and less spoilage of food products. These techniques assure that optimal conditions are maintained along the supply chain so that foodstuffs remain fresh and consumable(Revathi et al., 2024).

Optimized Storage Conditions: The conditions of storage—temperature and humidity—are very precisely controlled in real-time. If these parameters were to be continuously monitored, an operator would then be in a position to adjust the variables to maintain foodstuffs at the best possible condition. For example, in the case of a refrigeration unit failing, real-time alerts will prompt operators to take immediate action, such as turning on backup cooling systems or adjusting temperature settings.

Less Spoilage: Such predictive maintenance would avoid breakdowns of equipment that may result in spoilage. For example, predicting refrigeration units' problems and attending to them before they cause any change in temperature ranges would reduce a great deal of potential risks of spoilage. This proactive approach will ensure the maintenance of food products within safe temperature ranges to avoid its premature deterioration.

Extended Shelf Life: Creation of ideal conditions for storing food products and avoidance of equipment failure help the product remain good for a longer time. Real-time monitoring ensures rectification in real-time against deviations from the best conditions, and predictive maintenance minimizes risks associated with equipment-related problems that reduce shelf life.

Quality Assurance: Ensures a general quality with its continuous monitoring and maintenance, ensuring the foods to be delivered and sold to be safe and fresh. Real-time data and predictive insight enable operators to act ahead of time to ensure that any potential issue does not affect product quality, ensuring higher levels of food safety and greater customer satisfaction.

Cost Savings: Reducing spoilage and extending shelf life also provides cost savings for food manufacturers and distributors. Prevention of losses in the form of spoilage, along with reduced associated downtime because of predictive maintenance, can help optimize the production process for companies and thereby improve profitability.

Real-time monitoring and predictive maintenance will change the management of food quality. It would ensure operational efficiency, extended shelf life, and reduced wastage. Real-time data analysis gives a company the ability to respond timely to conditions that might affect foods' safety or freshness. On the other hand, predictive maintenance would work to prevent equipment breakdowns while ensuring that all conditions are kept within an optimal range. These would achieve the right handling and storage for both quality enhancement and reduction of wastages while securing customer satisfaction in foodstuffs.

CHALLENGES AND LIMITATIONS

Figure 5. Challenges and limitations of integrating IoT and deep learning for food quality monitoring

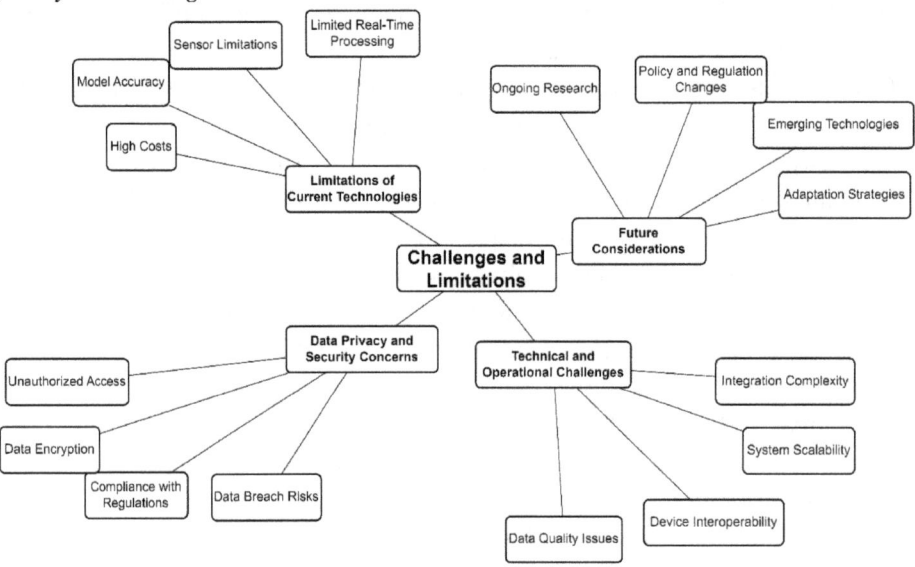

However, the integration of the IoT and deep learning into the sector of food quality monitoring faces several technical and operational issues, complexity, interoperability with the devices, data quality issues, scalability of the system, issues of privacy and security, risks of data breach, issues of unauthorized access, compliance with regulations, and encryption of data, as shown in Figure 5. Other limitations of current technologies include sensor limitations, model accuracies, and the cost of a high-volume sensor network. Future trends to be considered encompass emerging technologies, growing researches, adaptation strategies, and policy and regulation changes. Adaptation strategies are required for mitigating developing challenges and ensuring data privacy and security (Rebecca et al., 2024).

Technical and Operational Challenges

Integration of IoT devices and deep learning models with existing food quality-monitoring systems brings about significant challenges in technical integration. Diverse sensors and devices mostly use competing communication protocols; this may call for elaborate data integration and middleware solutions. All components must work together, which calls for the need for rigorous system design and testing.

Data management: dealing with the volume, velocity, and variety of data generated from the IoT sensors can be simply mind-boggling. Both managing and processing big data volumes in real time call for a strong, infrastructure-based data management approach with highly scalable cloud storage solutions. These data handling mechanisms really need to be efficient enough to make sure of this kind of data that can be collected, transmitted, and analyzed with simple delays or loss of information(Upadhyaya et al., 2024; Venkateswaran et al., 2024).

Scalability Issues: Maintaining system performance and reliability can be a daunting challenge as the count of IoT devices and data sources keeps increasing. Infrastructure can be scaled to deal with the increasing number of devices and increasing data streams through careful capacity planning and investment in high-performance computing resources. It is crucial to ensure that the system remains effective and responsive to continue operation.

Calibration and Maintenance of Devices: Sensors and their connected devices must be calibrated and maintained in a proper time cycle for providing the necessary level of accuracy and reliability. Inaccurate sensor reading can cause wrong analytics data and prove to be a poor prediction. Therefore, setting down great protocols about calibration, working with the right maintenance span, is important in the maintenance of such a monitoring system.

Model Performance: Deep learning models have to be aggressively trained on high-quality, diversified data to have good accuracy. Sustained high performance is generally achieved through retraining: the models are updated at every step with the changes in conditions and hence improved for higher accuracy.

Data Privacy and Security Concerns

Data Breaches: The collection of data and its subsequent transmission create concerns for privacy and security of the same. IoT devices and cloud servers are open to cyber-attacks that might result in data breaches through unauthorized access. Implementing strong encryption, rigorous access controls, and secure protocols is very instrumental in fending off malicious threats to data (Boopathi, 2024b; Nanda et al., 2024).

Compliance: It should adhere to various data protection laws, such as GDPR and CCPA. Both of these regulations have essentially imposed strict requirements on the practices of data collection, storage, and handling. Checking for compliance means having strong data governance practices in place, updated with respect to regulatory changes.

Data Anonymization: This becomes a big problem: how to use the data for analysis while protecting individual privacy. In this respect, methods of data anonymization must be used to ensure that PII is not exposed or misused. To that end, proper care must be taken to balance data utility with privacy considerations.

System Vulnerabilities: IoT devices and cloud platforms are found open to various security vulnerabilities, which attackers may exploit. Since risks of this sort may be incurred, regular security assessment, update of software, and vulnerability patches are cited as a result of such scenarios to curb potential weaknesses within the system and offer protection against threats. Ensuring that all components are secure and current is very important for the integrity of the data and the reliability of the system.

Limitations of Current Technologies

Sensor Limitations: Much as IoT sensors record and play a vital role in providing valuable data, they are still limited by their accuracy, sensitivity, and range. Some of the sensors may fail to pick up very minute changes to environmental conditions or food quality. Improvements in sensor technology will be needed to grow this capability in both increased measurement precision and a far broader range of parameters that can be detected.

Generalization: Deep learning models may have poor generalization across new or unseen conditions. The models learned on one dataset mostly fail to generalize very well in another environment or situations. This requires a large number of models that generalize across more diverse conditions.

Data Quality Issues: A deep learning model can provide as good results as the quality of the data fed to it. If the data is missing, noisy, or biased, the results will definitely be wrong. Great attention should therefore be paid to data quality, including proper collection, preprocessing, and validation to ensure high quality in order to achieve accurate and meaningful results.

Cost and Resource Constraints: IoT and deep learning technologies require huge investments in hardware, software, and computational resources. The cost of setting up and maintaining these technologies is often unbearable to some organizations, especially the SMEs. Great consideration and planning need to be done on the cost against the many benefits of advanced monitoring and analysis.

Technical Expertise: An integrated IoT and deep learning system requires specialized technical expertise in its development and maintenance. There may be problems of recruiting or retaining personnel with required knowledge and experience. In such a case, organizations must invest in training and developmental programs to build and sustain required competencies.

Challenges and limitations should be dealt with in the integration of IoT and deep learning technologies for the monitoring of food quality in order for the system performance to be optimized and effective. On the technical and operational levels, issues about integration complexity, data management, and device maintenance have to be managed through careful planning and investment. Data privacy and security concerns regarding possible data breaches or compliance with some provisions of regulatory concern require strong protective measures and incessant vigilance. Finally, constraints in today's technologies, such as sensor accuracy and model generalization, present the case for continued improvement and advancement. Addressing these ever-present challenges will potentially position an organization to have a much more certain, more accurate, and more effective food quality monitoring system for the delivery of better results in food safety and freshness.

FUTURE DIRECTIONS AND INNOVATIONS

Emerging Technologies and Trends

Advanced sensor technologies will be part of the future in relation to the use of IoT for food quality monitoring. New sensors that are more sensitive and accurate, with multiple functions, will provide better overall views of controllable parameters. Innovations such as nano sensors and flexible electronics may offer more precise and real-time data collection, letting food quality and safety be nearer to being under closer scrutiny (Ali et al., 2024; Venkateswaran et al., 2023).

5G and Edge Computing: Next-generation 5G networks and edge computing technologies are designed to take real-time data processing to new heights. In this regard, 5G technology will provide faster and more reliable data transmission, thus reducing latency and further increasing the responsiveness of IoT systems. At the same time, edge computing will enable much faster data processing at the source to enable faster decision-making without having to rely on any form of central data processing(Agrawal, Shashibhushan, et al., 2023; Sreedhar et al., 2024).

Blockchain Technology: This decentralized and immutable record of blockchain provides better trackability and visibility in the supply chain of food products. Each transaction and reading from the sensors are recorded on a blockchain, thus helping stakeholders to further ascertain authenticity and quality of food products while ensuring data tamper proofing for audits or quality checks(Sundaramoorthy et al., 2024; Tagesse et al., 2024).

Artificial Intelligence Integration: Coupling Artificial Intelligence with IoT and Deep Learning will drive still more sophisticated and autonomous monitoring systems. Establishment of the algorithm will then allow for predictive analytics and adaptive learning, allowing systems to become ever more accurate and efficient. Better decision-making based on AI-driven insights enables proactive management of food quality(Glady et al., 2024).

Smart Packaging: Next-generation smart packaging with edible sensors and RFID tags will give real-time monitoring of the quality of food directly from the pack. Such technologies can offer information on freshness, temperature history, and even identify the presence of spoilage, hence improving traceability and manageability of food products along the supply chain.

Potential Improvements in IoT and Deep Learning

Improved Data Fusion: In the future, most of the improvements will target better techniques in data fusion that can integrate multiple sources of data seamlessly. Improved data fusion algorithms will hence be able to better exploit multi-modal data emanating from sensors, images, and environmental factors for more accurate and actionable insights(Hema et al., 2023; Syamala et al., 2023).

More Solid Deep-Learning Models: Developments in deep learning would result in more robust and generalized models that handle a wide array of conditions and anomalies. Transfer learning, few-shot learning techniques, and others will optimize model performance with minimal data and generalize to new environments without heavy retraining.

Energy-efficient IoT devices: Innovations in low-power sensors and energy-efficient communication protocols will increase the battery life and operational efficiency of IoT devices. Developments in energy harvesting technologies—like solar and kinetic energy—will realize self-maintenance sensors, eliminating the need for frequent servicing and replacement of batteries.

Improved Security: Increasing data privacy and security-related issues require, in the future, even more advanced security for both IoT and deep learning systems; next-generation cryptography techniques will be applied, together with secure multi-party computation and decentralized security models for the protection of sensitive data and monitoring system integrity.

User-Friendly Interfaces: Further development of more intuitive and user-friendly interfaces for IoT and deep learning systems through which operators and stakeholders interact. Other improvements will be in visualization tools, real-time dashboards, and automated reporting features, ensuring the improvement of usability and access to monitoring systems.

FUTURE RESEARCH TRENDS

Cross-Domain Integration: The focus of the research will be on the integration of this food quality monitoring system with other domains, such as supply chain management, logistics, and consumer behavior. Through such an understanding of how food quality data can combine with supply chain and market data, new insights for the optimization of food production, distribution, and consumption can be obtained(Abass et al., 2024; Chhetri, 2024; Nerkar et al., 2023; Tutul et al., 2023).

Food Quality Management Personalization: On the other side, the coming lines of research will lie in the developing of personalized solutions in food quality management based on individual needs of preferences and dietary requirements. This is made possible through the use of data recorded from personal health devices and dietary logs, so that a system can automatically come up with specific recommendations and notifications on food safety and nutrition.

Advanced Anomaly Detection: The techniques for anomaly detection will continue to be perfected, much more towards perfecting the techniques for detecting subtle and emerging anomalies in food quality. It is related to the models of how to predict incipient spoilage or contamination that is otherwise hardly detectable currently with available technologies.

Sustainability and Environmental Impact: There will be long-term research with respect to how IoT and deep learning can help in supporting the sustainability aspect of the food industry, which will comprise in-depth analysis with regard to the reduction of wastage, optimization of resources, and reducing the environmental impact associated with food production and distribution(Vangeri et al., 2024; Vijaya Lakshmi et al., 2024).

It will also involve research into how human-machine interactions can be best improved in food quality monitoring systems. Understanding how operators interact with these systems and how to improve their experience will be critical in coming up with more effective and user-friendly solutions.

The future of food quality monitoring rests in continuously evolving IoT and deep learning technologies. Emerging technologies such as advanced sensors, 5G, and blockchain will drive innovation. Further steps in the development of data fusion, model robustness, and security could also be taken to enhance system performance. Future research into new frontiers of cross-domain integration, personalization, and sustainability will further drive more efficient, more accurate, and impactful food quality monitoring solutions. Mastering these will mean the food industry has achieved enhanced safety, freshness, and efficiency in the management of food products throughout their lifecycle.

CONCLUSION

It allows for a transformational leap into food quality monitoring and opens new opportunities for enhancing safety, freshness, and operational efficiency through the integration of IoT and deep learning technologies. Given the next generation of IoT sensors and real-time data analytics tools, firms will gain fine-grained environmental conditions and food quality details to permit timely intervention and better management throughout the supply chain.

Deep learning models then support these capabilities with a great degree of anomaly detection and predictive analytics so as to anticipate potential issues before they affect food quality. This interplay of technologies provides impetus toward more accurate monitoring, reduces spoilage, extends shelf life, and hence significantly improves food safety and operational cost savings.

Technical integration complexities, data privacy concerns, security concerns, and the limitations of current technologies are some of the key challenges that need to be addressed for the successful implementation of these technologies. Further innovations in the future—next-generation sensor technologies, 5G networks, and AI-driven systems—will push real-time monitoring and predictive maintenance to new heights by tackling these challenges.

In the near future, further research has to be done on enhancing cross-domain integration, the management of personalized food quality, and the achievement of sustainability. These developments enhance the efficiency and accuracy of monitoring food quality and contribute to sustainable practices, besides providing a personalized experience to consumers.

The future of food quality management will eventually be realized when these evolving technologies of IoT and deep learning move in a hand-in-hand fashion to create very powerful tools, ensuring the safety, freshness, and quality of all food products in an increasingly complex and demanding market.

REFERENCES

Abass, T., Itua, E. O., Bature, T., & Eruaga, M. A. (2024). Concept paper: Innovative approaches to food quality control: AI and machine learning for predictive analysis. *World Journal of Advanced Research and Reviews*, 21(3), 823–828. DOI: 10.30574/wjarr.2024.21.3.0719

Agrawal, A. V., Magulur, L. P., Priya, S. G., Kaur, A., Singh, G., & Boopathi, S. (2023). Smart Precision Agriculture Using IoT and WSN. In *Handbook of Research on Data Science and Cybersecurity Innovations in Industry 4.0 Technologies* (pp. 524–541). IGI Global. DOI: 10.4018/978-1-6684-8145-5.ch026

Agrawal, A. V., Shashibhushan, G., Pradeep, S., Padhi, S., Sugumar, D., & Boopathi, S. (2023). Synergizing Artificial Intelligence, 5G, and Cloud Computing for Efficient Energy Conversion Using Agricultural Waste. In *Sustainable Science and Intelligent Technologies for Societal Development* (pp. 475–497). IGI Global.

Ali, M. N., Senthil, T., Ilakkiya, T., Hasan, D. S., Ganapathy, N. B. S., & Boopathi, S. (2024). IoT's Role in Smart Manufacturing Transformation for Enhanced Household Product Quality. In *Advanced Applications in Osmotic Computing* (pp. 252–289). IGI Global. DOI: 10.4018/979-8-3693-1694-8.ch014

Atitallah, S. B., Driss, M., Boulila, W., & Ghézala, H. B. (2020). Leveraging Deep Learning and IoT big data analytics to support the smart cities development: Review and future directions. *Computer Science Review*, 38, 100303. DOI: 10.1016/j.cosrev.2020.100303

Bhardwaj, A., Dagar, V., Khan, M. O., Aggarwal, A., Alvarado, R., Kumar, M., Irfan, M., & Proshad, R. (2022). Smart IoT and machine learning-based framework for water quality assessment and device component monitoring. *Environmental Science and Pollution Research International*, 29(30), 46018–46036. DOI: 10.1007/s11356-022-19014-3 PMID: 35165843

Bikash Chandra Saha, M. S., Deepa, R., Akila, A., & Sai Thrinath, B. V. (2022). Iot Based Smart Energy Meter For Smart Grid.

Boopathi, S., Karthikeyan, K. R., Jaiswal, C., Dabi, R., Sunagar, P., & Malik, S. (2024). *IoT based Automatic Cooling Tower*.

Boopathi, S. (2023). Deep Learning Techniques Applied for Automatic Sentence Generation. In *Promoting Diversity, Equity, and Inclusion in Language Learning Environments* (pp. 255–273). IGI Global. DOI: 10.4018/978-1-6684-3632-5.ch016

Boopathi, S. (2024a). Advancements in Machine Learning and AI for Intelligent Systems in Drone Applications for Smart City Developments. In *Futuristic e-Governance Security With Deep Learning Applications* (pp. 15–45). IGI Global. DOI: 10.4018/978-1-6684-9596-4.ch002

Boopathi, S. (2024b). Balancing Innovation and Security in the Cloud: Navigating the Risks and Rewards of the Digital Age. In *Improving Security, Privacy, and Trust in Cloud Computing* (pp. 164–193). IGI Global.

Boopathi, S. (2024c). Sustainable Development Using IoT and AI Techniques for Water Utilization in Agriculture. In *Sustainable Development in AI, Blockchain, and E-Governance Applications* (pp. 204–228). IGI Global. DOI: 10.4018/979-8-3693-1722-8.ch012

Chhetri, K. B. (2024). Applications of Artificial Intelligence and Machine Learning in Food Quality Control and Safety Assessment. *Food Engineering Reviews*, 16(1), 1–21. DOI: 10.1007/s12393-023-09363-1

Dhanya, D., Kumar, S. S., Thilagavathy, A., Prasad, D., & Boopathi, S. (2023). Data Analytics and Artificial Intelligence in the Circular Economy: Case Studies. In *Intelligent Engineering Applications and Applied Sciences for Sustainability* (pp. 40–58). IGI Global.

Glady, J. B. P., D'Souza, S. M., Priya, A. P., Amuthachenthiru, K., Vikram, G., & Boopathi, S. (2024). A Study on AI-ML-Driven Optimizing Energy Distribution and Sustainable Agriculture for Environmental Conservation. In *Harnessing High-Performance Computing and AI for Environmental Sustainability* (pp. 1–27). IGI Global., DOI: 10.4018/979-8-3693-1794-5.ch001

Hema, N., Krishnamoorthy, N., Chavan, S. M., Kumar, N., Sabarimuthu, M., & Boopathi, S. (2023). A Study on an Internet of Things (IoT)-Enabled Smart Solar Grid System. In *Handbook of Research on Deep Learning Techniques for Cloud-Based Industrial IoT* (pp. 290–308). IGI Global. DOI: 10.4018/978-1-6684-8098-4.ch017

Kav, R. P., Pandraju, T. K. S., Boopathi, S., Saravanan, P., Rathan, S. K., & Sathish, T. (2023). Hybrid Deep Learning Technique for Optimal Wind Mill Speed Estimation. *2023 7th International Conference on Electronics, Communication and Aerospace Technology (ICECA)*, 181–186.

Khan, P. W., Byun, Y.-C., & Park, N. (2020). IoT-blockchain enabled optimized provenance system for food industry 4.0 using advanced deep learning. *Sensors (Basel)*, 20(10), 2990. DOI: 10.3390/s20102990 PMID: 32466209

Kumar, P. R., Meenakshi, S., Shalini, S., Devi, S. R., & Boopathi, S. (2023). Soil Quality Prediction in Context Learning Approaches Using Deep Learning and Blockchain for Smart Agriculture. In *Effective AI, Blockchain, and E-Governance Applications for Knowledge Discovery and Management* (pp. 1–26). IGI Global. DOI: 10.4018/978-1-6684-9151-5.ch001

Manikandan, R., Ranganathan, G., & Bindhu, V. (2023). Deep learning based IoT module for smart farming in different environmental conditions. *Wireless Personal Communications*, 128(3), 1715–1732. DOI: 10.1007/s11277-022-10016-5

Manisha, N., & Jagadeeshwar, M. (2023). BC driven IoT-based food quality traceability system for dairy product using deep learning model. *High-Confidence Computing*, 3(3), 100121. DOI: 10.1016/j.hcc.2023.100121

Nanda, A. K., Sharma, A., Augustine, P. J., Cyril, B. R., Kiran, V., & Sampath, B. (2024). Securing Cloud Infrastructure in IaaS and PaaS Environments. In *Improving Security, Privacy, and Trust in Cloud Computing* (pp. 1–33). IGI Global. DOI: 10.4018/979-8-3693-1431-9.ch001

Nayak, J., Vakula, K., Dinesh, P., Naik, B., & Pelusi, D. (2020). Intelligent food processing: Journey from artificial neural network to deep learning. *Computer Science Review*, 38, 100297. DOI: 10.1016/j.cosrev.2020.100297

Nerkar, P. M., Shinde, S. S., Liyakat, K. K. S., Desai, S., & Kazi, S. S. L. (2023). Monitoring fresh fruit and food using Iot and machine learning to improve food safety and quality. *Tuijin Jishu/Journal of Propulsion Technology, 44*(3), 2927–2931.

Rahamathunnisa, U., Subhashini, P., Aancy, H. M., Meenakshi, S., & Boopathi, S. (2023). Solutions for Software Requirement Risks Using Artificial Intelligence Techniques. In *Handbook of Research on Data Science and Cybersecurity Innovations in Industry 4.0 Technologies* (pp. 45–64). IGI Global.

Rebecca, B., Kumar, K. P. M., Padmini, S., Srivastava, B. K., Halder, S., & Boopathi, S. (2024). Convergence of Data Science-AI-Green Chemistry-Affordable Medicine: Transforming Drug Discovery. In *Handbook of Research on AI and ML for Intelligent Machines and Systems* (pp. 348–373). IGI Global.

Revathi, S., Babu, M., Rajkumar, N., Meti, V. K. V., Kandavalli, S. R., & Boopathi, S. (2024). Unleashing the Future Potential of 4D Printing: Exploring Applications in Wearable Technology, Robotics, Energy, Transportation, and Fashion. In *Human-Centered Approaches in Industry 5.0: Human-Machine Interaction, Virtual Reality Training, and Customer Sentiment Analysis* (pp. 131–153). IGI Global.

Sampath, B., Pandian, M., Deepa, D., & Subbiah, R. (2022). Operating parameters prediction of liquefied petroleum gas refrigerator using simulated annealing algorithm. *AIP Conference Proceedings*, 2460(1), 070003. DOI: 10.1063/5.0095601

Sangeetha, M., Kannan, S. R., Boopathi, S., Ramya, J., Ishrat, M., & Sabarinathan, G. (2023). Prediction of Fruit Texture Features Using Deep Learning Techniques. *2023 4th International Conference on Smart Electronics and Communication (ICOSEC)*, 762–768.

Saravanan, S., Khare, R., Umamaheswari, K., Khare, S., Krishne Gowda, B. S., & Boopathi, S. (2024). AI and ML Adaptive Smart-Grid Energy Management Systems: Exploring Advanced Innovations. In *Principles and Applications in Speed Sensing and Energy Harvesting for Smart Roads* (pp. 166–196). IGI Global. DOI: 10.4018/978-1-6684-9214-7.ch006

Sreedhar, P. S. S., Sujay, V., Rani, M. R., Melita, L., Reshma, S., & Boopathi, S. (2024). Impacts of 5G Machine Learning Techniques on Telemedicine and Social Media Professional Connection in Healthcare. In *Advances in Medical Technologies and Clinical Practice* (pp. 209–234). IGI Global. DOI: 10.4018/979-8-3693-1934-5.ch012

Sumathi, P., Subramanian, R., Karthikeyan, V., & Karthik, S. (2021). Retracted: Soil monitoring and evaluation system using EDL-ASQE: Enhanced deep learning model for IoT smart agriculture network. *International Journal of Communication Systems*, 34(11), e4859. DOI: 10.1002/dac.4859

Sundaramoorthy, K., Singh, A., Sumathy, G., Maheshwari, A., Arunarani, A., & Boopathi, S. (2024). A Study on AI and Blockchain-Powered Smart Parking Models for Urban Mobility. In *Handbook of Research on AI and ML for Intelligent Machines and Systems* (pp. 223–250). IGI Global.

Syamala, M., Komala, C., Pramila, P., Dash, S., Meenakshi, S., & Boopathi, S. (2023). Machine Learning-Integrated IoT-Based Smart Home Energy Management System. In *Handbook of Research on Deep Learning Techniques for Cloud-Based Industrial IoT* (pp. 219–235). IGI Global. DOI: 10.4018/978-1-6684-8098-4.ch013

Tagesse, T., Arulkumar, S., Mahalingam, S., Subbarao Tadepalli, N. V. R., Munjal, N., & Boopathi, S. (2024). Analyzing Fuel Cell Vehicles in India via the PESTLE Framework and Intelligent Battery Management Systems (BMS): Blockchain in E-Mobility. In *Advances in Mechatronics and Mechanical Engineering* (pp. 107–129). IGI Global. DOI: 10.4018/979-8-3693-5247-2.ch007

Tirlangi, S., Teotia, S., Padmapriya, G., Senthil Kumar, S., Dhotre, S., & Boopathi, S. (2024). Cloud Computing and Machine Learning in the Green Power Sector: Data Management and Analysis for Sustainable Energy. In *Developments Towards Next Generation Intelligent Systems for Sustainable Development* (pp. 148–179). IGI Global. DOI: 10.4018/979-8-3693-5643-2.ch006

Tutul, M. J. I., Alam, M., & Wadud, M. A. H. (2023). Smart food monitoring system based on iot and machine learning. *2023 International Conference on Next-Generation Computing, IoT and Machine Learning (NCIM)*, 1–6. DOI: 10.1109/NCIM59001.2023.10212608

Ugandar, R., Rahamathunnisa, U., Sajithra, S., Christiana, M. B. V., Palai, B. K., & Boopathi, S. (2023). Hospital Waste Management Using Internet of Things and Deep Learning: Enhanced Efficiency and Sustainability. In *Applications of Synthetic Biology in Health, Energy, and Environment* (pp. 317–343). IGI Global.

Upadhyaya, A. N., Saqib, A., Devi, J. V., Rallapalli, S., Sudha, S., & Boopathi, S. (2024). Implementation of the Internet of Things (IoT) in Remote Healthcare. In *Advances in Medical Technologies and Clinical Practice* (pp. 104–124). IGI Global. DOI: 10.4018/979-8-3693-1934-5.ch006

Vangeri, A. K., Bathrinath, S., Anand, M. C. J., Shanmugathai, M., Meenatchi, N., & Boopathi, S. (2024). Green Supply Chain Management in Eco-Friendly Sustainable Manufacturing Industries. In *Environmental Applications of Carbon-Based Materials* (pp. 253–287). IGI Global., DOI: 10.4018/979-8-3693-3625-0.ch010

Venkateswaran, N., Kiran Kumar, K., Maheswari, K., Kumar Reddy, R. V., & Boopathi, S. (2024). Optimizing IoT Data Aggregation: Hybrid Firefly-Artificial Bee Colony Algorithm for Enhanced Efficiency in Agriculture. *AGRIS On-Line Papers in Economics and Informatics*, 16(1), 117–130. DOI: 10.7160/aol.2024.160110

Venkateswaran, N., Kumar, S. S., Diwakar, G., Gnanasangeetha, D., & Boopathi, S. (2023). Synthetic Biology for Waste Water to Energy Conversion: IoT and AI Approaches. *Applications of Synthetic Biology in Health. Energy & Environment*, •••, 360–384.

Vijaya Lakshmi, V., Mishra, M., Kushwah, J. S., Shajahan, U. S., Mohanasundari, M., & Boopathi, S. (2024). Circular Economy Digital Practices for Ethical Dimensions and Policies for Digital Waste Management. In *Harnessing High-Performance Computing and AI for Environmental Sustainability* (pp. 166–193). IGI Global., DOI: 10.4018/979-8-3693-1794-5.ch008

KEY TERMS AND DEFINITIONS

AI: Artificial Intelligence
CCPA: California Consumer Privacy Act
CNN: Convolutional Neural Network
CO: Carbon Monoxide
DRL: Deep Reinforcement Learning
ERP: Enterprise Resource Planning
GAN: Generative Adversarial Network
GDPR: General Data Protection Regulation
LSTM: Long Short-Term Memory
NFC: Near Field Communication
PII: Personally Identifiable Information
QR: Quick Response (Code)
RFID: Radio Frequency Identification
RNN: Recurrent Neural Network
RTD: Resistance Temperature Detector
SME: Small and Medium Enterprises
SMS: Short Message Service

Chapter 5
Edible Energy Harvesting:
Powering the Future of Smart Food Technology

Dipan Kumar Das
https://orcid.org/0000-0003-0224-3295
Centurion University of Technology and Management, India

Padmaja Patnaik
https://orcid.org/0000-0003-4468-7871
Centurion University of Technology and Management, India

Sudip Kumar Das
https://orcid.org/0009-0003-0814-0884
Dr. C.V. Raman University, India

Mandakini Barala
Centurion University of Technology and Management, India

Nibedita Nayak
Centurion University of Technology and Management, India

ABSTRACT

Edible energy harvesting merges food science, biotechnology, and engineering to convert chemical energy from safe, ingestible substances into electrical power. This chapter explores the evolution of energy harvesting technologies, focusing on advancements in materials science and biocompatible electronics. It discusses the significance of edible energy harvesting in medical devices, sustainable technologies, and IoT applications. Safety and ethical considerations, along with recent innovations in fabrication techniques and regulatory frameworks, are analyzed. The chapter concludes with prospects for integrating energy harvesting components into

DOI: 10.4018/979-8-3693-5573-2.ch005

Copyright © 2025, IGI Global. Copying or distributing in print or electronic forms without written permission of IGI Global is prohibited.

food and addressing challenges for broader adoption in healthcare and environmental sustainability. Ethical and environmental considerations, including safety and biocompatibility, are crucial in advancing these technologies. The chapter also discusses challenges such as regulatory compliance and environmental impact mitigation, highlighting the pathway for future developments in edible energy harvesting.

1. INTRODUCTION

Edible energy harvesting is a growing field combining food science, biotechnology, and engineering to create sustainable energy solutions using edible substances. The goal is to convert chemical energy from food into usable electrical energy, preserving the nutritional value and safety of the food. Technologies being explored include biochemical fuel cells, piezoelectric materials, triboelectric devices, nanotechnology, and synthetic biology. Applications include medical devices, the food and beverage industry, and wearable technology. However, ethical and safety concerns remain, including these materials' long-term health and environmental impact. Prospects for edible energy harvesting include advancements in materials science, biocompatible electronics, and nanotechnology, which could drive its development from experimental prototypes to mainstream applications.

Energy harvesting from food waste, particularly edible organic matter, has been explored as a sustainable energy source (Youn et al., 2015; Corigliano et al., 2016). Anaerobic reactor dimensioning and the calculation of the amount of organic matter to be processed for user needs are made possible by the understanding of world energies, biogas consumption, and the organic fraction's characteristics. This approach aligns with the broader concept of energy harvesting, which involves converting ambient energy into electrical power (Blagg, 2011). The method's potential is further underscored by the need to replace or augment batteries in wireless electronic systems (Blagg, 2011). The use of vermicompost organic matter in Microbial Fuel Cells (MFC) has been shown to produce electricity while decomposing organic waste (Youn et al., 2015). This approach can potentially contribute to food waste recycling and provide sustainable electricity (Youn et al., 2015).

Recent studies have explored the potential of edible energy harvesting, focusing on sustainable electricity production. Kulkarni (2010) discussed the use of piezoelectric devices to capture ambient energy and convert it into electrical energy, with potential applications in wireless sensors (Kulkarni et al.,2010). Alimonti (2017) highlighted the need to balance energy inputs and waste in the food supply chain and proposed algae for biofuel production (Alimonti et al.,2017). Ambrożkiewicz (2022) presented new energy harvesting systems based on electromagnetic and piezoelectric effects, which could be power supplies for low-energy devices (Ambrożkiewicz &

Rounak,2022). These studies collectively underscore the potential of edible energy harvesting in addressing both energy and environmental challenges.

PAs worldwide provide various provisioning services and related goods, benefiting local communities, society, and the economy. These services include food, water, raw materials, and genetic resources. PAs can provide these services within or outside their sites, such as refuges or breeding places for species. Examples of related goods include food, clean water, raw materials, medical resources, ornamental resources, and genetic resources. PAs also provide resources, such as medicinal products, cosmetics, and pharmaceuticals (Kettunen, M. & D'Amato, D., 2013).

i. **Overview of Energy Harvesting Technologies and Their Evolution**

The process of obtaining minute quantities of energy from various environmental sources, such solar, thermal, wind, salinity gradients, or kinetic energy, and transforming it into electrical power that may be used is known as energy harvesting. This technology primarily aims to power small, wireless autonomous devices, such as those used in wearable electronics and wireless sensor networks.

Evolution of Energy Harvesting:

Early Developments: Initially, energy harvesting techniques, such as windmills and waterwheels, were simple and mostly mechanical. Over time, these evolved into more sophisticated technologies like solar panels and thermoelectric generators.

Expansion and Miniaturization: With the advent of the digital and information age, the focus shifted towards miniaturization and integration. Innovations such as piezoelectric materials and micro-electromechanical systems (MEMS) enabled the harnessing of vibrational and kinetic energy from the environment.

Recent Trends: Recent developments include advanced photovoltaics, enhanced energy storage systems, and the exploration of novel materials like nanogenerators and bio-harvesting devices, which have dramatically expanded the possibilities and efficiency of energy harvesting systems.

ii. **Definition of Edible Energy Harvesting and Its Significance**

Edible Energy Harvesting involves extracting energy from safe substances for humans or animals to consume. This energy can then be used to power ingestible medical devices, such as capsules for diagnostic purposes or drug delivery systems requiring a small amount of power.

Significance:

Medical Applications: Edible energy harvesters can power devices that monitor internal health metrics or deliver medications in precise dosages over extended periods.

Sustainability: Using biocompatible and often biodegradable materials reduces environmental impact, aligning with global sustainability goals.

Innovation in Wearables and IoT: Extending to edible contexts opens new avenues in the Internet of Things-IoT and wearable technology, potentially leading to innovations in how data integration and personal health management are approached.

iii. Historical Context and Recent Advancements

Recent advancements in edible energy harvesting have shifted towards piezoelectric devices to capture ambient energy, as discussed by Kulkarni (2010) (Kulkarni et al., 2010). This method and other miniature energy harvesters are particularly useful for powering electronic devices and wireless sensor networks, as highlighted by Yeatman (2009) (Yeatman, 2009). The potential of genetic breeding in developing energy plants that can withstand extreme environments and provide a sustainable energy supply is also a key area of focus, as noted by Saibi (2013) (Saibi et al., 2013). Jain (2012) further emphasizes the significance of energy harvesting in various forms, including thermal, solar, wind, and mechanical energy, and the need for improved harvesting capabilities and their application in useful systems (Jain, 2012).

Using biological processes for energy production isn't new and can be traced back to early experiments with bio-batteries and microbial fuel cells. The notion of integrating these concepts with ingestible technology began gaining traction as the miniaturization of electronic components progressed, allowing for the development of small, safe, and consumable devices that could operate internally.

Material Science: Advances in materials science have led to the development of more efficient and safer energy-harvesting materials that can function effectively within the human body.

Biomedical Engineering: Integrating energy harvesting with biomedical applications, such as glucose-powered bio-batteries that use body fluids to generate electricity.

Regulatory Progress: Increased attention from regulatory bodies has led to clearer guidelines and standards, which help develop and test ingestible energy harvesting devices.

Table 1. Some edible materials and their energy density, biodegradability, safety rating, and cost per kg (USD)(Approx.)

Material	Energy Density (kJ/g)	Biodegradability	Safety Rating	Cost per kg (USD)
Sugar	17.0	High	High	0.50
Starch	15.5	High	High	0.30
Cellulose	14.8	High	Medium	0.75
Fat	37.7	Medium	High	1.50
Protein (Casein)	23.4	Low	High	2.00
Edible Oil (Olive)	37.0	Low	High	3.00

All the figures are generated by Biblioshiny app programmed by R-programming utilizing and analysing the Scopus data of the topic edible energy harvesting from 1977-2024 year.

Figure 1. Annual Scientific Production of articles researched in edible energy harvesting (Scopus data of 1977-2024 year)(Biblioshiny Analysis)

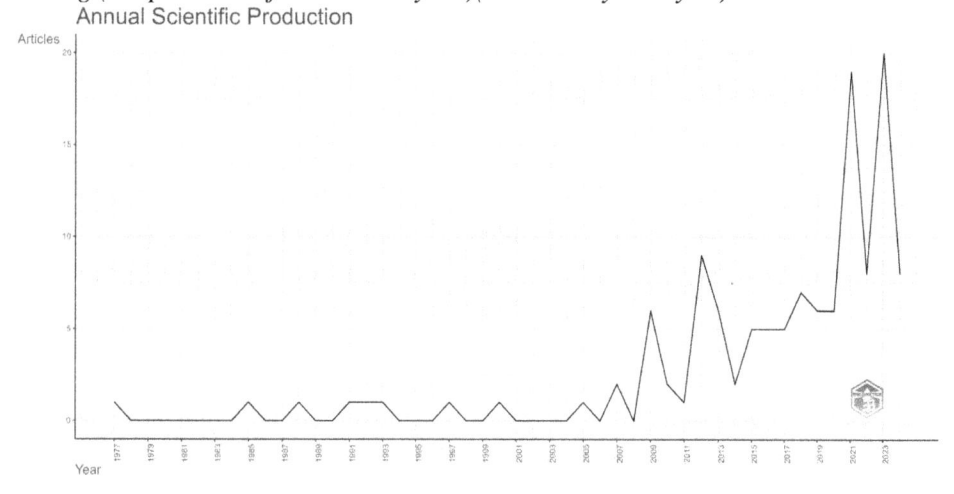

Figure 2. Most Relevant Sources researched in edible energy harvesting (Biblioshiny Analysis)

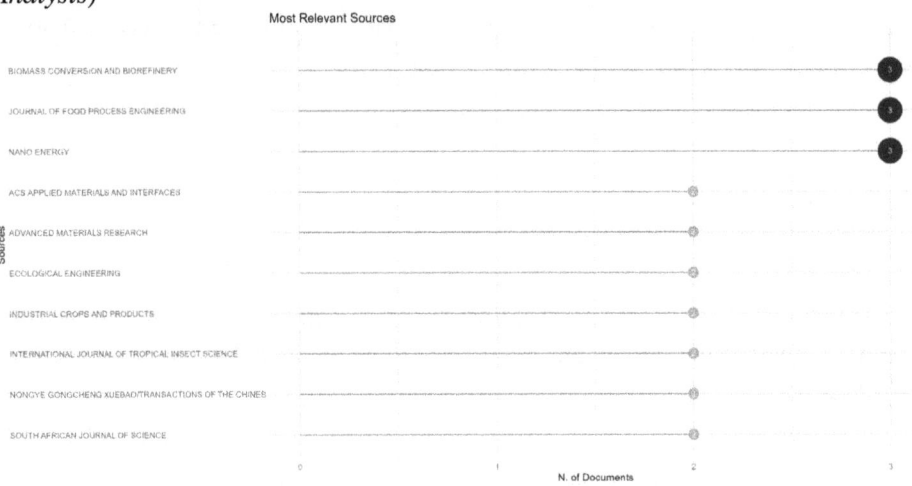

Implications of the Findings

The findings of our study on biologically active food products and edible energy harvesting have significant implications for several domains, including public health, environmental sustainability, and technological innovation. Firstly, the introduction of these advanced food products can revolutionize nutritional strategies by offering enhanced health benefits and personalized nutrition. However, the ethical considerations and potential health risks necessitate stringent regulatory frameworks and consumer education to ensure informed consent and safety.

Secondly, the effects of creating and discarding edible energy devices on the environment highlights the need for sustainable practices. Our findings suggest that biodegradable materials and eco-friendly manufacturing processes can mitigate negative environmental effects, promoting a more sustainable future. Lastly, the potential for these technologies to integrate with smart systems opens new avenues for real-time health monitoring and data collection, driving advancements in personalized medicine and wearable technology.

Methods Section

a. Study Design

The study was designed to explore the ethical, environmental, and practical aspects of biologically active food products and edible energy harvesting. It aimed to provide a comprehensive understanding of these emerging technologies by reviewing existing literature, analyzing regulatory frameworks, and assessing consumer acceptance and sustainability practices.

b. Study Execution

The study was carried out in several phases:

i.   **Literature Review**: Extensive literature review was conducted using Scopus to identify key research articles on edible energy harvesting from 1977 to 2024. Keywords such as "edible energy harvesting," "biologically active food products," and related terms were used.
ii. **Data Extraction and Analysis**: Articles from Scopus were extracted and analyzed using the Biblioshiny app programmed in R. The data covered various aspects of edible energy harvesting, including technological developments, regulatory issues, and consumer acceptance.
iii. **Case Studies and Comparative Analysis**: Specific case studies and comparative analyses were performed to understand the impact of different production methods, regulatory frameworks, and consumer perceptions across regions.

c. Data Analysis

The data analysis involved several steps:

i.   **Bibliometric Analysis**: The Biblioshiny app was utilized to perform bibliometric analysis, generating high-resolution figures to visualize trends, key authors, influential papers, and collaborations in the field.
ii. **Thematic Analysis**: Thematic analysis was performed on the retrieved data in order to find recurrent themes, challenges, and opportunities in the domain of edible energy harvesting.
iii. **Statistical Analysis**: Statistical methods were applied to evaluate the significance of findings, particularly in assessing the environmental impact and consumer acceptance of these technologies.

2. Design and Fabrication of Edible Energy Systems

Different drying treatments for Tenebrio molitor larvae flour production were investigated. The larvae were dried using various methods, including freeze-drying, ordinary drying, microwave drying, and microwave drying with 0.1% butylated hydroxytoluene. Results showed lower protein and higher fat content in the both ordinary drying and freeze-drying flours. Although the microwave drying process was quick and energy-efficient, it had a detrimental effect on the stability of lipids due to increased acidity and secondary oxidation products (Vlahova-Vangelova et al.,2024).

Walnuts are a nutritious food with numerous nutrients and minerals. However, their processing operations often produce by-products, which are often underutilized. By-products can be applied to create valuable items, increasing profitability and providing environmental and socioeconomic benefits. Understanding the value of these by-products and determining the technical feasibility of processing technologies is crucial for converting walnut waste into valuable food and non-food products (McGranahan & Leslie, 1991).

The effect of ozone treatment on the enzymatic and Tianjin-based non-enzymatic scavenging solutions for post-harvest "Jing Tao Xiang" strawberries. The strawberries were treated with different concentrations of ozone, varying from 2.144 to 15.008 mg/m3. The results showed that Ascorbic acid and phenolic components were greatly enhanced by the ozone treatment, preserving the overall flavonoid content. Nonetheless, the enzymatic antioxidant system demonstrated a rise in peroxidase activity; however, the treatment group receiving 15.008 mg/m3 had a rapid decline in this activity. The 2.144 mg/m3 concentration O3 treatment group showed minimal change in enzyme activity, but the 6.432 mg/m3 treatment group showed suppression of enzyme activity. SOD activity, APX activity, and POD enzyme activity were all markedly elevated in the 10.72 mg/m3 ozone treatment group. After harvesting strawberries, ozone treatment at a dose of 10.72 mg/m3 increased the enzymatic reaction activity of the redox system, increasing the strawberries' edible and commercial value (Chen et al., 2019).

Using egg white liquid, soy milk, and cow's milk as spin-coated positive layers in triboelectric nanogenerators (TENGs) for wearable electronics and green energy harvesting was investigated. The egg white liquid is found to be a superior comparable conductivity and high transparency in a liquid conductor. The EW-TENG, with its potential eco-friendly, flexible, and lightweight for electrophoretic deposition and energy harvesting (Kheirabadi et al.,2023).

Potato tubers are a popular food crop due to their high nutritional value and edible energy content. However, they are often affected by physiological and disease disorders due to poor handling and storage practices, resulting in losses ranging from 40-50%. These losses can be qualitative or quantitative. The chapter reviews these post-harvest diseases and disorders and suggests using classic and modern technologies to extend the shelf-life of potato tubers (Benkeblia, N., 2012).

The wolfberry, an important ingredient in Chinese medicine and cuisine, is usually dried after harvest and is subject to quick degradation. Drying, however, can reduce the material's bioactive components and use a lot of energy. Fresh wolfberry fruit may be well preserved in both CA and SA + CA storages, according to a study that looked at the impacts on wolfberry quality of salicylic acid (SA) and control atmosphere (CA). A possible post-harvest barrier technique for wolfberry preservation that might be helpful for the food and agricultural industries is the combination of SA and CA. The study emphasises how crucial it is to combine SA and CA in order to preserve wolfberry (Xiang et al.,2021).

The coconut tree is a vital resource in tropical regions, used for subsistence and economic activities. However, the significant amount of agro-industrial residue produced during harvesting raises questions. This study emphasises the lignocellulosic nature of coconut trash and its transformational potential. This examines the various uses for coconut goods with an emphasis on recycling trash to lessen its impact on the environment and promote a circular economy (Vieira et al.,2024).

Edible mushrooms are grown using lignocellulosic waste, but 70% of the substrate is still made up of discarded mushroom compost (SMC). SMC is useful in agriculture for prolonged cultivation, plant compost, animal feed, and mushroom media. Moreover, it is employed in the production of biogas, bioethanol, or biohydrogen for sustainable energy (Umor et al.,2021).

Degraded land resources caused by mining, erosion, and other activities can lead to soil degradation and challenges in producing healthy food. The goal of this research is to produce biomass energy on degraded land by planning from the upstream to the downstream sectors. A balanced industrial and land sectors as well as a well-functioning supply chain are necessary for the sustainable generation of biomass energy (Nurcholis et al., 2021).

The use of starch as a sustainable, eco-friendly, and biodegradable material for research was done on the creation of biodegradable triboelectric nanogenerators (TENGs). By introducing an edible Laver filler, the hydrophobicity of starch was enhanced, resulting in a composite with a contact angle of 107°, which can be utilised to power electronics and capture biomechanical energy (Khandelwal et al.,2021).

It is possible to market borage, an edible flower with a limited shelf life, by prolonging its shelf life. Three post-harvest procedures were assessed in this study: alginate edible coating, hot air convective drying, and freeze-drying. The flowers that were freeze-dried were more beautiful, however the flowers coated with alginate stayed looking fresher for a longer period of time. However, the flowers became fragile, making freeze-drying suitable for infusions and alginate coating promising for increasing shelf life (Pereira et al., 2018).

Figure 3. Treemap of keywords researched in edible energy harvesting (Biblioshiny Analysis)

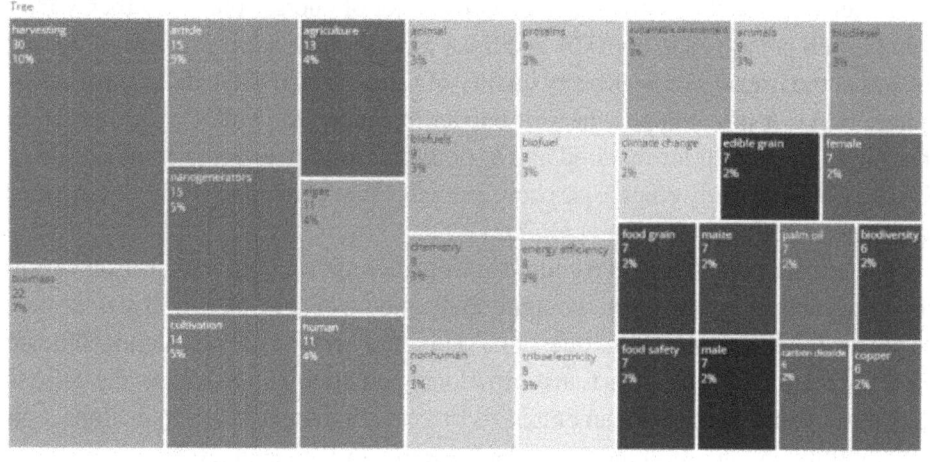

Table 2. Some edible materials preparation methods include efficiency (%), time required, scalability, and energy cost (approx.)

Method	Efficiency (%)	Time Required	Scalability	Energy Cost
Enzymatic Hydrolysis	80	2 hours	High	Low
Mechanical Pressing	70	30 minutes	Medium	Medium
Microwave-Assisted Extraction	85	15 minutes	Low	High
Supercritical CO2 Extraction	90	4 hours	Low	Very High
Solvent Extraction	75	3 hours	High	Medium
Fermentation	60	48 hours	High	Low

Table 3. Some edible materials preparation methods, energy output (joules), test conditions (approx.)

Material Used	Preparation Method	Energy Output (Joules)	Test Conditions
Sugar	Fermentation	1500	Temperature: 25°C, pH: 7
Starch	Enzymatic Hydrolysis	1300	Temperature: 30°C, pH: 6

continued on following page

Table 3. Continued

Material Used	Preparation Method	Energy Output (Joules)	Test Conditions
Cellulose	Acid Hydrolysis	1100	Temperature: 50°C, pH: 4
Fat	Mechanical Pressing	2000	Temperature: 20°C, Ambient Pressure
Protein (Casein)	Solvent Extraction	1600	Temperature: 60°C, pH: 7, High Pressure
Edible Oil (Olive)	Supercritical CO2 Extraction	2200	Temperature: 35°C, High Pressure

i. **Engineering Principles for Designing Edible Energy Systems**

When designing edible energy systems, several fundamental engineering principles must be adhered to, ensuring that the devices are effective, safe, and practical for ingestion. These principles include:

Energy Efficiency and Density: Edible energy systems should be optimized for high energy density and efficiency to maximize the power output from the limited volume and mass that can be ingested.

Material Selection: Materials used must be non-toxic, digestible, or biodegradable and able to function effectively in the variable pH and temperature conditions of the gastrointestinal tract.

Mechanical Integrity: The device must maintain structural integrity in the harsh mechanical environment of the gastrointestinal system but also disintegrate or pass safely through the body after its operational lifecycle is complete.

Scalability and Manufacturability: Designing with scalable manufacturing processes in mind is crucial for commercial viability, ensuring that the devices can be produced cost-effectively at scale.

ii. Techniques for Integrating Energy Harvesting Components into Food

Innovative techniques for integrating energy harvesting components into edible platforms include microencapsulation, edible conductive inks, and biocompatible interfaces. Microencapsulation protects components from digestive processes, while edible conductive inks enable electronic circuit printing on substrates. Biocompatible interfaces prevent harmful substances leaching.

iii. Safety and Biocompatibility Considerations

Safety and biocompatibility are paramount in the development of edible energy systems due to their intimate interaction with the human body:

Toxicological Assessments: Comprehensive testing to ensure that all materials and by-products are non-toxic and safe for ingestion.

Digestibility and Excretion: Materials should either be fully digestible or able to pass through the gastrointestinal tract without causing harm or discomfort.

Compliance with Regulations: Adhering to international safety standards and obtaining approvals from regulatory bodies like the FDA or EMA is essential to ensure safety and market acceptance.

iv. **Innovations in Fabrication Technology Specific to Edible Applications**

Fabrication technologies for edible applications have to consider the unique challenges posed by the need for biocompatibility and digestibility:

3D Printing of Edible Materials: Advances in 3D printing technologies allow for the creation of complex structures from edible materials, integrating energy harvesting components into these structures seamlessly.

Nanotechnology: Utilizing nanoscale materials and components can reduce the overall size of the energy harvesting device, making it easier to integrate into small edible forms.

Advanced Bioengineering: Techniques like synthetic biology can be employed to create biological systems that inherently produce or store energy, potentially bypassing the need for traditional electronic components.

3. Types of Edible Energy Harvesters

New evidence from Northton, a Mesolithic hunter-gatherer site in Scotland, shows the harvesting of edible plant roots and tubers. The excavations revealed abundant Ficaria verna Huds. Root tuber remains and the first evidence of Lathyrus linifolius (Reichard) Bässler tuber use at a European hunter-gatherer site. The findings highlight the importance of appropriate sampling on hunter-gatherer sites (Bishop et al., 2023).

The nutritional and phytochemical composition of Henicus whellani Chopard crickets, a Zimbabwean cricket, collected from Bikita District's four quadrants analysed. The results showed a high ash content, suggesting a rich source of minerals like calcium, iron, magnesium, phosphorus, and potassium. However, the presence of saponins, oxalates, and tannins may limit their benefits. Additional investigation is required to assess the bioaccessibility and safety of bioactive compounds for human consumption (Musundire et al., 2014).

The establishment and production of four native, edible, culturally significant forbs (Apios americana Medik., Helianthus annuus L., Helianthus tuberosus L., etc.) returned to a fallow farm field were compared to the impacts of three site preparation treatments and biomass harvesting. The findings indicate that plants cultivated from seed establish better after tillage, whereas plants cultivated from tubers establish populations more successfully following fire or mowing. According to the study, natural wild edibles might be returned to the old-field systems using low-energy input management techniques in order to preserve or improve ecosystem function (Law et al.,2020).

The dynamic alterations in morphological traits, oil content, fatty acid composition, and other components throughout the growth of Xanthoceras sorbifolium Bunge seeds were studied. According to the study, seed weight peaked 65 days after anthesis, while seed width expanded quickly within 50 days. Between 45 and 55 days, seed kernel oil quickly accumulated, reaching its maximal oil concentration at 70 days. The seed kernels' non-structural carbohydrate content dropped, giving the basic materials needed to produce seed oil (Zhang et al.,2020).

Global nutrition can be impacted by severe illness conditions that arise from food instability and malnutrition in children. Edible insects are commonly used in diets, but their nutritional information is higher than conventional products. Insect beverages and bars are more expensive than conventional commodities. Insects have environmental, economic, and nutritional advantages above other creatures. Large-scale production depends on pricing models for edible insect meals, and emerging and underdeveloped countries must work together to support this innovation through policies and coordinated efforts (Ojianwuna et al.,2023).

Gum polysaccharides are sustainable, renewable materials with adaptable qualities that may be used in a range of sectors. They are affordable, non-toxic, tasteless, colourless, chemically inert, and biocompatible. Plant gums have been employed as edible coating agents for fruits and vegetables as well as tragacanth, ghatti, acacia, and karaya gum. Moreover, they are employed in tissue engineering, medication delivery, nanofibers, nanoparticles, film creation, and wound healing dressings (Adnan et al.,2021).

The use of edible insects as a substitute for traditional agricultural methods has grown in favour due to concerns about climate change, rising carbon emissions, and the environmental damage they cause. Edible insects have several benefits, such as quick reproduction, high energy conversion efficiency, variety, wide dispersion, lower emissions of greenhouse gases and ammonia, potential for waste reduction, and high nutritional content (Musundire et al., 2021).

Auricularia polytricha's nutritional characteristics and transcriptome profiling (A. polytricha) between two harvesting periods, AP_S1 (first harvesting period) and AP_S2 (third harvesting period), were investigated. Results show that AP_S1 provides additional growth advantages because to its higher levels of protein, calcium content, amino acid content, biomass, and auricle area. Transcriptome profiling also shows that AP_S1 strengthens synthesis and metabolic processes, particularly in amino acid and protein synthesis and metabolism. This is the first study to compare these factors (Wang et al.,2021).

Figure 4. Word cloud keywords researched in edible energy harvesting (Biblioshiny Analysis)

i. **Photovoltaic Elements for Energy Harvesting in Foods**

Photovoltaic elements, traditionally used for solar energy harvesting, can be adapted for use in foods, primarily for powering small electronic devices within edible products. These elements convert light to electricity using materials that must be non-toxic and safe for incorporation into edible items.

Materials: Innovations might include the development of organic photovoltaic cells that utilize biodegradable and edible materials, such as modified natural dyes or conductive polymers derived from biological sources.

Application: Potential applications could be in smart packaging where photovoltaic cells power sensors that monitor food quality or in medical diagnostic devices that are activated by light after ingestion.

ii. Piezoelectric Components Derived from Edible Materials

Piezoelectric components generate electricity from mechanical stress, such as pressure or vibrations. Edible piezoelectric materials could harness energy from bodily movements or the mechanical forces exerted during digestion.

Materials: Research into biocompatible, edible piezoelectric materials has included modifications of substances like gelatin, which exhibits piezoelectric properties under certain conditions.

Application: Such components could be integrated into chewable tablets or food items that, when chewed, generate enough power to activate small diagnostic devices or release medication in controlled doses.

iii. Biochemical Energy Harvesters Utilizing Metabolic Processes

This approach involves harnessing the biochemical energy from metabolic processes directly within the body, similar to how cells convert glucose into energy.

Materials and Mechanism: Enzymatic biofuel cells, for instance, use enzymes as catalysts to extract electrons from metabolites like glucose or lactate directly in the human body.

Application: These devices could be particularly useful in continuously powered ingestible biosensors for monitoring metabolic functions or disease states in real time, leveraging the body's own biochemical activities as a power source.

iv. **Comparison of Efficiency and Practicality Among Different Types**

Photovoltaic Elements:
Efficiency is best in light-rich environments, but limited in ingestible applications. Practicality is best for external applications like smart packaging, unless activated by external light sources.

Piezoelectric Components:
Efficiency in edible forms depends on mechanical stress, while practicality involves consistent mechanical stress in applications like mastication or gastrointestinal movements.

Biochemical Energy Harvesters:
Efficiency and practicality of enzymes are limited by precise reactions and enzyme longevity, but are highly useful for internal applications like long-term monitoring and medical drug delivery systems.

Photovoltaic systems are generally more practical for external applications; piezoelectric systems find a niche in applications involving physical interactions, while biochemical harvesters offer a significant advantage for internal medical applications due to their ability to leverage the body's internal biochemical energy. The choice between these systems depends heavily on the specific application, required power levels, and environmental conditions of use.

4. Applications of Edible Energy Technologies

The study evaluates the environmental impacts of canned sardines in olive oil using the life cycle assessment (LCA) methodology. The canning factory in Portugal produces aluminum cans, contributing to high environmental loads. Canned sardines have a seven-fold higher environmental cost per kilogram than frozen and fresh sardines. To optimize their environmental performance, packaging should be replaced and olive oil losses minimized. Plastic packaging reduces greenhouse gas emissions by half. Frozen and fresh sardines have lower environmental impacts (Almeida et al., 2015).

Blueberries are popular for their flavor and nutritional content but have a short shelf life due to microbial contamination and water loss. Existing preservation methods, like irradiation and electrostatic fields, are effective but expensive. Edible composite films, created by incorporating functional substances, offer eco-friendly preservation technology by enhancing fruit storage quality. This article reviews the feasibility and potential applications of edible composite films (Shi et al., 2024).

The depletion of fossil fuels has led to research on alternative energy sources, with algae being used as a third-generation feedstock for biofuel production. However, algal-based biofuels face higher costs in cultivation, harvesting, and extraction. This review discusses the evolution of biofuel feedstocks, configurations of photobioreactor systems, and the importance of conditions like temperature, light intensity, inoculum size, CO_2, nutrient concentration, and mixing in bioreactor performance. Advancements in pretreatment methods, such as hydrothermal processing, are also discussed (Anto et al.,2020).

i. Smart Packaging Solutions with Built-in Energy Harvesters

Smart packaging integrates intelligent systems to provide functions such as tracking, freshness monitoring, and interactive communication. Incorporating energy harvesters into smart packaging enables autonomous operations without the need for external power sources.

Technology: Photovoltaic cells, piezoelectric materials, or thermoelectric generators can be used to harness ambient light, mechanical stress from handling, or temperature differences during transportation and storage.

Applications: These energy harvesters can power sensors that monitor temperature, humidity, and spoilage gases, or they can enable communication systems that provide product information or verify authenticity directly to consumers' smartphones.

Benefits: Enhances the safety and shelf life of food products, provides real-time data to both consumers and suppliers and supports proactive supply chain management.

ii. **Edible Electronics in Medical Diagnostics and Drug Delivery**

Edible electronics represent a revolutionary step in personalized medicine, allowing for internal diagnostic processes and targeted drug delivery systems powered by ingestible energy sources.

Technology: Biochemical energy harvesters using enzymatic reactions or small piezoelectric devices activated by gastrointestinal movements can be key players.

Applications: These devices can be designed to release drugs at specific locations within the gastrointestinal tract or to monitor conditions such as pH levels, temperature, or the presence of specific biomarkers.

Benefits: Provides precise, controlled therapy tailored to individual patient's needs, potentially improving therapeutic outcomes and reducing side effects.

iii. Enhancements in Food Safety Through Active Monitoring

The integration of energy harvesting technologies into food safety systems can enable continuous, active monitoring of food conditions, significantly enhancing safety protocols.

Technology: Sensors powered by energy harvesting (such as mechanical vibrations in transport or thermal differentials in storage areas) can continuously monitor food safety parameters.

Applications: These sensors can detect pathogens, spoilage indicators, or toxic compounds, transmitting data to stakeholders in real-time and enabling swift action to prevent foodborne illnesses.

Benefits: Increases consumer trust and compliance with regulatory standards, reduces waste by pinpointing spoilage, and optimizes logistics by monitoring food quality throughout the supply chain.

iv. **Sustainable and Self-Powered Food Systems**

Sustainable food systems aim to reduce environmental impact, and integrating self-powered systems can significantly contribute to these goals by minimizing energy consumption and waste.

Technology: Energy harvesting from solar, wind, or bioenergy within agricultural or processing settings can power sensors and actuators involved in precision agriculture or automated processing plants.

Applications: In agriculture, solar-powered sensors can monitor soil moisture and nutrient levels to optimize water and fertilizer use. In food processing, energy harvesters can power systems that monitor processing efficiency and automate energy-intensive processes.

Benefits: Reduces the carbon footprint of food production and processing, lowers operational costs through energy savings, and supports a more sustainable food supply chain.

Figure 5. Most Relevant Words researched in edible energy harvesting (Biblioshiny Analysis

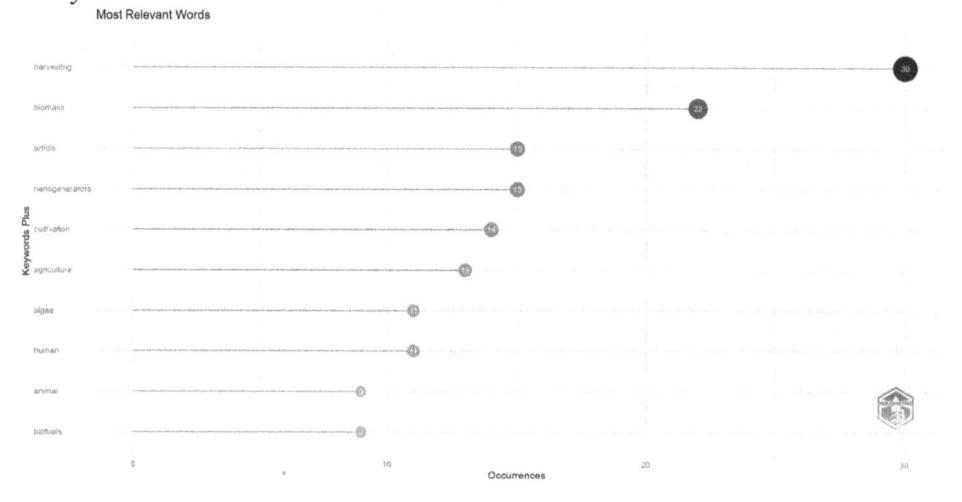

5. Ethical and Environmental Considerations

Insect consumption is a traditional practice in many countries, but due to a lack of standardized legislative rules and scientific data, many countries still consider it a grey area to introduce edible insects into food supply chains. The legal situation and consumer acceptance vary significantly across countries, influenced by geographical locations and cultural backgrounds. Safety concerns persist, especially in countries without insect consumption. However, commercial insect products like energy bars, burgers, and snack foods have emerged, and the European Union has issued a regulation item for insect-based foods (Siddiqui et al., 2023).

Sunflower oil, a crucial oil crop, is highly nutritious and has excellent nutritional value. Two samples of sunflower oil were analyzed for physicochemical properties and storage methods to minimize deterioration and prolong shelf life. The results showed that extracted oil after harvesting is more resistant to deterioration than stored oil. Glass containers were found to be more resistant than polythene, while galvanized containers were the worst. Brown containers showed the highest resistance to oxidative stability (Abdellah et al.,2012).

Common lands in Africa provide smallholder farmers with firewood, timber, and livestock feed, which are used to collect edible non-timber forest products (NTFPs) to supplement human diets. These NTFPs are crucial for coping with crop failure and are essential for wealthier and poor farmers during droughts. However, deforestation and illegal harvesting threaten their effectiveness. Farmers are now planning future land use to intensify crop production, cultivate firewood, maintain orchards, and improve grazing lands. Addressing issues like regeneration and conservation, access rules, and managing competing claims on common lands is crucial (Woittiez et al.,2013).

Africa needs to adapt to climate change by transforming its water and land resource management systems. This includes focusing on practical solutions for salty and freshwater resources, strengthening urban and rural communities, and addressing socioeconomic challenges. Soil and water conservation efforts should include commercializing small-scale farming, reforestation, and improving soil fertility. Urban wastewater treatment systems and cost-effective mechanization can also help. Food security programs should include mechanization, agronomy, and marketing drought-tolerant crops. Developing renewable energy technologies can improve rural communities' productivity and standard of life (Vushe, 2021).

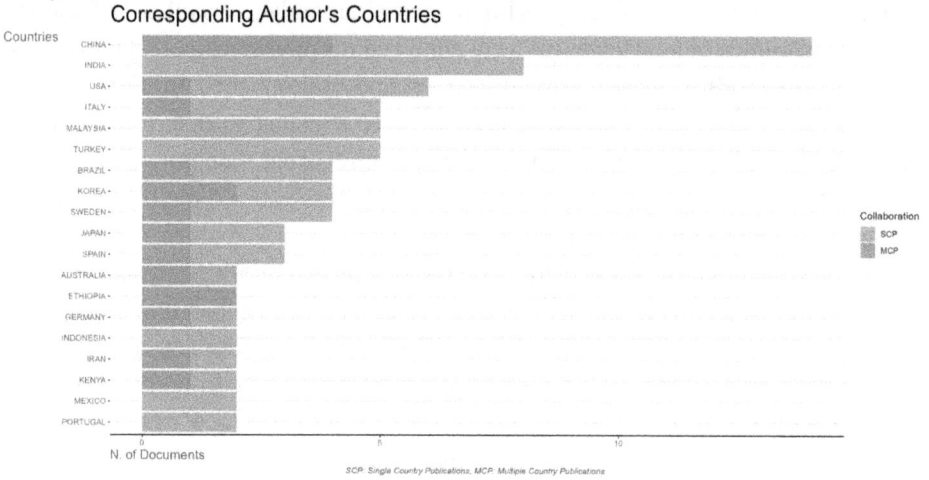

Figure 6. Most relevant countries researched edible energy harvesting (Biblioshiny Analysis)

The potential of upcycling food products for energy production is highlighted by Takemoto (2023), who emphasizes the need to consider the environmental impact of these processes (Takemoto et al.,2023). Pecunia (2023) further underscores the importance of innovative materials in converting ambient energy into electricity, which could be applied to edible energy harvesting (Pecunia et al.,2023). Corigliano (2016) and Singh (2012) both explore the energy recovery from organic biomass, with Corigliano focusing on residual organic biomass and Singh specifically on vegetable wastes (Corigliano et al.,2016 & Singh et al.,2012). These studies collectively underscore the potential of edible energy harvesting in addressing both ethical and environmental considerations.

Consumer Consent and Transparency: The introduction of biologically active components in food products raises concerns about informed consent. Consumers have the right to know if their food contains active elements that might interact with their bodies beyond traditional nutrition.

Health Risks: Potential health risks, especially for populations with specific dietary restrictions or allergies, must be thoroughly assessed and communicated. There's also the question of long-term effects, which may not be fully understood initially.

Data Privacy: Biologically active foods, particularly those that gather data, pose significant privacy issues. How the data is collected, stored, and used must adhere to stringent ethical standards.

Clear Labeling and Information: Providing consumers with clear, accessible information about the contents and effects of biologically active foods is essential.

Rigorous Safety Testing: Ensuring that these products undergo extensive testing to prevent adverse health impacts.

Secure Data Practices: Implementing robust data protection measures to safeguard personal health information.

i. Environmental Impact of Producing and Disposing of Edible Energy Devices

Environmental Concerns:
Resource Use: The production of edible energy devices involves both renewable and non-renewable resources, which could contribute to resource depletion if not managed sustainably.

Waste and Pollution: Depending on the materials used, the disposal of these devices might generate waste and pollution, especially if the devices contain components that do not biodegrade easily.

Mitigation Strategies:
Biodegradable Materials: Developing energy devices from fully biodegradable materials can reduce environmental impact.

Recycling and Reuse: Establishing methods for recycling or repurposing electronic and biochemical components can help minimize waste.

Eco-friendly Manufacturing Processes: Implementing greener manufacturing technologies that reduce environmental footprints.

ii. **Sustainability Practices and Their Promotion in the Industry**

Sustainability Initiatives:
Life Cycle Assessment (LCA): Performing LCAs to understand and minimize the environmental impact throughout a product's lifecycle, from sourcing materials to disposal.

Sustainable Sourcing: Using raw materials that are renewable, ethically sourced, and have a minimal environmental impact.

Energy Efficiency: Enhancing the energy efficiency of production processes and promoting the use of renewable energy sources.

Promotion in the Industry:
Certifications and Labels: Adopting and promoting sustainability certifications can help consumers make informed choices.

Corporate Responsibility Programs: Encouraging companies to integrate sustainability into their corporate ethos and business objectives.

Collaborations and Partnerships: Engaging with stakeholders across the supply chain to foster broader adoption of sustainable practices.

iii. **Potential Unintended Consequences and Management Strategies**

Unintended Consequences:
Market Disruption: The introduction of advanced biologically active and energy-harvesting food products could disrupt traditional markets, potentially harming industries that are slow to adapt.
Socioeconomic Impact: There could be a division between those who can afford these advanced products and those who cannot, potentially leading to inequality.
Dependency and Overreliance: Increased reliance on technology in food systems might make them more vulnerable to technical failures or cyber-attacks.
Management Strategies:
Risk Assessment and Planning: Continual evaluation of potential risks and development of contingency plans to address them.
Public Engagement and Education: Keeping the public informed and educated about the new technologies and their implications.
Regulatory Oversight: Ensuring that there is adequate governance to prevent abuse of the technology and to safeguard public interest.

Table 4. Country-wise safety standards and compliance requirements

Country/Region	Safety Standards	Compliance Requirements
USA	FDA Guidelines for Edible Devices	Biocompatibility Testing, Labeling
European Union	EU Regulation EC No 1935/2004	Migration Limits, Traceability
Japan	JFSA Standards for Food Safety	Toxicity Testing, Quality Control
China	GB Standards for Food Safety	Heavy Metal Testing, Packaging
Canada	Health Canada Food Guidelines	Allergen Control, GMP Certification
India	FSSAI Standards for Food Products	Composition Disclosure, Eco Labeling
Australia	FSANZ Food Standards Code	Nutritional Labelling, Contaminant Limits
Brazil	ANVISA Regulations	Registration, Microbial Contamination Tests

6. Challenges and Limitations

It investigated the impact of palm biodiesel on transport CO_2 emissions in Malaysia from 1990-2019 and predicted the impact of the B10 blending program. Using dynamic autoregressive distributed lag and Kernel-based regularized least squares, the study finds a one-way Granger causality between GDP from transport, diesel consumption, and motor petrol consumption to palm biodiesel consumption. A 10% increase in biodiesel consumption by 2030 could reduce road transport carbon emissions. Improving oil palm plantation and harvesting technologies can also reduce forest replacement and biodiversity loss (Solaymani, S., 2023).

A study in rural Burkina Faso found that 14% of children had a MUAC <125 mm, indicating acute undernutrition. The study also found a decline in MUAC below 3000 kcal/ae/d, with a mean MUAC of 2.49 mm less at 1000 kcal/ae/d. This suggests that In rural agricultural communities, worse nutritional status is linked to lower household crop output/per capita, making them vulnerable to adverse weather effects on the harvest of agriculture, particularly in light of climate change (Belesova et al., 2017).

The study explores the impact of copper contamination on rice growth and development. It found that high copper concentrations led to a decline in photosynthetic efficiency but increased Non-photochemical quenching and non-controllative energy release. This also delayed rice flowering. The study also found significant lipid peroxidation and leaf area reduction. The study concluded that excess copper inhibited rice photosynthetic capacity and growth, leading to reduced grain yield at harvesting (Htwe et al.,2022).

The use of microalgae for biofuel production has been a recent trend due to its high oil yield, short harvesting period, and lack of arable land requirements. These third-generation biofuels can be used as feedstock for biodiesel and other biofuels like bio-oil, bio-char, syngas, and bioethanol. The conversion technologies of microalgae into various biofuels have proven their feasibility and potential (Shuit et al., 2021).

The ability to prevent rivalry between food and fuel is drawing attention to the production of cellulosic bioethanol from inedible plants. Seasonality in feedstock availability, however, drives up prices and restricts commercialization. In Indonesia, partial harvesting of Napier grass shortens supply cycles and increases biomass yield, possibly due to the border effect (Sekiya et al., 2015).

Under Mediterranean circumstances, the effects on the environment of transitioning from conventional cereal, pasture, and horticulture systems to food/energy ones were investigated. Alternative scenarios include rapeseed introduction, artichoke residue valorization, and forage to biomass. Results show positive effects on environmental farming sustainability, with -32% and -8% burdens on land basis, respectively. The findings can help improve agricultural practices and land allocation options (Solinas et al., 2015).

Palm oil, a significant vegetable oil and biofuel, has potential sustainability but has emitted greenhouse gases, waste, and wastewater, affecting the environment. A sustainable palm oil industry promotes the bioeconomy, promoting biodiversity and environmental conservation. Solutions include introducing a green economy to coexist with the industry (Sakai et al., 2022).

Figure 7. Country annual scientific production researched in edible energy harvesting (Biblioshiny Analysis)
Country Scientific Production

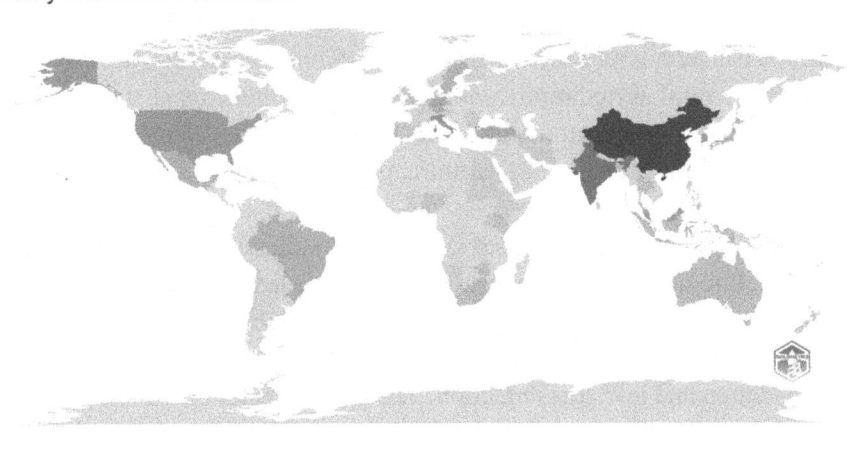

i. **Technical Hurdles in Enhancing Efficiency and Durability**

Challenges:
Material Limitations: Current materials used in edible electronics and energy harvesting may not provide the optimal balance between efficiency and safety required for prolonged use inside or in conjunction with the human body.

Power Management: Efficiently managing and storing the energy harvested in small-scale devices poses significant challenges, especially under the variable conditions found in food storage and human digestive systems.

Wear and Tear: Enhancing the durability of devices that are exposed to harsh environmental conditions, such as high acidity and varying temperatures within the gastrointestinal tract, is challenging.

Solutions:
Advanced Materials: Research into new biocompatible and more robust materials that can function under extreme conditions without degrading.

Innovative Design: Devices could be designed with multi-functional layers that protect core components while still performing their intended functions.

Improved Energy Storage: Development of better micro-scale energy storage solutions that can handle the intermittent and fluctuating power outputs typical of edible energy harvesters.

ii. **Cost Implications and Scalability Issues**

Challenges:

High Production Costs: Novel technologies, particularly those involving cutting-edge materials and methods, typically incur high initial production costs.

Scaling Manufacturing: Scaling manufacturing processes for highly specialized products like biocompatible energy devices requires significant capital investment and can introduce complexities in maintaining quality control.

Market Adoption: The cost of integrating new technologies into existing food products might be prohibitive for widespread market adoption, particularly in cost-sensitive markets.

Solutions:

Economies of Scale: As production volume increases, unit costs can decrease, making the technology more affordable.

Process Innovation: Developing more cost-effective manufacturing processes and automating parts of production can reduce costs.

Subsidies and Incentives: Governments and institutions could provide financial incentives to encourage the adoption and scaling of sustainable and innovative food technologies.

iii. Health and Safety Concerns Surrounding Novel Food Technologies

Challenges:

Unforeseen Reactions: Interactions between novel materials and the human body can have unpredictable results, including allergic reactions or toxicity.

Long-Term Health Effects: The long-term health effects of continuous exposure to new technologies, particularly ingestible or biologically active ones, are often unknown.

Regulatory Compliance: Ensuring that new food technologies comply with existing and future health and safety regulations can be challenging and slow down the introduction of innovative products.

Solutions:

Rigorous Testing: Comprehensive pre-market testing for safety and efficacy, including long-term impact studies.

Transparent Labeling: Clear labeling of all active and potentially reactive components to inform consumers and allow them to make safe choices.

Regulatory Engagement: Early and ongoing engagement with regulatory bodies to ensure that products meet all safety standards and facilitate smoother market entry.

iv. **Limitations in Current Technology and Potential Breakthroughs Needed**

Challenges:

Integration and Miniaturization: Integrating complex electronic functions into small, safe, and palatable formats remains a significant technical barrier.

Sensor Sensitivity and Specificity: Current sensor technologies may not yet achieve the levels of sensitivity and specificity needed for precise monitoring and control of food products.

Data Handling and Privacy: As devices become more connected and capable of collecting personal data, ensuring privacy and secure data handling becomes increasingly complex.

Breakthroughs Needed:

Nanotechnology: Advances in nanotechnology could lead to better integration and miniaturization of devices.

Bioengineering: Genetic and microbial engineering might offer new ways to produce energy or active components naturally within food products.

Artificial Intelligence: AI and machine learning could enhance sensor capabilities and data analysis, improving functionality and user interaction without compromising privacy.

7. The Future of Edible Energy Harvesting

Biorefineries can meet global energy, fuel, and chemical demands, but challenges remain in waste reduction, workflow optimization, and resource consumption. Microalgae, a non-edible feedstock with minimum land usage, constant harvesting, and high production issues, offers potential for integration into biofuels and petrochemical industries in Brazil. Current developments in microalgae large-scale production highlight these challenges (Brasil et al.,2017).

India produced 97.35 million tons of fruits in 2017-18, with less than 1% exported. Three percent of fruits are processed, and less than five percent are sold via organised supply chain management and e-commerce businesses. Over 90% of fruits follow the traditional route, with post-harvest losses occurring at each stage. Interventions to reduce PH losses include establishing pre-cooling facilities, primary processing, packaging, transportation, and storage facilities. Post-harvest management systems, farmer clusters, and mechanization can improve bargaining power and quality output. Transportation methods like evaporative cooling and phase change materials can improve shelf life. An integrated RFID system and sensors can help track and trace fresh produce. Establishing primary and secondary processing facilities can help transform farmers into primary processors (Oberoi & Dinesh, 2019).

The amount of Blad, a bioactive fungicide in Lupinus β-conglutin proteolysis, in seven(7) cultivars of sweet L. albus, was evaluated. The study found that BCO, a non-toxic bioactive fungicide, is most effective in inhibiting fungal growth When plantlets are growing and seeds are germinating. Additionally, the study discovered that keeping seeds at -20°C had no effect on BCO activity, indicating that prevailing edaphoclimatic conditions during seed development regulate variance in cotyledonary BCO accumulation. The study also found significant antifungal activity between 3 and 5-day-old plantlets, with no activity observed on eight-day-old plantlets or older. This research is crucial for maximizing BCO extraction from the cotyledons of sweet L. albus cultivars (Cruz et al., 2021).

i. Emerging Technologies and Research Frontiers

The field of edible energy harvesting and smart food technology is rapidly evolving with several promising research frontiers:

Biodegradable Electronics: Developing fully biodegradable and edible electronics that do not harm the environment or human health. Research is focusing on natural and synthetic materials that mimic the functionality of traditional electronics.

Energy Harvesting from Body Heat and Fluids: Utilizing thermoelectric generators or biochemical energy harvesters that convert body heat or fluids into electricity, powering internal devices such as health monitors and drug delivery systems.

Smart Probiotics: Engineering probiotics that can monitor gut health, treat diseases by delivering drugs directly where needed, or even detect and report on dietary intake and gut microbiome status.

Personalized Nutrition Devices: Devices that analyze real-time metabolic data to provide personalized dietary recommendations or release specific nutrients based on the individual's health data.

ii. Potential Collaborations Across Disciplines to Enhance Innovation

Cross-disciplinary collaborations are essential to drive innovation in the integration of technology into food and medicine:

Material Science and Bioengineering: Collaboration between these fields can lead to the development of new materials that are safe for consumption and capable of performing complex functions inside the body.

Data Science and Nutrition: Working together, these disciplines can harness big data from dietary habits and health outcomes to create smarter, more responsive dietary technologies.

Ethics and Technology: Ethicists, together with technologists, can ensure that new food technologies are developed and implemented in a way that respects consumer rights and data privacy.

Environmental Science and Food Technology: These collaborations can ensure that new technologies are sustainable and contribute positively to the environment, minimizing waste and reducing the carbon footprint.

iii. **Long-Term Visions for the Integration of Technology with Daily Diet**

Looking toward the future, the integration of technology with our daily diet could revolutionize how we think about and interact with food:

Interactive Food: Foods that change properties based on the eater's nutritional needs or preferences, possibly using real-time health data from wearable devices.

Automated Personal Chefs: Advanced AI systems that design and prepare perfectly balanced meals based on personal health data, taste preferences, and available ingredients.

Dietary Optimization Interfaces: Platforms that integrate seamlessly with users' health data and manage their nutritional intake automatically, adjusting as their health status or goals change.

iv. **Predictions for the Evolution of the Field Over the Next Decade**

In the next decade, several trends and innovations are likely to shape the field:

Mainstream Adoption of Smart Packaging: As costs decrease and technology advances, smart packaging will likely become commonplace, offering enhanced food safety, quality monitoring, and consumer interaction.

Advanced Biocompatible Energy Systems: We will see more sophisticated biocompatible and ingestible energy systems being used in clinical settings for monitoring and treatment, dramatically improving patient care and health monitoring.

Regulatory Evolution: Regulations will evolve to catch up with technological advances, focusing more on digital health, data security, and personalized nutrition.

Growth of Personalized Nutrition: Driven by a better understanding of individual health and metabolic differences, personalized nutrition will grow, supported by technology that makes real-time adjustments to dietary intake.

Table 5. Hypothetical example of future research directions

Suggested Area	Key Questions	Proposed Methods	Potential Impact
Nano-materials in Energy Storage	How can nano-materials enhance energy storage in edible forms?	Development of nano-enhanced biopolymers	Increase in energy density and efficiency; potential for new applications in wearable tech
Biodegradability Improvement	How can the biodegradability of energy-harvesting materials be improved?	Genetic modification of biomass sources; Advanced compostable materials research	Reduced environmental impact, enhanced sustainability

continued on following page

Table 5. Continued

Suggested Area	Key Questions	Proposed Methods	Potential Impact
Cost Reduction Strategies	What methods can reduce the cost of energy production from edible sources?	Optimization of extraction processes; Economies of scale in production	Lower cost, increased market adoption
Hybrid Energy Systems	Can edible materials be integrated with traditional energy systems for hybrid solutions?	System integration studies; Prototype development of hybrid systems	Enhanced system efficiency and reliability
Improved Energy Conversion	How can the efficiency of energy conversion from edible sources be enhanced?	Exploration of novel enzymatic pathways; Thermodynamic analysis	Higher output, broader usability, and practical application scope
Regulatory Compliance	What are the emerging regulatory needs for edible energy products?	Regulatory review and gap analysis; Continuous compliance monitoring	Faster market entry reduced legal and business risks

8. CONCLUSION

The integration of edible energy harvesting technologies offers ground-breaking possibilities in sustainable energy production and consumption. These innovations hold promise for applications in medical devices, wearable technology, and beyond, providing a unique intersection of food science, biotechnology, and engineering. However, realizing their full potential requires addressing ethical, environmental, and regulatory challenges. Rigorous safety testing, clear labelling, and secure data practices are essential to ensure consumer trust and acceptance. Moreover, adopting sustainable practices in production and disposal will mitigate environmental impacts. The cooperative efforts of industry stakeholders, legislators, and scientists will be pivotal in navigating the complexities and advancing the field. By fostering a holistic approach, edible energy harvesting can contribute significantly to global energy solutions, paving the way for a more sustainable and technologically advanced future.

Scope for Future Work

i. Advanced Materials Research: Investigate new biodegradable and biocompatible materials to enhance the efficiency and safety of edible energy harvesting devices.
ii. Long-term Health Studies: Conduct extensive longitudinal studies to assess the long-term health impacts of consuming biologically active food products, ensuring they are safe for diverse populations.

iii. Integration with Smart Technologies: Explore the integration of edible energy devices with Internet of Things (IoT) systems and wearable technology, enabling real-time health monitoring and data collection.
iv. Consumer Acceptance and Education: Study consumer perceptions and acceptance of these technologies. Develop educational programs to inform the public about the benefits and safety of biologically active and energy-harvesting foods.
v. Sustainability Improvements: Focus on enhancing the sustainability of production processes, including the use of renewable energy and reducing the environmental footprint of manufacturing edible energy devices.
vi. Regulatory Frameworks: Develop comprehensive regulatory frameworks that address safety, ethical considerations, and data privacy for biologically active foods and edible energy devices.
vii. Economic Impact Analysis: Analyze the economic impacts of widespread adoption of these technologies, including potential market disruptions and socioeconomic benefits or challenges.
viii. Global Collaboration: Foster international collaboration to standardize regulations, share research findings, and promote the global development and acceptance of edible energy harvesting technologies.
ix. Exploration of New Applications: Expand the potential applications of edible energy devices beyond medical and wearable tech, such as in disaster relief, military operations, and space exploration.
x. Environmental Impact Studies: Conduct detailed environmental impact assessments to ensure that the production, use, and disposal of these devices are environmentally sustainable. Develop strategies to mitigate any negative effects.

By addressing these areas, future research can significantly advance the field of edible energy harvesting, ensuring its safe, ethical, and sustainable integration into society.

Key Takeaways:

Technological Innovation: Developments in materials science have enabled the creation of biocompatible and even biodegradable materials that can safely operate within the human body or be incorporated into food items. These materials are crucial for the successful application of edible energy harvesting technologies in medical diagnostics, drug delivery, and interactive food products.

Health and Environmental Impact: Edible energy harvesters can lead to improved health monitoring and disease management by providing real-time, personalized medical data and therapeutic interventions. Environmentally, the push for biodegradable and non-toxic materials in these devices aligns with broader sustainability goals, reducing waste and the ecological footprint of medical and electronic waste.

Challenges and Future Directions: Despite its potential, the field faces significant challenges, including technical limitations related to energy efficiency and storage, safety and ethical concerns about biocompatibility and data privacy, and economic hurdles related to production costs and market adoption. Overcoming these challenges will require multidisciplinary collaborations and continued innovation.

Regulatory and Ethical Considerations: As these technologies enter mainstream markets, they will necessitate rigorous regulatory frameworks to ensure safety and efficacy, alongside ethical considerations to manage consumer data and ensure privacy.

Looking Ahead: The next decade will likely see increased research activity and commercial applications of edible energy harvesting technologies as these challenges are addressed and as the market for personalized healthcare and sustainable technologies grows. Innovations in this field could profoundly impact not only the medical field but also consumer electronics, the food industry, and environmental management practices.

REFERENCES

Abdellah, A. M., & Ishag, K. E. A. (2012). Effect of storage packaging on sunflower oil oxidative stability. *American Journal of Food Technology*, 7(11), 700–707. DOI: 10.3923/ajft.2012.700.707

Adnan, M., Oh, K. K., Cho, D. H., & Alle, M. (2021). Nutritional, pharmaceutical, and industrial potential of forest-based plant gum. Non-Timber Forest Products: Food, Healthcare and Industrial Applications, 105-128.

Alimonti, G., Brambilla, R., Pileci, R. E., Romano, R., Rosa, F., & Spinicci, L. (2017). Edible energy: Balancing inputs and waste in the food supply chain and biofuels from algae. *The European Physical Journal Plus*, 132(1), 132. DOI: 10.1140/epjp/i2017-11301-8

Almeida, C., Vaz, S., & Ziegler, F. (2015). Environmental life cycle assessment of a canned sardine product from Portugal. *Journal of Industrial Ecology*, 19(4), 607–617. DOI: 10.1111/jiec.12219

Ambrożkiewicz, B., & Rounak, A. (2022). ENERGY HARVESTING – NEW GREEN ENERGY. Journal of Technology and Exploitation in Mechanical Engineering.

Anto, S., Mukherjee, S. S., Muthappa, R., Mathimani, T., Deviram, G., Kumar, S. S., Verma, T. N., & Pugazhendhi, A. (2020). Algae as green energy reserve: Technological outlook on biofuel production. *Chemosphere*, 242, 125079. DOI: 10.1016/j.chemosphere.2019.125079 PMID: 31678847

Belesova, K., Gasparrini, A., Sié, A., Sauerborn, R., & Wilkinson, P. (2017). Household cereal crop harvest and children's nutritional status in rural Burkina Faso. *Environmental Health*, 16(1), 1–11. DOI: 10.1186/s12940-017-0258-9 PMID: 28633653

Benkeblia, N. (2012). Post-harvest diseases and disorders of potato tuber Solanum tuberosum L. *Potatoes Prod. Consum. Health Benefits*, 7, 99–114.

Bishop, R. R., Kubiak-Martens, L., Warren, G. M., & Church, M. J. (2023). Getting to the root of the problem: New evidence for the use of plant root foods in Mesolithic hunter-gatherer subsistence in Europe. *Vegetation History and Archaeobotany*, 32(1), 65–83. DOI: 10.1007/s00334-022-00882-1

Blagg, C.R. (2011). Preface. Hemodialysis International, 15.

Brasil, B. S. A. F., Silva, F. C. P., & Siqueira, F. G. (2017). Microalgae biorefineries: The Brazilian scenario in perspective. *New Biotechnology*, 39, 90–98. DOI: 10.1016/j.nbt.2016.04.007 PMID: 27343427

Chen, C., Zhang, H., Dong, C., Ji, H., Zhang, X., Li, L., Ban, Z., Zhang, N., & Xue, W. (2019). Effect of ozone treatment on the phenylpropanoid biosynthesis of post-harvest strawberries. *RSC Advances*, 9(44), 25429–25438. DOI: 10.1039/C9RA03988K PMID: 35530059

Corigliano, O., Florio, G., & Fragiacomo, P. (2016). Energy Valorization of Edible Organic Matter for Electrical, Thermal and Cooling Energy Generation: Part One. *Energy Procedia*, 101, 89–96. DOI: 10.1016/j.egypro.2016.11.012

Cruz, F., Batista-Santos, P., Monteiro, S., Neves-Martins, J., & Ferreira, R. B. (2021). Maximizing Blad-containing oligomer fungicidal activity in sweet cultivars of Lupinus albus seeds. *Industrial Crops and Products*, 162, 113242. DOI: 10.1016/j.indcrop.2021.113242

Htwe, T., Chotikarn, P., Duangpan, S., Onthong, J., Buapet, P., & Sinutok, S. (2022). Integrated biomarker responses of rice associated with grain yield in copper-contaminated soil. *Environmental Science and Pollution Research International*, 29(6), 1–10. DOI: 10.1007/s11356-021-16314-y PMID: 34498193

Jain, A. K. (2012). Emerging Dimensions in the Energy Harvesting. *IOSR Journal of Electrical and Electronics Engineering*, 3(1), 70–80. DOI: 10.9790/1676-0317080

Kettunen, M., & D'Amato, D. (2013). PROVISIONING SERVICES AND RELATED GOODS. In Social and Economic Benefits of Protected Areas: An assessment guide. DOI: 10.4324/9780203095348-10

Khandelwal, G., Joseph Raj, N. P. M., Alluri, N. R., & Kim, S. J. (2021). Enhancing hydrophobicity of starch for biodegradable material-based triboelectric nanogenerators. *ACS Sustainable Chemistry & Engineering*, 9(27), 9011–9017. DOI: 10.1021/acssuschemeng.1c01853

Kheirabadi, N. R., Karimzadeh, F., Enayati, M. H., & Kalali, E. N. (2023). Green flexible triboelectric nanogenerators based on edible proteins for electrophoretic deposition. *Advanced Electronic Materials*, 9(2), 2200839. DOI: 10.1002/aelm.202200839

Kulkarni, V., Mrad, R.B., El-Diraby, T.E., & Prasad, E. (2010). Energy Harvesting Using Piezoceramics.

Law, E. P., Arnow, E., & Diemont, S. A. (2020). Ecosystem services from old-fields: Effects of site preparation and harvesting on restoration and productivity of traditional food plants. *Ecological Engineering*, 158, 105999. DOI: 10.1016/j.ecoleng.2020.105999

McGranahan, G., & Leslie, C. (1991). Walnuts (Juglans). *Genetic Resources of Temperate Fruit and Nut Crops*, 290, 907–974.

Musundire, R., Ngonyama, D., Chemura, A., Ngadze, R. T., Jackson, J., Matanda, M. J., Tarakini, T., Langton, M., & Chiwona-Karltun, L. (2021). Stewardship of wild and farmed edible insects as food and feed in Sub-Saharan Africa: A perspective. *Frontiers in Veterinary Science*, 8, 601386. DOI: 10.3389/fvets.2021.601386 PMID: 33681322

Musundire, R., Zvidzai, C. J., Chidewe, C., Samende, B. K., & Manditsera, F. A. (2014). Nutrient and anti-nutrient composition of Henicus whellani (Orthoptera: Stenopelmatidae), an edible ground cricket, in south-eastern Zimbabwe. *International Journal of Tropical Insect Science*, 34(4), 223–231. DOI: 10.1017/S1742758414000484

Nurcholis, M., Ahlasunnah, W., Utami, A., Krismawan, H., & Wibawa, T. (2021, November). Management of degraded land for developing biomass energy industry. In AIP Conference Proceedings (Vol. 2363, No. 1). AIP Publishing. DOI: 10.1063/5.0061179

Oberoi, H. S., & Dinesh, M. R. (2019). Trends and innovations in value chain management of tropical fruits. *Journal of Horticultural Sciences*, 14(2), 87–97. DOI: 10.24154/jhs.v14i2.773

Ojianwuna, C. C., Enwemiwe, V. N., Esiwo, E., Orji, G. O., & Nkeze, A. J. (2023). Food and Feed Additive of Insects: Economic and Environmental Impacts. *International Journal of Child Health and Nutrition*, 12(3), 107–119. DOI: 10.6000/1929-4247.2023.12.03.5

Pecunia, V., Silva, S. R. P., Phillips, J. D., Artegiani, E., Romeo, A., Shim, H., Park, J., Kim, J. H., Yun, J. S., Welch, G. C., Larson, B. W., Creran, M., Laventure, A., Sasitharan, K., Flores-Diaz, N., Freitag, M., Xu, J., Brown, T. M., Li, B., & Joshi, A. P. (2023). Roadmap on energy harvesting materials. *JPhys Materials*, 6(4), 042501. DOI: 10.1088/2515-7639/acc550

Pereira, J. A., Saraiva, J. A., Casal, S., & Ramalhosa, E. (2018). The effect of different post-harvest treatments on the quality of borage (Borago officinalis) petals. *Acta Scientiarum Polonorum. Technologia Alimentaria*, 17(1), 5–10. PMID: 29514420

Saibi, W., Brini, F., Hanin, M., & Masmoudi, K. (2013). Development of energy plants and their potential to withstand various extreme environments. *Recent Patents on DNA & Gene Sequences*, 7(1), 13–24. DOI: 10.2174/1872215611307010004 PMID: 22779438

Sakai, K., Hassan, M. A., Vairappan, C. S., & Shirai, Y. (2022). Promotion of a green economy with the palm oil industry for biodiversity conservation: A touchstone toward a sustainable bioindustry. *Journal of Bioscience and Bioengineering*, 133(5), 414–424. DOI: 10.1016/j.jbiosc.2022.01.001 PMID: 35151536

Sekiya, N., Abe, J., Shiotsu, F., & Morita, S. (2015). Effects of partial harvesting on napier grass: Reduced seasonal variability in feedstock supply and increased biomass yield. *Plant Production Science*, 18(1), 99–103. DOI: 10.1626/pps.18.99

Shi, D., Zhao, B., Zhang, P., Li, P., Wei, X., & Song, K. (2024). Edible composite films: Enhancing the post-harvest preservation of blueberry. *Horticulture, Environment and Biotechnology*, 65(3), 1–19. DOI: 10.1007/s13580-023-00581-4

Shuit, S. H., Tee, S. F., Sim, L. C., & Lim, S. (2021). Biofuels Production from Microalgae: Processes and Conversion Technologies. In Biofuel Production from Microalgae, Macroalgae and Larvae: Processes and Conversion Technologies.

Siddiqui, S. A., Tettey, E., Yunusa, B. M., Ngah, N., Debrah, S. K., Yang, X., Fernando, I., Povetkin, S. N., & Shah, M. A. (2023). Legal situation and consumer acceptance of insects being eaten as human food in different nations across the world–A comprehensive review. *Comprehensive Reviews in Food Science and Food Safety*, 22(6), 4786–4830. DOI: 10.1111/1541-4337.13243 PMID: 37823805

Singh, A., Kuila, A., Adak, S., Bishai, M., & Banerjee, R. (2012). Utilization of Vegetable Wastes for Bioenergy Generation. *Agricultural Research*, 1(3), 213–222. DOI: 10.1007/s40003-012-0030-x

Solaymani, S. (2023). Biodiesel and its potential to mitigate transport-related CO2 emissions. *Carbon Research*, 2(1), 38. DOI: 10.1007/s44246-023-00067-z

Solinas, S., Fazio, S., Seddaiu, G., Roggero, P. P., Deligios, P. A., Doro, L., & Ledda, L. (2015). Environmental consequences of the conversion from traditional to energy cropping systems in a Mediterranean area. *European Journal of Agronomy*, 70, 124–135. DOI: 10.1016/j.eja.2015.07.008

Takemoto, M., Yunoki, A., Miao, S., & Dowaki, K. (2023). Environmental Impact Analysis of Food Considering Upcycling. *IOP Conference Series. Earth and Environmental Science*, 1187(1), 1187. DOI: 10.1088/1755-1315/1187/1/012032

Umor, N. A., Ismail, S., Abdullah, S., Huzaifah, M. H. R., Huzir, N. M., Mahmood, N. A. N., & Zahrim, A. Y. (2021). Zero waste management of spent mushroom compost. *Journal of Material Cycles and Waste Management*, 23(5), 1726–1736. DOI: 10.1007/s10163-021-01250-3

Vieira, F., Santana, H. E., Jesus, M., Santos, J., Pires, P., Vaz-Velho, M., Silva, D. P., & Ruzene, D. S. (2024). Coconut Waste: Discovering Sustainable Approaches to Advance a Circular Economy. *Sustainability (Basel)*, 16(7), 3066. DOI: 10.3390/su16073066

Vlahova-Vangelova, D., Balev, D., Kolev, N., Dragoev, S., Petkov, E., & Popova, T. (2024). Comparison of the Effect of Drying Treatments on the Physicochemical Parameters, Oxidative Stability, and Microbiological Status of Yellow Mealworm (Tenebrio molitor L.) Flours as an Alternative Protein Source. *Agriculture*, 14(3), 436. DOI: 10.3390/agriculture14030436

Vushe, A. (2021). Proposed research, science, technology, and innovation to address current and future challenges of climate change and water resource management in Africa. Climate Change and Water Resources in Africa: Perspectives and Solutions Towards an Imminent Water Crisis, 489-518.

Wang, W., Wang, Y., Gong, Z., Yang, S., & Jia, F. (2021). Comparison of the nutritional properties and transcriptome profiling between the two different harvesting periods of Auricularia polytricha. *Frontiers in Nutrition*, 8, 771757. DOI: 10.3389/fnut.2021.771757 PMID: 34765633

Woittiez, L. S., Rufino, M. C., Giller, K. E., & Mapfumo, P. (2013). The use of woodland products to cope with climate variability in communal areas in Zimbabwe. *Ecology and Society*, 18(4), art24. DOI: 10.5751/ES-05705-180424

Xiang, W., Wang, H. W., Tian, Y., & Sun, D. W. (2021). Effects of salicylic acid combined with gas atmospheric control on post-harvest quality and storage stability of wolfberries: Quality attributes and interaction evaluation. *Journal of Food Process Engineering*, 44(8), e13764. DOI: 10.1111/jfpe.13764

Yang, K., Bai, C., Liu, B., Liu, Z., & Cui, X. (2023). Self-Powered, Non-Toxic, Recyclable Thermogalvanic Hydrogel Sensor for Temperature Monitoring of Edibles. *Micromachines*, 14(7), 1327. DOI: 10.3390/mi14071327 PMID: 37512638

Yeatman, E.M. (2009). Energy harvesting: small scale energy production from ambient sources. Smart Structures and Materials + Nondestructive Evaluation and Health Monitoring.

Youn, S.B., Yeo, J., Joung, H., & Yang, Y. (2015). Energy harvesting from food waste by inoculation of vermicomposted organic matter into Microbial Fuel Cell (MFC). 2015 IEEE SENSORS, 1-4.

Zhang, Z., Ao, Y., Su, N., Chen, Y., Wang, K., & Ou, L. (2022). Dynamic changes in morphology and composition during seed development in Xanthoceras sorbifolium Bunge. *Industrial Crops and Products*, 190, 115899. DOI: 10.1016/j.indcrop.2022.115899

Chapter 6
Edible Electronics for Smart Detection of Food Spoilage

S. N. Kumar
https://orcid.org/0000-0002-2530-1454
Amal Jyothi College of Engineering, India

Nikki John Kannampilly
Amal Jyothi College of Engineering, India

Jomin Joy
Amal Jyothi College of Engineering, India

ABSTRACT

This chapter describes an innovative approaches which employs edible electronics sensors to monitor food in real time to prevent spoilage. In today scenario, food spoilage detection plays a vital role, since many packaged food items are there in the market. Edible sensors role is inevitable in food industry for spoilage detection of liqud and solid foods. This chapter discuss about various edible electronics sensors for food spoilage detection. Biosensors are widely used in many applications, the sensor values are processed by an IoT based system comprising of a microcontroller and internet connectivity enables the transfer of data through cloud network. Food spoilage detection enables the buyers to predict the shelf life and minimizes the wastage of food items.

DOI: 10.4018/979-8-3693-5573-2.ch006

Copyright © 2025, IGI Global. Copying or distributing in print or electronic forms without written permission of IGI Global is prohibited.

1. INTRODUCTION

Food spoilage is one of the greatest difficulties faced in food industry. Acidic food spoilage is one important among them. Spoilage can be termed as deterioration of food to a condition that is unfit for human consumption, (Sowmya, 2017). Spoilage of food is an intricate process which causes undesirable changes in foods such as texture, appearance or organoleptic changes. These changes associated within the food render it unsafe and inedible, (Sperber & Doyle, 2009). Changes in microbial, chemical, enzymatic activity and physical damage results in deterioration of food, (Ray & Bhunia, 2013). Therefore, it is very essential to develop rapid and innovative techniques to extend food shelf life, ensure food safety, and minimize economic losses, (Luning & Marcelis, 2006). Reduced nutritive value and serious health risks are associated with intake of spoiled food.

Spoilage of food leads to decline in sensory, nutritional quality and safety of the food. This deterioration is due to certain factors such as microbial growth, chemical reactions, physical damage or enzymatic reactions in the food. Various microorganism responsible for spoilage of food include bacteria, yeasts and molds. *Pseudomonas spp.*, *Listeria spp.*, and coliforms are common bacteria causing spoilage. Species of *Aspergillus* and *Penicillium* are common molds causing food spoilage, (Batt & Tortorello, 2014). Intrinsic and extrinsic factors such as characteristics of the food and storage and handling of the food products are major factors responsible for microbial food spoilage, (Jay, 2000). The most common chemical changes associated to food is oxidation, causing rancidity predominantly occurring in fats and oils. Pigment degradation and browning of fruits and vegetables owing to enzymatic activity are are also chemical changes associated with food spoilage, (Beuchat, 2006). Any physical damage to food can lead to spoilage and more prone to microbiological contamination. Major physical damages occur due to harvesting, processing, transportation, or storage, (Campbell-Platt, 2009). Enzymatic spoilage occurs basically due to the reaction of naturally present enzymes in the food. Few common enzyme reactions are breakdown of proteins by protease, fat breakdown by lipase and carbohydrate breakdown by amylase, (Labuza & Sinskey, 2006).

Food safety is essential for public health and everyday life, economic development, social stability, and the reputation of both the government and the nation. Consumption of spoiled food can have an adverse effect on socio-economic impact and also possess a severe health impact. Food borne illness, gastrointestinal discomfort to severe dehydration, kidney failure, or even death are the common issues related with intake of spoiled foods. Food poisoning outbreaks caused by pathogenic organisms like Salmonella, E. coli, and Clostridium botulinum are risky and deadly, (Doyle & Buchanan, 2013). Certain mycotoxins produced by molds are very toxic and can cause severe health effects, (Frias & Vidal-Valverde, 2008). The economic impact of food spoilage is significant. There is a huge loss and wastage of food wordwide to the food produced from human consumption, (Food and Agricultural Organization, 2011). A substantial economic loss, food wastage and also food insecurity are of major concern. This substantial loss arising due to spoilage of food also plays an important affect in food uncertainty and wastage of food.

2. NEED / REQUIREMENT FOR SMART DETECTION SYSTEMS:

Detection of food spoilage through traditional method have been the chief method of ensuring the quality and safety of a food product for decades. The visual check, assessment of odour and checking expiry dates of foods are the easy methods of evaluation of ensuring food safety, but not reliable and accurate. These traditional methods mostly fail to detect spoilage at an initial stage. Food spoilage indicators like microbial growth or chemical changes cannot be detected initially until an advanced level is reached, whereby the consumption of food becomes unsafe, (Gram *et al.*, 2002). Real-time monitoring of food, lack of sufficient and skilled manpower and resources, especially for large-scale operators are the major drawback in traditionally available method of quality or spoilage detection of foods. Therefore, there is an immediate requirement to shift to a rapid, reliable and advanced method or techniques to determine quality and safety of a food.

Modern/ smart detection techniques give an accurate reliable result, real-time monitoring, earlier detection as well as good operational efficiency, (Zou *et al.*, 2019). Thus, a modern technology driven method could improvise the safety or quality of foods, minimise food wastage and also aid in food supply chain management. By implementing these innovative environmentally friendly techniques, a sustainable economy with food safety assurance ca be achieved. Recent such advancement includes use of sensors to determine food spoilage, integration of IoT with food monitoring systems, Machine learning and AI and Blockchain technology. These technologies to the field of food spoilage detection offer several merits such as enhanced food security, economic stability and minimises environmental impact.

3. DETECTION OF FOOD SPOILAGE USING INFORMATION AND COMMUNICATION TECHNOLOGY

Various types of sensors such as gas sensors, temperature and humidity sensors, pH sensors and biosensors have improved capability to detect food quality and spoilage. Smart packages are one such approach incorporating sensors in real-time monitoring of food spoilage. These packages indicate or alert customers about the quality of food product inside a package, thus ensuring safe food and also prevent food wastage, (Kuswandi *et al.,* 2011). Commonly available sensors are:

Gas-sensors: Volatile organic compounds (VOCs) releasing by the foods spoiled can be monitored using different gas sensors. Specific gases released such as ethylene gas, ammonia and hydrogen sulfide can be identified by use of these gas sensors, indicating the spoilage of food. Fruits and vegetables when ripen releases ethylene, which can be sensed by ethylene sensors, thereby indicating the freshness of the fruit. Early timely indication can help in spoilage prevention and reduction in food wastage, (Arshak *et al.,* 2007).

Humidity and Temperature Sensors: It is very important to monitor the humidity and temperature inside and outside the food package. Temperature sensors like thermocouples and resistance temperature detectors (RTDs) help in monitoring environmental conditions and thus ensure the safe temperature limits to store the food commodity to maintain a good quality product. A higher moisture level promotes microbial growth; thus, it is very essential to check the moisture levels using humidity sensors to arrest mold growth and prevent spoilage, (Fitzgerald *et al.*, 2001).

Biosensors: Certain devices with help of a transducer can integrate biologically to indicate spoilage by detecting specific biochemical reactions associated to food. Enzymatic activity, microbiological activity and biological activities can be identified by these biosensors. Specific bacteria can be detected by biosensors, thereby indicating the microbiological safety of the foods, (Velusamy *et al.,* 2010). An early spoilage detection is possible using these biosensors which makes it ideal to monitor food spoilage.

pH sensors: Food spoilage associated with change in acidity or alkalinity of foods can be determined by the pH sensors. Monitoring pH levels in milk, meat and seafoods can determine the freshness and indicate the microbial activity of the food, and hence assess the food quality and avoid intake of unsafe foods, (Pacquit *et al.,* 2007).

Internet of Things (IoT): A significant impact by integrating IoT with food monitoring systems has paved its way to improve the quality and safety of the foods and also improved efficiency of supply chain management. These devices can track and monitor the condition of food throughout the supply chain. This data can update the storage conditions of the food commodity during shipping and storage and thus minimize food spoilage, (Tao *et al.,* 2018). IoT enabled food packages can monitor freshness of the stored product and intimates whether the product is safe or unsafe for consumption.

Machine Learning and Artificial Intelligence (AI): These technologies can predetermine spoilage before they occur by the AI- driven power systems. Larger datasets can be analysed such as storage conditions, transportation routes, and historical spoilage data throughout the supply chain management. Reduction in food wastage and efficiently in food supply chain are the major challenges in this area. Visual detection using convolutional neural networks (CNNs) can identify food products with signs of spoilage and give rapid accurate results. Early warning/spoilage detection system are more sensitive and reliable than traditional methods, (Zou *et al.*, 2019).

Blockchain Technology: Food product freshness can be tracked by the implementing the blockchain technology systems. Food traceability and transparency are the major aspects in supply chain management of a food industry, (Balamurugan, Ayyasamy, & Joseph, 2022). Blockchain system provides a decentralized and absolute record of all the transactions in the supply chain, thereby precise data regarding the production of food, processing and distribution of food is attained. This data can help maintain the quality standards and also detect possible spoilage problems. Advancements in blockchain technology will transform agri-food supply chains by preserving value, enhancing information security and traceability, (Yin *et al.*, 2017).

4. EDIBLE ELECTRONICS SENSORS FOR SPOILAGE DETECTION

Edible electronics is an innovative emerging field to develop sensors and electronic devices that are edible for consumption. This is basically an inter-disciplinary study combining concepts of electronic engineering and food science, which focuses on devices made from edible polymers, biodegradable substance and food grades substances. These devices help monitor the activities inside the food and also interact with biological processes inside the body, and helps in real-time monitoring of the foods. This system focusses on creating a system to provide actual data like pH levels, temperature and humidity levels in food product. These edible sensors are directly incorporated into the food products or packaging, giving updated data of the property of food and hence indirectly intimates the presence of microbial level and the spoilage indicators. This instant data helps in determining the actual quality of the food product inside the package and thereby preventing intake of contaminated food and also decrease the prevalence of foodborne illnesses.

One major problem faced in the supply chain management of food processing industry is the risk of spoilage of food due to changes in storage and climatic conditions in the warehouses. There are millions of illness and death occurring due to foodborne illness and this is of public health concern. Integration of the edible sensors into the food packages can ensure a safe food supply chain from processing, storage to the food consumption. These sensors can reduce health risks and also improve food safety practices, by alerting consumers about the food quality and safety. A sustainable and ecologically friendly approach are the major merits of using these edible sensors than traditional electronic sensors. Biodegradable materials, naturally available materials used for designing edible electronics, are used thereby increasing environmental awareness and the need for green technologies. These edible sensors are developed to monitor interaction of food with the GI tract of the humans, (Ganesan et al., 2022).

Edible electronics also have a huge potential in health monitoring for personalized medicine. Chronic diseases can be monitored and managed by certain data using edible electronics such as information regarding vital signs, track the release and absorption of nutrients and medications, (Patel & Sharma, 2013). Personalised nutritional requirements and needs can be tailored easily using these recommendations for medical and health enhancement. Real-time monitoring of the foods can embark a development in the field of food safety and health care using edible electronics.

Biological recognition is the key feature of the biosensors. The enzymatic biosensor detects the enzymes present in the deteriorated food; Proteases, lipases, and amylases are some of the typical enzymes released during the food decay. Microbial biosensors detect the microorganism present in the decayed food and chemical biosensor detects the chemical compounds present in the decayed food. Gas sensors and pH sensors are the best examples of the chemical sensors. Gas sensors detect the gas generated from the decayed food and is widely based on metal oxide semiconductor technology. pH sensor measures the acidity or alkalinity nature of the food, that may change when the food deteriorates. The optical biosensor utilizes the optical properties in sensing food decay susch as color and turbidity. The electrochemical biosensor detects the changes in electrical parameters produced by the chemical reactions. The conductivity sensors are widely used in the liquid food spoilage detection and measures the change in electrical conductivity generated by the ions in the spoiled food.

4.1 Gas Sensors for Food Spoilage Detection

The characterstics of gas sensor is depicted in Figure 1.

Figure 1. Characteristics of gas sensor

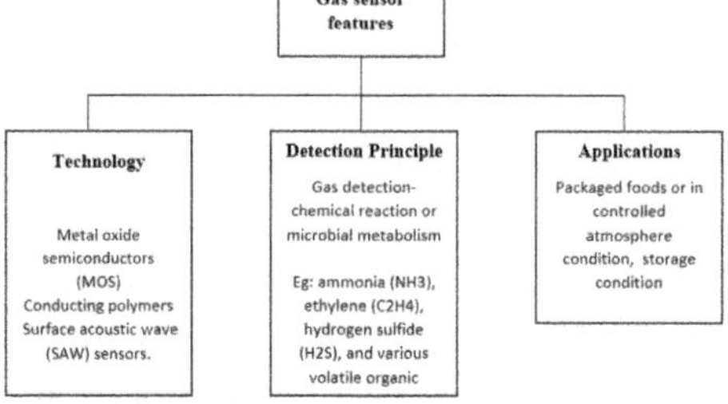

A critical review of gas sensors and features of two dimesional materials based gas sensors for food spoilage detection are proposed in Joshi, Pransu, & Conte-Junior (2023). In Matindoust et al. (2021), a detailed reiew on different types of gas sensors based on conductive polymers for meat spoilage detection are discussed. The various techniques involved in the food spoilage detection followed by the ammonia and various toxic gases detecion are also covered in Matindoust et al. (2021). The fabrication of ammonia gas detection sensor based on polyaniline was put forward in Matindoust *et al.* (2017), the gas sensor with a substrate thickness of 0.25mm exhibits superior performance. A electronic nose system was prposed in Luo *et al.* (2023) for the detection of gases from spoiled food, the images are generated from the fourier series computation of sensors and processed by convolution neural network. Four gas sensors are utilized in the e nose and the performance of the CNN architecture was found to be superior, when compared with the classical machine learning models, (Luo *et al.*, 2023). For low temperature application, specialized gas sensors are required and in Nguyen et al. (2019), Polydiacetylene based sensor was proposed for meat spoilage detection. The ammonia and hydrogen sulphide are the posionous gases generated during food spoilage and Preethichandra *et al.* (2023) discusses about conducting polymer based sensor for these gases detection. The design and fabrication of nanostructured conducting polymer wireeless gas sensor is highlited in Ma *et al.* (2018), can be integrated with smart phone for analysis.

4.2 pH Sensors for Food Spoilage Detection

The characterstics of pH sensor is depicted in Figure 2.

Figure 2. Characteristics of pH sensor

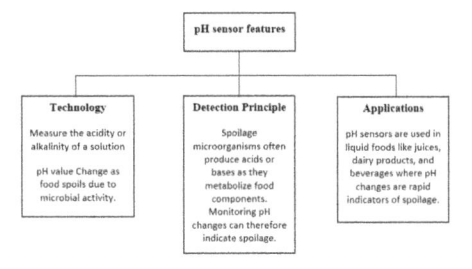

A low-cost wireless pH sensor was proposed in Waimin *et al.* (2022), and passive resonant frequency sensing was coupled with the pH-responsive polymer for spoilage detection in meat. The wireless pH sensor comprising IrO x /AgCl sensing electrodes was proposed in Huang *et al.* (2011) for spoilage detection in fish meat.

A simple halochromic sensor was put forward in Liu, Gurr, & Giao (2020) for the spoilage detection in meat, the sensor was found to be proficient for protein-rich foods spoilage detection. The wireless pH sensor with hydrogel coating was proposed in Mu *et al.* (2022) for spoilage detection in fish meat and had a sensitivity of −49.184 mv/pH. The sensor generates efficient results with varying temperature ranges from 4 °C to 20 °C. the wireless pH sensor system comprised of indium tin oxide (ITO) electrode, an Ag/AgCl reference electrode was proposed in Mu *et al.* (2021) for spoilage detection in fish meat, continuous monitoring of spoilage at 28 °C for 36 hours was highlighted in this work. Miniature wireless sensor that can be incorporated into packaged foods was put forward in Istif *et al.* (2023) for meat spoilage detection, the size of the sensor was $2 \times 2 \, cm^2$.

4.3 Biosensors for Food Spoilage Detection

The characterstics of biosensor is depicted in Figure 3.

Figure 3. Characteristics of biosesnor

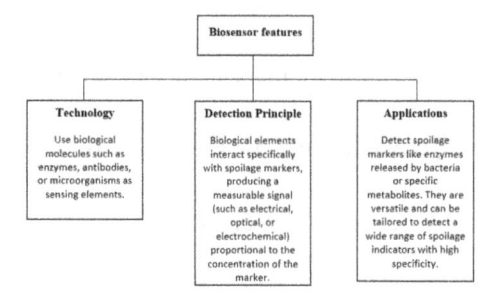

A detailed review of various types of biosensors for the detection of fungi and mycotoxins was proposed in Oliveira *et al.* (2019). In Schaertel and Firstenberg-Eden (1988), a detailed survey on biosensors, their types features, and evolution are highlighted. Different types of biosensing methods like label-free sensors, Immuno sensors, and fluorescence sensor are described in Poltronieri *et al.* (2024) and lab on chip-based biosensors was also discussed in this work. An impedimetric biosensor that utilizes pig odorant-binding protein (pOBP) as the biorecognition element was proposed in Calabrese *et al.* (2023). A detailed review describing different types of non destructive biosnsors are putforward in Turasan and Kokini (2021). Enzyme based electrochemical biosensor for biogenic amine detection in food samples was putforward in Masód, Azhari, and Sathishkumar (2022). The schematic of biosensor is depicted in Figure 4.

Figure 4. Schematic of biosensor

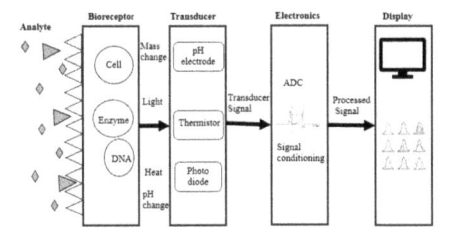

4.4 Electrochemical Sensors for Food Spoilage Detection

The characteristics of electrochemical sensor is depicted in Figure 5.

Figure 5. Characteristics of electrochemical sensor

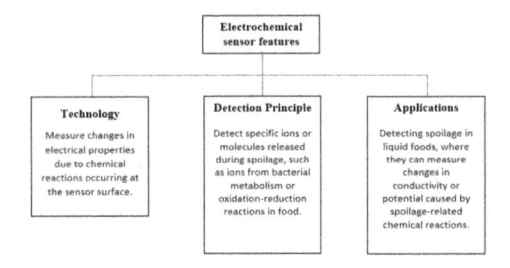

The microbial growth and enzyme activity change the impedance of food sample and a measure of this indicates the quality of the food. Real time spoilage detection is possible by incorporating in the packaged and storage foods. The microbial activity and chemcial reaction during spoilage condition alters the ionic composition of food and the conductometry approach measures this activity.

Figure 6. (a) Impedimetry approach, (b) conductometry approach

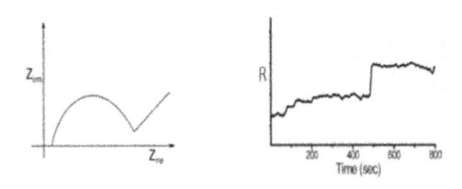

The potential difference between the active electrode and reference electrode is measured in the potentiometry approach and is an indication of the ions in the food sample, ionic concentration changes due to microbial action and chemical reaction. In the voltammetry approach, the chemical composition is measured by measuring the current response concerning the voltage applied and different types of voltammetry approaches are there.

Figure 7. (a)Potentiometry approach, (b) Voltammetry approach

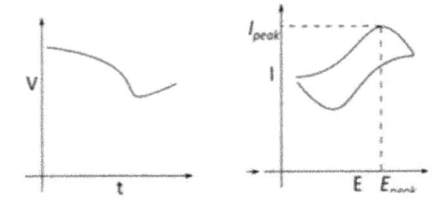

A fixed voltage is applied to the active electrode and the resultant current is measured, it indicates the composition of species in the spoiled food.

Figure 8. Amperometry approach for food spoilage dtection

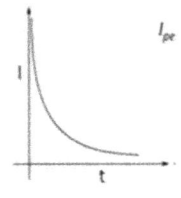

4.5 Optical Sensors for Food Spoilage Detection

The characterstics of optical sensor is depicted in Figure 9.

Figure 9. Characteristics of optical sensor

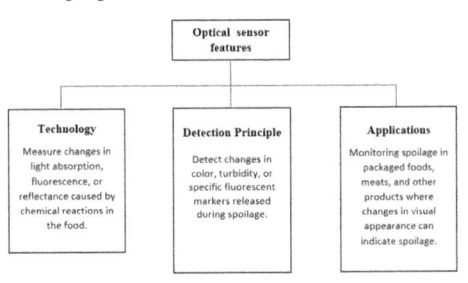

Optical sensors gains promience in food spoilage detction and in Sun *et al.* (2023), the authors proposes a dual mode optical sensor for spoilage detction in meat. A single optical gas sensor was putforward in Semeano *et al.* (2018) for the spoilage detection in tilapaia fish. Optical sensors for the detection of biogenic amines in food was proposed in Danchuk *et al.* (2020). Optical sensor for the detection of oxygen and carbon dioxide in spoilage food was proposed in McEvoy *et al.* (2003). A detailed review on optical sensors for food spoilage detection was highlighted in Narsaiah *et al* (2012). Label free optical biosensor was proposed in Khansili, Rattu, and Krishna (2018) for food spoilage detection.

5. REGULATORY AND SAFETY ISSUES CONCERNING THE USE OF EDIBLE ELECTRONICS IN FOOD SAFETY

The chief regulatory concern in edible electronics is to ensure that safe materials are used for consumption. The materials used must be food-grade and comply with food safety regulatory authorities like FDA and EFSA. Toxicity and biodegradability studies should be carried out to ensure the adverse effect when consumed and also the materials will safely breakdown in GI tract respectively. Functional testing should be performed to check reliability of the edible sensor on adverse circumstances of the digestive system. The product should be labelled clearly to intimate customers regarding the edible electronics in the food package and also any risks associated with their consumption, (Johnson & Brown, 2020). The safety aspects associated with the edible electronics is the possible risk of allergic reactions to material and the interaction of it to medicines. Comprehensive studies should be carried out to identify side-effect associated with the use of it. The dosage level and use of edible electronics in long-term safety on human health should be studied. Though these materials are biodegradable, studies on impact of these materials to the environment should be studied.

6. CHALLENGES OF EDIBLE ELECTRONICS

Durability of edible electronics is the chief parameter where more research should be carried out. Conventional electronics in stable environmental condition, are intended to be used for a prolonged period. However, the durability of edible electronics in food or health care in acidic condition with digestive enzymes in gastrointestinal (GI) tract is of substantial challenge. Studies on several food-grade biodegradable materials and its reaction in GI track should be focused, to maintain an equilibrium between durability and biodegradability, (Miller, 2014). Sensitivity of edible electronics is another vital task. The edible sensor should possess an ability to monitor and record every change in incipient spoilage parameters like pH, temperature, humidity change and moisture of the food. Edible sensors development that are safe for consumption and sensitive to variable environment condition is a major challenge in this area of research. Accuracy is also dependent on sensitivity of the edible sensor. A consistent data with high precision and performance is of utmost requirement. Another factor of concern is the size of the edible sensor, which must be suitable of intake and accurate to give sufficient data. Transmission of data within the body

7. CONCLUSION AND FUTURE PROSPECTUS OF EDIBLE ELECTRONICS

Edible electronics is an emerging field which finds its application in monitoring food quality, assuring safe food, monitoring health etc. Certain challenges such as sensitivity, durability and the accuracy are of concern. The possible future application of edible electronics is to replace traditional methods of monitoring food safety to modern rapid smart method. Personalized nutrition and health monitoring in real-time is one promising application in the field of edible electronics . A wider application in field of food packaging, food safety, monitoring food quality and providing personalized nutrient requirements are few aspects in edible electronics. An effective edible sensor must be capable to transmit data in the body to an external commodity with accuracy and good signal strength. This is one major challenge in the field of edible electronics. The material selected, sensors used and bioengineering are major aspects to be studied. Hence, advanced research in harvesting energy without signal breakage and reliable power sources are essential to study the potential of edible electronics.

REFERENCES

Arshak, K., Moore, E., Lyons, G. M., Harris, J., & Clifford, S. (2007). A review of gas sensors employed in electronic nose applications. *Sensor Review*, 27(1), 7–20.

Balamurugan, S., Ayyasamy, A., & Joseph, K. S. (2022). IoT-Blockchain driven traceability techniques for improved safety measures in food supply chain. *International Journal of Information Technology : an Official Journal of Bharati Vidyapeeth's Institute of Computer Applications and Management*, 14(2), 1087–1098. DOI: 10.1007/s41870-020-00581-y

Batt, C. A., & Tortorello, M. L. (2014). *Encyclopedia of food microbiology*. Elsevier.

Beuchat, L. R. (2006). Control of foodborne pathogens. In Doyle, M. P., & Beuchat, L. R. (Eds.), *Springer*.

Calabrese, A., Battistoni, P., Ceylan, S., Zeni, L., Capo, A., Varriale, A., D'Auria, S., & Staiano, M. (2023). An impedimetric biosensor for detection of volatile organic compounds in food. *Biosensors (Basel)*, 13(3), 341. DOI: 10.3390/bios13030341 PMID: 36979553

Campbell-Platt, G. (2009). *Food science and technology*. Wiley-Blackwell.

Danchuk, A. I., Komova, N. S., Mobarez, S. N., Doronin, S. Y., Burmistrova, N. A., Markin, A. V., & Duerkop, A. (2020). Optical sensors for determination of biogenic amines in food. *Analytical and Bioanalytical Chemistry*, 412(17), 4023–4036. DOI: 10.1007/s00216-020-02675-9 PMID: 32382967

Doyle, M. P., & Buchanan, R. L. (2013). *Food microbiology: Fundamentals and frontiers*. ASM Press.

Fitzgerald, J. E., Wang, C., Jamieson, R. P., & Coughlan, A. P. (2001). Temperature and humidity sensors for use in food storage. *Sensors and Actuators. B, Chemical*, 81(1), 115–119.

Food and Agriculture Organization (FAO). (2011). Global food losses and food waste – Extent, causes and prevention. Rome.

Frias, J., & Vidal-Valverde, C. (2008). *Advances in food and nutrition research*. Elsevier.

Ganesan, V., Manoj, C., Ramaswamy, V., Tejaswi, T., Akilan, T., & Kumar, G. A. (2022, December). Food safety checking measures by artificial intelligent IOT. In 2022 4th International Conference on Advances in Computing, Communication Control and Networking (ICAC3N) (pp. 1385-1389). IEEE. DOI: 10.1109/ICAC3N56670.2022.10074055

Gram, L., Ravn, L., Rasch, M., Bruhn, J. B., Christensen, A. B., & Givskov, M. (2002). Food spoilage—Interactions between food spoilage bacteria. *International Journal of Food Microbiology*, 78(1-2), 79–97. DOI: 10.1016/S0168-1605(02)00233-7 PMID: 12222639

Huang, W. D., Deb, S., Seo, Y. S., Rao, S., Chiao, M., & Chiao, J. C. (2011). A passive radio-frequency pH-sensing tag for wireless food-quality monitoring. *IEEE Sensors Journal*, 12(3), 487–495. DOI: 10.1109/JSEN.2011.2107738

Istif, E., Mirzajani, H., Dağ, Ç., Mirlou, F., Ozuaciksoz, E. Y., Cakır, C., Koydemir, H. C., Yilgor, I., Yilgor, E., & Beker, L. (2023). Miniaturized wireless sensor enables real-time monitoring of food spoilage. *Nature Food*, 4(5), 427–436. DOI: 10.1038/s43016-023-00750-9 PMID: 37202486

Jay, J. M. (2000). *Modern food microbiology*. Springer. DOI: 10.1007/978-1-4615-4427-2

Johnson, M., & Brown, T. (2020). Ingestible Sensors for Chronic Disease Management. *Journal of Medical Devices*.

Joshi, N., Pransu, G., & Conte-Junior, A. (2023). Critical review and recent advances of 2D materials-based gas sensors for food spoilage detection. *Critical Reviews in Food Science and Nutrition*, 63(30), 10536–10559. DOI: 10.1080/10408398.2022.2078950 PMID: 35647714

Khansili, N., Rattu, G., & Krishna, P. M. (2018). Label-free optical biosensors for food and biological sensor applications. *Sensors and Actuators. B, Chemical*, 265, 35–49. DOI: 10.1016/j.snb.2018.03.004

Kuswandi, B., Wicaksono, Y., Abdullah, A., Heng, L. Y., & Ahmad, M. (2011). Smart packaging: Sensors for monitoring of food quality and safety. *Sensing and Instrumentation for Food Quality and Safety*, 5(3-4), 137–146. DOI: 10.1007/s11694-011-9120-x

Labuza, T. P., & Sinskey, A. J. (2006). *Chemical deterioration and physical instability of food and beverages*. Academic Press.

Liu, B., Gurr, P. A., & Qiao, G. G. (2020). Irreversible spoilage sensors for protein-based food. *ACS Sensors*, 5(9), 2903–2908. DOI: 10.1021/acssensors.0c01211 PMID: 32869625

Luning, P. A., & Marcelis, W. J. (2006). A techno-managerial approach in food quality management research. *Trends in Food Science & Technology*, 17(7), 378–385. DOI: 10.1016/j.tifs.2006.01.012

Luo, J., Zhu, Z., Lv, W., Wu, J., Yang, J., Zeng, M., Hu, N., Su, Y., Liu, R., & Yang, Z. (2023). E-nose system based on Fourier series for gases identification and concentration estimation from food spoilage. *IEEE Sensors Journal*, 23(4), 3342–3351. DOI: 10.1109/JSEN.2023.3234194

Ma, Z., Chen, P., Cheng, W., Yan, K., Pan, L., Shi, Y., & Yu, G. (2018). Highly sensitive, printable nanostructured conductive polymer wireless sensor for food spoilage detection. *Nano Letters*, 18(7), 4570–4575. DOI: 10.1021/acs.nanolett.8b01825 PMID: 29947228

Majer-Baranyi, K., Székács, A., & Adányi, N. (2023). Application of electrochemical biosensors for determination of food spoilage. *Biosensors (Basel)*, 13(4), 456. DOI: 10.3390/bios13040456 PMID: 37185531

. Masód, N. H., Azhari, S., & Sathishkumar, P. (2022). Food spoilage: Detection of biogenic amines in food samples by enzyme-based electrochemical biosensors. Malaysian Journal of Chemistry, 24(3), 74-87. ttps://DOI: 10.55373/mjchem.v24i3.947

Matindoust, S., Farzi, A., Baghaei Nejad, M., Shahrokh Abadi, M. H., Zou, Z., & Zheng, L. R. (2017). Ammonia gas sensor based on flexible polyaniline films for rapid detection of spoilage in protein-rich foods. *Journal of Materials Science Materials in Electronics*, 28(11), 7760–7768. DOI: 10.1007/s10854-017-6471-z

Matindoust, S., Farzi, G., Nejad, M. B., & Shahrokhabadi, M. H. (2021). Polymer-based gas sensors to detect meat spoilage: A review. *Reactive & Functional Polymers*, 165, 104962. DOI: 10.1016/j.reactfunctpolym.2021.104962

McEvoy, A. K., Von Bueltzingsloewen, C., McDonagh, C. M., MacCraith, B. D., Klimant, I., & Wolfbeis, O. S. (2003). Optical sensors for application in intelligent food-packaging technology. In Opto-Ireland 2002: Optics and Photonics Technologies and Applications (Vol. 4876, pp. 806-815). SPIE. DOI: 10.1117/12.464210

Miller, R. (2014). *Green Technologies in Electronics*. Sustainable Technology Journal.

Mu, B., Cao, G., Zhang, L., Zou, Y., & Xiao, X. (2021). Flexible wireless pH sensor system for fish monitoring. *Sensing and Bio-Sensing Research*, 34, 100465. DOI: 10.1016/j.sbsr.2021.100465

Mu, B., Dong, Y., Qian, J., Wang, M., Yang, Y., Nikitina, M. A., Zhang, L., & Xiao, X. (2022). Hydrogel coating flexible pH sensor system for fish spoilage monitoring. *Materials Today. Chemistry*, 26, 101183. DOI: 10.1016/j.mtchem.2022.101183

Narsaiah, K., Jha, S. N., Bhardwaj, R., Sharma, R., & Kumar, R. (2012). Optical biosensors for food quality and safety assurance—A review. *Journal of Food Science and Technology*, 49(4), 383–406. DOI: 10.1007/s13197-011-0437-6 PMID: 23904648

Nguyen, L. H., Naficy, S., McConchie, R., Dehghani, F., & Chandrawati, R. (2019). Polydiacetylene-based sensors to detect food spoilage at low temperatures. *Journal of Materials Chemistry. C, Materials for Optical and Electronic Devices*, 7(7), 1919–1926. DOI: 10.1039/C8TC05534C

Oliveira, I. S., da Silva, A. G.Junior, de Andrade, C. A., & Oliveira, M. D. (2019). Biosensors for early detection of fungi spoilage and toxigenic and mycotoxins in food. *Current Opinion in Food Science*, 29, 64–79. DOI: 10.1016/j.cofs.2019.08.004

Pacquit, A., Lau, K. T., McLaughlin, H., Frisby, J., Quilty, B., & Diamond, D. (2007). Development of a smart packaging for the monitoring of fish spoilage. *Food Chemistry*, 102(2), 466–470. DOI: 10.1016/j.foodchem.2006.05.052

Patel, K., & Sharma, V. (2013). Edible sensors in personalized medicine. *Journal of Personalized Medicine*.

Poltronieri, P., Mezzolla, V., Primiceri, E., & Maruccio, G. (2014). Biosensors for the detection of food pathogens. *Foods*, 3(3), 511–526. DOI: 10.3390/foods3030511 PMID: 28234334

Preethichandra, D. M., Gholami, M. D., Izake, E. L., O'Mullane, A. P., & Sonar, P. (2023). Conducting polymer based ammonia and hydrogen sulfide chemical sensors and their suitability for detecting food spoilage. *Advanced Materials Technologies*, 8(4), 2200841. DOI: 10.1002/admt.202200841

Ray, B., & Bhunia, A. (2013). *Fundamental food microbiology*. CRC Press. DOI: 10.1201/b16078

Schaertel, B. J., & Firstenberg-Eden, R. (1988). Biosensors in the food industry: Present and future. *Journal of Food Protection*, 51(10), 811–820. DOI: 10.4315/0362-028X-51.10.811 PMID: 28398862

Semeano, A. T., Maffei, D. F., Palma, S., Li, R. W., Franco, B. D., Roque, A. C., & Gruber, J. (2018). Tilapia fish microbial spoilage monitored by a single optical gas sensor. *Food Control*, 89, 72–76. DOI: 10.1016/j.foodcont.2018.01.025 PMID: 29503510

Sowmya, Y. (2017). A short review on milk spoilage. *Journal of Food Dairy Technology*, 5, 1–5.

Sperber, W. H., & Doyle, M. P. (2009). *Compendium of the microbiological spoilage of foods and beverages*. Springer. DOI: 10.1007/978-1-4419-0826-1

Sun, R., Li, Y., Du, T., & Qi, Y. (2023). Recent advances in integrated dual-mode optical sensors for food safety detection. *Trends in Food Science & Technology*, 135, 14–31. DOI: 10.1016/j.tifs.2023.03.013

Tao, F., Qi, Q., Liu, A., & Kusiak, A. (2018). Data-driven smart manufacturing. *Journal of Manufacturing Systems*, 48, 157–169. DOI: 10.1016/j.jmsy.2018.01.006

Turasan, H., & Kokini, J. (2021). Novel nondestructive biosensors for the food industry. *Annual Review of Food Science and Technology*, 12(1), 539–566. DOI: 10.1146/annurev-food-062520-082307 PMID: 33770468

Velusamy, V., Arshak, K., Korostynska, O., Oliwa, K., & Adley, C. (2010). An overview of foodborne pathogen detection: In the perspective of biosensors. *Biotechnology Advances*, 28(2), 232–254. DOI: 10.1016/j.biotechadv.2009.12.004 PMID: 20006978

Waimin, J., Gopalakrishnan, S., Heredia-Rivera, U., Kerr, N. A., Nejati, S., Gallina, N. L., Bhunia, A. K., & Rahimi, R. (2022). Low-cost nonreversible electronic-free wireless pH sensor for spoilage detection in packaged meat products. *ACS Applied Materials & Interfaces*, 14(40), 45752–45764. DOI: 10.1021/acsami.2c09265 PMID: 36173396

Yin, S., Bao, J., Zhang, Y., & Huang, X. (2017). M2M security technology of CPS based on Blockchains. *Symmetry*, 9(9), 193. DOI: 10.3390/sym9090193

Zou, Q., Lin, G., Jiang, X., Liu, X., & Zeng, W. (2019). Smart technologies for food quality assurance and safety: Status and perspectives. *Trends in Food Science & Technology*, 91, 18–28.

Zou, Q., Lin, G., Jiang, X., Liu, X., & Zeng, W. (2019). Smart technologies for food quality assurance and safety: Status and perspectives. *Trends in Food Science & Technology*, 91, 18–28.

Chapter 7
Advancements in Edible Electronics and Robotics for Smart Food Packaging Processes

Sudhakara Rao J.
https://orcid.org/0009-0005-4176-4837
Department of Life Science (Microbiology and Food Science & Technology), GITAM University (Deemed), Visakhapatnam, India

Gitanjali Behera
https://orcid.org/0000-0002-0444-6304
Department of Life Science (Microbiology and Food Science & Technology), GITAM University (Deemed), Visakhapatnam, India

Shreenidhi K. S.
https://orcid.org/0000-0002-6844-0454
Department of Biotechnology, Rajalakshmi Engineering College, Chennai, India

Harishchander Anandaram
https://orcid.org/0000-0003-2993-5304
Department of Artificial Intelligence, Amrita Vishwa Vidyapeetham, Coimbatore, India

G. Durai Muthu Mani
Department of Biochemistry, SRM Arts and Science College, Kattankulathur, India

DOI: 10.4018/979-8-3693-5573-2.ch007

Copyright © 2025, IGI Global. Copying or distributing in print or electronic forms without written permission of IGI Global is prohibited.

ABSTRACT

This chapter explores recent advancements in edible electronics and robotics for smart food packaging, focusing on sustainability, food safety, and consumer interaction. Edible electronics, designed for human consumption, offer new applications in food quality monitoring, including real-time detection of spoilage, monitoring nutritional content, and product authentication, thereby enhancing food safety Robotics and other technologies will enable innovative functionalities like dynamic packaging responses, automated freshness indicators, and active interventions to extend shelf life. The chapter delves into the intricacies of edible sensors and robotic systems, examining their materials, technologies, design, regulatory, safety, and ethical implications. The study explores the potential of IoT in packaging for supply chain management, personalized nutrition, and waste reduction, highlighting its integration into digital ecosystems, presenting a shift in food distribution, safety, and sustainability.

INTRODUCTION

The integration of edible electronics and robotics in smart food packaging shows a jump in evolution. Such solutions have never been more important, with global challenges like food waste, sustainability, and food safety gaining prominence. Although traditional methods of packaging ensure food stays longer and also travels safely, they cannot increasingly meet modern consumers' demands for more transparent, safe, and ecologically friendly packaging. Adding to this, the rapid growth in the number of connected devices and smart technologies opens new frontiers of innovation within the food industry. In this respect, edible electronics and robotics might propose a novel way of food packaging and distribution (Sharova et al., 2021).

Edible electronics represents a pretty new, fast-developing area which deals with the designing of such electric devices that can be safely ingested without causing harm to the consumer. Such devices, made up of food-grade material, will definitely embed sensors directly into food products or their packaging for real-time monitoring of various parameters related to food safety, including but not limited to temperature, humidity, spoilage, and contamination. Pragmatically, this might insinuate that both the consumer and the manufacturer will have immediate information about the freshness or quality of a product. This therefore will lead to foodborne illnesses being minimized and also reduce waste. With foodborne diseases running amok in almost every region of the world and being one of the major concerns of public health, this ability in tracking and verifying food quality from production to consumption is a major stride forward (Floreano et al., 2024).

On the other hand, robotics embeds such smart packaging systems with active functionality. Embedding robotic systems in packaging might be starting a new age in the monitoring of the safety and quality of food. Robotics can provide interactive packaging that could respond either to environmental conditions or to consumer interaction. For example, packaging might change its properties dynamically-change colors upon spoilage, activating mechanisms that would act to alter the internal conditions of storage to extend the shelf life. These intelligent packaging systems will not only inform consumers about the condition of the product but can intervene by acting to maintain food for a longer period (Idumah et al., 2020).

Edible electronics and robotics can greatly increase the functionality of food packaging together. Printable sensors, made with biodegradable and food-safe materials, could be printed directly on the package to continuously monitor those factors affecting food quality. These sensors can, in turn, trigger robotic mechanisms inside the packaging to alter atmospheric conditions, such as oxygen or humidity, inside the package. It can even send notifications to distributors and retailers when the products are near their expiration dates. This could also include tracking and authentication of the product, identifying whether or not the product inside is indeed genuine and that it has not been tampered with during transport (Annese et al., 2023).

Another key driver for smart food packaging technologies is sustainability. Food packaging is thought to be highly wasteful, relying on single-use plastics and non-biodegradable materials. This has been evident in the emerging crisis of waste production globally, and especially food packaging having to do with such challenges through the disposal of used food containers. These could be avoided by the incorporation of edible electronics and robotics in the food packaging using sustainable materials that are biodegradable. These technologies can reduce reliance on conventional plastics and allow packaging that either degrades naturally or is consumed with the food product. Over the long term, such innovations can reduce the ecological footprint of the food industry while also offering an added degree of functionality to meet the modern consumer's needs (Bharti et al., 2023).

The development of smart food packaging also largely involves the Internet of Things. By embedding sensors and other electronic components in packaging, food products will be incorporated into a larger digital environment connecting food manufacturers, distributors, retailers, and consumers. IoT-imbued packaging makes sharing real-time information throughout the supply chain possible, right from the status of the product condition during transport, storage, and retail. With this, logistics efficiency can be enhanced, spoilage reduced, and consumers can be ensured of getting only the freshest of products. It can also interact with IoT-enabled packaging, smartphones, or other devices to provide consumers with personalized information on nutritional content, sourcing, and even ways of preparation (Vashishth et al., 2022).

On the other hand, edible electronics and robotics in food packaging are not without their challenges. This transition in technology raises several safety and regulatory concerns. Because these are technologies touching food directly, special concern for safety considerations and regulations must be given to make the materials non-toxic and biodegradable for human consumption. Long-term effects of the electronics ingested also need to be carefully assessed since the field is still very young and still requires many more researches to assess some risks. There are also ethical considerations around data privacy and consumer trust, particularly in the data to be collected through IoT-connected packaging. Industry players will have to consider and build out a framework to address these issues so that the rights of consumers are preserved while fostering innovation (Mahalik, 2009a).

In conclusion, the part that edible electronics and robotics together play in smart food packaging presages a sea change in the way foods will be packaged, tracked, and consumed. The chapter shall look at the basic technologies that make this integration possible, materials used, and design principles. It will also consider the broader ramifications of such innovations on sustainability, food safety, and global food supply chains. Since food industries are answering consumer needs with high-quality foods and concern for the environment, active and smart packaging technologies are continuously changing. This gives a promising vision for food packaging in the future. While some of this has started to be developed through real-time monitoring, dynamic responses, and sustainable design, the edible electronics and robotics have great potential to challenge current thinking on food packaging in creating a far safer, far more efficient, and highly sustainable system for delivering food to consumers worldwide (Nayik et al., 2015).

Scope

This chapter covers the scope of edible electronics and robotics in smart food packaging for reviewing the disruptive potential in food safety, sustainability, and consumer experience. The present work provides a comprehensive review on the progress in the development and application of edible electronic sensors and robotic mechanisms integrated into food packaging to play a role in the real-time monitoring of food quality and spoilage detection and consumer interaction. The chapter covers key technological advances, materials used, and design considerations in creating the ingestible, safe, and functional creation of electronic components. Besides, the chapter explores how these innovations lead to waste minimization and environmental sustainability due to the application of biodegradable materials. Further, the chapter addresses the regulatory and ethical considerations of the emphasis on standards with regard to safety and concern about data privacy. This chapter attempts to give a broad overview of both existing and future technologies that can transform the food packaging industry through the combined use of edible electronics and robotics.

Objectives

The chapter is focused on a description of edible electronics and robotics, integrated into smart food packaging, and assessment for their potential applicability to change the face of the food industry. A more basic understanding of how these technologies enhance food safety through real-time monitoring of quality, freshness, and spoilage. The chapter explores the potentiality of robotics to develop dynamic packaging systems that, for example, will be able to keep up with environmental conditions and extend the lives of various products. Also, this work is focused on sustainable, biodegradable materials in edible electronics with the purpose of reducing packaging waste and enhancing the sustainability of the environment. Another objective is solving challenges like regulatory, safety, and ethical issues to ensure such innovations meet industrial standards. This chapter surveys existing applications and future possibilities to paint a forward-looking vision of how these technologies might shape food distribution, safety, and consumer engagement in the coming years.

EDIBLE ELECTRONICS: TECHNOLOGIES AND APPLICATIONS

The development of edible electronics heralds a revolutionary advance in the field of food technology, which enables new ways to improve food safety and quality monitoring, and engage consumers. Edible electronics are electronic devices and components that are developed from biocompatible, biodegradable, and food-grade materials and are intended to be safely consumed. Once this field has reached full maturity, applications include smart food packaging and monitoring systems with significant benefits for consumers, manufacturers, and distributors in almost equal measure. Edible electronics, types of edible sensors, their application in real-time monitoring of food, and the challenges and limitations relating to this new technology are discussed in this section (Licardo et al., 2024). Figure 1 visually illustrates the connection between technologies and applications in edible electronics.

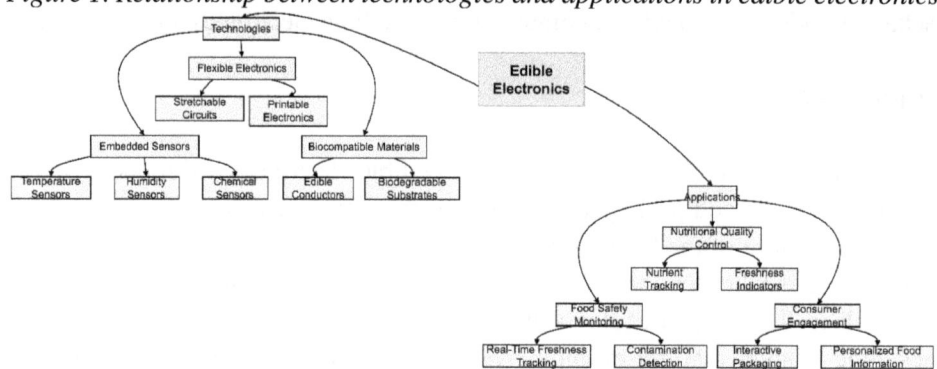

Figure 1. Relationship between technologies and applications in edible electronics

Basic Aspects of Edible Electronics

Edible electronics are a new generation of electronic devices that, after serving a purpose, are designed to be consumed and digested by the human body without causing any harm. The device will, therefore, be fabricated from nonpoisonous, biodegradable materials capable of passing through the digestive system or being completely metabolized in the body. Key constituting elements of edible electronics will be sensors, batteries, and circuits; all these need to be prepared from nontoxic materials that could be safely ingested (Prasad, 2017).

Instead, edible electronics are made of food-grade materials like silk proteins, cellulose, and gelatin. For edible electronics, the conductive materials could be carbon-based compounds, such as graphene, or metal ions that occur naturally in foods. These materials, when structured, are able to create devices with a wide range of functionalities, including the capability to sense environmental changes or to provide data about the food product they are imbedded in.

Accordingly, enabling packaging integration requires the development of thin, flexible, and degradable electronics. For instance, edible electronics can be integrated into packaging films, labeling, or even directly into the food product. As these kinds of materials continue to evolve, their functionality and range for edible electronics will also continue to expand.

Types of Edible Sensors and Their Functions

The main functions involve monitoring the quality and safety of food, thus rendering edible sensors the basis of edible electronics. There have been various types of edible sensors developed for diverse roles and functions in smart food packaging and real-time monitoring systems of foods(S., 2024; Upadhyaya et al., 2024; Venkateswaran et al., 2024) . A few of the most important kinds include:

Temperature Sensors

These sensors monitor the temperature of food products at all stages in the supply chain to guarantee that the best storage conditions are maintained. Edible temperature sensors are made of materials like gelatin and cellulose, and have the ability to track temperature changes that could affect the quality or safety of food. Such sensors can be crucial in the expiration of temperature-sensitive items like dairy products and meat through real-time data indicating whether they have been exposed to unsafe temperatures.

pH Sensors

pH sensors feature among the significant sensors that detect acidity or alkalinity in food products, forming a major indication about the product's freshness and safety. These sensors are often integrated into packaging in order to monitor changes in pH levels of food. This, particularly in food items that easily spoil, such as seafood or beverages, can easily be observed. Generally, edible pH sensors are made from nontoxic polymers and signal when pH levels start to become outside the threshold of safety for foods in order to help the consumer and retailer avoid ingesting or selling spoiled products.

Gas Sensors

Among the main functions of gas detectors is the detection of the emergence of gases that evidence spoilage or contamination. The rise of carbon dioxide or ammonia inside the packaging normally evidences bacterial proliferation or food spoilage. Thus, packages can be fitted with gas sensors to track the level of gasses throughout their lifetime and immediately raise warnings if the food is at risk of contamination. These sensors help reduce waste by detecting spoilage early, thereby enabling better inventory management and decision-making.

Biochemical Sensors

Biochemical sensors can sense a wide range of biological markers or contaminants, from pathogenic agents to allergens and harmful chemicals. These sensors play a significant role in enhancing food safety and preventing foodborne diseases. They undergo a reaction with the target molecule, whether in food or the environment within the food packaging, and provide a detectable signal to indicate an alarm for contamination. Integrating edible biochemical sensors into food packaging allows for continuous product monitoring regarding harmful substances.

Applications of Real-Time Food Monitoring

Integrated edible electronics and sensors into smart food packaging have a variety of practical uses in real-time food monitoring. Their immediate benefit is the ability to continuously trace and estimate the quality of food along the entire supply chain, from production to consumption. It thus provides, in real time, information on the condition of a packaged food item to both the manufacturer and consumer, thereby helping to reduce waste and improve food safety (Prasad, 2017).

Edible sensors have major applications in the detection of spoilage. Sensors embedded inside the package can monitor temperature, humidity, and gas composition that affect food freshness. An edible gas sensor, for example, could detect the generation of gases associated with spoilage, such as ammonia, and send an alarm to show food is no longer safe for consumption. This type of monitoring is particularly useful in cold chains, where improper storage conditions can render the product spoiled and lead to losses. Other applications are nutritional monitoring. The edible sensors can be used to measure nutrients within food with a view to availing detailed information of the product being consumed to the consumer. This could be very useful for people on special diets or with allergies, as one could be in a better position to track their intake more precisely. Real-time monitoring of nutritional content can extend to personalized nutrition where sensors could deliver data on meal nutritional values and help consumers make wiser choices. Other areas that might be transformative include product authentication. There is a heightened incidence of counterfeiting in the food industry, especially in high-value commodities such as premium meats, wines, and specialty foods. Such foods may utilize edible sensors to authenticate their genuineness and make sure that whatever the consumers get is what it claims to be. Authentication features integrated with packaging shall help manufacturers protect their products against tampering and counterfeiting.

Challenges and Limitations

In turn, though edible electronics have a lot of promises, several challenges and limitations need to be overcome before wide-scale use is achieved. One of the main challenges to be faced will be over safety and the degradable nature of materials within these devices. Significant advances have been made in terms of food-safe material development; still, proper and continuous research is needed to ensure that such material would not be harmful and does not result in any long-term health or ecological effect.

Another challenge is the manufacturing and scalability of edible electronics. The fabrication of electronic components from biodegradable, food-grade materials is more challenging compared to conventional electronics manufacturing, and presently there are limits in terms of the size, power, and functionality of edible devices. Further material science and improvements in manufacturing techniques will be needed beyond those the field is developing to overcome these limitations and allow mass-manufacturing of edible electronics. Besides, some other regulatory challenges are also there. Ingestible electronics raise safety, consumer protection, and privacy questions, especially with devices integrated into IoT-enabled smart packaging. The regulatory bodies will have to work out sufficient guidelines on the use of these technologies within the food industry since demands for their safe and ethical deployment are increasing.

In any case, edible electronics stands out as an exciting novelty that is going to transform the way food is traced, authenticated, and consumed. Though there are still a number of challenges, improvement in this area promises great perspectives for the future when it comes to enhancement of food safety, reduction of waste, and improvement of consumer experiences.

ROBOTICS IN SMART PACKAGING

Robotics will play an important role in the revolution of food packaging, as it automates processes, improves efficiency, and allows for better food safety. Due to the increasing demands of consumers to make packaging more intelligent and sustainable, robotics technology creates advanced systems with dynamic functionality, extended shelf life, and real-time monitoring. The areas covered in the section are robotics for packaging automation, the characteristics and advantages of a dynamic packaging system, how robotics enhances Shelf Life and Food Quality, and the future directions of robotics in smart packaging (Abimbola & Okpara, 2023).

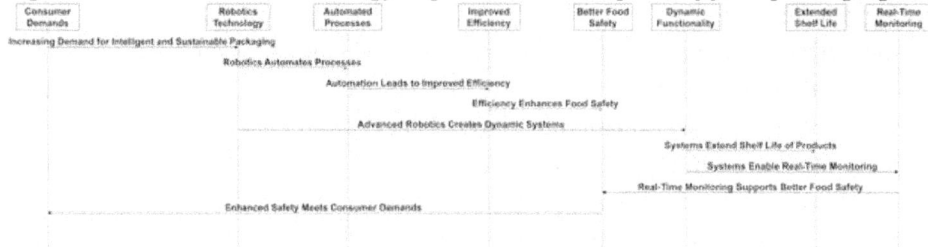

Figure 2. How robotics technology impacts various aspects of food packaging

Consumer Demands	Robotics Technology	Automated Processes	Improved Efficiency	Better Food Safety	Dynamic Functionality	Extended Shelf Life	Real-Time Monitoring

Increasing Demand for Intelligent and Sustainable Packaging
Robotics Automates Processes
Automation Leads to Improved Efficiency
Efficiency Enhances Food Safety
Advanced Robotics Creates Dynamic Systems
Systems Extend Shelf Life of Products
Systems Enable Real-Time Monitoring
Real-Time Monitoring Supports Better Food Safety
Enhanced Safety Meets Consumer Demands

Robotics technology is revolutionizing food packaging by automating processes, improving efficiency and food safety. It also extends product shelf life through dynamic functionality and real-time monitoring of food products as showun in Figure 2. These advancements meet consumer demands for intelligent and sustainable packaging, ensuring better food safety and extending product shelf life. The technology also supports real-time monitoring, further enhancing its impact.

Role of Robotics in Packaging Automation

It has transformed the food industry by presenting solutions to enhance the speed, accuracy, and consistency of packaging processes. From the initial stage of food production to the last stage of food distribution, robotic systems serve the sorting, packing, labeling, sealing, and palletization processes of the products. These are designed to reduce the need for human intervention while ensuring that packaging is accurately done with minimal wastage (Chisenga et al., 2020).

One of the key advantages of robotics in packaging automation is the ability to work under harsh or sterile conditions. In food packaging plants, often a high level of sterility is required for a food packet or package to avoid its contamination. With sensors and algorithms in place, robotic systems can operate in such environments with no fatigue or errors while upholding high sanitation standards. Also, robotic arms fitted with a vision system can ensure that the right amount of product is placed in each package, therefore improving accuracies while reducing human error.

Other key packaged goods industry uses for automation include the growing adoption of collaborative robots, or "cobots." These work right alongside human workers, extending labor capacity without the need to include safety barriers. In smaller or mixed-product packaging lines where flexibility and precision are more crucial, for example, cobots prove particularly valuable. Automation of repetitive tasks—such as product handling or case packing—with robotics frees human operators to attend to more complex tasks and thus can enhance efficiency in overall operations.

Dynamic Packaging Systems: Features and Benefits

Dynamic packaging systems are part of a new frontier in food packaging where robotics and smart technologies combine to give interactive packaging with the environment. Whereas traditional packaging remains static after it is produced, dynamic packaging systems change due to such external factors as temperature, moisture, or gas content. Robotics is at the heart of how these systems work, making real adjustments possible in conditions of packaging to optimize preservation and safety (Sadeghi et al., 2022).

The most distinctive features of dynamic packaging are that they regulate permeability: for example, systems that make use of robotic mechanisms in controlling the gas exchange between a package and its environment. By being able to control the rates of oxygen and carbon dioxide, packaging can retard the spoilage processes that take place in perishable products such as fruits and vegetables, extending their shelf life and reducing wastage by ensuring that consumers are provided with fresh goods.

Another beauty of dynamic packaging is its real-time monitoring and reporting on the quality of food. On that note, embedding sensors and Internet-of-Things technologies within this type of packaging enables dynamic packaging systems to pick up data with regard to the condition of the food product and, upon need, send out notices in case the quality is approaching spoilage. For instance, a robotically integrated system within the packaging may detect changes in temperature or humidity that will affect the safety for consumption of food. This information can subsequently be relayed to the manufacturer, retailer, or consumer so that proactive actions are initiated with a view toward the prevention of spoilage and contamination. These advantages of dynamic packaging systems have great implications: They enhance the safety of food by providing real-time information about the status of the product, reduce waste of food by extending shelf life, and increase consumer satisfaction by allowing the product to reach them in an optimized state. Many dynamic packaging systems also have an added dimension of sustainability through the use of recyclable or biodegradable materials that reduce the environmental impact of packaging waste.

Improving Shelf Life and Food Quality

Perhaps the greatest use of robotics in active packaging is in longer shelf lives and higher qualities in food items. Using robotic systems and sensors, smart packaging can actively monitor the state of the food and manipulate the internal environment of the package for the express purpose of prolonging freshness (Hwa & Te Chuan, 2024).

For example, one of the treatment methods, MAP, seeks to change the internal atmosphere composition of a package to retard natural product degradation. This may involve robotic systems automatically adjusting the internal gases of the package based on real-time sensor data to maintain the optimum environment to extend shelf life. This becomes all the more relevant to products like meats, cheeses, and fresh produce, in which minor changes in atmospheric conditions might strongly impact product quality.

Robotics could also serve in the regulation of moisture inside the packaging. Several products do have a certain demand for humidity in maintaining their texture and flavor. The employment of robotic systems that will inspect and control the level of moisture helps smart packaging maintain the sensorial properties of food, such as the crispiness of baked goods or the juiciness of fresh fruits. These systems also prevent spoilage due to excess moisture or dryness, which leads to bacterial attack or product degradation.

Another potential area in which robotics may play a role is active packaging, since the package interacts with the food for the purpose of improving its quality. For example, some smart packaging systems are utilizing robotic-controlled oxygen scavengers, which absorb excess oxygen from the package to prevent oxidation and spoilage. Others may release preservatives or antimicrobial agents based on changes in the state of the food before actual deterioration occurs. Further extensions of shelf life and improvements in food safety result accordingly.

Future Directions in Packaging Robotics

In the future, robotics in smart packaging will be even more sophisticated, improving food safety, sustainability, and engagement by consumers. As robotics technology continues to evolve, one wouldn't be surprised to see innovative, intelligent, totally autonomous packaging systems created. Another development area for the future is the field of artificial intelligence applied to packaging robotics. AI algorithms can be installed in robotic systems to make the most of packaging processes, predict the shelf life of a product, and detect possible defects not visible to human vision. For instance, machine learning models analyze historical data from sensors embedded in the packaging and use that information to predict when a product will spoil, enabling better inventory management and reducing food waste(Sekhar et al., 2024; Teja et al., 2024).

Another probable development is the continuous use of robotics to personal package goods. It is envisioned that, in years to come, the robotic systems may be utilized for the elaboration of personalized packages for each single consumer, taking into consideration his/her preferences or dietary needs. For example, dynamically changing packaging according to specific nutritional information or even changing its composition due to the end-customer's lifestyle could take place. Sustainability will be one of the major driving forces for developing future packaging robotics. The use of robotic systems is likely to become one of the main factors in reducing the environmental footprint since it allows for more ecological materials and reduces packaging waste. An autonomous robotic system will make logistics and supply chains optimal, reducing the carbon footprint from food distribution.

Finally, robotics in smart packaging will move toward more integration with IoT and blockchain for better transparency and traceability in the value chain of food supply. Consumers will have real-time information provided concerning the origin, quality, and condition of their food products, which builds confidence and enables purchasing decisions to be more informed. It is a conclusion where robotics will change the face of food packaging and give great incentives in automation, thereby assuring food safety and improving the life of the product on the shelf. By dynamic packaging systems, the robotic technologies will enable, in real time, to monitor and adapt to the environmental conditions that keep food products fresh and safe for longer periods. While the subject is getting advanced, the future innovations in robotics, AI, and sustainability are bound to take this further in smart packaging and open new vistas to manufacturers, retailers, and consumers.

SUSTAINABILITY IN SMART FOOD PACKAGING

Food packaging has now become of great importance in view of the growing environmental concerns. Growing consumer awareness encourages the adoption of suitable eco-friendly packaging solutions, not limited to reducing waste but also contributing to resource conservation. Smart food packaging, by virtue of the advanced technologies and materials incorporated, is a critical player in the whole shift towards sustainability, reducing environmental impact to a minimum. This section looks at the environmental impact of conventional packaging, the rise of biodegradable and edible materials, and the contribution smart technologies are making to reduce waste (Mahalik, 2009b; Prasad, 2017). Figure 3 illustrates the transition from conventional to eco-friendly packaging, emphasizing its advantages in reducing waste and conserving resources.

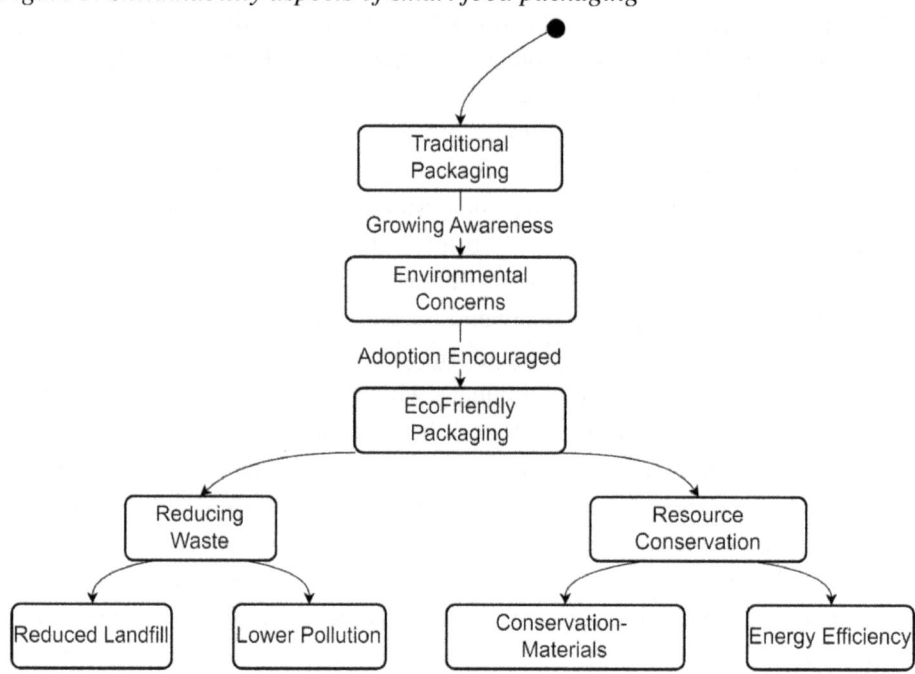

Figure 3. Sustainability aspects of smart food packaging

```
                    ●
                    ↓
            ┌─────────────┐
            │ Traditional │
            │  Packaging  │
            └─────────────┘
                    │
            Growing Awareness
                    ↓
            ┌─────────────┐
            │Environmental│
            │  Concerns   │
            └─────────────┘
                    │
           Adoption Encouraged
                    ↓
            ┌─────────────┐
            │ EcoFriendly │
            │  Packaging  │
            └─────────────┘
              ↙         ↘
    ┌──────────┐     ┌──────────┐
    │ Reducing │     │ Resource │
    │  Waste   │     │Conservation│
    └──────────┘     └──────────┘
      ↙     ↘          ↙      ↘
┌────────┐ ┌────────┐ ┌──────────┐ ┌──────────┐
│Reduced │ │ Lower  │ │Conservation-│ │  Energy  │
│Landfill│ │Pollution│ │ Materials │ │Efficiency│
└────────┘ └────────┘ └──────────┘ └──────────┘
```

Environmental Impact of Conventional Packaging

Traditional food packaging, a lot of which involves plastic, aluminum, and other non-biodegradable materials, comes with quite a good deal of environmental impact. Plastics especially are major contributors to pollution on both land and in the water. An approximate amount of millions of tonnes of plastic waste is generated every year throughout the world, to which a great part can be attributed to single-use packaging. These materials take several hundred years to decompose and most often land in landfills or the environment, forming a threat to wildlife and ecosystems(Suresh et al., 2024; Vangeri et al., 2024).

Besides, conventional packaging is one of the most resource-intensive productions. Production processes related to materials like plastic and aluminum are extremely energy- and water resource-intensive; they contribute to a great extent to greenhouses gases and deplete natural resources. This has posed a great dilemma for food manufacturers who rely on packaging methods for quality preservation and extending the shelf life of food but contribute to environmental degradation.

However, traditional packaging has its own problems with disposal. In most cases, the recycling rates continue to stay low, and a large portion of this packaging waste is either incinerated or deposited into landfills. Incineration and landfills emit harmful chemicals and gases in the air, further contributing to environmental degradation. These pose urgent needs for alternative materials far more sustainable than traditional food packaging.

Biodegradable and Edible Materials

Biodegradable and edible material development in packaging is a very promising direction in seriously decreasing the environmental impact of food packaging. Biodegradable packaging materials are naturally degraded into fragments within a relatively short period of time without causing harm to the environment. Biodegradable materials include plant-based plastics, compostable films, and paper-based packaging. Biodegradable packaging material can decompose in a few months after proper disposal, in contrast to hundreds of years that are generally estimated for conventional plastics(Sengeni et al., 2023).

Bioplastics, or plant-based plastics, come from renewable resources including corn, sugarcane, and potato starch. These materials are a more viable option to the normal petroleum-based plastic because they do not require much in production and also emit less greenhouse gas in production. Bioplastics can be used for all kinds of food-packaging applications ranging from flexible wraps to rigid containers. Although bioplastics are rarely fully biodegradable, increasing development to increase composting ability may make them less harmful to the environment.

Edible packaging materials extend this a step further by providing packaging that can be consumed along with the product. Made from natural ingredients, such as seaweed, rice, or potato starch, edible packaging is an innovative solution that completely eliminates packaging waste. Certain companies, for example, have been able to create from seaweed a packaging material used to wrap foods, such as sandwiches or burgers, instead of wrapping them in plastic. This material does not harm the consumer and can easily degrade if it is not consumed.

Edible coatings and films can also be directly applied to the food products themselves, extending shelf life and negating the need for packaging altogether. These act like barriers to moisture, oxygen, and contaminants, reducing food spoilage and wastes. Although edible packaging is still a nascent technology, its potential to drastically cut down packaging waste makes it an exciting alternative for sustainable food packaging.

Smart Technologies Reduce Waste

This helps reduce waste by improving the storage of food, generally enhances efficient supply chain, and provides more environmentally friendly materials. Technologies involved in smart packaging, such as sensors, indicators, and active packaging, will also help prevent food spoilage and, consequently minimize packaging wastes(Boopathi, 2024; Hanumanthakari et al., 2023).

One of the key ways through which smart packaging reduces waste involves extended shelf life for food products. In some cases, active packaging incorporates parts that might interact either with food itself or the external atmosphere to delay spoilage. Other active ingredients in packaging might involve oxygen scavengers, moisture absorbers, and antimicrobial agents that, when incorporated into foods, can delay microbial growth as well as oxidation or moisture buildup responsible for loss of quality in foodstuffs. With active packaging, foods will remain fresh longer; this reduces the overall quantity of foods that are spoiled during production. It therefore indirectly cuts down on packaging as fewer foods will need packaging for consumption at later dates.

It is assumed that in turn, sensors and indicators in smart packaging may reduce food waste by informing on the real-time status of the food. TTIs and freshness indicators record temperature and freshness history of a product during its lifetime and warn the consumer and retailer about possible spoilage before it actually takes place. This, in turn, enables better inventory management and reduces the possibility of discarding food that is still good. The presence of smart packaging helps to better monitor food quality, which, in turn, reduces the amount of food and packaging waste.

Besides that, smart technologies play a role in the optimization of the logistical side of the supply chain. Smart packaging can be fitted with RFID tags and IoT devices for product tracking along the supply chain. This level of visibility therefore allows the management of stock by both the manufacturer and retailer for efficient delivery and sale before the products spoil. These technologies therefore reduce overall waste by reducing the number of products that go to waste due to improper handling or storage.

Besides, it is a trend nowadays that the design of smart packaging systems has increasingly taken into consideration the aspect of sustainability. Most of the smart packaging materials are biodegradable, such as bioplastics or compostable films, which highly reduces the environmental impact caused by package discard. These can be combined with smart technologies to develop functional and ecologically friendly packaging solutions.

Food packaging within the concept of smart packaging is driven by trying to reduce environmental impacts emanating from conventional packaging, increasing the use of biodegradable and edible materials, and using smart technologies in waste reduction. With further innovation within the food industry, a combination of eco-friendly material with intelligent packaging systems may be considered one promising path toward a sustainable future. Accordingly, smart packaging can increase the food and packaging waste reduction considerably with the extension of shelf life, reduction of spoilage, and streamlining of supply chains, thus aiding the environment and consumer.

IOT AND SMART PACKAGING INTEGRATION

Internet of Things-integrated smart packaging opens completely new insights into the innovative food supply chain for improving visibility, traceability, and consumer connect. IoT-enabled smart packaging interlinks physical packaging with the latest digital technologies for sharing real-time data that enables monitoring and management of the product in a far better way. This section explains the role of IoT in food supply chain management, how real-time data sharing and tracking change the face of the industry, and how personalization, consumer engagement, and automation open new avenues of efficiency (Bharti et al., 2023; Hassoun et al., 2023; Licardo et al., 2024).

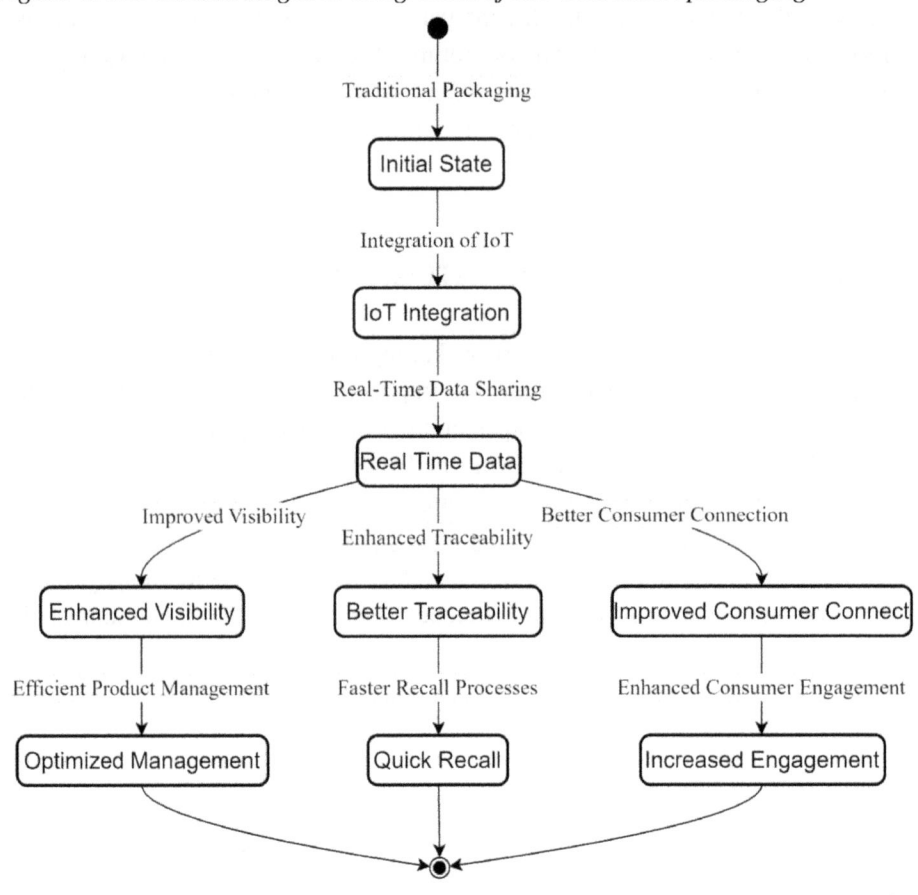

Figure 4. The various stages in integration of IoT with smart packaging

Figure 4 illustrates the transition from traditional packaging to an IoT-integrated smart packaging system, emphasizing the advantages of real-time data sharing for visibility, traceability, and consumer engagement.

IoT in Food Supply Chain Management

IoT has emerged as a game-changing technology in managing food supply chains, offering complete connectivity and visibility into processes all the way from production, through transportation at retail, up to consumer usage. IoT-powered devices make data seamlessly flow between the different stakeholders of food supply chains. This real-time connectivity is quite critical to optimize operations that ensure food safety, reduce waste, and thereby enhance overall efficiency in the supply chain(Mohanty et al., 2023; Vangeri et al., 2024; Verma et al., 2024).

Smart packaging with inbuilt IoT sensors can monitor temperature, humidity, and storage conditions in real time to enable food preservation and transportation under optimum conditions. For instance, perishable products such as dairy, seafood, or frozen foods need to maintain certain temperatures for freshness. IoT sensors embedded in packaging can monitor such conditions constantly and send an alert if the conditions go beyond safety levels. This reduces spoilage, improves the quality of food, and helps avoid food-borne illnesses.

Also, IoT provides more traceability throughout the supply chain. Tagging food items with RFID or NFC technology allows the manufacturer or retailer to know at all points in time from farm to table where the product has been. This increases accountability and transparency in the tracking and capturing of issues related to contamination or product recalls.

Real-Time Data Sharing and Tracking

IoT-enabled smart packaging considers the sharing and tracking of real-time data to be of great importance. With sensors and communication technologies imbedded in the package, continuous collection, storage, and transmission offer real-time information about status-location contingencies of food products (Agrawal et al., 2024). These enable a supply chain manager or retailer to trace the condition of food and predict disruptions in supply for making proactive decisions.

For example, IoT-integrated TTIs can provide immediate feedback whether the product has been under undesirable conditions during transportation or storage. If the temperature of the product exceeds the threshold value beyond which the product is no longer safe, notifications can be sent via the IoT system to all relevant parties for immediate interference and reduction of loss of food.

Real-time product tracking thus allows for better monitoring of inventory. Automatic data collection of stock levels and conditions of packaged products by IoT-enabled packaging automatically guarantees that inventory stocking will be better controlled, therefore reducing problems of overstocking or understocking. Consequently, storage space is used more efficiently, there is a reduction in wastage, and there will be efficient distribution of food products.

Second, IoT-based tracking systems improve the delivery accuracy and timeliness of retailers. For food products, GPS-enabled packaging can track goods throughout their journey by continuously updating on location and estimated delivery time. This creates a supply chain with greater visibility, hence allowing the retailer to ensure that the product reaches a destination in optimal condition.

Personalization and Consumer Engagement

IoT in smart packaging extends beyond supply chain optimization to new avenues of personalization and consumer engagement. Through the integration of IoT, packaging will be able to communicate with consumers' smartphones or other devices and provide personalized information on the product or suggest ways of usage, upon consumer preference. Examples include QR codes or NFC tags on packaging that connect consumers to digital platforms, which provide detailed information on the origin of the product, its nutritional content, or other sustainability practices. This level of transparency and personalization builds trust with consumers and can potentially enhance brand loyalty.

It will also make packaging interactive: intelligent packaging integrated with IoT will suggest recipes regarding the consumer's eating habits, offer usage tips, or track consumption to suggest similar products next time. Brands can also promote loyalty programs or promotional offers connected with the packaging for deeper customer engagement. IoT personalization also enables companies to deliver products matching specific consumer needs. For example, packaging can monitor the freshness of a product and send notifications to the consumer when it is time to use it, therefore further reducing food waste and enhancing customer satisfaction. Such levels of engagement turn passive packaging into an active tool of engagement, creating value for both consumers and brands.

Automating Opportunities and Efficiencies

Moreover, the integration of IoT into smart packaging allows immense opportunity for automation and operational efficiency along the food supply chain. IoT-based automation allows streamlining many of the supply chain activities which are labor-intensive in nature; therefore, operations become speedier and more accurate(Gnanaprakasam et al., 2023).

Another such important area is the automation of inventory management, wherein IoT is making a difference. Smart packaging fitted with sensors can automatically update inventories in real time and reduce manual counting, hence minimizing errors. It enables warehouses and retailers to maintain optimal levels of stock and avoid stockouts, reducing risks of expired products.

The automation driven by IoT in logistics can do route optimization by real-time data so that the delivery of the products will be more efficient. Such IoT devices can consider the patterns of traffic, the state of the weather, and the consumption of fuel to come up with the best routes that the delivery trucks can take to their destination. This will consume less fuel and lower costs, reducing the overall ecological footprint of food distribution.

IoT also increases automation in quality control methods. Sensors embedded in the packaging could track food items for any sort of quality changes continuously, hence preventing the need to do human inspections. If there are any abnormalities regarding temperature fluctuations or contamination, the automatic systems can flag the problem so that immediate corrective measures may be taken.

Moreover, IoT-driven automation could help the sustainable enterprise by reducing food waste. Smart packaging informs a company in real time of the condition regarding its products and, through this, optimizes the supply chain to prevent spoilage and deliver products in the best possible condition to the consumer. This could reduce waste but also contribute to more sustainability within the food system.

Integration of IoT into smart packaging empowers each node in the food supply chain by creating more visibility and better product quality, apart from opening the doors to new consumer interaction methods. IoT allows sharing and tracking information in real-time, hence entrepreneurs or managers can make more informed decisions and manage operations more effectively. IoT-driven personalization increases the bond between brands and consumers, and automation on the back of IoT enhances efficiency along the supply chain. With further evolvement, IoT technologies are expected to contribute immensely to forming the future of smart food packaging, driving both operational excellence and sustainability.

CASE STUDY AND INDUSTRY APPLICATIONS

The smart packaging technologies will break into many industries, improving the safety of the product, enhancing efficiency in the supply chain, and creating new ways for consumer engagement. Pilot projects and industrial initiatives further demonstrate how such technologies have the potential to drive step changes in packaging practice and meet growing demands for sustainability and transparency. While these innovations are still evolving, they may well continue to be an integral part of food packaging in the future, playing a great role in creating an efficient, safe, and more consumer-oriented industry (Hwa & Te Chuan, 2024; Jagtap et al., 2020; Sadeghi et al., 2022).

Novel Applications in Food Packaging

The incorporation of advanced technologies in food packaging is a major driver for innovation in various segments. A very good example of such innovative applications relates to smart packaging systems adopted in the dairy industry. The companies manufacturing packaging solutions, such as Amcor and Tetra Pak, have introduced intelligent packaging that includes sensors using IoT for monitoring and maintaining freshness in dairy products. These sensors monitor temperature and humidity in the supply chain from the time of production to retail. Smart packaging systems reduce spoilage and waste, hence improving the quality of the product by availing real-time information. For instance, active packaging solutions developed by Tetra Pak are integrated with an oxygen and moisture barrier to extend shelf life while retaining nutritional value in dairy products.

Winnow Solutions has introduced smart packaging into fruit and vegetables to reduce food waste. The packaging works in concert with AI systems that track the freshness of fruits and vegetables. Sensors embedded in the packaging offer information on ripeness and spoilage, analyzed to work out inventory management to decrease food loss. Besides bringing lots of benefits to the retailers, it improves customer satisfaction in the way of buying fresh produce.

Another famous application is in the meat industry through smart packaging at companies like Cargill. It will increase the safety of food because Cargill's packaging will be integrated with TTI and RFID technology for tracking conditions passed by the meat products during handling and storage along the value chain. All this information will contribute to keeping the products at temperatures that will not affect their safety and give full traceability in case of a recall. Not only does this increase the confidence of the consumer, but it also helps to satisfy the increasingly strict regulation requirements for food safety.

Industry Adoption and Pilot Projects

Smart packaging technologies have started to be adopted by various industries. Several companies have started pilot projects to test innovations in smart packaging and to enhance them further. PepsiCo, for example, is one of the biggest beverage firms that has undertaken a smart packaging pilot using NFC tags for consumer engagement. Consumers are going to scan these tags on their smartphones, through which they would get information about the origin of the product and nutritional content. They also get promotional offers because this project ensures better consumer interactions and collects important data for sales trends regarding customer preferences and behavior.

The PharmaPack pilot project has mandated the integration of IoT with packaging for medication adherence in the pharmaceutical industry. Smart packaging with sensors and features of connectivity reminds patients about the timely intake of medication, hence monitoring adherence. This will be developed to enhance patient outcomes and smooth medication management through the access of real-time alerts and monitoring.

Nestlé, too, is very aggressive with regards to using smart packaging technologies. The company's various pilot projects include use of biodegradable and compostable materials for the purpose of packaging so that environmental degradation can be reduced. Nestlé tried out materials from plant sources, such as corn and sugarcane, to make packaging that would degrade faster compared to conventional plastics. Such projects reflect Nestlé's commitment to sustainability, while providing a certain degree of understanding about the feasibility of large-scale adoption of eco-friendly materials. Simultaneously, innovative activities are in common and proceed actively in the beverage industry: an example could be Coca-Cola's experiment with smart labels, which convey-in real-time-the freshness of the product and conditions of the supply chain. A Smart Label fitted with sensors can give a detailed picture of the journey of the product right from manufacturing to consumption, thus enabling transparency and reducing the risk of quality issues.

Figure 5. Future of smart food packaging with the convergence of edible electronics and robotics

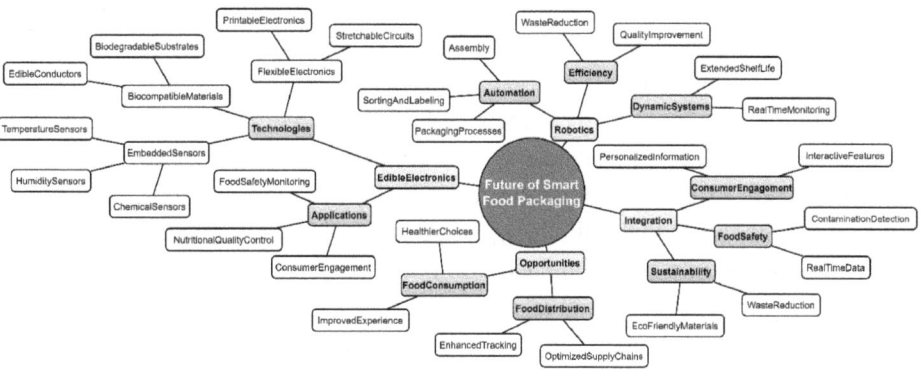

FUTURE OUTLOOK

Imagine the smart food packaging of the future: a convergence of emerging edible electronics and robotics. Integration in the future will stand to redefine the ways of food packaging, distribution, and consumption for more valued opportunities concerning food safety, sustainability, and consumer engagement (Chisenga et al., 2020; Mahalik, 2009a; Prasad, 2017). Figure 5 presents a comprehensive analysis of the potential impact of edible electronics and robotics on future food packaging, outlining key technologies, applications, and opportunities.

Edible Electronics and Robotics-Emerging Technologies: The work of edible electronics is one of the fastest-evolving areas, where research is directed to the elaboration of such advanced materials which, while being ingested, would be able to perform certain functions critical for the human organism. Future developments in edible electronics are very likely to be related to highly sophisticated sensors and complicated circuits embedded into food packaging, able to monitor a wide range of parameters from nutritional content to freshness and safety. These emerging technologies will, for the first time, enable the direct and real-time delivery of data directly to the consumer, offering the opportunity for new insight into the quality and safety of the food placed in their bodies. Meanwhile, robotics continue to transform the packaging, with increasingly more autonomous systems perform various tasks with assembly, sorting, and labeling. AI-driven robotics comes in to increase precision and efficiency, while reducing waste and offering higher overall quality in packaged foods.

The Future of Smart Food Packaging: Smart food packaging will continue to interconnect increasingly with the greater set of digital ecosystems. In this future, the integration of higher levels of IoT capabilities in packaging is going to enable seamless communication among products, consumers, and supply chain stakeholders. This high degree of interconnectedness can further enable much improved tracking and management of food products from production to consumption. Innovations such as dynamic labels that will change with either the environment or consumer preference will become commonplace. Similarly, packages will change in form to extend the life of a product. Integration of blockchain technology could mean further improvements in transparency and traceability, inclusive of comprehensive details for consumers about where their food has come from and how it arrived.

Long-term effects on food distribution and safety: The near future is sure to make good use of advanced technologies in food packaging. Supply chains will be better managed due to enhanced monitoring and real-time data, which will reduce the incidents of spoilage and waste. Smart packaging improves traceability and thus quicker responses to any situation of contamination or recall, hence improving the safety of food in general. Also, real-time monitoring and control of food conditions will contribute to better compliance with safety regulations and standards. This, in turn, will make food distribution systems increasingly reliable and effective in delivering quality products to consumers under the best possible conditions.

Vision for a Sustainable Food System: The long-term view toward a truly sustainable food system is directly related to understanding and adopting green or environmentally friendly packaging solutions and technologies. There will also be greater emphasis on the usage of biodegradable, compostable, and edible materials that minimize environmental impact in the drive towards sustainability. Innovations in smart packaging will support these goals through reductions in food waste by means of enhanced preservation and real-time condition monitoring. In this way, while the industry is moving toward more sustainable practices, driving the adoption of solutions will require collaboration among technology developers, manufacturers, and policymakers. The ultimate vision is such that the food system would meet today's consumer demand, but also ensure that future generations are healthy and well. This vision will have to call for a holistic approach from production to distribution-a balance of technological advancement against environmental stewardship towards a more resilient and sustainable food supply chain.

CONCLUSION

Edible electronics, robotics, and smart technologies in food packaging mark one giant leap toward improving food safety and sustainability and also consumer interaction. It is the edible electronics that introduce new ways of monitoring food quality and its nutritional content in real time, while robotics smoothen the process of packaging, hence being more efficient with less waste. As such knowledge continues to evolve, it will form part of smart packaging solutions that shall offer more accurate tracking, elongated shelf life, and dynamic interaction with consumers.

The future of smart food packaging is bright, steered by advances in IoT, blockchain, and sustainable materials toward a more transparent, efficient, and ecological food system. Such developments will enable better management not only at the food supply chain level but also in reducing the environmental impact, which will assure consumers receive high-quality, safe products. By embracing these technologies, the food industry will be well on its way to a resilient, sustainable future—tackling today's challenges while preparing for tomorrow's needs. Continuous development and adoption of such cutting-edge solutions will be critical to driving the next generation of food packaging and distribution forward.

REFERENCES

Abimbola, O. F., & Okpara, M. O. (2023). Artificial Intelligence in the Food Packaging Industry. In *Sensing and Artificial Intelligence Solutions for Food Manufacturing* (pp. 165–171). CRC Press. DOI: 10.1201/9781003207955-12

Agrawal, A. V., Bakkiyaraj, M., Das, S., Reddy, C. M. S., Kiran, P. B. N., & Boopathi, S. (2024). Digital Strategies for Modern Workplaces and Business Through Artificial Intelligence Techniques. In *Multidisciplinary Applications of Extended Reality for Human Experience* (pp. 231–258). IGI Global., DOI: 10.4018/979-8-3693-2432-5.ch011

Annese, V. F., Coco, G., Galli, V., Cataldi, P., & Caironi, M. (2023). Edible Electronics and Robofood: A Move Towards Sensors for Edible Robots and Robotic Food. *2023 IEEE International Conference on Flexible and Printable Sensors and Systems (FLEPS)*, 1–4. DOI: 10.1109/FLEPS57599.2023.10220412

Bharti, S., Jaiswal, S., & Sharma, V. (2023). Perspective and challenges: Intelligent to smart packaging for future generations. In *Green sustainable process for chemical and environmental engineering and science* (pp. 171–183). Elsevier. DOI: 10.1016/B978-0-323-95644-4.00015-2

Boopathi, S. (2024). Minimization of Manufacturing Industry Wastes Through the Green Lean Sigma Principle. *Sustainable Machining and Green Manufacturing*, 249–270.

Chisenga, S., Tolesa, G., & Workneh, T. (2020). Biodegradable food packaging materials and prospects of the fourth industrial revolution for tomato fruit and product handling. *International Journal of Food Sciences*, 2020(1), 8879101. DOI: 10.1155/2020/8879101 PMID: 33299850

Floreano, D., Kwak, B., Pankhurst, M., Shintake, J., Caironi, M., Annese, V. F., Qi, Q., Rossiter, J., & Boom, R. M. (2024). Towards edible robots and robotic food. *Nature Reviews. Materials*, •••, 1–11.

Gnanaprakasam, C., Vankara, J., Sastry, A. S., Prajval, V., Gireesh, N., & Boopathi, S. (2023). Long-Range and Low-Power Automated Soil Irrigation System Using Internet of Things: An Experimental Study. In *Contemporary Developments in Agricultural Cyber-Physical Systems* (pp. 87–104). IGI Global.

Hanumanthakari, S., Gift, M. M., Kanimozhi, K., Bhavani, M. D., Bamane, K. D., & Boopathi, S. (2023). Biomining Method to Extract Metal Components Using Computer-Printed Circuit Board E-Waste. In *Handbook of Research on Safe Disposal Methods of Municipal Solid Wastes for a Sustainable Environment* (pp. 123–141). IGI Global. DOI: 10.4018/978-1-6684-8117-2.ch010

Hassoun, A., Boukid, F., Ozogul, F., Aït-Kaddour, A., Soriano, J. M., Lorenzo, J. M., Perestrelo, R., Galanakis, C. M., Bono, G., Bouyahya, A., Bhat, Z., Smaoui, S., Jambrak, A. R., & Câmara, J. S. (2023). Creating new opportunities for sustainable food packaging through dimensions of industry 4.0: New insights into the food waste perspective. *Trends in Food Science & Technology*, 142, 104238. DOI: 10.1016/j.tifs.2023.104238

Hwa, L. S., & Te Chuan, L. (2024). A brief review of artificial intelligence robotic in food industry. *Procedia Computer Science*, 232, 1694–1700. DOI: 10.1016/j.procs.2024.01.167

Idumah, C., Zurina, M., Ogbu, J., Ndem, J., & Igba, E. (2020). A review on innovations in polymeric nanocomposite packaging materials and electrical sensors for food and agriculture. *Composite Interfaces*, 27(1), 1–72. DOI: 10.1080/09276440.2019.1600972

Jagtap, S., Bader, F., Garcia-Garcia, G., Trollman, H., Fadiji, T., & Salonitis, K. (2020). Food logistics 4.0: Opportunities and challenges. *Logistics (Basel)*, 5(1), 2. DOI: 10.3390/logistics5010002

Licardo, J. T., Domjan, M., & Orehovački, T. (2024). Intelligent robotics—A systematic review of emerging technologies and trends. *Electronics (Basel)*, 13(3), 542. DOI: 10.3390/electronics13030542

Mahalik, N. P. (2009a). Processing and packaging automation systems: A review. *Sensing and Instrumentation for Food Quality and Safety*, 3(1), 12–25. DOI: 10.1007/s11694-009-9076-2

Mahalik, N. P. (2009b). Processing and packaging automation systems: A review. *Sensing and Instrumentation for Food Quality and Safety*, 3(1), 12–25. DOI: 10.1007/s11694-009-9076-2

Mohanty, A., Venkateswaran, N., Ranjit, P., Tripathi, M. A., & Boopathi, S. (2023). Innovative Strategy for Profitable Automobile Industries: Working Capital Management. In *Handbook of Research on Designing Sustainable Supply Chains to Achieve a Circular Economy* (pp. 412–428). IGI Global.

Nayik, G. A., Muzaffar, K., & Gull, A. (2015). Robotics and food technology: A mini review. *Journal of Nutrition & Food Sciences*, 5(4), 1–11.

Prasad, S. (2017). Application of robotics in dairy and food industries: A review. *International Journal of Science, Environment and Technology*, 6(3), 1856–1864.

S., B. (2024). Advancements in Optimizing Smart Energy Systems Through Smart Grid Integration, Machine Learning, and IoT. In *Advances in Environmental Engineering and Green Technologies* (pp. 33–61). IGI Global. DOI: 10.4018/979-8-3693-0492-1.ch002

Sadeghi, K., Kim, J., & Seo, J. (2022). Packaging 4.0: The threshold of an intelligent approach. *Comprehensive Reviews in Food Science and Food Safety*, 21(3), 2615–2638. DOI: 10.1111/1541-4337.12932 PMID: 35279943

Sekhar, K. Ch., Ingle, R. B., Banu, E. A., Rinawa, M. L., Prasad, M. M., & Boopathi, S. (2024). Integrating VR and AR for Enhanced Production Systems: Immersive Technologies in Smart Manufacturing. In *Advances in Computational Intelligence and Robotics* (pp. 90–112). IGI Global. DOI: 10.4018/979-8-3693-6806-0.ch005

Sengeni, D., Padmapriya, G., Imambi, S. S., Suganthi, D., Suri, A., & Boopathi, S. (2023). Biomedical waste handling method using artificial intelligence techniques. In *Handbook of Research on Safe Disposal Methods of Municipal Solid Wastes for a Sustainable Environment* (pp. 306–323). IGI Global. DOI: 10.4018/978-1-6684-8117-2.ch022

Sharova, A. S., Melloni, F., Lanzani, G., Bettinger, C. J., & Caironi, M. (2021). Edible electronics: The vision and the challenge. *Advanced Materials Technologies*, 6(2), 2000757. DOI: 10.1002/admt.202000757

Suresh, P., Paul, A., Kumar, B. A., Ramalakshmi, D., Dillibabu, S. P., & Boopathi, S. (2024). Strategies for Carbon Footprint Reduction in Advancing Sustainability in Manufacturing. In *Environmental Applications of Carbon-Based Materials* (pp. 317–350). IGI Global., DOI: 10.4018/979-8-3693-3625-0.ch012

Teja, N. B., Kannagi, V., Chandrashekhar, A., Senthilnathan, T., Pal, T. K., & Boopathi, S. (2024). Impacts of Nano-Materials and Nano Fluids on the Robot Industry and Environments. In *Advances in Computational Intelligence and Robotics* (pp. 171–194). IGI Global. DOI: 10.4018/979-8-3693-5767-5.ch012

Upadhyaya, A. N., Saqib, A., Devi, J. V., Rallapalli, S., Sudha, S., & Boopathi, S. (2024). Implementation of the Internet of Things (IoT) in Remote Healthcare. In *Advances in Medical Technologies and Clinical Practice* (pp. 104–124). IGI Global. DOI: 10.4018/979-8-3693-1934-5.ch006

Vangeri, A. K., Bathrinath, S., Anand, M. C. J., Shanmugathai, M., Meenatchi, N., & Boopathi, S. (2024). Green Supply Chain Management in Eco-Friendly Sustainable Manufacturing Industries. In *Environmental Applications of Carbon-Based Materials* (pp. 253–287). IGI Global., DOI: 10.4018/979-8-3693-3625-0.ch010

Vashishth, R., Pandey, A. K., Kaur, P., & Semwal, A. D. (2022). Smart technologies in food manufacturing. In *Smart and Sustainable Food Technologies* (pp. 125–155). Springer. DOI: 10.1007/978-981-19-1746-2_5

Venkateswaran, N., Kiran Kumar, K., Maheswari, K., Kumar Reddy, R. V., & Boopathi, S. (2024). Optimizing IoT Data Aggregation: Hybrid Firefly-Artificial Bee Colony Algorithm for Enhanced Efficiency in Agriculture. *AGRIS On-Line Papers in Economics and Informatics*, 16(1), 117–130. DOI: 10.7160/aol.2024.160110

Verma, R., Christiana, M. B. V., Maheswari, M., Srinivasan, V., Patro, P., Dari, S. S., & Boopathi, S. (2024). Intelligent Physarum Solver for Profit Maximization in Oligopolistic Supply Chain Networks. In *AI and Machine Learning Impacts in Intelligent Supply Chain* (pp. 156–179). IGI Global. DOI: 10.4018/979-8-3693-1347-3.ch011

Chapter 8
From Ingestible to Edible:
Quantifying and Analyzing Edible Electronics' Role in Environment Monitoring and Future Advancement

Snehasis Dey
https://orcid.org/0000-0002-0490-8628
College of Engineering Bhubaneswar, BPUT University, India

ABSTRACT

Edible electronics symbolize a departure from the traditional concept of ingestible electronics, characterized by devices that are not only tailored for ingestion but also completely disintegrate in the body and can be safely released into the environment without the need for retrieval. In recent years, there has been a swift progression in the field of Green Electronics, marked by a significant rise in the number of research groups showcasing remarkable achievements in the realm of edible and biodegradable electronics. On this context this chapter caters to the study of edible electronics, its physical characterization and their role in environment monitoring. Thus, researchers are putting hard time out to discover the development of electronic components from food waste which controls the wastage and creation of proper sustainable approach towards edible electronics which also monitors the environment. This chapter dives into the edible electronics role in environment monitoring as well as its future advancement and current research.

DOI: 10.4018/979-8-3693-5573-2.ch008

I. INTRODUCTION:

Edible electronics represent a progression from the conventional notion of ingestible electronics. These devices are designed to be ingested, completely broken down within the body, and released into the environment without requiring retrieval. This advancement paves the way for an alternative, optimized, self-monitored medication, which leads to edible electronics and its invention. Time and again in recent years, the agricultural sector and food production sector have emerged as the best innovation fields in their respective areas, as huge studies have proven the best results for AI integration in the agricultural sector, green technology advancement, and edible electronics application. A novel approach must be incorporated to design and develop a sustainable environment in these fields as well. Edible electronics is an emerging technology that works on food waste and exploits food-oriented materials to develop electronic biodegradable components. These components are environment-friendly and non-hazardous. They have the unique characteristics of being safe to integrate, safe to use, and safe to ingest. Global studies have proved that food supply chain loss is alarming, estimated to be around 30% approximately. Precisely to incorporate those wastes for better usage from cultivation to consumption, there must be an efficient way for handling those wastes and developing some productive material from them. According to UNEP's food wastage index report, India independently wastes around 68.7 million metric tons of food annually, as compared to 1.3 billion metric tons of global waste. This statistic is quite alarming if we consider the amount of environmental imbalance that is happening due to food waste and its side effects, as industry contributes approximately 51.7% to this waste from its end. The sole solution to the ongoing problem is efficient utilization of these wastes as well as optimization and control of the wastes. This chapter caters to the proper utilization side of the waste by advancing towards edible electronics and global environmental monitoring. Also, we discussed the future advancements and current research on edible electronics.

Figure 1. Food wastage statistics

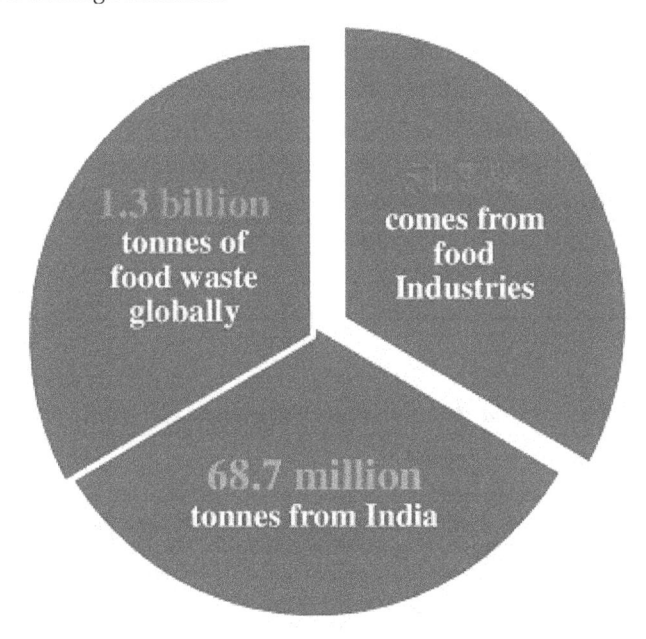

II. LITERATURE SURVEY:

Edible electronics represent an evolution beyond traditional ingestible electronics. These devices, designed for ingestion, can be completely digested by the body, allowing for safe environmental release without the necessity of retrieval. The field of edible electronics is concentrating on the advancement of ingestible electronics that can be safely swallowed, digested, and excreted. This area aims to offer a secure and efficient approach for remote health monitoring, aligning with the growing movement towards self-managed healthcare and personalized medicine. Ingestible electronics, which wirelessly send data from within the body, offer prospects for revolutionary healthcare advancements. These devices possess the capability to revolutionize the basic monitoring, optimization, and treatment of medical conditions, offering a more convenient and less intrusive option compared to conventional healthcare methods.

Edible electronics are pivotal in environmental monitoring; offering innovative solutions while also confronting significant challenges. These include the endurance of bioelectronics sensors under harsh conditions, vulnerability to interference from external electrical signals, the necessity for power sources in isolated areas (Ilic et al., 2023), and the creation of stable, biodegradable sensors (Prasanna et al., 2022). Nevertheless, recent progress offers optimism for overcoming these obstacles. For example, the invention of low-voltage edible transistors and circuits, crafted from consumable materials like chitosan and gold electrodes, provides a scalable and economical production method (Sharova et al., 2023). Moreover, the development of a rechargeable edible battery using everyday food components such as riboflavin and quercetin introduces a renewable energy source for edible electronics, facilitating safer medical diagnostics and food quality assessments (Teixeira et al.,2023). The prospective advancements in edible electronics are set to revolutionize environmental monitoring and enhance the food and medical industries (Zvezdin et al.,2020).

Table 1. Recent articles and its contribution in edible electronics field

Article	Insights	Contributions
Sharova et al., 2023	• Edible electronics are pivotal in environmental monitoring, facilitating the use of ingestible biosensors for food quality assurance and anti-counterfeiting measures. • The challenges lie in creating appropriate materials that can be manufactured at scale.	The study investigates the use of an alternative Au nanoparticle formulation for inkjet print electrodes on edible substrates, highlighting the advantages of low-temperature drying, green water-based composition, stability, and commercial availability of the colloidal gold ink.
Ilic et al., 2023	• Edible electronics, like the edible rechargeable battery, offer potential for environmental monitoring due to their ingestible nature, enabling innovative approaches for sustainable monitoring solutions.	Edible electronics field growing, focusing on ingestible devices using food. Edible battery developed using riboflavin and quercetin for power source.
Prasanna et al., 2022	• Edible electronics, like EM-TENG using rice flour, offer sustainable energy generation. Challenges include scalability and biodegradability testing, crucial for environmental monitoring applications.	This paper discusses the use of edible materials in triboelectric nanogenerators (TENGs) for sustainable energy generation and human joint movement monitoring
Teixeira et al.,2023	• The article emphasizes the trend of adopting biodegradable polymeric materials for creating flexible and sustainable electronic device components. It provides a summary of the latest scientific advancements and approaches in acquiring environmentally friendly polymers for such uses.	The paper outlines strategies for the incorporation of biodegradable polymers and methods for creating portable bioelectronic devices with flexible and sustainable substrates, that aims at the development of eco-friendly electronics.

continued on following page

Table 1. Continued

Article	Insights	Contributions
Zvezdin et al., 2020	• The research paper suggests solutions such as designing products for durability, reparability, and safe recycling, as well as promoting closed-loop systems for collection and reuse of electronic devices	Introduces the concept of organic electronics as a sustainable ultimate solution to the issue of waste electrical and electronic equipment (WEEE) by utilizing biodegradable materials and devices that has prolonged life span.
Radovanovic et al.,2023	• The paper presents the design, production, and testing of an interdigital capacitor exclusively made from edible materials such as biscuits, edible wax, and buckwheat tiles. • This innovative approach demonstrates the feasibility of creating electronic components from food and its wastes	Edible electronics components, like interdigital capacitors made from food materials, show promise for biomedical applications, particularly in oral cavity pathology detection, showcasing the potential and challenges in this innovative field.

II. Problem Statement:

The vivid literature study provides enough gaps to look into, like the prospects of edible electronics role in Industry 4.0, the healthcare sector, environmental monitoring, etc. Though articles (Teixeira et al., 2023) studied some parts of biodegradable electronics devices for a sustainable environment, the scope for vivid study is always there. Article (Zvezdin et al.,2020) proposed methodologies to control environmental imbalances through proper use of waste and provided solutions for the collection and reuse of electronic devices, but their continuous role in environmental monitoring is missing. Article (Radovanovic et al., 2023) presented the way ahead for edible electronics towards their application in the biomedical field and healthcare sector as a very innovative scope, but solutions for environmental monitoring are also missing there. This chapter dives into the gap in this study where we can properly use and monitor edible electronics in environmental monitoring.

The continuous and systematic literature study provides the following problem formulation from the research gap found in the study:

- Quantification of the role of the role of edible electronics in environmental monitoring.
- Analyzing the role of edible electronics in future industry advancement.
- Maintenance of an adequate balance between the rise of green electronics and the current environmental scenarios.

IV. Ingestible to Edible Electronics: The Rise Ahead.

Figure 2. The rise of edible electronics

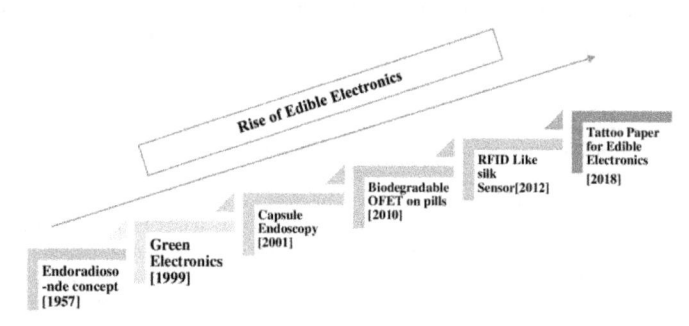

Rise of Edible Electronics

- Endoradiosonde concept [1957]
- Green Electronics [1999]
- Capsule Endoscopy [2001]
- Biodegradable OFET on pills [2010]
- RFID Like silk Sensor [2012]
- Tattoo Paper for Edible Electronics [2018]

Edible electronics have shown tremendous growth in the last decade or so. Figure 3 shows the rise of edible electronics from 1952, when the endoradiosonde concept first came into play, to date, where we are advancing towards edible robots, robotic food, etc.

Figure 3. Recent edible electronics growth

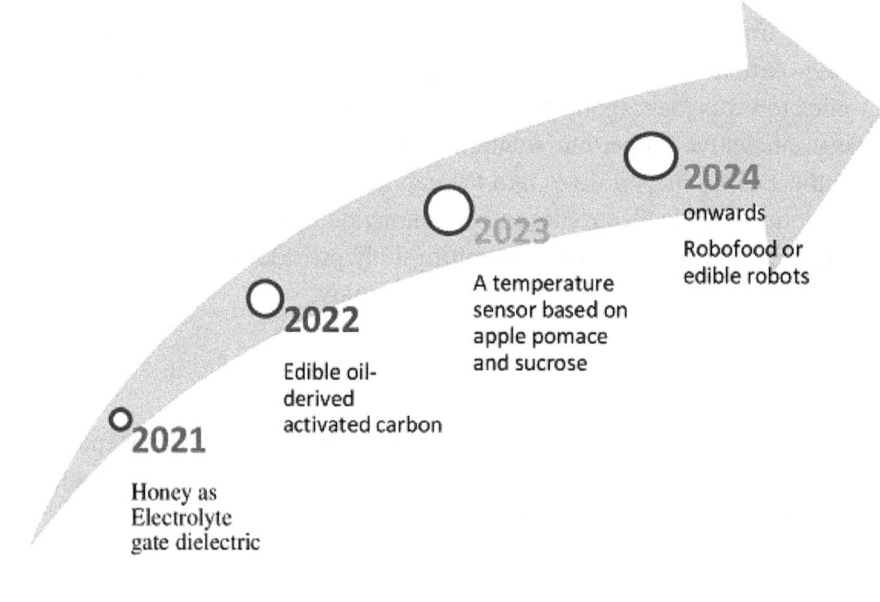

- 2021: Honey as Electrolyte gate dielectric
- 2022: Edible oil-derived activated carbon
- 2023: A temperature sensor based on apple pomace and sucrose
- 2024 onwards: Robofood or edible robots

Sharova et al. (2021) proposed edible transistors with humidity responsiveness for potential moisture control of dried or dehydrated food. He accumulated honey as a key material for the electrolyte gate dielectric. Similarly, Tyagi et al. (2022) used edible biocompatible capacitors and super-capacitors based on the molecular growth of hydrocarbons. The fabrication process includes burning the mustard oil and activating zinc chloride, followed by the use of a tubular furnace with a controlled temperature of 900 °C. A temperature sensor based on apple pomace and sucrose was developed for use in the oral cavity by Stojanovic, G. M. et al. (2023), and now the rise of edible electronics has reached robofood and edible robots in 2024. So, the rise of edible electronics from ingestible to edible is quite evident and impressive. Edible electronics are now quite substantial in providing solutions to the entire production chain in the agricultural field. As these have impressive characteristics of being biodegradable in nature, they possess the advantages of being safe to ingest and apply to anything without the need for recollection or safe disposal.

V. *Quantification of Edible Materials and Edible Electronics: Statistical Perspective of Edible Electronics.*

Edible electronics are basically generated and developed from edible materials, which can be of three types depending upon their electrical properties and conductivity.

- Conductor
- Semiconductor
- Insulator

Conductors are integral components within electronic devices, manifesting in interconnections, electrodes, vias, and other relevant features based on their designated purpose. It is noteworthy that edible materials have the capacity to demonstrate inherent electronic or ionic conductivity, often combining both types of conductivity, with biological systems frequently leveraging ionic transport as a fundamental mechanism. Conductors like gold, silver, magnesium, zinc, copper, calcium, and iron with higher conductivity can be termed electronic conductors. But ionic conductivity is a fundamental contributor to electrical conductivity within the realm of nature. Marmite/vegemite, hydrogel, etc. are typical examples of these types of ionic conductors. β-Carotene, indigo, and perylene diimide with less conductivity or minimal conductivity act as semiconductors, whereas powder infant milk, albumen, glucose, aloe vera, starch, natural rubber, etc. behave like typical insulators with zero conductivity (Betlej et al., 2022). These edible materials are considered the building blocks of the edible electronics that we design. Figure 5 below shows the food materials and their characterization.

Figure 4. Edible material and its respective characterization

[Pyramid diagram:
- Top (Insulator): Albumen, Glucose, Aloe vera, Starch and Natural rubber — Conductivity zero or negligible
- Middle (Semiconductor): β-Carotene, Indigo, Perylene diimide — Conductivity minimal
- Bottom left (Ionic conductor): Ex: Marmite/Vegemite, Hydrogel etc — Conductivity high but less than Electronic conductor
- Bottom right (Electronic conductor): Ex: Gold, Silver, Magnesium, Zinc, Copper, Calcium and Iron — Conductivity very high]

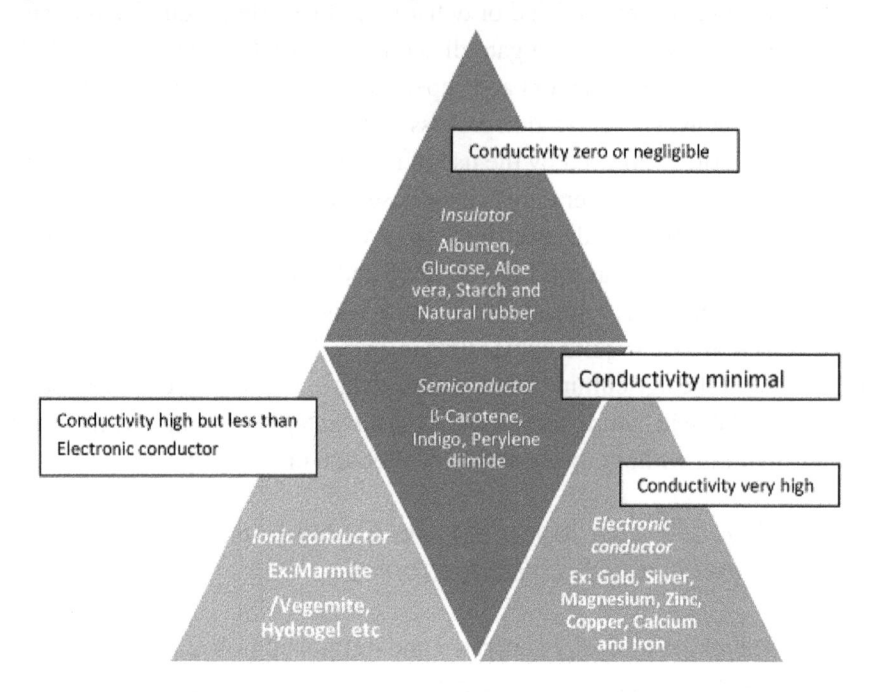

The study considers natural foods and foodstuffs as potential candidates for materials in the field of electronics. Subsequently, additional edible processed foods and food components were selected based on specific requirements. Additionally, nontoxic levels of electronic materials were incorporated to a limited extent in order to develop comprehensive electronic structures. The elements, like resistors, capacitors, inductors, and antennas, are interconnected and supported by a substrate, enclosed by an encapsulating layer for protection. Insulators, also referred to as dielectric materials, necessitate a conductivity "σ" of $<10^{-8}$ S m^{-1}, while conductors require a conductivity of $>10^6$ S m^{-1}. Typically, insulators are used for encapsulation and as dielectric materials in capacitors, with capacitance values ranging from 1 pF to 100 nF. The development of functional devices is essential for the creation of sensors tailored to specific characteristics for particular applications (Hafezi et al., 2015).

Table 2. Quantification of edible materials in edible electronics

Edible Material	Electrical Property	Conductivity [S cm^{-1}] Or Carrier mobility [cm^2 V^{-1}s^{-1}]	Electrical function and applications
Iron	Electronic conductor	1×10^5	Used in designing ingestible and implantable sensors,
Calcium	Electronic conductor	2.98×10^5	Interconnects/wires, power connections as high conductivity
Gold	Electronic conductor	4.10×10^5	It is accepted in food industry as garnish to various food items, and is assigned with food additivecode.The substance in a form of gold leaf or processedthrough sputtering/thermal evaporation is widely exploited indifferent electronic devices, at the macro and microscale, targetingthe concept of edibility
Hydrogel	Ionic conductor	1-40	Building blocks for ingestible and implantable devices due to its good conductivity
Vegemite/Marmite	Ionic conductor	$20 \pm 3 / 13 \pm 1$	Edible energy storage devices such as batteries, super-capacitors, and fuel cells.
β-Carotene	Semiconductor	4×10^{-4}	β-carotene belonging to the group of carotenoids, exhibits poor charge carrier field-effect mobility of 4×10^{-4} cm^{2}V^{-1}s^{-1}
Perylene diimide	Semiconductor	2×10^{-2}	Regarded as good electron transporting semiconductor and usually adopted to realize n-type OFETs (μ = 0.01–0.02 cm^2 V^{-1}s^{-1}). Application in lipsticks, nail polishes, and hair dyes.
Aloe vera	Insulator	σ = 0 but Relative dielectric constant =3.39	Used as substrates in biodegradable electronic devices.
Glucose	Insulator	σ = 0 but Relative dielectric constant =6.35	Used as substrates in biodegradable electronic devices.
Natural rubber	Insulator	σ = 0 but Relative dielectric constant =3.35-3.38	Used both as constituent elements in substrates, insulating and encapsulating layers and for the active role that dielectric polarization plays in the operation of energy harvesters, resonating circuits and transistors.

Table 1 specifies the statistical value of different edible materials for edible electronics. Generally; any electrical circuit consists of resistors, capacitors, inductors, antennas, connecting wires, and sensors etc.

VI. Edible Electronics Role in Environment Monitoring &Sustainability:

Edible electronics has the potential to provide solutions for environmental monitoring by utilizing biodegradable materials like edible food ingredients to create sensors and devices that are eco-friendly and sustainable (Ilic et al., 2023) (Prasanna et al., 2022) (Teixeira et al.,2023). The materials, such as rice flour and cellulose, can be incorporated into flexible films for sensor fabrication, reducing electronic waste and pollution (Prasanna et al., 2022). However, challenges persist, including the durability, interference, power supply, and cost issues associated with bioelectronic sensors, which can limit their widespread deployment for environmental monitoring applications (Teixeira et al.,2023). Despite these challenges, the integration of edible electronics in environmental monitoring holds promises for sustainable and environmentally friendly technological advancements. Edible electronics present a promising avenue for mitigating e-waste through the introduction of devices that can be ingested, digested, or metabolized safely by the human body, thereby preventing the buildup of electronic waste. These devices, made from food ingredients and additives, can degrade in the body or be digested, contrasting with traditional electronics that pose environmental challenges upon disposal. Additionally, the development of edible power sources, such as rechargeable batteries using common food ingredients like riboflavin and quercetin, further enhances the sustainability of edible electronics, enabling safer medical diagnostics and treatments and innovative ways to monitor food quality. Furthermore, the concept of recyclable organic flexible (ROF) electronic devices showcases a closed-loop recycling approach, where entire materials are recaptured and reused, and demonstrating reliable electrical properties even after multiple recycling cycles, thus contributing to a sustainable future for wearable electronic systems (Chai et al., 2015).

- Edible electronics offer non-invasive monitoring, advancing environmental monitoring by providing eco-friendly, biodegradable materials like pea protein biofilms with apple pomace extract, enhancing point-of-care diagnostics, and patient health tracking. Edible electronics, such as stable organic polymer sensors, have the potential to facilitate economical and extensive environmental monitoring in difficult marine settings, presenting an innovative method for sensing applications.
- Edible electronics, utilizing biodegradable materials, offer low-cost, disposable devices for environmental monitoring. They can be integrated into food packaging, plastic bags, and biomedical implants, revolutionizing monitoring techniques.

- Edible electronics, like field-effect transistor-based biosensors, offer precise, low-cost, and real-time environmental monitoring, revolutionizing current techniques by enabling in situ, continuous, and selective measurements.
- Edible electronics have the potential to bring about a significant transformation in the field of environmental monitoring through the enhancement of real-time tracking capabilities for variables such as temperature, humidity, illumination, and levels of CO_2 within the context of edible fungi cultivation. This advancement holds promise for elevating the quality of output and concurrently diminishing labor costs.
- Edible electronics, such as conductive oleogel paste, provide sustainable options for point-of-care diagnostics. These biodegradable and ingestible materials are crafted from food-grade substances, contributing to environmental sustainability.
- Biodegradable transparent substrates composed of conductive nanocomposites derived from starch and chitosan provide a sustainable approach to edible electronics, mitigating environmental issues by diminishing e-waste contamination.
- Edible electronics on fruit-waste paper substrates offer sustainable circular electronics with biocompatibility and eco-friendly disposal options, promoting a positive impact on the environment.
- Edible electronics, utilizing materials like chitosan and ethyl cellulose, offer sustainable solutions for ingestible devices, bioelectronics, and eco-friendly electronic components, contributing to a greener environment.
- "Green" electronics represents not only a novel scientific term but also an emerging area of research aimed at identifying compounds of natural origin and establishing economically efficient routes for the production of synthetic materials that have applicability in environmentally safe (biodegradable) and/or biocompatible devices. Thus, the ultimate goal of current research is to create paths for the production of human- and environmentally friendly electronics in general and the integration of such electronic circuits with living tissue in particular. Research into the emerging class of "green" electronics may help fulfill not only the original promise of organic electronics, which is to deliver low-cost and energy-efficient materials and devices, but also achieve unimaginable functionalities for electronics, for example, benign integration into life and the environment.
- A variety of natural substrates and dielectrics, including guanine, adenine, and caramelized glucose, are currently being researched, characterized, and refined for use in organic electronic devices. Their biocompatibility and biodegradability, along with their suitability for bio-implantable and bioresorbable applications, render them promising candidates for future develop-

ments in the realm of edible electronics and to maintain proper environmental sustainability.
- Edible electronics, utilizing materials from common diets, offer sustainable power sources for non-invasive medical implants, contributing to a more environmentally friendly approach to healthcare technology development.
- Edible electronics, like the edible rechargeable battery using food ingredients, offer biodegradable alternatives, potentially reducing e-waste by providing digestible and environmentally friendly electronic devices.
- Edible electronics, like the interdigital capacitor made from food materials, offer biodegradable alternatives, potentially reducing e-waste by utilizing edible substrates and conductive segments.

VII. Edible Electronics Role in Future Industry Advancement

Figure 5. Six-dimensional application of edible electronics

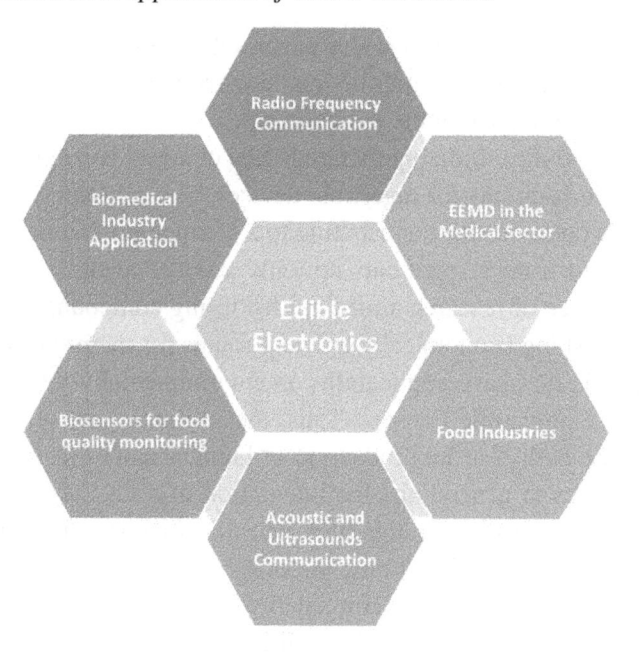

As shown in the figure above, the basic six-dimensional applications of edible electronics are confined to the food, pharma, medical, electronics, communication, and electrical industries. The application in different medical fields and the food industry is also broadly shown in:

- Detection of GI fluid pH and temperature
- Monitoring of the body's internal temperature
- Discovery of acidity in the stomach and intestines
- Examination of glucose concentration in GI fluid
- Identification of the swelling and dissolution properties of the intestinal mechanism
- Simulation of a food and drink consumption model using stomach temperature
- Diagnosis of diseases of the digestive tract
- Assessment of gastrointestinal bleeding
- Detection of an adenoma, a non-cancerous tumor
- Analysis of gas sensor enzymes and microbial communities
- Gastrorrhagia-determining capsules
- Edible carbon paste biodevices help in examining the acid level of the stomach and intestine.
- Edible hydrogel device with temperature sensor for detecting food and drink intake patterns
- X-ray imaging capsule for detecting colon impairment

Though edible electronics is a growing field and the most researched field in the current scenario, the proper balancing of its growth with environmental monitoring is quite essential. Moreover, by combining edible circuits and sensors driven by researchable edible batteries made from everyday common food components, we assure enhanced safety in medical testing, therapies, and creative methods for ultimate food standards, along with our food supply and distribution networks (Berean et al., 2018). It can monitor health and deliver nutrients effectively. Edible electronics have the potential to reshape the medical and food industries efficiently. Organic semiconductors that are safe to eat, such as circuits crafted from squid ink and everyday edible ingredients, hold great potential for eco-conscious and sustainable electronic devices in the upcoming era (Kim et al.,2017).

VII. Advantages of Edible Electronics

- Cost-effective
- Environmentally friendly
- Robustness
- Eliminate additional packaging in food produce
- Harmless
- Remote health care Monitoring
- Safe Ingestion
- Digestible/ eliminate device recollection

- Real-time Communication

IX. Maintenance of Adequate Balance Between the Rises of Green Electronics with the Current Environment Scenarios

The emergence of eco-friendly electronics offers a pivotal chance to tackle existing environmental hurdles through the advocacy of sustainable methods in crafting, producing, and discarding technological gadgets (Xu et al.,2017). By incorporating materials that decompose naturally, like eco-friendly semiconductors, into electronic parts, the buildup of e-waste could be lessened, thus reducing the ecological footprint of discarded tech products. Moreover, the integration of eco-conscious communication technologies, such as solar power and intelligent energy grids, could further boost efficiency in energy usage and resource management within electronic devices, thereby contributing to a healthier ecosystem (Meta et al., 2023). Embracing a circular economy model in the realm of electronics, which entails effective recycling methods and the adoption of eco-friendly components, can aid in upholding a harmonious equilibrium between the expansion of eco-friendly electronics and the conservation of the environment for forthcoming generations. Using green electronics reduces e-waste overflow in landfills, limits harmful substances, and creates a safer environment (Sharova et al.,2023). It helps in providing a sustainable and harmless environment for future generations.

Figure 6. Green electronics with smart optimization for the sustainable environment scenarios

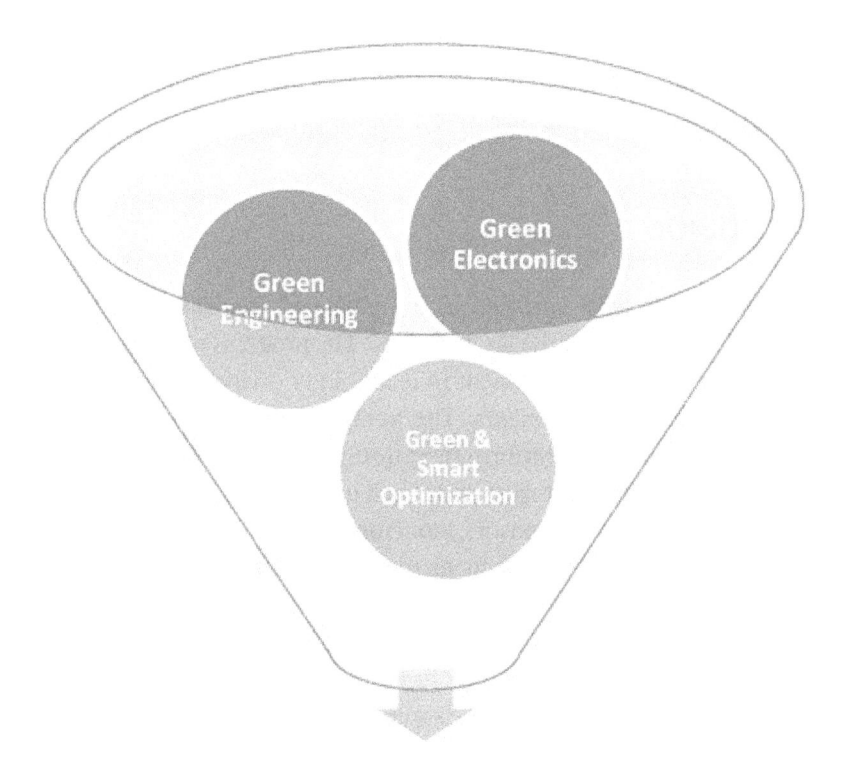

Sustainable Envionment

As shown in the figure above, green electronics not only rely on green engineering for a sustainable environment, but they also depend on green and smart optimization. These optimization techniques are the building blocks of smart edible electronics, which are contributing to the healthcare sector, the food processing sector, and different industries of necessity (Dey, 2023). So, maintaining the right balance between these three ingredients leads to a sustainable environment. Green electronics contribute to the reduction of e-waste through the implementation of eco-conscious technology during manufacturing (Zeng et al., 2017). This effort not only fosters a safer environment for forthcoming generations but also plays a crucial role in sustainable practices by minimizing the adverse effects of electronic waste. The employment of eco-friendly electronics, such as the environmentally degradable computer mouse examined in the research, lessens the carbon footprint on the environment by 60.2% in contrast to conventional electronics, advocating for sustainability (Bettinger et al., 2019). So, the maintenance of an adequate balance between the rise of green electronics and the current environmental scenarios is quite essential. The inclusion of artificial intelligence and different machine learning algorithms in edible electronics will definitely provide a new horizon to the edible electronics field in the coming days (Dey, 2024).

X. CONCLUSION

Edible electronics have surfaced as a significant innovation in the realm of healthcare technology during the last ten years. The progress of these edible gadgets represents a noteworthy advancement in technology owing to their use of natural or artificial food-based components. The human body can easily digest, absorb, and process these gadgets to perform their designated tasks. Included in the realm of edible electronics are an array of parts such as resistors, conductors, transistors, capacitors, batteries, antennas, sensors, inductors, conductive binders' substrates, and dielectrics. The basic framework acts as the cornerstone of electronic equipment, covering both functionally operative and dormant elements, circuits, sensors, power sources, and communication mechanisms. Quite often, edible materials display inherent electronic or ionic conductivity, sometimes even both. The communication approach utilized in edible electronics features a passive device resembling a radio frequency system, built with edible conductive materials employed to energize basic circuits. This chapter studies the basic need for edible electronics, its relevance in environmental monitoring, and future industrial growth as a whole.

REFERENCES

Berean, K. J., Ha, N., Ou, J., Chrimes, A. F., Grando, D., Yao, C. K., Muir, J. G., Ward, S. A., Burgell, R. E., Gibson, P. R., & Kalantar-zadeh, K. (2018). The safety and sensitivity of a telemetric capsule to monitor gastrointestinal hydrogen production in vivo in healthy subjects: A pilot trial comparison to concurrent breath analysis. *Alimentary Pharmacology & Therapeutics*, 48(6), 646–654. DOI: 10.1111/apt.14923 PMID: 30067289

Betlej, K., Rybak, M., Nowacka, A., Antczak, S., Borysiak, B., Krochmal-Marczak, K., Lipska, P., & Boruszewski, P. (2022). Pomace from Oil Plants as a New Type of Raw Material for the Production of Environmentally Friendly Biocomposites. *Coatings*, 13(10), 1722. Advance online publication. DOI: 10.3390/coatings13101722

Bettinger, C. J. (2019). Edible hybrid microbial-electronic sensors for bleeding detection and beyond. *Hepatobiliary Surgery and Nutrition*, 8(2), 157–160. Advance online publication. DOI: 10.21037/hbsn.2018.11.14 PMID: 31098367

Chai, P. R., Castillo-Mancilla, J. R., Buffkin, E., Darling, C. E., Rosen, R. K., Horvath, K. J., Boudreaux, E. D., Robbins, G. K., Hibberd, P. L., & Boyer, E. W. (2015). Utilizing an Ingestible Biosensor to Assess Real-Time Medication Adherence. *Journal of Medical Toxicology; Official Journal of the American College of Medical Toxicology*, 11(4), 439–444. DOI: 10.1007/s13181-015-0494-8 PMID: 26245878

Dey, S. (2023). Design & Development of new age IOT based smart garbage monitoring & controlling system for smart city. *Journal of Emerging Technologies and Innovative Research*, 10(6), 694–700.

Dey, S. (2024). Phenomenon of Excess of Artificial Intelligence: Quantifying the Native AI, Its Leverages in 5G/6G and beyond. In *Radar and RF Front End System Designs for Wireless Systems* (pp. 245–274). IGI Global. DOI: 10.4018/979-8-3693-0916-2.ch010

Hafezi, H., Robertson, T. L., Moon, G. D., Au-Yeung, K. Y., Zdeblick, M. J., & Savage, G. M. (2015). An Ingestible Sensor for Measuring Medication Adherence. *IEEE Transactions on Biomedical Engineering*, 62(1), 99–109. DOI: 10.1109/TBME.2014.2341272 PMID: 25069107

Ilic, I.K., Galli, V., Lamanna, L., Cataldi, P., Pasquale, L., Annese, V.F., Athanassiou, A. & Caironi, M., (2023) An Edible Rechargeable Battery. *Adv Mater*. 2023 May;35(20): e2211400. doi: . Epub. PMID: 36919977.DOI: 1002/adma.202211400

Kim, J., Jeerapan, I., Ciui, B., Hartel, M. C., Martin, A., & Wang, J. (2017). Edible Electrochemistry: Food Materials Based Electrochemical Sensors. *Advanced Healthcare Materials*, 6(22), 1700770. Advance online publication. DOI: 10.1002/adhm.201700770 PMID: 28783874

Mete, B., Durukan, D., Keskin, Y., Dinçer, O., Ogeday, M., Cicek, B., Yildiz, S., Aygün, B., Ercan, H., & Unalan, E. (2023). An Edible Supercapacitor Based on Zwitterionic Soy Sauce-Based Gel Electrolyte. *Advanced Functional Materials*. Advance online publication. DOI: 10.1002/adfm.202307051

Prasanna, A. P. S., Vivekananthan, V., Khandelwal, G., Alluri, N. R., Maria Joseph Raj, N. P., Anithkumar, M., & Kim, S. J. (2022). Green energy from edible materials: Triboelectrification-enabled sustainable self-powered human joint movement monitoring. *ACS Sustainable Chemistry & Engineering*, 10(20), 6549–6558. DOI: 10.1021/acssuschemeng.1c08030

Radovanović, M. R., Stojanović, G. M., Simić, M., Suvara, D., Milić, L., Kojic, S. P., & Škrbić, B. D. (2023). Edible Electronic Components Made from Recycled Food Waste. *Advanced Electronic Materials*.

Sharova, A., & Caironi, M. (2021). Sweet Electronics: Honey-Gated Complementary Organic Transistors and Circuits Operating in Air. *Advanced Materials*, 33(40), 2103183. Advance online publication. DOI: 10.1002/adma.202103183 PMID: 34418204

Sharova, A., Modena, F., Luzio, A., Melloni, F., Cataldi, P., Viola, F., Lamanna, L., Zorn, N., Sassi, M., Ronchi, C., Zaumseil, J., Beverina, L., Antognazza, M., & Caironi, M. (2023). Chitosan gated organic transistors printed on ethyl cellulose as a versatile platform for edible electronics and bioelectronics. *Nanoscale*, 15(25), 10808–10819. Advance online publication. DOI: 10.1039/D3NR01051A PMID: 37334549

Teixeira, S. C., Gomes, N. O., Oliveira, T. V., Fortes-Da-Silva, P., Soares, N. D., & Raymundo-Pereira, P. A. (2023). Review and Perspectives of sustainable, biodegradable, eco-friendly and flexible electronic devices and (Bio)sensors. *Biosensors and Bioelectronics: X.*, Volume 14,2023,100371, ISSN 2590-1370, https://doi.org/ DOI: 10.1016/j.biosx.100371

Xu, W., Yang, H., Zeng, W., Houghton, T., Wang, X., Murthy, R., Kim, H., Lin, Y., Mignolet, M., Duan, H., Yu, H., Slepian, M., & Jiang, H. (2017). Food-Based Edible and Nutritive Electronics. *Advanced Materials Technologies*, 2(11), 1700181. Advance online publication. DOI: 10.1002/admt.201700181

Zeng, Z., Piao, S., & Li, L. (2017). *Climate mitigation from vegetation biophysical feedbacks during the past three decades.* Nature Clim Change., DOI: 10.1038/nclimate3299

Zvezdin, A., Mauro, E. D., Rho, D., Santato, C., & Khalil, M. S. (2020). En route toward sustainable organic electronics. *MRS Energy & Sustainability : a Review Journal*, 7(1), 1–8. DOI: 10.1557/mre.2020.16

Chapter 9
Edible Electronics' Role in Healthcare:
Application, Challenges, and Future Research

Snehasis Dey
https://orcid.org/0000-0002-0490-8628
College of Engineering Bhubaneswar, India

Kadambini Himanshu
College of Engineering Bhubaneswar, India

ABSTRACT

Edible electronics have emerged as a prominent technology within the healthcare industry over the past decade. The advancement of edible electronics signifies a notable progression in the realm of technology due to their composition of natural or synthetic food-based materials. These devices can be easily digested, absorbed, and processed by the human body to carry out their intended functions. Within edible electronics are a variety of components including resistors, conductors, transistors, capacitors, batteries, antennas, sensors, inductors, conductive binders' substrates, and dielectrics. The fundamental structure serves as the foundation of electronic devices, encompassing functionally active and passive elements, circuits, sensors, power supplies, and communication mechanisms. In many instances, edible materials showcase inherent electronic or ionic conductivity, or both.

DOI: 10.4018/979-8-3693-5573-2.ch009

INTRODUCTION

Edible electronics have emerged as a prominent technology within the healthcare industry over the past decade. Edible electronics combine technology with food. The rise of edible electronics represents a significant progression in the field of technology because these devices are prepared from natural food-based materials that can easily be digested, absorbed, and processed within the human body to perform their functions. Edible electronics are designed from basic electronics components like resistors, capacitors, sensors, and many more. They can be made up of batteries, antennas, and some dielectric, depending on the user's requirements. Generally, these are the building blocks of active and passive elements, power supplies, and communication technologies. In the majority of cases, edible substances may reveal intrinsic electronic, ionic, or both conductivities.

The quality of the material must meet high standards that are used to evaluate its suitability for consumption and safety, as well as its biochemical and sensory attributes. Components such as conductors, which form part of the connections and electrodes, play a vital role in the system. There are two types of conductors used in editable electronics. It's electronic or ionic. Edible substances can exhibit either electrical or ionic conductivity, which are the fundamental processes of ionic motion. For an electrical service, a specific list of editable materials is recommended. Cellulose and chitin-activated carbon shellac are basically supplemented by tiny gold, silver, magnesium, and zinc elements.

At this moment, innovative electronic devices have emerged for the continuous observation, detection, and management or healing of the human body. Presently, there are four divisions of medical electronics that have been recognized, encompassing wearable, skin-based (epidermal), implantable, and ingestible. When compared to implantable electronics, ingestible electronics are less intrusive but are capable of navigating near vital organs through the gastrointestinal (GI) tract, tracking a broad array of biomarkers and therapeutic objectives, and functioning as efficient clinical instruments for diagnosis and treatment. Edible electronics components, like the interdigital capacitor made from food materials, show promise in biomedical applications for detecting oral cavity conditions, highlighting the potential for future health care innovations.

As edible electronics have been growing stronger and stronger, their early concept dates back to the 1950s. It was an ingestible endoradiosonde concept. Then there was an in vitro test for body temperature monitoring in 1988. A revolutionary achievement in this field is the contribution of the smart pill, which is for real-time position tracking and can fetch data about intestinal gases with the help of smart sensors.

This chapter dives into the basics of edible electronics, its promising role in different fields, including health care applications, on-board challenges, and the future ahead.

Figure 1. Typical field of application of edible electronics

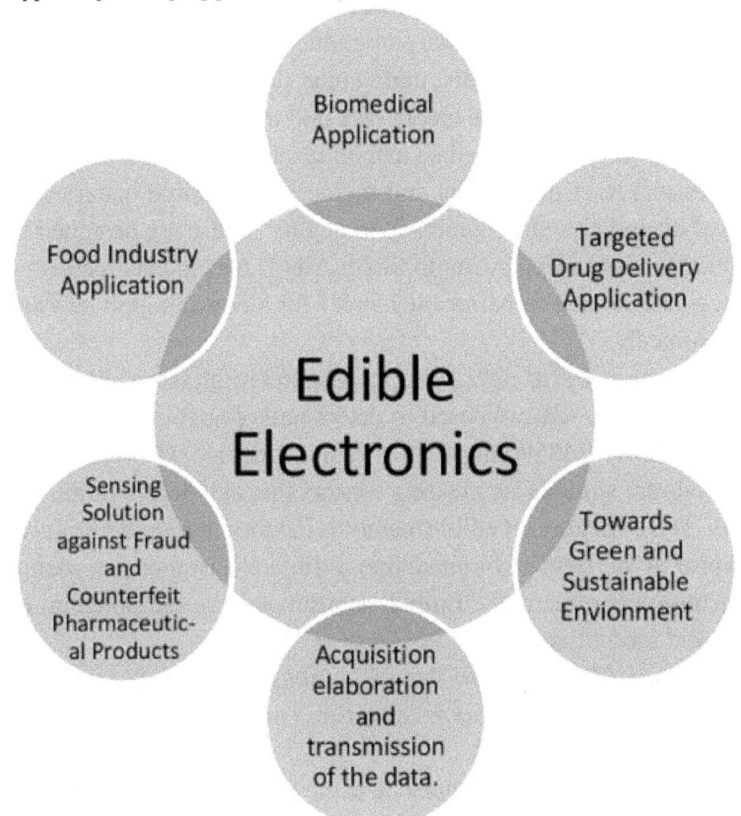

The figure above describes the diversity of the of the field where edible electronics roles are quite prominent. These fields range from the biomedical sector, pharmaceutical industry, food industry, data acquisition and transmission applications, smart healthcare sectors, etc. Recent research has revealed that the most frequent applications of edible electronics are generally in the biomedical field. For smart health care and monitoring, technologies like edible electronics and the internet of things are taking over others very aggressively.

Literature Survey

Edible electronics' evolution is just commendable, starting in the early 1950s and continuing to date. Different independent studies and collective research have taken place to fully explore the creditability and necessity of this technology. Recent technologies like artificial intelligence, machine learning, and IOT have taken center stage, but the popularity and necessity of edible electronics make it a key field to explore. In this section, we vividly study some of the key contributions made in recent times in the field of edible electronics and their current relevance.

Ivan K., Ilic et al. (2023) described in their research the use of electronic devices for gastrointestinal tract monitoring and therapeutics. Edible batteries can enable safer medical diagnostics and monitor food quality. These are designed from daily edible materials. It is realized by immobilizing riboflavin and quercetin on activated carbon. They fabricated a rechargeable battery for a power supply of 48μA at 0.65 volts, approximately.

Milan, Radovanovic et al. (2023) proposed an interdigital capacitance made from edible materials (food) that was used to detect pathological conditions in the oral cavity using a HIOKI IM3590 impedance analyzer device. His study specified that edible materials are suitable for creating sensors that can detect various conditions in the mouth. The utilization of edible materials in fabricating interdigital capacitors shows promise in the field of biomedicine. Edible electronics can monitor health and deliver nutrients effectively. Edible electronics have the potential to reshape the medical and food sciences.

F., Annese et al. (2023) studied that edible sensors reduce the risk of poisoning and surgical extractions. Food-derived materials enable gastrointestinal monitoring with digestible sensors.

Gargi, Konwar, et al. (2022) presented the dielectric composition of natural biopolymers, a unique way to tune and improve the electrical stability of simple carbon transistors. Transistors using this food-grade dielectric compound exhibit low voltage, high electrical conductivity, circuit adaptability, and long life, making them ideal for food and biomedical applications.

Saravanavel, G. et al. (2022) in their paper, focused on the development of green, biodegradable electronics using isomalt, a form of sugar, for healthcare applications. It discussed the design and development of inductor-capacitor (LC) circuits based on isomalt, showcasing resonators like capacitors, planar inductors, helical inductors, and inductors with an edible core.

Sharova, A.S. et al. (2021) proposed in their study edible transistors using materials like honey as an electrolyte gate dielectric. It also investigated future eco-friendly technologies and the relevant complexities of this process. Major issues have been discussed related to communication, human-body interaction, and healthcare applications.

Pietro, Cataldi., et al. (2022) in their research focused on the development of edible electronics for point-of-care testing and the establishment of safe devices that can be either digested in the body or degraded in the environment after being used. According to the study, the admission of a variety of materials is critical to the basis of edible platforms. Such materials shall be fabricated from green methods, have to be ingestible without causing danger to health in large quantities, and be made of food derivatives.

Bettinger C. J. (2019) described in their research paper the development of edible hybrid microbial-electronic sensors, which have potential applications in bleeding detection and beyond, showing the better integration of biological components with electronic systems for healthcare appointments.

Zeng, Wei, et al. (2017) presented the concept of edible supercapacitors, which prior research has shown can have applications in medical stimulation, powering microrobots, and others, such as bacteria elimination inside the human body.

The literature study specifies that there is enough research regarding edible electronics and its subsequent role in the health care sector, but a vivid study needs to be done regarding the challenges and future research ahead. There is a progressive study that needs to be done regarding the edible materials that can help in building those edible electronics for sustainable utility in the health care sector. This chapter builds on the research gap drawn from this literature study and provides a progressive study towards mitigating that.

Optimization of Green Materials to Edible Electronics: Background and Way Ahead.

Objective 1

The green system is the major inspiration for edible electronics. "Green" strategies were previously adopted by the researcher community in pursuit of a secure and sustainable future. Now green engineering is gaining more attention because it minimizes environmental impact and reduces risks to human health. Green Chemistry aims to reduce waste throughout the life cycle. To form the green triangle, electronics took an essential part of the "green n triangle." Green electronics are strongly connected to green chemistry and green technology. It is a non-toxic and environmentally friendly electronic; research is also going on worldwide. For the fabrication of green electronics devices, natural, biodegradable, and biocompatible organic materials are used. The aim is to utilize the semiconductor properties of food and edible materials to form ingestible electronic devices and systems. Edible materials for electronic purposes are very vast. Organic, inorganic, and semiconductor materials are used for edible electronics. Materials derived from organics and semiconductors are biopolymers in nature; cellulose is an essential nutrient for human health and is safe when consumed. Numerous benefits, like less processing required, more economic, and less environmental harm, make the effort put into the legislative procedure go beyond the scope of edible electronics selectivity. For designing new technology, control over electronic waste (e-waste) has become more important.

Figure 2. Categorization of edible materials

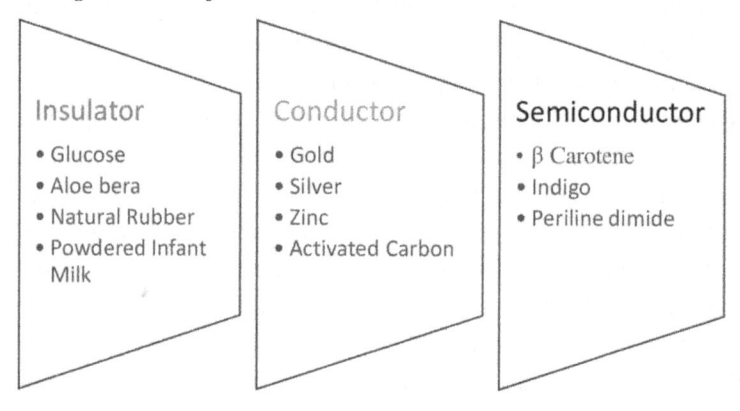

Insulator
- Glucose
- Aloe bera
- Natural Rubber
- Powdered Infant Milk

Conductor
- Gold
- Silver
- Zinc
- Activated Carbon

Semiconductor
- β Carotene
- Indigo
- Periline dimide

The below table shows different edible materials with their typical electrical values, like conductivity, carrier mobility, and relative dielectric constant. It is evident that these materials exhibit these electrical characteristics, on which their sustainability relies. Silver, gold, and copper are the best edible conductors, and therefore they are generally used more in the daily intake in comparison to others.

Table 1. Edible insulator with relative dielectric constant value

Edible Insulator Material	Relative Dielectric constant
Glucose	6.35
Aloe vera	3.39
Natural Rubber	4–5 (≈5)
Starch	2.2–3.20
Cellulose	1.3–6

Table 2. Edible conductor with specified conductivity vale

Edible Conductor Material	Conductivity [S cm⁻¹]
Gold	4.10×10^5
Silver	6.30×10^5
Zinc	1.69×10^5
Copper	5.96×10^5
Iron	1.00×10^5

Table 3. Edible semiconductor with carrier mobility value

Edible Semiconductor Material	Carrier mobility [cm²V⁻¹s⁻¹]
β-Carotene	$\mu_h = 4 \times 10^{-4}$
Indigo	$\mu_{e,h} = 1 \times 10^{-2}$
Perylene diimide	$\mu_e \leq 2 \times 10^{-2}$

As shown in the below figure, different edible materials are plotted according to their conductivity, carrier mobility, and relative dielectric constant value. The figure shows conducting edible materials with higher values of conductivity. Silver as an edible conductor shows the highest conductivity amongst all, and zinc is the edible conductor with the least conductivity amongst these. Similarly, edible semiconductors show their promising carrier mobility in the provided list.

Figure 3. Carrier mobility [$cm^2V^{-1}s^{-1}$] values of some edible semiconductor.

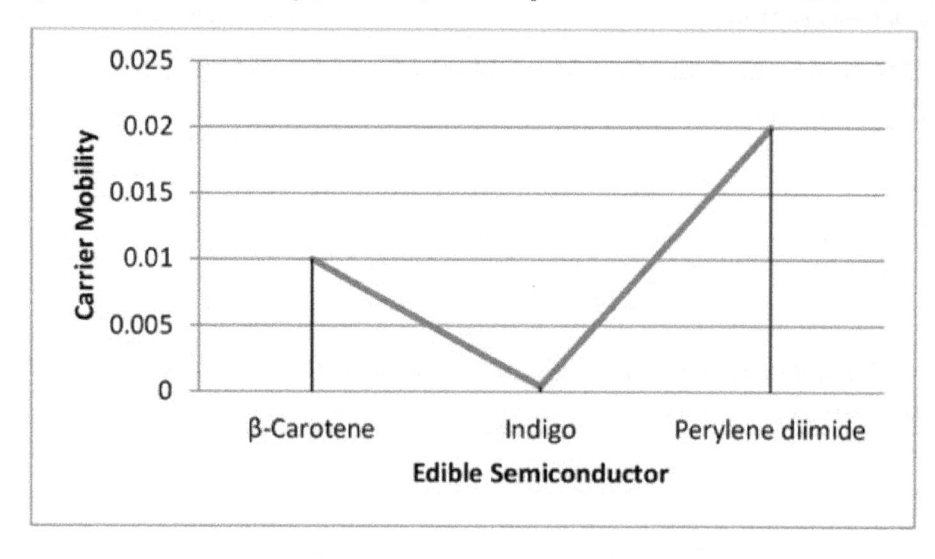

Figure 4. Minimum relative dielectric constant value of edible insulators

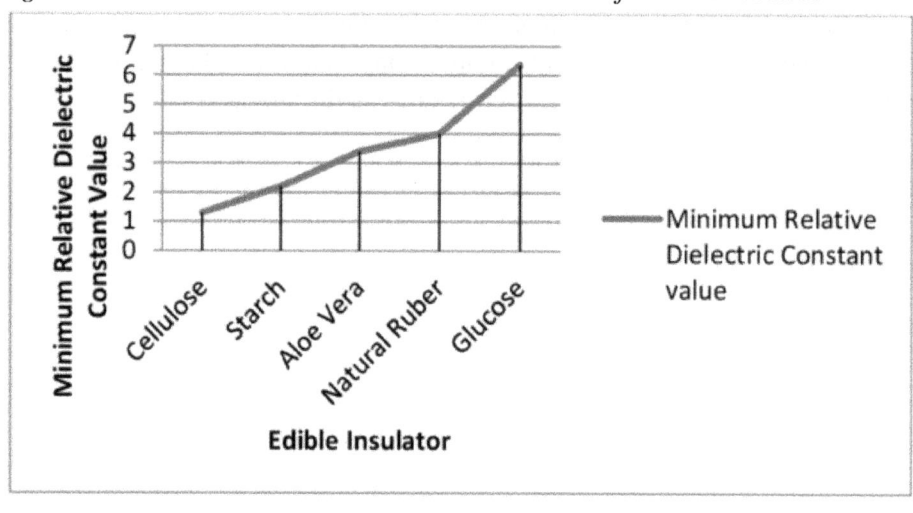

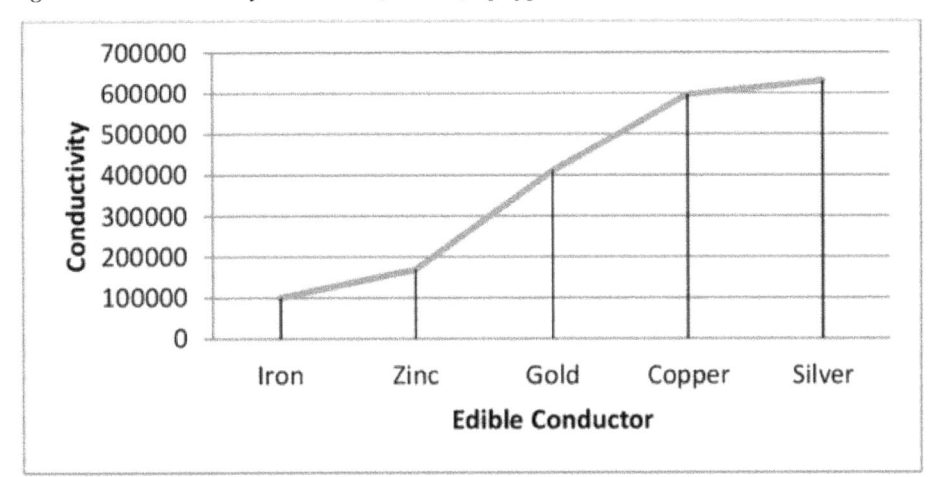

Figure 5. Conductivity values in [S cm^{-1}] of typical edible conductors

Edible Electronics and Health Care Monitoring

Objective 2

Although part of the important breakthrough medicine has ever witnessed in terms of diagnosis and treatment for many diseases, some intrinsic limitations characterize electronic medical implants. These implants are associated with invasive surgical procedures, operate in very challenging microenvironments, and are prone to bacterial infections or perpetual inflammation. Moreover, there are other issues like biocompatibility and immune reactions that have added on to this list, thus diminishing their applications. The integration of new materials and unconventional techniques into device manufacturing can change the interaction between electronic devices and the human body. In comparison to implantable devices, ingestible electronic devices present various advantages that could enhance the management of a wide range of pathologies, such as gastrointestinal infections and diabetes. This topic provides an overview of current technologies and emphasizes recent advancements in edible electronics applications in various fields that are considered safe, non-hazardous, and possibly biodegradable. Furthermore, in this topic, we discuss future obstacles and future scope in this field. Depending on their use, edible electronic systems are categorized into two main groups: devices that function within the gastrointestinal (GI) tract and those designed to perform tasks externally. The latter aims to introduce technologies for monitoring the GI system through the production of ingestible devices made of materials that are naturally safe for use. The first-ever ingestible device was invented in 1950–1957, known as endoradiosonde. It measures pressure, pH levels, and temperature within GI tract. In 1981, a video capsule endoscope was developed, but it did not come to market until the 2001s. In the 2000s, the Smart Pill (wireless motility capsule (WMC)) was developed. It is a small capsule that can be swallowed by a person and contains the sensing transducer and radio transmitter. The "device successfully operated in the gastrointestinal tract," which was not only developed but also approved and introduced into the market to monitor pressure, pH, and temperature crucial GI parameters for diagnosing individuals with delayed gastric emptying. In the 21st century, microelectronic circuits, enhancements in powering techniques, progress in biomaterials, and revolutions in drug delivery techniques enable researchers in the biomechanical industry to produce products with a variety of functions like temperature assessment, pH surveillance, motility detection, and capsule endoscopy. biopsy technique, GI gas monitoring, identification of inflammation, drug administration, and adherence monitoring. Despite the advances, there are still many problems with oral capsules, such as the fact that most ingestible devices have a limited time in the digestive tract. These have lower sensitivity. Scientists have developed several ingestible devices with ingenious techniques to increase residence time in the digestive tract.

Edible electronics have the potential to broaden their use into the realm of the food industry, focusing on overseeing and tracing the quality of food throughout the supply chain, serving as the main goals.

The typical health care applications of edible electronics are as follows:

- Edible electronics have promising healthcare applications, as demonstrated by various research papers. These innovative devices, made from edible materials like biscuits, edible wax, and natural waxes (Ivan *et al.*, 2023) (Floreano *et al.*, 2024) can be utilized for monitoring physiological functions, detecting pathological conditions in the oral cavity, and even gastrointestinal (GI) monitoring, (Trajkovska *et al.*, 2021).
- The development of edible sensors, such as interdigital capacitors and pressure sensors, not only eliminates the risk of toxicity from traditional materials but also reduces the need for surgical extractions in cases of malfunctioning devices (Ivan *et al.*, 2023) (Trajkovska *et al.*, 2021).
- Additionally, the integration of edible electronics with robotics opens up possibilities for applications in food monitoring, healthcare, and search and rescue scenarios, (Trajkovska *et al.*, 2021).
- These advancements in edible electronics hold the potential to revolutionize healthcare by providing safe, ingestible devices that can perform specific functions and transmit vital information for point-of-care testing.

- Recognizing the swelling and dissolution properties of the intestinal mechanism.
- Simulation of a food and drink consumption model using stomach temperature.
- Diagnosis of diseases of the digestive tract.
- Assessment of gastrointestinal bleeding.
- Detection of an adenoma, a non-cancerous tumor.
- Analysis of gas sensor enzymes and microbial communities.

Edible Electronics and Environment Sustainability

Objective 3

Today, environmental pollution like air, water, soil, noise, radioactive, and thermal pollution is the is the most significant threat to the environment and all living organisms. It impacts, directly or indirectly, human health and the ecosystem. From above, pollution is the most significant threat to the environment. For air quality monitoring Generally, disposable sensors are not used. Disposable sensors are widely employed to monitor contaminants present in water and soil. Water pollutants can be divided into three categories. i) inorganic; ii) organic; iii) biological. Inorganic contaminants, generally heavy metals, in water and soil samples are a significant application of disposable sensors in environmental analysis. Inorganic contaminants can be detected, and optical and electrochemical techniques are mostly employed. In the utilization of apparatuses constructed from paper, polymer, silicon, or glass substrates with carbon or metallic electrodes. As compared to colorimetric sensors, electrochemical sensors designed for environmental monitoring typically have higher analytical capabilities, but they are more complex and expensive.

Recent Challenges in Edible Electronics

Objective 4

Recent challenges in the field of edible electronics include the need to expand the whole field of edible electronic materials with suitable properties for scalable production. Although progress has been made in the development of edible circuits and sensors, the creation of fully edible electronic devices requires innovations in edible power supplies, which are scarce in examples Dey (2024) and Hafezi et al. (2014). In addition, the development of very low-power wireless interface electronics for edible electronic devices is crucial to enable non-intrusive monitoring of the gastrointestinal tract, (Chai et al., 2015). In addition, improving the functionality and usability of edible electronics, analyzing food components to build sensors, and ensuring consumer acceptance are important aspects. The study of biomedical applications of edible materials in sensor design, such as the use of interdigital capacitors to detect conditions in the oral cavity, shows the innovative and practical application possibilities of edible electronics in health care, (Kalantar-Zadeh *et al.,* 2018). It is important to know how an orally administered drug interacts with our body mechanisms. The gastrointestinal tract is composed of various organs, each playing a crucial role in digesting food, absorbing nutrients, and expelling waste from the body. The major difficulties a microdevice will present with after oral ingestion, entry into the stomach and its passage through the GIT to the large intestine much have to do with its minute size. Therefore, it is of importance to minimize accidents. Its size determines safe passage via the esophagus to the large intestine, the GI tract, with a maximum residence time of 1-2 days. But the journey between the two is too risky. Not to mention the low pH in the stomach, which puts the material at risk of undesired leaching, and various enzymes that are present at various parts of the digestive tract. Tissue wall localization is another challenge due to the continuous motility and peristalsis of the gastrointestinal tract. Both the safety profile and therapeutic accuracy of any device that is orally digestible are controlled by the device size itself and associated design. Challenges come in the way of size, material used, power consumption, and communication.

Limiting the size of the ingestible capsule is a major challenge in the development of any ingestible device. There are two main factors to consider when deciding on the size of the tank: is it in an ingestible area and does it create obstructions? The instrument camera measures 26 mm x 13 mm and contains several integrated components, including a power supply unit, a communication unit and a sensor unit, but has a short standby time. Certain implantable medical device materials are generally not regulated by the FDA in the United States. instead, it regulates the entire system. In the design of consumable medical devices, biocompatibility of materials is a key factor, along with chemical stability, mechanical properties and product toxicity. Typically, consumable medical devices are temporary, reducing the potential for chronic toxicity and biofouling. One of the biggest design challenges for wearable devices is powering the electronic components due to size limitations, battery life and potential risk to patients. Coins or button batteries are the most suitable solution for swallowing capsules. Lithium-ion batteries have a higher power density; they are not recommended for clinical use because they increase the pH of the gastric fluid. Commercial vehicles use silver oxide batteries that produce 20 MW of power and last for eight hours. Biodegradable batteries can replace traditional batteries. A biodegradable paper-based mg-moo3 battery (2 cm x 2 cm x 0.5 cm) can produce 1.6 volts for 13 days. The SmartPill (WMC) has a transmitter that can receive data at 434 MHz and a battery life of at least five days. In general, input devices communicate using wireless radio frequency telemetry, which typically consists of a transmitter, receiver, or radio receiver with an antenna. The size of the required antenna is a quarter of the wavelength. As a result, shorter wavelength radio frequency signals are generated. A smaller antenna is required for higher frequencies. However, the increased attenuation and dispersion results in a greater loss of power absorption in the human body. Li et al. used water as a medium to reproduce the attenuation of RF signals from 2.45 GHz to 135 GHz.

Beginning of a New Era: Research Perspective

Objective 5

In recent years, the development of edible electronics has been promising. One of the latest innovations is the development of smart pills. Equipped with smart sensors, the pill can detect intestinal gas and provide real-time location tracking. The capsule developed by Khan Lab is specially designed to detect gases associated with gastritis and stomach cancer. Current innovations in AI and different machine learning techniques, like artificial neural networks (ANN), are creating a new horizon in the field of edible electronics. ML techniques like supervised, unsupervised, and reinforced learning will definitely foster the scenario of edible electronics altogether, (Kim *et al.,* 2017). The benefit of the inclusion of ANN is that it will locate the exact position of the capsule in the body with the help of a magnetic field created through wearable sensors. Another area of ongoing research in the field of edible electronics is edible robots, which could be life-saving food in cases of emergency in the field of medical science. There is also tremendous achievement in the field of current endoscopy where doctors prefer "PillBot," which is a stomach endoscopy method via telemedicine, in place of traditional endoscopy methods. Generally, if we coin the future of edible electronics, then it must be coined "STEPS."

Figure 6. Edible electronics future ahead

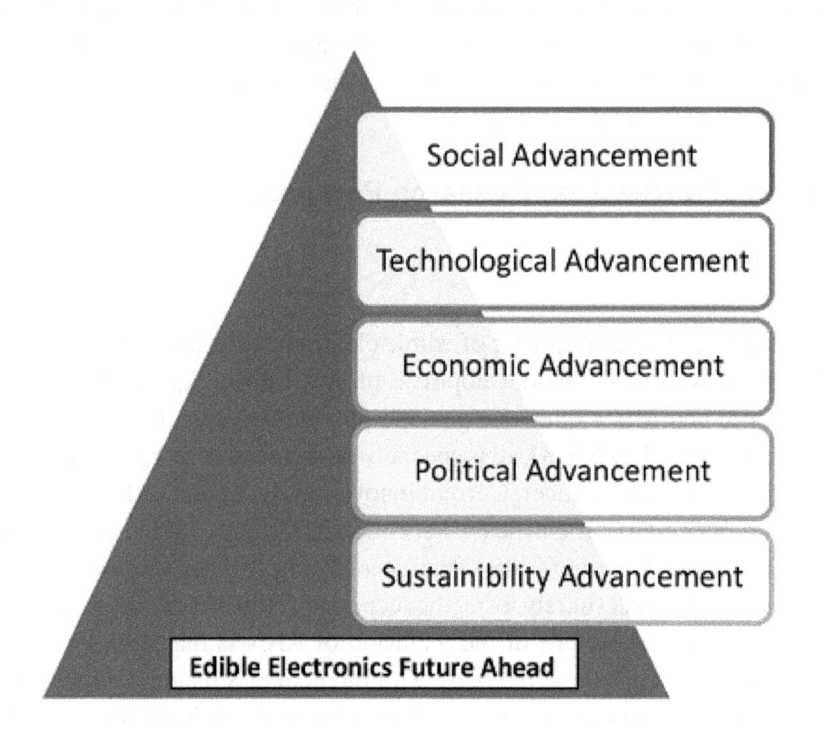

[Pyramid diagram with layers from top to bottom: Social Advancement, Technological Advancement, Economic Advancement, Political Advancement, Sustainability Advancement, Edible Electronics Future Ahead]

Social Advancement: Medical science research is increasingly concentrating on the development of edible electronic devices for less invasive and more accurate diagnostic procedures. A specific focus is on stomach cancer diagnosis, a condition that impacts close to 15,000 individuals in Italy each year.

Technological Advancement: The progression of digital solutions in medicine is propelled by ingestible electronic devices, which play a significant role in the emergence of Medicine 4.0. This new era integrates digital technologies, data analytics, and AI systems into healthcare. The Internet of Medical Things is increasingly prominent, evident in the growing network of connected medical devices and applications that interface with healthcare IT systems over the internet.

Economic Advancement: Edible electronic devices for telemedicine and diagnostics carry substantial economic implications, potentially lowering healthcare costs through early diagnosis and targeted, personalized medical treatments.

Political Advancement: Edible electronic devices gained initial approval from the US Food and Drug Administration as early as 2001, starting with an ingestible diagnostic camera, and later in 2015, with the first integrated circuit micro-sensor designed for daily ingestion by patients. In Europe, the European Food Safety Authority (EFSA) and the European Medicines Agency (EMA) have set acceptable daily intake (ADI) limits for substances that may cause side effects due to chronic or excessive consumption, which also reflects progress in regulatory policies.

Sustainability Advancement: Sustainability is a critical factor in developing edible electronic devices, with significant implications for patient health, environmental impact, and food safety. The focus is on solutions that prevent side effects by utilizing natural and entirely biocompatible materials. Edible electronics seek to incorporate food-grade materials into sophisticated systems like robots, thus contributing to the reduction of electronic waste. In addition to sustainability, biocompatibility, and biodegradability, employing food-grade materials in electronics offers the benefit of reduced toxicity, especially if ingested.

Current advancement also needs to be diverted towards some more critical as well as important aspects of the edible electronics field. Still, edible electronics and its surrounding fields, like nutritive electronics, are in their infancy, but their potential applications and solutions in different fields make them a researcher's paradise. As shown in the figure below, edible electronics can have potential applications in clinical medicine and food technology, like biosensors and glucose sensors for measuring acidity in the stomach and intestine. Flexible and edible electronics have a role in wound healing as well. Some critical diseases, like cardiovascular problems and diabetic diseases, are easily monitored and controlled with the help of edible electronics. The monitoring of pH values in the GI tract can also be solved by the application of edible electronics in the form of edible sensors. The future of edible electronics lies in searching for and optimizing different semiconductors, conductors, and insulators that are really edible and in the utilization of nutritive electronics as a whole that are fabricated from these materials.

Figure 7. Future of edible electronics in some critical fields

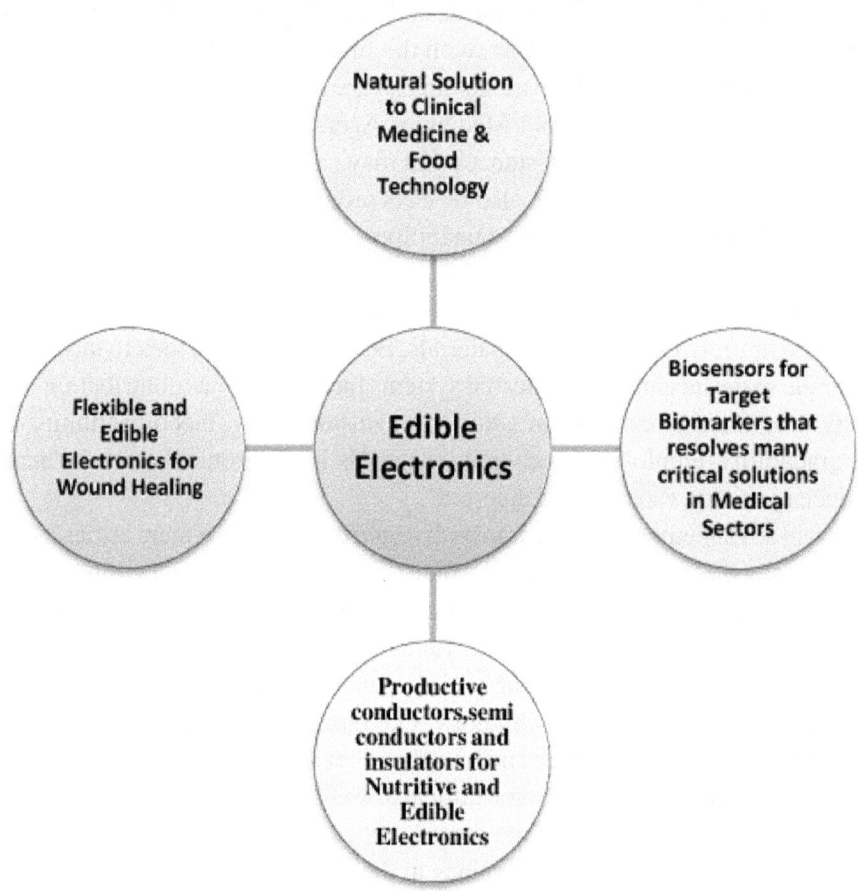

- Natural Solution to Clinical Medicine & Food Technology
- Flexible and Edible Electronics for Wound Healing
- Edible Electronics
- Biosensors for Target Biomarkers that resolves many critical solutions in Medical Sectors
- Productive conductors, semi conductors and insulators for Nutritive and Edible Electronics

Some key edible devices and their possible applications have been listed in the below figure. The pH value measurement device fabricated from carbon and olive oil is used to test the acidity in the stomach. A temperature sensor can be applied to evaluate the real-time temperature of an animal. An ingestible micro-bioelectronic device (IMBED) was developed as a diagnostic platform for the examination of biomarkers and gathering diverse sorts of information on gastrointestinal diseases. The sensitivities at 60 minutes and 120 minutes for the blood-sensing IMBED were both as high as 83.3%, while the specificity was 100%. So, IMBEDs achieve high specificity and sensitivity in detecting trace amounts of an analyte in these harsh gastric surroundings. So now, researchers are very eager to develop edible and nutritive electronics for a biodegradable environment with effective sustainability.

Figure 8. Some key edible devices

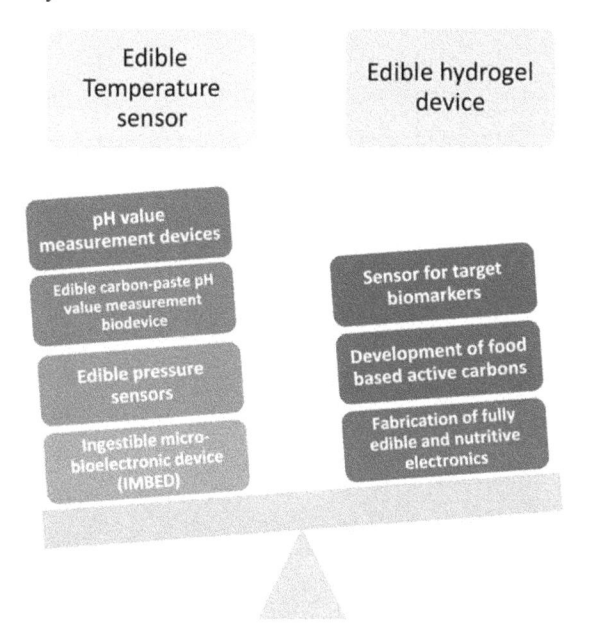

- Edible Temperature sensor
- Edible hydrogel device
- pH value measurement devices
- Edible carbon-paste pH value measurement biodevice
- Edible pressure sensors
- Ingestible micro-bioelectronic device (IMBED)
- Sensor for target biomarkers
- Development of food based active carbons
- Fabrication of fully edible and nutritive electronics

CONCLUSION

This chapter studies the basics of edible electronics and their subsequent applications in the biomedical field. Today, the field of edible electronics is growing from strength to strength. So, the right balance between environmental sustainability and edible electronics must be there. In this chapter, we study the challenges in the upcoming days for edible electronics as well as their subsequent solutions. Edible electronics future research areas along with progressive advancements have also been equally studied. Recently, some advanced biosensor and molecular imprinting technologies have been utilized to develop high-efficiency sensors for specific biomarker detection. Various ethical and legal issues related to edible food and food electronics are receiving a lot of attention. Finally, more practical gastrointestinal simulators are widely used to quickly assess the bioabsorbable safety of their electronics in the degradation process. One major step forward in edible electronics is the development of fully edible and nutritional electronics. The development of edible and nutritional electronics will further improve the biosafety and bio-absorbability of both ingestible and implantable electronics. So, it is clear that the production of fully edible and nutritious electronics is possible.

REFERENCES

. Alina, S., Sharova., Alina, S., Sharova., Mario, Caironi. (2021). Sweet Electronics: Honey-Gated Complementary Organic Transistors and Circuits Operating in Air. *Advanced Materials,* 33(40):2103183-. .DOI: 10.1002/adma.202103183

Bettinger, C. J. (2019). Edible hybrid microbial-electronic sensors for bleeding detection and beyond. *Hepatobiliary Surgery and Nutrition*, 8(2), 157–160. DOI: 10.21037/hbsn.2018.11.14 PMID: 31098367

Chai, P. R., Castillo-Mancilla, J. R., Buffkin, E., Darling, C. E., Rosen, R. K., Horvath, K. J., Boudreaux, E. D., Robbins, G. K., Hibberd, P. L., & Boyer, E. W. (2015). Utilizing an Ingestible Biosensor to Assess Real-Time Medication Adherence. *Journal of Medical Toxicology; Official Journal of the American College of Medical Toxicology*, 11(4), 439–444. DOI: 10.1007/s13181-015-0494-8 PMID: 26245878

Dey, S. (2024). Phenomenon of Excess of Artificial Intelligence: Quantifying the Native AI, Its Leverages in 5G/6G and beyond. In *Radar and RF Front End System Designs for Wireless Systems* (pp. 245–274). IGI Global. DOI: 10.4018/979-8-3693-0916-2.ch010

Floreano, D., Kwak, B., Pankhurst, M., Shintake, J., Caironi, M., Annese, V. F., Qi, Q., Rossiter, J., & Boom, R. M. (2024). Towards edible robots and robotic food. *Nature Reviews. Materials*, 9(8), 589–599. Advance online publication. DOI: 10.1038/s41578-024-00688-9

Hafezi, H., Robertson, T. L., Moon, G., Au-Yeung, K.-Y., Zdeblick, M., & Savage, G. (2014). An Ingestible Sensor for Measuring Medication Adherence. *IEEE Transactions on Biomedical Engineering*, 62(1), 99–109. Advance online publication. DOI: 10.1109/TBME.2014.2341272 PMID: 25069107

Ivan, K. (2023, May). Ilic., Valerio, Galli., Leonardo, Lamanna., Pietro, Cataldi., Lea, Pasquale., V., F., Annese., Athanassia, Athanassiou., Mario, Caironi. (2023). An Edible Rechargeable Battery. *Advanced Materials*, 35(20), 2211400. Advance online publication. DOI: 10.1002/adma.202211400

Kalantar-Zadeh, K., Berean, K. J., Ha, N., Chrimes, A. F., Xu, K., Grando, D., Ou, J. Z., Pillai, N., Campbell, J. L., Brkljača, R., Taylor, K. M., Burgell, R. E., Yao, C. K., Ward, S. A., McSweeney, C. S., Muir, J. G., & Gibson, P. R. (2018). A human pilot trial of ingestible electronic capsules capable of sensing different gases in the gut. *Nature Electronics*, 1(1), 79–87. DOI: 10.1038/s41928-017-0004-x

Kim, J., Jeerapan, I., Ciui, B., Hartel, M. C., Martin, A., & Wang, J. (2017). Edible Electrochemistry: Food Materials Based Electrochemical Sensors. *Advanced Healthcare Materials*, 6(22), 1700770. Advance online publication. DOI: 10.1002/adhm.201700770 PMID: 28783874

Mete, B. (2024, February). Durukan., Deniz, Keskin., Yiğithan, Tufan., Orçun, Dinçer., Melih, Ogeday, Cicek., Bayram, Yildiz., Simge, Çınar, Aygün., Batur, Ercan., Husnu, Emrah, Unalan. (2023). An Edible Supercapacitor Based on Zwitterionic Soy Sauce-Based Gel Electrolyte. *Advanced Functional Materials*, 34(6), 2307051. Advance online publication. DOI: 10.1002/adfm.202307051

Saravanavel, G., John, S., Wyatt-Moon, G., Flewitt, A., & Sambandan, S. (2022). Edible resonators. *arXiv preprint arXiv*:2202.13782.

Sharova, A., Modena, F., Luzio, A., Melloni, F., Cataldi, P., Viola, F., Lamanna, L., Zorn, N. F., Sassi, M., Ronchi, C., Zaumseil, J., Beverina, L., Antognazza, M. R., & Caironi, M. (2023). Chitosan gated organic transistors printed on ethyl cellulose as a versatile platform for edible electronics and bioelectronics. *Nanoscale*, 15(25), 10808–10819. Advance online publication. DOI: 10.1039/D3NR01051A PMID: 37334549

Trajkovska Petkoska, A., Daniloski, D., D'Cunha, N. M., Naumovski, N., & Broach, A. T. (2021). Edible packaging: Sustainable solutions and novel trends in food packaging. *Food Research International*, 140, 109981. DOI: 10.1016/j.foodres.2020.109981 PMID: 33648216

Xu, W., Yang, H., Zeng, W., Houghton, T., Wang, X., Murthy, R., Kim, H., Lin, Y., Mignolet, M., Duan, H., Yu, H., Slepian, M., & Jiang, H. (2017). Food-Based Edible and Nutritive Electronics. *Advanced Materials Technologies*, 2(11), 1700181. DOI: 10.1002/admt.201700181

Zeng, Z., Piao, S., Li, L., Zhou, L., Ciais, P., Wang, T., Li, Y., Lian, X., Wood, E. F., Friedlingstein, P., Mao, J., Estes, L. D., Myneni, R. B., Peng, S., Shi, X., Seneviratne, S. I., & Wang, Y. (2017). Climate mitigation from vegetation biophysical feedbacks during the past three decades. *Nature Climate Change*, 7(6), 432–436. DOI: 10.1038/nclimate3299

Zvezdin, C., Di Mauro, E., Rho, D., Santato, C., & Khalil, M. (2020). En route toward sustainable organic electronics. *MRS Energy & Sustainability : a Review Journal*, 7(1), 16. Advance online publication. DOI: 10.1557/mre.2020.16

Chapter 10
Integrating Internet of Things (IoT) With Edible Electronics

Himadri Sekhar Das
https://orcid.org/0000-0002-3509-3388
Haldia Institute of Technology, India

Subir Maity
KIIT University, India

ABSTRACT

The unification of Internet of Things (IoT) accompanying edible electronic devices marks a meaningful advancement in two together science and healthcare. This synergy integrates the relatedness and data processing wherewithal of IoT accompanying the innovative potential of edible electronics, offering unprecedented opportunities for monitoring and improving human health. By implanting IoT sensors and actuators within succulent substrates, in the way that biocompatible materials or ingestible sensors, a smooth and non-obtrusive monitoring whole is worked out. This abstract explores the arising flows, challenges, and potential applications of integrating IoT with edible electronics. From embodied cure to real-occasion well-being monitoring, this unification holds promise for transforming healthcare delivery and improving the feature of life. However, righteous concerns concerning privacy and dossier protection must be addressed to guarantee the responsible deployment concerning this transformational technology.

DOI: 10.4018/979-8-3693-5573-2.ch010

1. INTRODUCTION:

The Internet of Things (IoT) is revolutionizing various sectors, including healthcare, agriculture, and manufacturing, by enabling the interconnection of everyday objects to the internet. Meanwhile, edible electronics is an emerging field that focuses on the development of ingestible electronic devices for health monitoring and therapeutic applications. This chapter explores the convergence of IoT and edible electronics, discussing the potential benefits, challenges, and future directions of this interdisciplinary integration.

The convergence of Internet of Things (IoT) technology with edible electronics represents a transformative leap in the realm of healthcare and medical diagnostics. Edible electronics involve the creation of ingestible devices that are biocompatible, biodegradable, and capable of performing various electronic functions within the human body (Abdulmalek S et al, 2022). When these devices are integrated with IoT, they can transmit real-time data to external systems, enabling continuous monitoring, precise diagnostics, and personalized treatment. This chapter explores the synergies between IoT and edible electronics, discussing their applications, materials and design considerations, power supply solutions, communication protocols, and the challenges and future directions of this innovative field.

Edible electronics are designed to be safe for ingestion and operation within the human body. These devices are constructed from biocompatible and biodegradable materials to ensure they do not harm the body and can be naturally excreted or absorbed. Applications of edible electronics include gastrointestinal monitoring, drug delivery systems, and nutrient absorption tracking (M. Radovanović et al,2023). The integration of IoT with edible electronics enhances their functionality by enabling real-time data transmission and analysis. This synergy allows for continuous health monitoring, early detection and prevention of diseases, precise drug delivery, and improved dietary management. IoT-enabled edible electronics can communicate with external devices such as smartphones, tablets, or dedicated medical equipment, facilitating remote patient monitoring and personalized healthcare (W. A. Jabbar et al.,2019).

The design of IoT-enabled edible electronics involves careful selection of materials and components to ensure biocompatibility, functionality, and safety. Biodegradable polymers, silk fibroin, and magnesium are commonly used materials. The integration of sensors, power sources, and communication modules is crucial for the device's performance. Miniaturization and energy efficiency are key considerations to ensure the devices can operate effectively within the body's environment. Providing a reliable power supply for ingestible devices is a significant challenge. Potential solutions include using biocompatible batteries, energy harvesting from bodily fluids, and wireless power transfer. Advances in miniaturization and energy efficiency are crucial for developing practical IoT-enabled edible electronics (Luke A. Beardslee et al, 2020).

Effective communication between ingestible devices and external IoT systems requires robust and low-power communication protocols (Ramy Ghanim et al,2023). Near Field Communication (NFC), Bluetooth Low Energy (BLE) and Radio Frequency Identification (RFID) are potential options. The choice of protocol depends on few factors such as range, power consumption and data transmission rates. While the integration of IoT with edible electronics holds great promise, several challenges must be addressed. These include ensuring long-term biocompatibility and safety, developing reliable power sources, and creating secure and efficient communication protocols. Future research and development will focus on overcoming these challenges, paving the way for broader adoption and new applications of IoT-enabled edible electronics.

The integration of IoT with edible electronics represents a significant advancement in healthcare technology. By combining the connectivity and data processing capabilities of IoT with the innovative potential of ingestible devices, we can achieve new levels of health monitoring, disease management, and personalized medicine. This interdisciplinary approach has the potential to transform various sectors and improve human health and well-being significantly.

2. OVERVIEW OF IOT AND EDIBLE ELECTRONICS

2.1 Internet of Things (IoT)

The Internet of Things (IoT) refers to the network of physical devices embedded with sensors, software, and other technologies to connect and also exchange data with other devices and through the internet. IoT enables real-time monitoring, data collection and automation, offering significant advantages in various domains such as smart homes, smart cities, and industrial automation. Figure.1. Traditional IoT devices

Figure 1. Traditional IoT devices

2.2 Edible Electronics

Edible electronics involves the creation of electronic devices that are safe to ingest and operate within the human body. These devices are designed using biocompatible and biodegradable materials, ensuring they do not harm the body and can be naturally excreted or absorbed. Applications include gastrointestinal monitoring, drug delivery systems, and nutrient absorption tracking. Edible electronics involves the creation of electronic devices that are safe to ingest and operate within the human body. These devices are designed using biocompatible and biodegradable materials, ensuring they do not harm the body and can be naturally excreted or absorbed. Applications include gastrointestinal monitoring, drug delivery systems, and nutrient absorption tracking. Figure 2. shows the IoT-enabled environmental sensing and Agriculture and different applications.

Figure 2. IoT-enabled environmental sensing and agriculture and different applications.

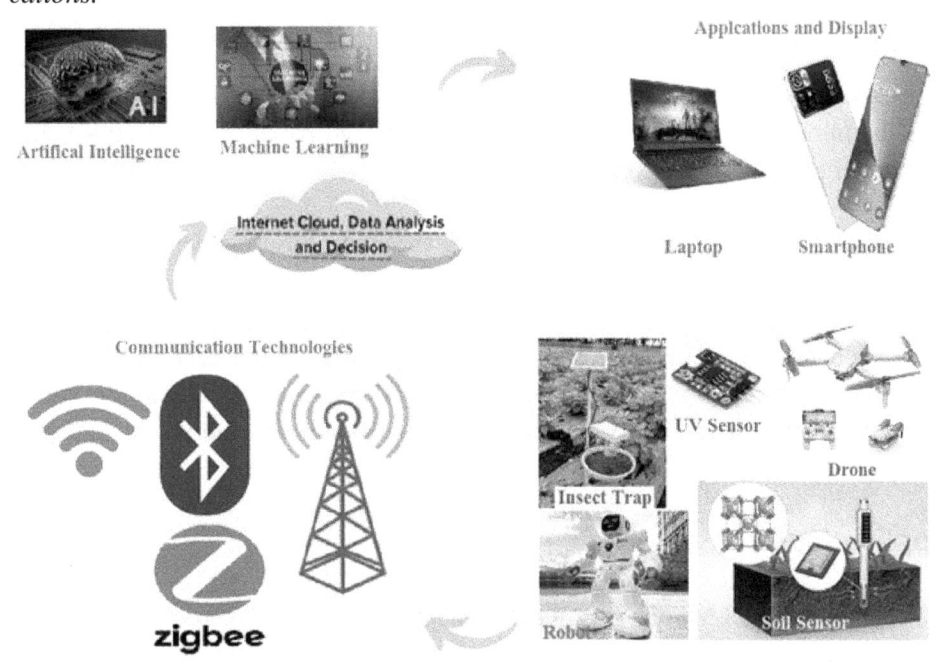

The field of edible electronics is driven by the need for minimally invasive diagnostic and therapeutic tools. Traditional medical devices often require surgical implantation or external attachment, which can be uncomfortable and carry risks of infection. Edible electronics, by contrast, offer a non-invasive alternative that can provide continuous monitoring and treatment from within the body.

2.3. Materials and Design

Biocompatible Materials: The success of edible electronics hinges on the use of materials that are both safe for ingestion and capable of performing electronic functions (Bai, L et al, 2023). Commonly used materials include:

- **Biodegradable Polymers**: polyglycolic acid (PGA) and Polylactic acid (PLA) are often used for their biocompatibility and ability to degrade naturally within the body.
- **Silk Fibroin**: Derived from silkworms, silk fibroin is a protein-based material that is both biocompatible and mechanically robust, making it suitable for flexible electronic devices.

- **Magnesium**: This metal is used in transient electronics due to its biocompatibility and ability to dissolve safely in the body.

Electronic Components: Integrating electronic components into edible devices poses unique challenges (Danielly B. Avancini, et al, 2021). These components must be miniaturized and designed to function in the harsh environment of the gastrointestinal (GI) tract. Key components include:

- **Sensors**: Edible sensors can measure a variety of physiological parameters, such as pH, temperature, and pressure.
- **Power Sources**: Options for powering edible electronics include biocompatible batteries, energy harvesting systems, and wireless power transfer.
- **Antennas**: Flexible and biocompatible antennas are essential for wireless communication between the ingestible device and external monitoring systems.

Applications

Gastrointestinal Monitoring: One of the primary applications of edible electronics is in the monitoring of gastrointestinal health. These devices can provide real-time data on various conditions, such as:

- **pH Levels**: Monitoring pH can help diagnose and manage conditions like acid reflux and ulcers.
- **Temperature**: Changes in temperature can signify inflammation or contamination.
- **Pressure**: Pressure sensors can help diagnose motility disorders by tracking the movement of food through the GI tract.

Drug Delivery Systems: Edible electronics can also be used to enhance drug delivery. By integrating sensors and control systems, these devices can release medication in response to specific physiological conditions, ensuring targeted and timely treatment. Benefits include:

- **Controlled Release**: Medication can be released at the optimal location and time, improving efficacy and reducing side effects.
- **Dose Adjustment**: The system can adjust the dosage based on real-time monitoring of the patient's condition.

Nutrient Absorption Tracking: Edible electronics can monitor the absorption of nutrients, providing valuable data for dietary management. Applications include:

- **Nutrient Tracking**: Monitoring levels of key nutrients can help manage dietary deficiencies and optimize nutrition.
- **Digestive Health**: Data on how well the body is absorbing nutrients can indicate the presence of digestive disorders.

Challenges and Considerations:

Safety and Biocompatibility: Ensuring the safety and biocompatibility of edible electronics is paramount. Materials must be carefully selected to avoid any toxic effects, and the devices must be designed to degrade safely within the body. Regulatory approval from bodies like the FDA is crucial for ensuring these standards are met.

Power Supply: Providing a reliable and safe power supply for ingestible devices is a significant challenge. Potential solutions include:

- **Biocompatible Batteries**: These must be small, efficient, and capable of providing sufficient power for the device's operation.
- **Energy Harvesting**: Systems that can harvest energy from the body's own processes, such as stomach acid or movement, offer a promising alternative.
- **Wireless Power Transfer**: Techniques such as inductive coupling can be used to transfer power wirelessly to the device.

Data Communication: Effective communication between the ingestible device and external monitoring systems is essential for the functionality of edible electronics (Bettinger CJ et al., 2019). This demands the development of strong and depressed-power ideas protocols, in the way that:

- **Bluetooth Low Energy (BLE)**: Offers low power consumption and sufficient range for many applications.
- **Near Field Communication (NFC)**: Suitable for short-range communication, useful for devices that need to transmit data when they are close to a reader.
- **Radio Frequency Identification (RFID)**: Useful for tracking and identification purposes.

Future Directions

Advances in Materials Science: Continued research in materials science will drive the development of more advanced and biocompatible materials for edible electronics. Innovations in biodegradable polymers, organic electronics, and nanomaterials will enable the creation of more sophisticated and reliable devices.

Integration with Artificial Intelligence (AI): The unification of AI with IoT-allowed edible electronics can improve data analysis and in charge. AI algorithms can process vast amounts of health data to identify patterns, predict health issues, and recommend personalized interventions. This synergy can significantly improve healthcare outcomes.

Broader Applications

While healthcare is the primary focus, IoT-enabled edible electronics have potential applications in other fields such as:

- **Agriculture**: Monitoring the health and nutrition of livestock.
- **Food Safety**: Detecting contaminants in food products.
- **Sports**: Tracking athlete nutrition and hydration.

The integration of IoT with edible electronics represents a promising frontier in technology and healthcare. By combining the connectivity and data processing capabilities of IoT with the innovative potential of ingestible devices, we can achieve unprecedented levels of health monitoring, disease management, and personalized medicine. However, addressing technical, ethical, and regulatory challenges is essential to realize the full potential of this interdisciplinary innovation. As research and development continue, IoT-enabled edible electronics have the potential to transform various sectors, offering new opportunities for enhancing human health and well-being.

2.4. Synergies Between IoT and Edible Electronics

The integration of IoT with edible electronics presents unique opportunities for enhancing healthcare and dietary management. By leveraging IoT's connectivity and data processing capabilities, edible electronics can transmit real-time data from within the body to external devices, enabling continuous health monitoring and timely interventions.

The integration of IoT with edible electronics presents unique opportunities for enhancing healthcare and dietary management. By leveraging IoT's connectivity and data processing capabilities, edible electronics can transmit real-time data from within the body to external devices, enabling continuous health monitoring and timely interventions. Real-Time Health Monitoring: Edible electronics equipped with sensors can monitor physiological parameters such as pH levels, temperature, and nutrient absorption (Wu JY et al, 2023). When joined accompanying IoT, this data maybe sent to healthcare providers for real-time study, allowing for early discovery of health issues and personalized situation plans.

> **Continuous Monitoring:** IoT-enabled edible electronics can provide continuous, real-time monitoring of a patient's physiological state. This is particularly beneficial for managing chronic diseases such as diabetes or hypertension, where ongoing data is critical for effective management. For instance, ingestible glucose sensors can continuously track blood sugar levels, sending alerts to patients and healthcare providers when levels become abnormal.
>
> **Early Detection and Prevention:** By continuously monitoring internal health metrics, IoT-enabled edible electronics can detect early signs of potential health issues before they become critical. For example, subtle changes in gastric pH or temperature could indicate the onset of gastrointestinal disorders. Early discovery allows for prompt intervention, conceivably preventing more harsh health questions.
>
> **Drug Delivery and Management:** IoT-enabled edible electronics can facilitate precise drug delivery systems. By monitoring the body's response to medication in real-time, these devices can adjust dosages automatically, ensuring optimal therapeutic outcomes. This integration can also help track medication adherence and manage chronic conditions more effectively.

Smart Pills: Smart pills are an excellent example of IoT-enabled drug delivery systems. These ingestible devices can release medication at specific times or in response to certain physiological conditions. For instance, a smart pill could release insulin when it detects high blood glucose levels. The pill can communicate with external devices to confirm the delivery and effectiveness of the medication, providing valuable data for healthcare providers.

> **Medication Adherence:** One of the challenges in chronic disease management is ensuring patients adhere to their medication schedules. IoT-enabled edible electronics can monitor and record when a patient takes their medication, send-

ing reminders or alerts if a dose is missed. This data can also be shared with healthcare providers to track adherence and adjust treatment plans as necessary.

Dietary and Nutrient Tracking: Edible electronics can monitor nutrient intake and digestive health, providing valuable data for dietary management (Ulfa M et al, 2022). IoT integration allows this data to be synced with fitness trackers and health apps, offering personalized dietary recommendations and promoting better nutritional habits.

Personalized Nutrition: IoT-enabled edible electronics can track the absorption of specific nutrients and identify deficiencies. This data can be used to create personalized nutrition plans tailored to an individual's needs. For example, if the device detects low levels of vitamin B12, it can recommend dietary adjustments or supplements to address the deficiency.

Fitness and Wellness Integration: By integrating data from edible electronics with fitness trackers and health apps, individuals can gain a comprehensive view of their overall health. This holistic approach allows for better management of diet and exercise, leading to improved fitness and wellness outcomes. For instance, tracking hydration levels and nutrient intake can help athletes optimize their performance and recovery.

Advanced Diagnostics and Treatment: The synergy between IoT and edible electronics can lead to advanced diagnostic and treatment capabilities that are less invasive and more accurate than traditional methods.

Gastrointestinal Diagnostics: Edible electronics can provide detailed information about the gastrointestinal tract, which is often difficult to access with conventional methods. IoT integration allows for real-time data transmission, enabling dynamic analysis and more accurate diagnostics. This maybe particularly valuable for labeling conditions like irritable bowel syndrome (IBS), Crohn's disease, and different GI disorders.

Remote Patient Monitoring: IoT authorized edible electronics can speed detached patient monitoring, lowering the need for frequent hospital visits and permissive care in more convenient scenes. This is specifically advantageous for elderly victims or those accompanying mobility issues. Real-time data transmission allows healthcare providers to monitor subjects' energy remotely and happen when inevitable.

Enhancing Clinical Research: The integration of IoT and edible electronics can significantly enhance clinical research by providing continuous, real-time data collection and improving the accuracy of clinical trials.

Continuous Data Collection: Traditional clinical trials often rely on periodic data collection, which can miss critical information between visits. IoT-enabled edible electronics can provide continuous data collection, offering a more comprehensive view of a patient's response to a treatment or intervention.

Improved Compliance and Accuracy: IoT-enabled edible electronics can improve patient compliance in clinical trials by ensuring that medication is taken as prescribed and accurately recording when it is ingested. This leads to more reliable data and better outcomes in clinical research (Chai PR et al, 2022). The integration of IoT with edible electronics represents a significant advancement in healthcare and dietary management. By combining the real-time monitoring and data processing capabilities of IoT with the innovative potential of ingestible devices, we can achieve new levels of health monitoring, disease management, and personalized medicine. The synergies between IoT and edible electronics offer numerous benefits, including continuous health monitoring, precise drug delivery, personalized nutrition, advanced diagnostics, and enhanced clinical research. However, to fully realize these benefits, it is essential to address the technical, ethical, and regulatory challenges associated with this interdisciplinary innovation. As research and development continue, the integration of IoT and edible electronics has the potential to transform various sectors and improve human health and well-being significantly.

3. TECHNICAL CONSIDERATIONS

3.1 Materials and Design

The design of IoT-enabled edible electronics requires careful consideration of the materials and components used, ensuring they are safe for ingestion and capable of performing necessary electronic functions. This section covers the essential aspects of materials selection and device design, focusing on biocompatibility, power supply, and communication protocols.

Biocompatible Materials: The success of edible electronics hinges on the use of materials that are both safe for ingestion and capable of performing electronic functions. These materials must be biocompatible, biodegradable, and able to withstand the harsh environment of the gastrointestinal (GI) tract (Alam F et al, 2024).

Biodegradable Polymers: Biodegradable polymers are commonly used in edible electronics due to their ability to safely degrade within the body. Examples include:

- **Polylactic Acid (PLA)**: A biodegradable thermoplastic derived from renewable resources such as corn starch. PLA is biocompatible and suitable for various medical applications.
- **Polyglycolic Acid (PGA)**: A biodegradable polymer known for its high strength and fast degradation rate. PGA is often used in surgical sutures and can be adapted for edible electronics.
- **Polycaprolactone (PCL)**: Another biodegradable polymer with excellent biocompatibility, PCL degrades more slowly, making it suitable for applications requiring longer device stability.

Silk fibroin, derived from silkworms, is a protein-based material that offers biocompatibility, mechanical robustness, and biodegradability. It can be processed into various forms, such as films and hydrogels, making it versatile for use in flexible electronic devices. Magnesium is used in transient electronics due to its biocompatibility and ability to dissolve safely in the body. Magnesium-based components can serve as electrodes or structural elements in edible electronics, providing necessary functionality while ensuring safe degradation.

Integrating electronic components into edible devices poses unique challenges. These components must be miniaturized and designed to function in the harsh environment of the gastrointestinal (GI) tract. Key components include sensors, power sources, and antennas. Edible sensors can measure various physiological parameters, such as pH, temperature, and pressure. These sensors must be designed to operate reliably in the GI tract and provide accurate data for health monitoring and diagnostics.

- **pH Sensors**: Essential for monitoring the acidity levels in the stomach and intestines, helping diagnose conditions like acid reflux and ulcers.
- **Temperature Sensors**: Used to discover changes in internal body temperature, which can signify swelling or infection.
- **Pressure Sensors**: Useful for tracking the movement of food through the GI tract, aiding in the diagnosis of motility disorders.
    Power Sources: Providing a reliable power supply for ingestible devices is a significant challenge. Potential solutions include biocompatible batteries, energy harvesting systems, and wireless power transfer.
- **Biocompatible Batteries**: These batteries must be small, efficient, and capable of providing sufficient power for the device's operation. Zinc-based

and magnesium-based batteries are commonly explored for their safety and biodegradability.
- **Energy Harvesting**: Systems that can harvest energy from the body's own processes, such as stomach acid or movement, offer a promising alternative. For example, glucose-based biofuel cells can generate power from the glucose present in the digestive tract.
- **Wireless Power Transfer**: Techniques such as inductive coupling can be used to transfer power wirelessly to the device. This approach eliminates the need for onboard batteries, reducing the device's size and complexity.

Antennas: Flexible and biocompatible antennas are essential for wireless communication between the ingestible device and external monitoring systems. These antennas must be designed to operate effectively within the body's environment, ensuring reliable data transmission (Aliqab K et al, 2023).

Flexible Printed Antennas: Made from biocompatible conductive inks and printed onto flexible substrates, these antennas can conform to the body's internal structures.

Miniaturized Coils: Used for near-field communication and wireless power transfer, these coils must be designed to maximize efficiency while remaining small enough to be safely ingested.

Communication Protocols: Effective ideas between ingestible devices and external IoT systems demands strong and low-capacity communication codes. The choice of contract depends on factors in the way that range, capacity consumption, and data transmission rates.

Bluetooth Low Energy (BLE) offers low power consumption and sufficient range for many applications. BLE is suitable for continuous monitoring applications, where the device needs to transmit data frequently but must conserve power. Near Field Communication (NFC) is suitable for short-range communication, useful for devices that need to transmit data when they are close to a reader. NFC is ideal for applications where the ingestible device is designed to interact with a smartphone or dedicated reader device during specific time intervals. Radio Frequency Identification (RFID) is useful for tracking and identification purposes. Passive RFID tags can operate without an onboard power source, drawing energy from the reader's signal. This makes RFID ideal for applications where the device only needs to transmit data occasionally, such as during excretion.

The design of IoT-enabled edible electronics requires a multidisciplinary approach, combining expertise in materials science, electronics, and biomedical engineering. By selecting biocompatible and biodegradable materials, integrating miniaturized electronic components, and employing robust communication protocols, it is possible to create ingestible devices that are safe, effective, and capable of providing valuable health data. These devices have the potential to revolutionize healthcare by enabling continuous monitoring, precise drug delivery, and advanced diagnostics, finally improving the patient outcomes and quality of life.

The design of IoT-enabled edible electronics requires careful selection of materials that are safe for ingestion, biocompatible, and capable of performing the desired electronic functions. Common materials include biodegradable polymers, silk fibroin, and magnesium. These materials must also support the integration of sensors, antennas, and power sources.

3.2 Power Supply and Energy Harvesting

Providing a reliable power supply for ingestible devices is a significant challenge due to the need for biocompatibility, safety, and sufficient energy density. Various solutions are being explored to address these challenges, including biocompatible batteries, energy harvesting from bodily fluids, and wireless power transfer. Advances in miniaturization and energy efficiency are crucial for developing practical IoT-enabled edible electronics (Yang S Y et al, 2021).

Biocompatible Batteries: Biocompatible batteries must be safe for ingestion, capable of providing sufficient power, and ideally biodegradable to prevent any long-term impact on the body.

- **Zinc-based Batteries**: Zinc is a biocompatible material commonly used in transient electronics. Zinc-air batteries, which generate electricity through the reaction of zinc with oxygen, offer high energy density and are safe for ingestion.
- **Magnesium-based Batteries**: Magnesium is another biocompatible material that can be safely ingested. Magnesium batteries can dissolve in bodily fluids, making them suitable for temporary use in ingestible devices.
- **Solid-state Batteries**: These batteries use solid electrolytes, which can be made from biocompatible materials. Solid-state batteries are safer than liquid electrolyte batteries as they are less prone to leakage, making them more suitable for ingestible applications.

Energy Harvesting: Energy harvesting involves capturing and converting energy from the body's natural processes or environment into electrical energy to power ingestible devices. This approach can extend the operational life of the device and reduce the need for large onboard power sources. Figure.3. shows the different types of Energy Harvesting Techniques.

Figure 3. Different types of energy harvesting techniques

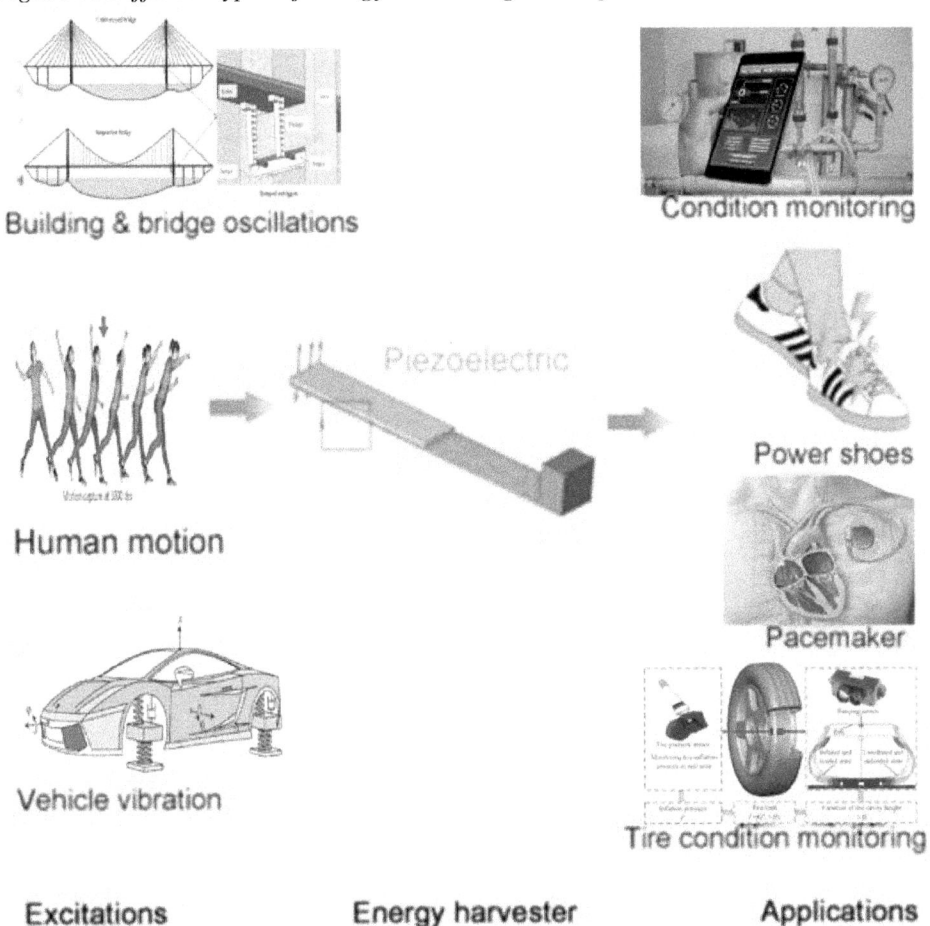

Building & bridge oscillations

Condition monitoring

Human motion

Power shoes

Pacemaker

Vehicle vibration

Tire condition monitoring

Excitations   Energy harvester   Applications

Biofuel Cells: Biofuel cells generate electricity by exploiting biochemical reactions in the body, such as glucose oxidation. Glucose biofuel cells can harness energy from the glucose present in the digestive system, providing a continuous power source for ingestible electronics. Figure.4. shows the Cycle of biomass energy.

Figure 4. Cycle of biomass energy

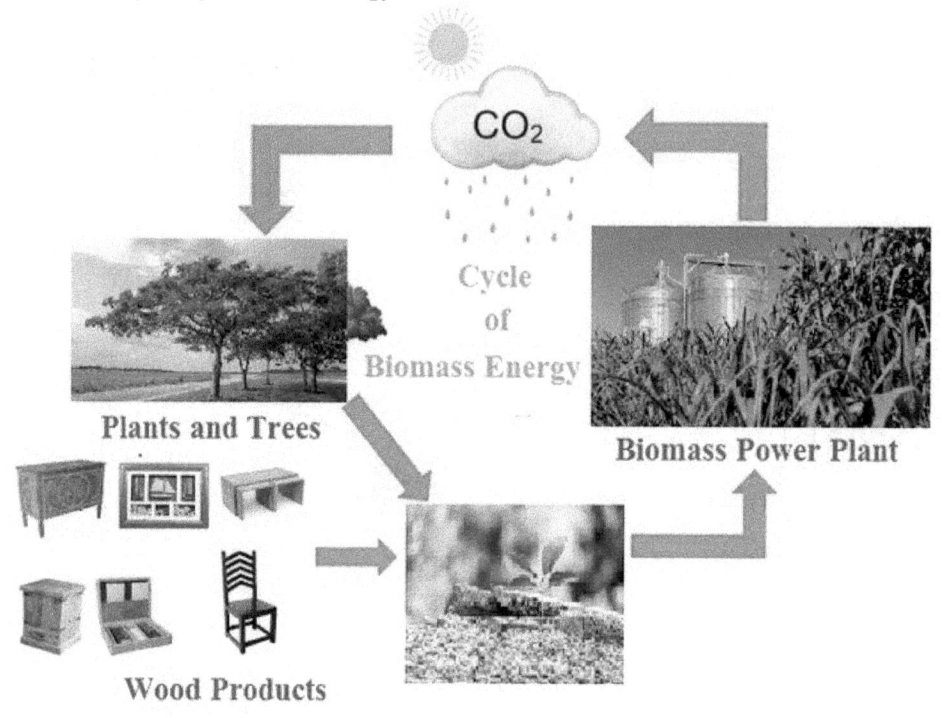

Thermoelectric Generators (TEGs): TEGs convert heat gradients into electrical energy. The temperature difference between the body's internal environment and the ambient surroundings can be harnessed by TEGs to generate power for ingestible devices.

Mechanical Energy Harvesting: Mechanical energy from bodily movements, such as peristalsis in the GI tract, can be converted into electrical energy using piezo-electric materials. These materials generate electricity in response to mechanical stress, providing a potential power source for ingestible electronics.

Wireless Power Transfer: Wireless power transfer involves transmitting power to ingestible devices without direct electrical contacts. This method can power devices over short distances and is particularly useful for applications requiring intermittent data transmission or operation.

- **Inductive Coupling**: Inductive coupling transfers power through electromagnetic induction between coils in the external transmitter and the ingestible device. This method is commonly used in wireless charging technologies and can be adapted for short-range power transfer to ingestible electronics.

- **Radio Frequency (RF) Energy Harvesting**: RF energy harvesting captures and converts ambient RF signals into electrical power. External RF transmitters can be used to deliver power to ingestible devices, which then use rectifying circuits to convert RF energy into usable electrical power.
- **Near Field Communication (NFC)**: NFC-based power transfer uses magnetic field induction to power devices at very close ranges. This method is suitable for applications where the ingestible device needs to communicate with or be powered by a device in close proximity, such as a smartphone.

Advances in Miniaturization and Energy Efficiency: Advances in miniaturization and energy efficiency are critical for the development of practical IoT-enabled edible electronics. These advances allow for smaller, more energy-efficient components that can fit into ingestible formats without compromising functionality.

Microfabrication Techniques: Microfabrication techniques, such as photolithography and 3D printing, enable the creation of tiny, precise components that can be integrated into ingestible devices. These techniques are essential for producing miniaturized sensors, circuits, and power sources.

Low-power Electronics: Developing low-power electronic components, such as microcontrollers and communication modules, helps extend the battery life of ingestible devices. Advances in semiconductor technology and power management algorithms contribute to reducing the overall power consumption of these devices.

Energy-efficient Communication Protocols: Implementing energy-efficient communication protocols, such as Bluetooth Low Energy (BLE) and Near Field Communication (NFC), reduces the power required for data transmission. These protocols are designed to minimize energy consumption while maintaining reliable communication between the ingestible device and external systems.

Providing a reliable power supply for IoT-enabled edible electronics involves a combination of innovative materials, energy harvesting techniques, and wireless power transfer methods. Advances in biocompatible batteries, biofuel cells, thermoelectric generators, and wireless power transfer are essential for developing safe and effective ingestible devices. Additionally, miniaturization and energy efficiency improvements are critical for integrating these technologies into practical, functional IoT-enabled edible electronics. These advancements will enable continuous health monitoring, precise drug delivery, and other transformative healthcare applications, ultimately improving patient outcomes and quality of life.

Providing a reliable power supply for ingestible devices is a significant challenge. Potential solutions include using biocompatible batteries, energy harvesting from bodily fluids, and wireless power transfer. Advances in miniaturization and energy efficiency are crucial for developing practical IoT-enabled edible electronics.

3.3 Communication Protocols

Effective communication between ingestible devices and external IoT systems is crucial for the functionality and utility of IoT-enabled edible electronics. The choice of communication protocol is influenced by factors such as range, power consumption, and data transmission rates. Key communication protocols include Bluetooth Low Energy (BLE), Near Field Communication (NFC), and Radio Frequency Identification (RFID) (Bettinger C J et al, 2019).

Bluetooth Low Energy (BLE): Bluetooth Low Energy (BLE) is a Wi-Fi communication obligation devised for short-range ideas with depressed capacity consumption. It is widely used in wearable devices and IoT applications due to its balance between power efficiency and data transmission capabilities.

Range: BLE typically supports communication ranges up to 100 meters, although practical ranges are often shorter in real-world conditions.

Power Consumption: BLE is designed to operate with minimal power, making it suitable for battery-powered ingestible devices. The low energy consumption extends the device's operational life.

Data Transmission Rates: BLE can achieve data transmission rates up to 2 Mbps, which is sufficient for most health monitoring applications. The protocol supports periodic data transmission, allowing ingestible devices to send data at regular intervals.

BLE is ideal for continuous health monitoring applications where the ingestible device needs to transmit data frequently but must conserve power. It allows for seamless integration with smartphones and other BLE-enabled devices, facilitating real-time data access and analysis.

Near Field Communication (NFC): Near Field Communication (NFC) is a temporary ideas agreement that enables data exchange between devices over distances of any centimetres It is commonly used in contactless payment systems and identification cards.

Range: NFC operates at very short ranges, typically within 4 cm. This limitation ensures secure communication but requires the external reader to be in close proximity to the ingestible device.

Power Consumption: NFC is highly energy-efficient and can operate in passive mode, where the ingestible device does not need its own power source. The external reader generates the necessary electromagnetic field to power the device and facilitate communication.

Data Transmission Rates: NFC supports data rates up to 424 kbps. While lower than BLE, these rates are adequate for applications that require sporadic data transmission, such as confirming the ingestion of a pill or retrieving data from an ingestible sensor.

NFC is particularly useful for applications where the ingestible device needs to communicate with a reader only at specific times, such as during a doctor's visit or when paired with a smartphone for data retrieval.

Radio Frequency Identification (RFID): Radio Frequency Identification (RFID) is a electronics that uses electromagnetic fields to recognize and path objects. It is established in management, inventory administration, and approach control.

Range: RFID ranges vary depending on the type of RFID system used. Passive RFID tags have short ranges, typically up to a few meters, while active RFID systems can achieve ranges of several meters.

Power Consumption: Passive RFID tags do not have an onboard power source and are powered by the electromagnetic field generated by the RFID reader. This makes passive RFID tags suitable for ingestible applications where long battery life is essential. Active RFID tags contain a battery and can transmit signals over longer distances but at the cost of higher power consumption.

Data Transmission Rates: RFID data transmission rates are generally lower than those of BLE and NFC, but they are sufficient for identification and tracking purposes.

RFID is useful for applications that require tracking and identification of ingestible devices. Passive RFID tags are particularly beneficial for disposable ingestible devices that do not require continuous data transmission but need to communicate data during specific events, such as passing through an RFID reader.

Choosing the Right Communication Protocol: Selecting the appropriate communication protocol for IoT-enabled edible electronics involves considering the specific requirements of the application, including range, power consumption, and data transmission needs.

Range: For applications requiring longer communication ranges, BLE may be the preferred option. For applications with very short-range communication requirements, such as data retrieval during a medical examination, NFC is suitable. RFID can be chosen based on the desired range and power constraints, with passive RFID suitable for short-range applications and active RFID for longer ranges.

Power Consumption: If minimizing power consumption is critical, NFC and passive RFID are ideal due to their low energy requirements. BLE, while efficient, consumes more power compared to NFC and passive RFID but offers greater range and data transmission rates.

Data Transmission Rates: For applications needing higher data transmission rates, such as continuous health monitoring, BLE is appropriate. NFC and RFID, with lower data rates, are suitable for applications that require less frequent data transmission.

Effective communication between ingestible devices and external IoT systems is essential for the functionality of IoT-enabled edible electronics. The choice of communication protocol—BLE, NFC, or RFID—depends on the specific needs of the application, including range, power consumption, and data transmission rates. BLE is well-suited for continuous monitoring and longer-range communication, while NFC is ideal for short-range, low-power applications. RFID offers flexibility with passive and active options, catering to a range of distances and power requirements. By selecting the appropriate communication protocol, developers can ensure reliable and efficient data transmission, enhancing the overall performance and utility of IoT-enabled edible electronics.

Effective communication between ingestible devices and external IoT systems requires robust and low-power communication protocols. Bluetooth Low Energy (BLE), Near Field Communication (NFC), and Radio Frequency Identification (RFID) are potential alternatives. The choice of agreement depends on determinants in the way that range, capacity devouring, and dossier broadcast rates.

4. APPLICATIONS IN HEALTHCARE

4.1 Gastrointestinal Monitoring

IoT-enabled edible electronics can revolutionize gastrointestinal (GI) health monitoring. Devices equipped with pH, temperature, and pressure sensors can track digestive processes and detect abnormalities such as ulcers, inflammation, and motility disorders. Real-time data transmission to healthcare providers enables early diagnosis and personalized treatment. IoT-enabled edible electronics represent a cutting-edge innovation poised to revolutionize gastrointestinal (GI) health monitoring (Chong K P et al, 2021). One of the pioneering features of such devices is their ability to measure pH levels within the digestive tract, offering invaluable insights into the acidity or alkalinity of various regions within the gastrointestinal system.

Traditional methods of pH monitoring in the GI tract often involve uncomfortable or invasive procedures, such as inserting catheters or swallowing capsules attached to external monitoring equipment. However, IoT-enabled edible electronics eliminate the need for such invasive measures by integrating miniature sensors directly into digestible devices. These devices can be ingested like ordinary pills, allowing for non-intrusive and hassle-free monitoring of pH levels throughout the digestive system. By continuously tracking pH levels in real-time, these devices can provide clinicians and patients with a comprehensive understanding of GI health. For example, fluctuations in pH levels can indicate conditions such as acid reflux, gastritis, or even the presence of certain pathogens. Early discovery of aforementioned abnormalities can further up-to-the-minute interventions and embodied situation plans, eventually improving patient effects and lowering the risk of complications.

Moreover, IoT-enabled edible electronics offer unparalleled convenience and accessibility. Patients can simply swallow a small, unobtrusive device and carry on with their daily activities while the device wirelessly transmits data to a smartphone or other connected device. This seamless integration with existing technology platforms enables remote monitoring and data sharing, allowing healthcare providers to remotely track patients' GI health and make informed decisions in real-time.

Furthermore, the biocompatible materials used in the construction of these devices ensure safe ingestion and passage through the digestive system without causing harm or discomfort. Once the monitoring period is complete, the device can be naturally excreted from the body, eliminating the need for retrieval procedures and minimizing patient inconvenience. In addition to clinical applications, IoT-enabled edible electronics hold promise for research purposes, enabling scientists to gather valuable data on digestive processes and disease mechanisms. The wealth of information obtained from these devices can contribute to the development of novel therapies, diagnostic tools, and preventive strategies for a wide range of GI disorders.

4.2 Chronic Disease Management

For patients with chronic conditions like diabetes, hypertension, or heart disease, constant monitoring is vital. Edible electronics can provide real-time data on glucose levels, blood pressure, and cardiac function. Integrating this data with IoT platforms allows for remote monitoring and timely interventions, improving patient outcomes (Peyroteo M et al,2021).

Continuous monitoring is paramount for managing chronic conditions such as diabetes, hypertension, and heart disease. IoT-enabled edible electronics offer a groundbreaking solution by providing real-time data on key health metrics like glucose levels, blood pressure, and cardiac function. By integrating these devices with IoT platforms, healthcare providers can remotely monitor patients' health status and intervene promptly when necessary, significantly enhancing patient outcomes (K M A et al, 2024).

For individuals with diabetes, the ability to monitor glucose levels non-invasively and continuously through edible electronics eliminates the need for frequent finger pricks or invasive procedures. Real-time data on blood glucose levels enables patients and clinicians to adjust medication dosages, dietary habits, and lifestyle factors promptly, thereby improving glycemic control and reducing the risk of complications. Similarly, continuous monitoring of blood pressure and cardiac function is essential for managing hypertension and heart disease effectively. Edible electronics equipped with sensors can track these vital signs in real-time, alerting healthcare providers to any deviations from normal ranges. Timely interventions, such as medication adjustments or lifestyle modifications, can be initiated remotely, helping to prevent adverse events such as heart attacks or strokes.

Integration with IoT platforms enables seamless data transmission, storage, and analysis, allowing for personalized and proactive healthcare interventions. Patients benefit from improved disease management, reduced hospitalizations, and enhanced quality of life, while healthcare providers can deliver more efficient and cost-effective care.

4.3 Post-Surgical Monitoring

After surgical procedures, monitoring the patient's recovery is critical to prevent complications. Edible electronics can track vital signs and detect early signs of infection or complications. IoT integration ensures that this data is available to healthcare providers in real-time, facilitating prompt medical responses (Amin T et al, 2021).

Post-surgical monitoring plays a pivotal role in ensuring patients' safe recovery and early detection of complications. Edible electronics represent a groundbreaking approach to this task, as they can continuously track vital signs and detect early signs of infection or other complications following surgical procedures. By integrating these devices with IoT platforms, healthcare providers can access real-time data on patients' recovery progress, enabling prompt medical responses and interventions when necessary. Edible electronics embedded with sensors can monitor vital signs such as temperature, heart rate, and respiratory rate, providing clinicians with comprehensive insights into patients' post-operative status. Any deviations from normal parameters can be promptly identified, allowing healthcare providers to intervene quickly and mitigate potential risks.

Moreover, these devices can also detect early signs of infection, inflammation, or other post-surgical complications, such as leakage or bleeding at the surgical site. By continuously monitoring biochemical markers or physiological changes associated with these conditions, edible electronics enable early intervention, reducing the likelihood of complications and improving patient outcomes. Integration with IoT platforms ensures seamless data transmission and accessibility for healthcare providers, regardless of their location. Real-time monitoring allows for timely medical responses, such as adjusting medication regimens, initiating additional diagnostic tests, or scheduling follow-up appointments, thereby optimizing patient care and recovery.

5. ETHICAL AND REGULATORY CONSIDERATIONS

5.1 Privacy and Data Security

The accumulation and broadcast of health data through IoT-authorized edible electronic devices raise significant solitude and protection concerns. Ensuring that patient data is shielded from unauthorized approach and breaches is principal. Robust encryption and secure communication codes are essential to safeguard patient news (Butpheng, C et al,2020).

The utilization of IoT-enabled edible electronics for health monitoring introduces important considerations regarding privacy and data security. Collecting and transmitting sensitive health data necessitates robust measures to safeguard patient privacy and prevent unauthorized access or breaches. Central to protecting patient data is the implementation of robust encryption techniques and secure communication protocols. By encrypting data at both the device and transmission levels, healthcare providers can ensure that patient information remains confidential and protected from interception or tampering. Additionally, incorporating authentication mechanisms, such as user authentication and device authentication, helps verify the identities of authorized users and devices, further enhancing security.

5.2 Safety and Biocompatibility

The safety of ingestible devices is a critical consideration. Regulatory bodies such as the FDA and EMA must establish stringent guidelines for the biocompatibility, degradation, and excretion of these devices. Continuous monitoring and evaluation of the long-term effects of ingestible electronics on human health are necessary (Thwaites P A et al, 2024).

Ensuring the security and biocompatibility of ingestible schemes is superior to their extensive enactment and use in healthcare. Regulatory physique like the FDA (Food and Drug Administration) and EMA (European Medicines Agency) play a important duty in beginning tight directions and guidelines to rule the incident, production, and arrangement of these ploys (Joppi R et al, 2019). One of the primary concerns is biocompatibility, ensuring that the materials used in the construction of ingestible electronics are non-toxic and compatible with the human body. Regulatory agencies require thorough testing to assess the potential risks of allergic reactions, tissue damage, or other adverse effects.

Additionally, the degradation and excretion of ingestible devices are important considerations. Devices must degrade safely within the digestive tract and be excreted from the body without causing harm or obstruction. Studies evaluating the degradation kinetics and biocompatibility of materials over time are essential to ensure patient safety. Continuous monitoring and evaluation of the long-term effects of ingestible electronics on human health are also necessary. Post-market surveillance and clinical studies enable researchers to assess the safety and efficacy of these devices in real-world settings and identify any unforeseen risks or complications.

5.3 Ethical Issues

The use of IoT-enabled edible electronics raises ethical questions regarding consent, autonomy, and the potential for misuse. Ensuring informed consent and maintaining patient autonomy in the use of these devices is essential. Additionally, guidelines must be established to prevent the misuse of health data and technology.

6. FUTURE DIRECTIONS

6.1 Advances in Materials Science

Continued research in materials science is crucial for developing more advanced and biocompatible materials for edible electronics. Innovations in biodegradable polymers, organic electronics, and nanomaterials will enable the creation of more sophisticated and reliable devices.

6.2 Integration with Artificial Intelligence (AI)

The unification of AI accompanying IoT-enabled succulent VCRs can enhance data reasoning and decision-making. AI algorithms can process far-reaching amounts of strength data to label patterns, foresee energy issues, and recommend embodied invasions. This synergy can considerably better healthcare outcomes.

6.3 Broader Applications

While healthcare is the primary focus, IoT-enabled edible electronics have potential applications in other fields such as agriculture (e.g., monitoring livestock health), food safety (e.g., detecting contaminants), and sports (e.g., tracking athlete nutrition and hydration).

7. CONCLUSION

The integration of IoT with edible electronics represents a promising frontier in technology and healthcare. By combining the connectivity and data processing capabilities of IoT with the innovative potential of ingestible devices, we can achieve unprecedented levels of health monitoring, disease management, and personalized medicine. However, addressing technical, ethical, and regulatory challenges is essential to realize the full potential of this interdisciplinary innovation. As research and development continue, IoT-enabled edible electronics have the potential to transform various sectors, offering new opportunities for enhancing human health and well-being.

REFERENCES

Abdulmalek, S., Nasir, A., Jabbar, W. A., Almuhaya, M. A. M., Bairagi, A. K., Khan, M. A.-M., & Kee, S.-H. (2022, October 11). IoT-Based Healthcare-Monitoring System towards Improving Quality of Life: A Review. *Healthcare (Basel)*, 10(10), 1993. DOI: 10.3390/healthcare10101993 PMID: 36292441

Alam, F., Ashfaq Ahmed, M., Jalal, A., Siddiquee, I., Adury, R., Hossain, G., & Pala, N. (2024, March 30). Recent Progress and Challenges of Implantable Biodegradable Biosensors. *Micromachines*, 15(4), 475. DOI: 10.3390/mi15040475 PMID: 38675286

Aliqab, K., Nadeem, I., & Khan, S. R. (2023, July 21). A Comprehensive Review of In-Body Biomedical Antennas: Design, Challenges and Applications. *Micromachines*, 14(7), 1472. DOI: 10.3390/mi14071472 PMID: 37512782

Amin, T., Mobbs, R. J., Mostafa, N., Sy, L. W., & Choy, W. J. (2021, July 20). Wearable devices for patient monitoring in the early postoperative period: A literature review. *mHealth*, 7, 50. DOI: 10.21037/mhealth-20-131 PMID: 34345627

Avancini, D. B.. (2021). A new IoT-based smart energy meter for smart grids, Volume 45, Issue 1, Special Issue: Smart. *Energy Technology (Weinheim)*, •••, 189–202.

Bai, L., Liu, M., & Sun, Y. (2023). Overview of Food Preservation and Traceability Technology in the Smart Cold Chain System. *Foods*, 2023(12), 2881. DOI: 10.3390/foods12152881 PMID: 37569150

Beardslee, L. A., Banis, G. E., Chu, S., Liu, S., Chapin, A. A., Stine, J. M., Pasricha, P. J., & Ghodssi, R. (2020). Ingestible Sensors and Sensing Systems for Minimally Invasive Diagnosis and Monitoring: The Next Frontier in Minimally Invasive Screening. *ACS Sensors*, 5(4), 891–910. DOI: 10.1021/acssensors.9b02263 PMID: 32157868

Bettinger, C. J.. (2019, April). Edible hybrid microbial-electronic sensors for bleeding detection and beyond. *Hepatobiliary Surgery and Nutrition*, 8(2), 157–160. DOI: 10.21037/hbsn.2018.11.14 PMID: 31098367

Bettinger, C. J.. (2019, April). Edible hybrid microbial-electronic sensors for bleeding detection and beyond. *Hepatobiliary Surgery and Nutrition*, 8(2), 157–160. DOI: 10.21037/hbsn.2018.11.14 PMID: 31098367

Butpheng, C., Yeh, K.-H., & Xiong, H. (2020). Security and Privacy in IoT-Cloud-Based e-Health Systems—A Comprehensive Review. *Symmetry*, 2020(12), 1191. DOI: 10.3390/sym12071191

Chai, P. R., Vaz, C., Goodman, G. R., Albrechta, H., Huang, H., Rosen, R. K., Boyer, E. W., Mayer, K. H., & O'Cleirigh, C. (2022, February 28). Ingestible electronic sensors to measure instantaneous medication adherence: A narrative review. *Digital Health*, 8, 20552076221083119. DOI: 10.1177/20552076221083119 PMID: 35251683

Chong, K. P., & Woo, B. K. P. (2021, March 28). Emerging wearable technology applications in gastroenterology: A review of the literature. *World Journal of Gastroenterology*, 27(12), 1149–1160. DOI: 10.3748/wjg.v27.i12.1149 PMID: 33828391

Jabbar, W. A., Kian, T. K., Ramli, R. M., Zubir, S. N., Zamrizaman, N. S. M., Balfaqih, M., Shepelev, V., & Alharbi, S. (2019). Design and Fabrication of Smart Home With Internet of Things Enabled Automation System. *IEEE Access : Practical Innovations, Open Solutions*, 7, 144059–144074. DOI: 10.1109/ACCESS.2019.2942846

Joppi R et al, (2019) Food and Drug Administration vs European Medicines Agency: Review times and clinical evidence on novel drugs at the time of approval. Br J Clin Pharmacol. 2020 Jan;86(1):170-174. . Epub 2019 Dec 16. PMID: 31657044; PMCID: PMC6983504.DOI: 10.1111/bcp.14130

K M A et al. (2024, March 14). Internet of Things enabled open source assisted real-time blood glucose monitoring framework. *Scientific Reports*, 14(1), 6151. DOI: 10.1038/s41598-024-56677-z PMID: 38486038

Peyroteo, M., Ferreira, I. A., Elvas, L. B., Ferreira, J. C., & Lapão, L. V. (2021, December 21). Remote Monitoring Systems for Patients With Chronic Diseases in Primary Health Care: Systematic Review. *JMIR mHealth and uHealth*, 9(12), e28285. DOI: 10.2196/28285 PMID: 34932000

Radovanović, M., ... (2023), "Edible electronics components for biomedical applications," 2023 22nd International Symposium INFOTEH-JAHORINA (INFOTEH), East Sarajevo, Bosnia and Herzegovina, 2023, pp. 1-4, DOI: 10.1109/INFOTEH57020.2023.10094196

Ramy Ghanim et al,(2023),Communication protocols integrating wearables, ingestibles, and implantables for closed-loop therapies,Device,1, 3,100092,ISSN 2666-9986,DOI: 10.1016/j.device.2023.100092

Thwaites, P. A., Yao, C. K., Halmos, E. P., Muir, J. G., Burgell, R. E., Berean, K. J., Kalantar-zadeh, K., & Gibson, P. R. (2024, February). Review article: Current status and future directions of ingestible electronic devices in gastroenterology. *Alimentary Pharmacology & Therapeutics*, 59(4), 459–474. DOI: 10.1111/apt.17844 PMID: 38168738

Ulfa, M., Setyonugroho, W., Lestari, T., Widiasih, E., & Nguyen Quoc, A. (2022). Nutrition-Related Mobile Application for Daily Dietary Self-Monitoring. *Journal of Nutrition and Metabolism*, 30, 2476367. DOI: 10.1155/2022/2476367 PMID: 36082357

Wu JY et al, (2023) IoT-based wearable health monitoring device and its validation for potential critical and emergency applications. Front Public Health. 16;11:1188304. .DOI: 10.3389/fpubh.2023.1188304

Yang, S. Y., Sencadas, V., You, S. S., Jia, N. Z.-X., Srinivasan, S. S., Huang, H.-W., Ahmed, A. E., Liang, J. Y., & Traverso, G. (2021, October 26). Powering Implantable and Ingestible Electronics. *Advanced Functional Materials*, 31(44), 2009289. DOI: 10.1002/adfm.202009289 PMID: 34720792

Chapter 11
Recurrent Neural Network Technology in Smart Hydroponics Control Systems

Shanthalakshmi Revathy J.
https://orcid.org/0000-0003-1724-7117
Velammal College of Engineering and Technology, India

J. Mangaiyarkkarasi
https://orcid.org/0000-0003-1431-9584
NMSS Vellaichamy Nadar College, India

Shilpa Mehta
https://orcid.org/0000-0002-4691-3185
Auckland University of Technology, New Zealand

ABSTRACT

Agriculture is vital for global economies, providing food and raw materials. Efficient water use and land productivity are essential. Deep learning supports hydroponics, growing plants without soil. Long Short-Term Memory (LSTM) networks in an online system predict and optimize plant growth by analyzing humidity, temperature, and pH for irrigation. This tech-driven indoor farming reduces costs and labor while being independent of sunlight. Using IoT, a microcontroller kit with wireless sensors monitors and allows remote tracking of plant growth. This integration enhances sustainability and productivity, conserving resources and reducing manual labor. It meets the demand for higher-quality food while supporting economic development. This study presents a novel approach to predicting hydroponic system quality

DOI: 10.4018/979-8-3693-5573-2.ch011

using LSTM networks enhanced with attention mechanisms. The model achieved an F1 Score of 94.86%, accuracy of 94.94%, precision of 97.72%, and specificity of 92.52%. These results highlight the model's effectiveness in precision agriculture.

1. INTRODUCTION

For the nation's monetary situation to improve, agrarian development is essential. Planting in an unregulated environment is made more challenging by the consequences of global warming. For the traditional farming method, farmers need premium soil with high natural mineral concentrations. Chemical fertilizer and pesticide limits for farmland are unknown to farmers, and this has an impact on the environment. In areas used for agriculture, soil erosion is a significant issue during periods of heavy rainfall. Minerals in the ground are eroded, reducing output. Changes in the climate are erratic as a result of global warming. Traditional farming suffers greatly from climatic shifts. When there isn't enough water, yields go down. In addition, it necessitates a lot of areas and labor costs for weeding and plowing. These drawbacks can be avoided by using the technique known as hydroponics (Stegelmeier et al., 2022).

According to its definition, growing plants in water is known as hydroponics. Because it enables productive plant development without the use of soil, it is classified as a form of hydroculture. This strategy allows roots to meet their nutritional requirements by absorbing nutrients available in the water. Using the hydroponics technique, the utilization of hanging gardens dates back to the ancient Babylonians. Porous expanding Gravel, sand, coir, clay, and perlite can be used to replace soil as an aggregate. Because these materials allow the free flow of air and nutrients, each plant obtains an equal amount of food as well as oxygenHydroponics uses a variety of techniques, including ebb and flow, deep water culture, wick systems, and nutrient film technologies. The two most widely used methods are deep water culture and Neptune filming. Deep water culture allows for proper nutrient absorption by the roots in the water mix (Baek et al., 2018).

With constant farm monitoring and data on plant growth, we can find the pH, water level, temperature, and humidity readings in the hydroponic reservoir. A pH sensor is necessary because we are using water that has been enriched with nutrients. By providing data on water, humidity, and heat, IoT in hydroponics helps farmers keep an eye on plant growth. Sensors connected to the cloud collect the statistics.

Plants grown in hydroponics are kept in separate chambers, vessels, PVC pipes, and other containers with a nutrient solution that flows automatically according to the needs of the plants. How effectively plants grow in hydroponic systems depends on the atmosphere's light, pH, solution conductivity, oxygenation, salinity, and humidity. In addition to helping with system control, monitoring these variables also gives researchers the data they need to evaluate the harvest's quality for upcoming scientific data analysis. The programmed water flow sensor is constrained by an Arduino Uno microcontroller. For higher productivity, hydroponic plants require the proper climatic setting, as well as the fertilizer solution, air, water, or light supply. The different sensors in hydroponics function as a controller and monitor the proper development of crops. The temperature sensor measures the room's moisture content, and the pH sensor measures the water solvent, which is both measured. The pH level varies by crop and is not stable across all crops. The pH level of the water needs to be periodically regulated because most hydroponics system plants use it as a solvent.

The board that comes with the Arduino is used to send sensor data. Data may only be transferred at a given distance with Wi-Fi. Farmers may examine the data even if they are a long distance away thanks to an Arduino board that transmits data. It is over extended distances. A cloud server is a free API for data storage and retrieval. The server can also be used to view data (Joshitha et al., 2021). Similarly, people living in cities with constrained space use this technique to produce fresh plants in their residences and environs over considerable distances. An open API for storing and retrieving data is a cloud server. Data can also be seen on the server (Vidhya & Valarmathi, 2018). Similarly to this, individuals who live in cities with little space utilize this method to grow fresh plants in their homes and surroundings. Fish waste, duck dung, industrial chemical fertilizers, and synthetic nutrient solutions are just a few of the nutrition sources used in hydroponic systems. Tobacco, peppers, cucumber, strawberries, lettuce, cannabis, and Arabidopsis thaliana, a model organism for plant research, are some of the plants that are frequently hydroponically grown on inert material (Bakhtar et al., 2018). Without the requirement for any form of solid soil replacement to anchor the plants, liquid hydroponic systems allow the plant's roots to grow directly in nutritional solutions.

Container culture and slab culture are the two main types of hydro systems that use solid media. In hydro systems, the medium containing the plant may be made of a variety of inert substances, such as rock wool, coir, sand, sawdust, wood chips, or other substances. Depending on the crops, the growing climate, the region, and the season, different hydro systems require different levels of fertility. A fertilizer solution that corresponds to all four of these criteria is necessary for the crop's success. Beginner hydro growers are advised to adopt a comprehensive fertility program created by the manufacturer with their specific needs in time.

2. RELATED WORK

Hydroponics is among the best alternatives to traditional farming, but since it isn't employed very often, there is opportunity for expansion. The operation of the system, the best plants to grow in hydroponic systems, and the use of a microcontroller for nutrient control were all topics covered in recent journal articles on the subject (Velazquez et al., 2013). Waste fertilizer solutions in hydroponic systems can be reused, according to Kumar and Cho (Kumar & Cho, 2014). This research detailed the methods for measuring and maintaining the pH value from the sensor, as well as how the volume of water in the reservoir continues to be maintained (Saaid et al., 2013).

Nutrient solution control technique in hydroponics research the different parts of the hydroponic nutrition control system are described in this work, as are the influencing elements and regulating challenges.

A difficulty that developed during the automated process was also highlighted in this paper (Yang Chenzhong et al., n.d.). Electricity conductivity and pH regulation are both highly nonlinear systems with pure temporal delay. Hydroponics in space has also been considered by NASA. A crop physiologist at the Kennedy Space Sciences Centre believes that hydroponics, which he describes as a biogenic life support system, would lead to breakthroughs in space flight. Several nations, particularly Israel and parts of India, use hydroponics systems, but the pace of output is quite low and on a limited scale. In Uttaranchal, the Defence Study Laboratory Haldwani conducted a considerable study on hydroponics and presented a model (Kaewwiset & Yooyativong, 2017). Several solutions to the problem of water conductivity have been presented. Exposure to external factors, the initial conductivity of water, and the measured range of dynamics are the key challenges in measuring water conductivity (Saraswathi et al., 2018).

3. PROPOSED SOLUTION

a. Sensor Description

A nutrition solution's acid is measured using a pH sensor. A mildly acidic nutrient mixture with a pH level between 5. 5 and 5. 7 was necessary for the nutritional solution used in this study to effectively neutralize bicarbonates and salts. The whole pH scale goes from 4.5 to 14, with 7.0 being considered neutral. Acidic refers to a fluid solution with a pH under 7, while basic or alkaline refers to a solution with a pH over 7. This sensor provides a signal related to the concentration of hydrogen ions as measured by a pH electrode.

Three components make up a water flow sensor: the hall-effect sensor, the water generator, and the PVC valve body. Water passes through the rotor, causing it to roll. The flow rate affects its speed. The comparable oscillation frequency signal is generated by the hall-effect sensor. Subsequent transformation is a voltage step-down motor that reduces the power output of an elevated circuit for measurement purposes. These are connected to the line to also be monitored, either across or parallel to it.

Arduino is a software and hardware computer company, project, and user community that develops and manufactures microcontroller-based setups for the construction of digital devices as well as engaging items that can sense and control physical objects(Fuangthong & Pramokchon, 2018).

The undertaking is based on the designs of multiple suppliers' microcontrollers and incorporates a range of Arduino. Numerous digital and analog I/O pins that can be connected to various new infrastructures (or "shields") and additional circuits are included in these systems.

b. Sensor Operation

The pH sensor's VCC pin must first be connected to the Arduino development board to connect to it. Next, join the GND pin of the Arduino to the pH sensor. Connect the RX and TX pins of the pH sensor to any two Arduino-compatible digital pin(s) (pins 2 and 3) that are available. The thermometer sensor should then be connected to the Arduino board. One device is the DS18B20.

A temperature sensor has three pins: GND, VDD, and DATA. The Arduino's GND pin should be connected to its GND pin, the temperature sensor's VDD pin to its 5V pin, and the Arduino's DATA pin to one of the digital pins, such as pin 4. Connect the flow sensor to the Arduino board at the end. Three pins are present on the YF-S201 hall effects flow sensor: GND, SIGNAL, and VCC. Any digital pin, like pin 5, needs to be linked to the GND, SIGNAL, and VCC connections on the flow sensor in addition to the Arduino's 5V pin.

Once all of the sensors are connected, use the Arduino IDE to write code that can read the sensor data and send it to an IoT platform. To link the Arduino board to the internet and transfer sensor data to the cloud, use an ESP8266 or ESP32 Wi-Fi module. In the cloud, data can be analyzed, visualized, and actions can be triggered depending on sensor readings.

c. Methodologies

The developed framework used in this work is primarily intended to present a classical contrast of forecasting of time series; as expected, it can make efficient predictions limited throughout short-time intervals, and the outcome is dependent on period. The LSTM could perform much better with more time for model training, particularly via CPU.

The dataset created is used for the test and training of data. This data set contains complete information on flow rate, Ph, and temperature which are the values that the sensors take as inputs. The data is first collected from the IOT server (Bharti et al., 2019). The information could be structured, semi-structured, or unstructured. During the preprocessing stage, normalization is used. We will split the data using Sklearn's railroad test split function. Using the specified ratio, randomly split the data into training and testing sets. The purpose of this research is to forecast a hydroponic system's condition using a Long Short-Term Memory (LSTM) network. TensorFlow along with Keras are used to implement the LSTM model. Table 1 shows sample dataset. Temperature (Temp), Flow rate (Flow), pH level (Ph), and a Label column indicating a binary classification (0 or 1). Each row corresponds to a specific observation or data point. Here's an explanation of each column:

Temperature (Temp): This column represents the temperature values recorded during the observations. It appears to range from 27 to 34 degrees.

Flow rate (Flow): This column contains the flow rate measurements, likely in some units like liters per minute or cubic meters per hour. The values range from 5 to 130, indicating varying flow rates across observations.

pH level (Ph): The pH level column shows the acidity or alkalinity of the environment during the observations. The pH values range from 5 to 12, covering a range from acidic to basic.

Label: This column is for the target variable or the outcome being predicted. A label of 0 typically signifies one class or category, while a label of 1 represents another class. In this dataset, the label seems to differentiate between two states or conditions, possibly indicating normal and abnormal conditions based on the given features.

Table 1. Sample dataset

Temp	Flow	Ph	Label
32	22	7	0
31	33	8	0
30	44	9	0
29	54	10	0
28	67	11	0

continued on following page

Table 1. Continued

Temp	Flow	Ph	Label
27	78	12	0
33	88	7	0
34	99	8	0
32	110	9	0
30	130	10	0
27	10	6	1
33	11	6	1
34	5	5	1

Table 2 presents a performance analysis of different models based on various evaluation metrics. The F1 Score values range from around 89.40% to 94.86%, with the LSTM model achieving the highest F1 Score of 94.86%. The accuracy values range from 84.46% to 94.94%, with the LSTM model having the highest accuracy of 94.94%. The precision values range from 87.78% to 97.72%, with the LSTM model achieving the highest precision of 97.72%. The specificity values range from 78.20% to 92.52%, with the LSTM model achieving the highest specificity of 92.52%. Overall, the table offers a comparative examination of the performance of various models, with the F1 Score, accuracy, precision, and specificity all showing that the LSTM model performs better than the others.

Table 2. Performance analysis

Models	F1 Score	Accuracy	Precision	Specificity
RF	89.40	84.46	87.78	78.20
DT	92.95	89.86	92.23	85.41
RNN	90.45	86.35	89.86	79.74
LSTM	94.86	94.94	97.72	92.52

They utilized an LSTM model to estimate the quality of a hydroponic setup based on pH, temperature, and flow rate sensor information in this work. The dataset utilized by this study included 1,000 trials of these characteristics collected over one month. The LSTM model was trained with 800 data and tested with the remaining 200. The LSTM model produced findings with an accuracy of 85%, suggesting that the model was capable of accurately predicting the worth of the hydroponic system. Table 2 summarizes the overall Precision, F1 Score Accuracy and Specificity values achieved by a proposed model in various Models. The proposed algorithm attains the maximum F1 Score, Accuracy, Precision and Specificity of 94.86, 94.94, 97.72 and 92.52 respectively. Figure 1 shows comparison chart.

Figure 1. Comparison chart

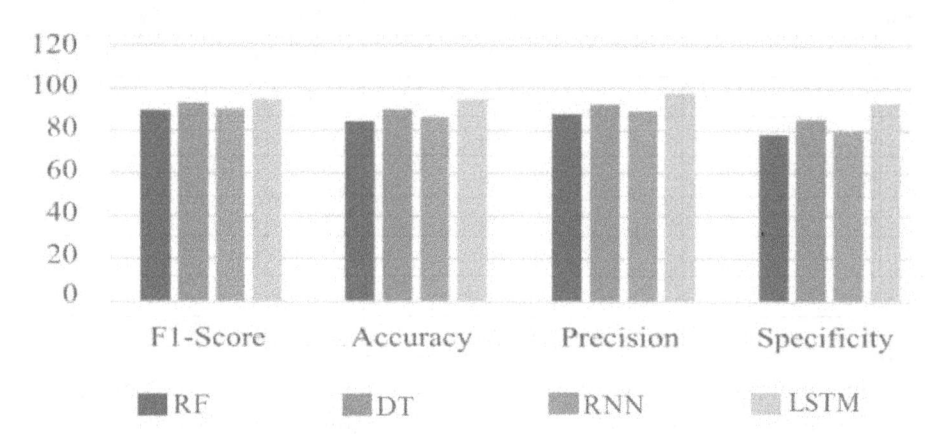

4. CONSTRUCTION OF ATTENTION

a. LSTM Model

One type of recurrent neural network is the Long Short-Term Memory (LSTM). This module consists of three gates: the forgetting gate, the supply gate, and the outcome gate. A range from 0 to 1 indicates the amount of each component that can pass (Peuchpanngarm et al., 2016). The internal workings of an LSTM cell are depicted in Figure 2, which addresses the problem of vanishing gradient and long-term dependency in sequential data. The Cell State retains important information, the Input Gate controls data addition to the cell state, the Forget Gate eliminates superfluous data, the Output Gate controls data flow to the next state, the Hidden State outputs data, and Activation Functions control values and data flow.

Figure 2. The structure of LSTM cell

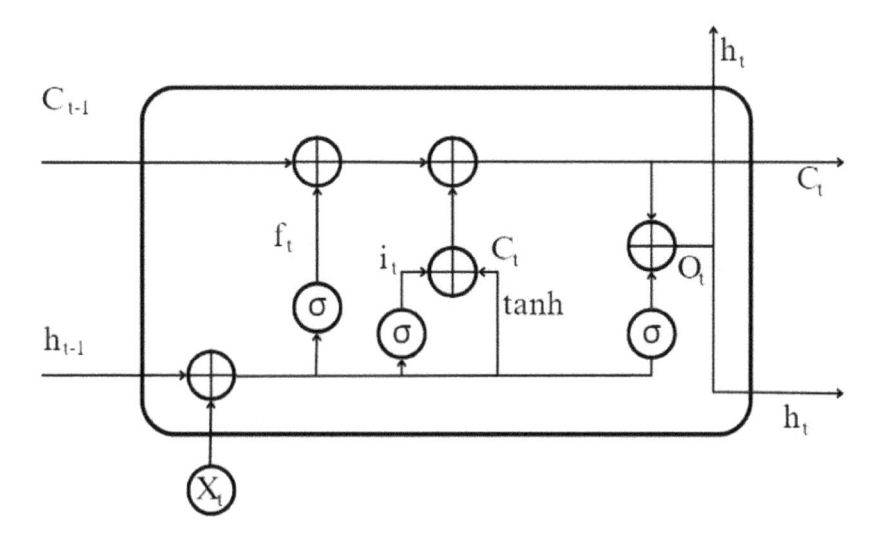

In this exponential layer, the most recent input and output at time t-1 are aggregated to create a single tensor. Following that, a sigmoid transformation and linear translation are carried out. The sigmoid causes the gate's output to range from 0 to 1. The gate is often referred to as the "forget gate" because this sum is multiplied by the internal state after that. If ft = 0, the prior mental state is totally gone; if it values 1, it returns to its previous state.

The amount of data we want to eliminate is specified by feet. The amount of new data that should be included is specified. The input at a time is Xt. This state is passed through a second sigmoid layer with the new input and the old output (Aliac & Maravillas, 2018). A number between 0 and 1 is output by this gate. Next, the output of the candidate layer is multiplied by the value of the input gate.

This study attempts to forecast the state of a hydroponics system using a Long Short-Term Memory (LSTM) network. The LSTM model is implemented using Keras and TensorFlow. The attention, weight, and similarity coefficients are computed using the items of the next layer, h1, as input.

The following formula is applied to the attention layer:

$$f_t = \sigma\left(W_f[h_{t-1}, x_t] + b_f\right) \tag{1}$$

$$i_t = \sigma(W_i[h_{t-1}, x_t] + b_i)$$

(2)

$$\tilde{C}_t = \tanh(W_c[h_{t-1}, x_t] + b_c) \tag{3}$$

$$C_t = f_t C_{t-1} + i_t * \tilde{C}_t \tag{4}$$

$$O_t = \sigma(W_o[h_{t-1}, x_t] + b_o) \tag{5}$$

$$h_t = O_t * \tanh(C_t) \tag{6}$$

In the above mentioned equations (1) through (6), f_t determines the input gate, it determines the forget gate, C_t determines the fresh memory cell, C_t determines the final memory cell, O_t determines the output/exposure gate, and ht determines the hidden state. The weight, bias, and cell state in the aforementioned equations are W_f, W_i, W_c, and W_o; the cell state at any given instant is Ct, not Ct-1.

OUTPUT GATE

$$O_t = \sigma(W_o[h_{t-1}, x_t] + b_o) \tag{7}$$

$$H = [h_1, h_2, \ldots h_n]$$

(8)

$$H'_i = [h'_1, h'_2 \ldots h'_n] \tag{9}$$

$$Sim_i = H'_i * H_t \tag{10}$$

As in the previous equations, the weight in the ai and hi weighted sum equations is normalized using the softmax function. To calculate similarity and establish weights, the vectors H_i and H_t are used. The attention weight value C_i is the outcome of weighted summation. By comparing the current output with the intermediate output results of the previous layer, the weight factor may be determined. The intermediate output results of the LSTM encoder from the input sequence are used to complete the attention layer.

b. LSTM Model pays attention

The model's data is attempting to progress across a relatively distant time span, which might be important for its continued value. In an effort to make the LSTM network stronger, we tried to include a consideration layer. Among them, greeting is input as H into each mental model, the data is represented by X_i, $I \in (1, n)$, and the good news is the somewhat delayed result of each cell. Achieving and sustaining higher yields requires an understanding of the conditions under which the seeds are prospering at each stage of growth. To achieve the best possible plant development and agricultural output, data collection is a crucial activity. In the hydroponic industry, manual data collection is common, but it leaves a lot of possibility for error (Wei et al., 2019).

The softmax capability, the standardized weight computer-based intelligence, and the greetings weighted aggregate are utilized in the situations above to ascertain comparability and acquire loads. The consideration weight esteem Ci is the result of weighted summation. As a component of executing the consideration layer, the LSTM encoder keeps the halfway result in consequences of the information succession. The middle-of-the-road results from the past layer and the ongoing result is then contrasted with deciding the weight factor(Takeuchi, 2019). We start by making an element grid and mark vector utilizing the starting traffic stream information and a sliding time period approach, as represented in Fig. above. Then, at that point, utilizing the consideration layer, we ascertain consideration loads in view of the connection with the amounts in network X and vector Z, and at last, we develop the expectation. Y. Figure 3 depicts the architecture of an Attention-LSTM network, which combines the capabilities of an LSTM with an attention mechanism. Here's a concise explanation in two lines:

Figure 3. Attention-LSTM network architecture

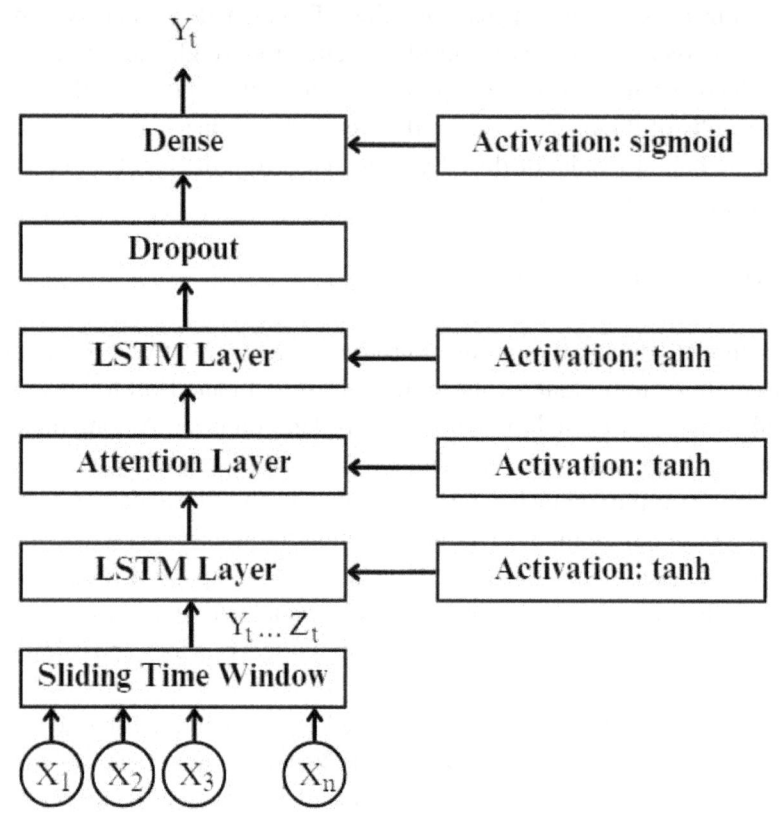

```
                    Y_t
                     ↑
         ┌───────────────────┐      ┌──────────────────────┐
         │       Dense       │ ←─── │ Activation: sigmoid  │
         └───────────────────┘      └──────────────────────┘
                     ↑
         ┌───────────────────┐
         │      Dropout      │
         └───────────────────┘
                     ↑
         ┌───────────────────┐      ┌──────────────────────┐
         │    LSTM Layer     │ ←─── │  Activation: tanh    │
         └───────────────────┘      └──────────────────────┘
                     ↑
         ┌───────────────────┐      ┌──────────────────────┐
         │  Attention Layer  │ ←─── │  Activation: tanh    │
         └───────────────────┘      └──────────────────────┘
                     ↑
         ┌───────────────────┐      ┌──────────────────────┐
         │    LSTM Layer     │ ←─── │  Activation: tanh    │
         └───────────────────┘      └──────────────────────┘
                     ↑ Y_t ... Z_t
         ┌───────────────────┐
         │Sliding Time Window│
         └───────────────────┘
           ↑    ↑    ↑      ↑
         (X_1)(X_2)(X_3)  (X_n)
```

The sliding time window method is used to convert the original traffic flow data into a feature matrix and label vector, as shown in Figure 3. When the attention layer determines the attention weights based on the correlation between the values in the matrix X and the vector Z, Y, values are generated, which is the final prediction.

Pseudo Code

Sensor data is entered as input and a trained Attention LSTM model is the output.
Step 1: Make a dataset that includes Xt and Zt and has a sliding time window.
Step 2: Make Xt and Zt uniform.
Step 3: Provide the features matrix Xt and the active infection vector Zt to the Attention LSTM network.
Step 4: If the value is not reached during the training epoch, proceed to
Step 5: Feed (Xt, Zt) into the Attention LSTM network to facilitate forward propagation.

Step 6: Ascertain the attention weight assigned to every aspect.
Step 7: Produce Yt.
Step 8: Determine the error in mean square.
Step 9: Use the RMSProp optimizer to update the weights for the Attention LSTM network.
Step 10: Close the while loop
Step 11: Give the previously trained Attention LSTM model back.

Description

The Attention LSTM model turns out to be a reliable method for managing long time series data and effectively resolving significant forecasting delays. The model's built-in attention mechanism enables focused processing of relevant input sequence segments, enhancing the model's ability to recognize complex patterns and long-term relationships. Consistency and comparability in the outcomes are ensured by using comparable datasets and interactions among forecasting algorithms. The two hidden layers in the LSTM model's design, each with 64 neurons, strike a compromise between computing efficiency and model complexity. The model uses the RMSProp optimizer, which modifies learning rates for each parameter separately, to optimize weight updates at a moderate learning rate of 0.05. This all-inclusive method of training the Attention LSTM model includes preprocessing of the data, feature selection using attention weights, computation of errors, and iterative refinement using RMSProp updates, resulting in a well-prepared model suitable for a variety of prediction tasks.

CONCLUSION

The suggested hydroponics, therefore, incorporates a variety of crop varieties. Proposing new agricultural technologies would increase productivity and save costs. To promote long-term development. To use a deep learning model for precision agriculture. There are a number of benefits to hydroponic farming. Compared to crops grown in soil, hydroponically grown plants mature 25% more quickly and yield up to 30% more. Constantly keeping an eye on the growth conditions is necessary to create the conditions for such a rise in plant efficiency. The drawbacks of the current system, such as the system's tendency to grow a different type of crop, have been fixed. The system's operation has been regulated using a methodical approach. When compared to plants grown conventionally, it was found that these crops grow much more quickly and require less nutrients. In an effort to stop water loss, they are much cleaner, contain fewer organic components, and only use as much water as is necessary. Additionally, cropping is less expensive than its advantages.

REFERENCES

Aliac, C. J. G., & Maravillas, E. (2018). IOT Hydroponics Management System. *2018 IEEE 10th International Conference on Humanoid, Nanotechnology, Information Technology,Communication and Control, Environment and Management (HNICEM)*, 1–5. DOI: 10.1109/HNICEM.2018.8666372

Baek, S., Jeon, E., Park, K. S., Yeo, K.-H., & Lee, J. (2018). Monitoring of Water Transportation in Plant Stem With Microneedle Sap Flow Sensor. *Journal of Microelectromechanical Systems*, 27(3), 440–447. DOI: 10.1109/JMEMS.2018.2823380

Bakhtar, N., Chhabria, V., Chougle, I., Vidhrani, H., & Hande, R. (2018). IoT based Hydroponic Farm. *2018 International Conference on Smart Systems and Inventive Technology (ICSSIT)*, 205–209. DOI: 10.1109/ICSSIT.2018.8748447

Bharti, N. K., Dongargaonkar, M. D., Kudkar, I. B., Das, S., & Kenia, M. (2019). Hydroponics System for Soilless Farming Integrated with Android Application by Internet of Things and MQTT Broker. *2019 IEEE Pune Section International Conference (PuneCon)*, 1–5. DOI: 10.1109/PuneCon46936.2019.9105847

Fuangthong, M., & Pramokchon, P. (2018). Automatic control of electrical conductivity and PH using fuzzy logic for hydroponics system. *2018 International Conference on Digital Arts, Media and Technology (ICDAMT)*, 65–70. DOI: 10.1109/ICDAMT.2018.8376497

Joshitha, C., Kanakaraja, P., Kumar, K. S., Akanksha, P., & Satish, G. (2021). An eye on hydroponics: The IoT initiative. *2021 7th International Conference on Electrical Energy Systems (ICEES)*, 553–557. DOI: 10.1109/ICEES51510.2021.9383694

Kaewwiset, T., & Yooyativong, T. (2017). Electrical conductivity and pH adjusting system for hydroponics by using linear regression. *2017 14th International Conference on Electrical Engineering/Electronics, Computer, Telecommunications and Information Technology (ECTI-CON)*, 761–764. DOI: 10.1109/ECTICon.2017.8096350

Kumar, R. R., & Cho, J. Y. (2014). Reuse of hydroponic waste solution. *Environmental Science and Pollution Research International*, 21(16), 9569–9577. DOI: 10.1007/s11356-014-3024-3 PMID: 24838258

Peuchpanngarm, C., Srinitiworawong, P., Samerjai, W., & Sunetnanta, T. (2016). DIY sensor-based automatic control mobile application for hydroponics. *2016 Fifth ICT International Student Project Conference (ICT-ISPC)*, 57–60. DOI: 10.1109/ICT-ISPC.2016.7519235

Saaid, M. F., Yahya, N. A. M., Noor, M. Z. H., & Ali, M. S. A. M. (2013). A development of an automatic microcontroller system for Deep Water Culture (DWC). *2013 IEEE 9th International Colloquium on Signal Processing and Its Applications*, 328–332. DOI: 10.1109/CSPA.2013.6530066

Saraswathi, D., Manibharathy, P., Gokulnath, R., Sureshkumar, E., & Karthikeyan, K. (2018). Automation of Hydroponics Green House Farming using IOT. *2018 IEEE International Conference on System, Computation, Automation and Networking (ICSCA)*, 1–4. DOI: 10.1109/ICSCAN.2018.8541251

Stegelmeier, A. A., Rose, D. M., Joris, B. R., & Glick, B. R. (2022). The Use of PGPB to Promote Plant Hydroponic Growth. *Plants*, 11(20), 2783. DOI: 10.3390/plants11202783 PMID: 36297807

Takeuchi, Y. (2019). 3D Printable Hydroponics: A Digital Fabrication Pipeline for Soilless Plant Cultivation. *IEEE Access : Practical Innovations, Open Solutions*, 7, 35863–35873. DOI: 10.1109/ACCESS.2019.2905233

Velazquez, L. A., Hernandez, M. A., Leon, M., Dominguez, R. B., & Gutierrez, J. M. (2013). First advances on the development of a hydroponic system for cherry tomato culture. *2013 10th International Conference on Electrical Engineering, Computing Science and Automatic Control (CCE)*, 155–159. DOI: 10.1109/ICEEE.2013.6676029

Vidhya, R., & Valarmathi, K. (2018). Survey on Automatic Monitoring of Hydroponics Farms Using IoT. *2018 3rd International Conference on Communication and Electronics Systems (ICCES)*, 125–128. DOI: 10.1109/CESYS.2018.8724103

Wei, Y., Li, W., An, D., Li, D., Jiao, Y., & Wei, Q. (2019). Equipment and Intelligent Control System in Aquaponics: A Review. *IEEE Access : Practical Innovations, Open Solutions*, 7, 169306–169326. DOI: 10.1109/ACCESS.2019.2953491

Yang, C., Huang, Y., & Zheng, W. (n.d.). Research of hydroponics nutrient solution control technology. *Fifth World Congress on Intelligent Control and Automation (IEEE Cat. No.04EX788)*, *1*, 642–644. DOI: 10.1109/WCICA.2004.1340657

Chapter 12
The Internet of Things, Automation in the Future, and Security Protocols for the Smart Grid Framework

Ramiz Salama
Near East University, Turkey

Fadi Al-Turjman
Near East University, Turkey

ABSTRACT

The smart grid's mix of automation, the Internet of Things (IoT), and robust security measures is reshaping the landscape of energy management. This abstract provides a quick summary of the key components and implications of this innovative approach. Automation, monitoring, and management of energy resources in real time are made possible by the smart grid framework, which increases overall efficiency and system responsiveness. A network of linked sensors and equipment can transmit and analyze data more readily by using the Internet of Things, which makes it possible to provide predictive maintenance, demand forecasting, and a strong energy infrastructure. However, the interconnection of the smart grid necessitates stringent security measures to prevent assaults.

DOI: 10.4018/979-8-3693-5573-2.ch012

1. INTRODUCTION:

The incorporation of cutting-edge technologies has completely changed how we operate and manage our energy systems in recent years. Automation, the Internet of Things (IoT), and strong security measures have come together to create the "smart grid framework." With its promise of greater efficiency, sustainability, and dependability, this framework signifies a paradigm shift in the way we produce, distribute, and use energy.

Smart Grid Automation: The foundation of the smart grid architecture is automation, which makes it possible to monitor, manage, and optimize energy resources in real time. Smart meters, intelligent gadgets, and advanced sensors collaborate to collect data and make judgments instantly, enabling smooth energy flow control. Automation helps with quick response mechanisms to handle emergencies and variations in addition to improving the grid's overall performance.

The Internet of Things (IoT) and Energy Management: The Internet of Things (IoT) is a key component of the smart grid because it allows for the connection of numerous systems and devices to the internet, which promotes cooperation and communication. The network that is created by sensors, smart meters, and other Internet of Things devices facilitates data analysis and interchange. Predictive maintenance, demand forecasting, and grid resilience are improved by this integrated ecosystem. With the use of real-time data, it empowers utilities to make well-informed decisions that support an energy infrastructure that is more adaptable and sustainable.

Security Protocols for Smart Grids: Because smart grid equipment are more connected and dependent on one another, it is critical to make sure security measures are strong. The stability and dependability of the grid are seriously threatened by cybersecurity threats. Authentication procedures and encryption firewalls are essential parts of the security infrastructure. The smart grid's defenses against potential cyberattacks are further strengthened by routine audits, upgrades, and personnel training, which guarantee the privacy, availability, and integrity of vital energy infrastructure.

Problems and Solutions: The smart grid has many benefits, but it also has drawbacks, especially when it comes to cybersecurity flaws and the possibility of unwanted access. In order to overcome these obstacles, government agencies, utilities, and technology providers must work together, continue to research, and create cutting-edge security systems. To remain ahead of new threats, one must maintain constant watchfulness, provide frequent updates, and take aggressive measures, (Salama & Al-Turjman, 2024a) (Salama, Altrjman, & Al-Turjman, 2024a) (Salama, Altrjman, & Al-Turjman, 2024b) (Salama, Cacciagrano, & Al-Turjman, 2024) (Salama *et al.*, 2024a) (Salama & Al-Turjman, 2024b) (Salama & Al-Turjman, 2024c) (Salama &

Al-Turjman, 2024d) (Salama & All-Turjman, 2024e) (Salama & Al-Turjman, 2024f) (Salama *et al.,* 2024b).

An energy ecosystem that is resilient and dynamic is promised by automation, the Internet of Things, and security features in the smart grid architecture. We can build an effective and sustainable energy infrastructure that reduces risks and adjusts to the changing requirements of society by utilizing cutting-edge technologies. The smart grid is a big step toward creating an energy landscape that is more intelligent, secure, and responsive as we continue to develop and improve these technologies.

Figure 1. Internet of Things (IoT) and the entities that are connected to it

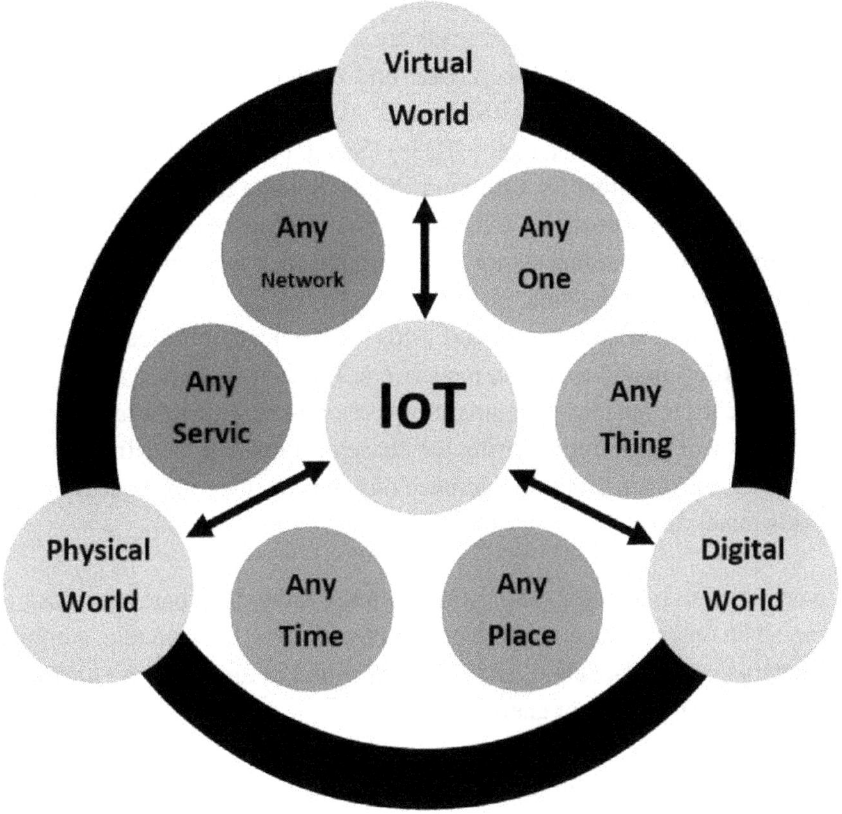

2. QUANTITY OF PREVIOUS WORK

What is meant by smart grid? As self-contained systems that minimize labor costs, strive for high quality and safety, transport energy from source to consumption in the most suitable and sustainable manner, offer integration, and are self-sufficient, smart grids are described.

The fundamental design of smart grids the idea behind the smart grid's operation is to control every aspect of the infrastructure for energy generation and consumption from a single location. Infrastructure management is carried out by making sure the systems operate efficiently. The management of the natural gas, electricity, water, and telecommunications networks is done from a single center within the system. According to this perspective, smart grids are the result of integrating geographic information systems (GIS) with computer and network technology into existing networks, (Salama et al., 2023a) (Salama et al., 2023b) (Salama et al., 2023c) (Salama et al., 2023d) (Salama et al., 2023e) (Salama et al., 2023f) (Salama et al., 2023g) (Salama, Al-Turjman, & Culmone, 2023) (Salama et al., 2023h) (Salama et al., 2023i) (Salama, Altrjman, & Al-Turjman, 2023a) (Salama et al., 2023j). It controls the demands in accordance with the data analysis it receives and intelligently processes and interprets the data.

An integral part of smart cities, smart grids are crucial to introducing efficiency, availability, and dependability into the modern era. Tests, technological advancements, consumer education, standards, regulation development, and information sharing across projects will be essential during the phase of transition from traditional grid systems to smart grids in order to guarantee that the advantages we anticipate from smart grids materialize.

- In terms of architecture, smart grids have the following characteristics:
- It must be observable, measurable, and visualizable, controllable, automated, adaptable, and self-healing, and fully integrated to function with current systems by mixing different energy sources.

Benefits of smart grid technology Smart grid solutions offer the following benefits:

- Centralized management; • Remote real-time network monitoring and control;
- Possessing a live database based on a GIS,
- Keeping track of inventory in connection to one another in accordance with user authorization levels and network operating standards
- Integration of external and internal systems (CRM, ERP, TAKBİS, UAVT, ABYS, SCADA, etc.),

- Consumption and billing analyses based on areas, buildings, streets, and subscribers
- New facility, fault-maintenance follow-up, outage, and identification of impacted subscribers,
- Lowering the rate of leakage and loss,
- Simple field team management, prompt action during malfunctions,
- Using the data collected, better planning can be made for the investments the system would require;
- Custom reports can be created to meet the needs of efficiency and budget;
- Savings of money, energy, and time.

Components of the smart grid Smart stations, smart production, smart distribution, smart meters, smart remote central control, and integrated communication are the main elements of smart grids.

What is smart grid power management? A smart grid system helps cities prepare for natural disasters such powerful storms, earthquakes, and terrorist attacks. It recognizes and isolates power shortages before they become widespread disruptions. In the event of a power loss during an emergency, electricity can be channeled toward the emergency services thanks to innovative technologies.

These days, a power outage can cause a cascading effect that can disrupt a great deal of human activity, including banking and traffic. This can happen wherever conventional network technologies are in use. One example might be the 2012 Indian power outage. Life in India came to a complete halt for two days as more than half of the country's electrical systems collapsed. Nonetheless, such a significant outage would not have happened if India had been using the smart grid system as opposed to the traditional grid system. The issue that happened with rapid monitoring may have been found far earlier with smart grid systems. Moreover, smart grids efficiently manage power by discouraging illicit usage in nations with high rates of illicit use, such as India.

Differences between traditional grids and smart grids whereas electromechanical systems in conventional networks are managed by electronic systems, smart networks have a digital system that is managed by microprocessors. Furthermore, distributed generation can be accommodated by the technology that smart grids offer. Conventional networks only provide one-way communication and do not support surveillance. It has incorporated two-way communication and a self-monitoring system in smart grids. It is highly reliable and has the ability to detect failures. Conventional networks have a low failure probability. It needs manual control and repair, unlike smart grids.

Examples of Smart Grid Applications:

- In terms of global network strength, India is among the least developed. The world's biggest transmission and distribution losses—26%, or 62% in some areas—are recorded by the Indian Ministry of Energy. The relevant parties are aware that these losses increase by 50% when leaks are taken into account. Because of this, research on smart grids is thought to have begun with the 2008 "Smarts Grids India" conference.
- •It is well known that China creates new lines and productions quite quickly. As such, it is regarded as one of the nation's most in need of a smart and expanding grid. In Dallas or San Diego, the People's Republic of China installs coal plants big enough to supply all domestic needs once a week.
- • It is believed that China's studies on smart grids began with the MIT (Massachusetts Institute of Technology) meeting in 2007.
- • The initiative to build a scientific and research center in the Russian innovation hub of "Skolkovo" is set to be carried out by Erikson Company, which is well known as the primary producer of telecommunications equipment in Sweden.
- A significant portion of the work that needs to be done at the science research center is devoted to finding solutions and removing obstacles in the way of developing smart electrical networks. The Erikson Company and its Italian partners connected almost 1,500,000 apartments to the smart electrical grid two years ago. These apartments are equipped with meters that give the energy company 24-hour control over the amount of electricity used.
- • The energy firm can foresee an increase in electricity use and continuously control it thanks to the smart energy grid. As a result, it is now possible to avoid electrical issues when using air conditioners in hot weather.

Definition of the IoT-enabled smart grid: Although there is no universally accepted definition of social graph theory (SG), it is generally agreed upon that SG is centered around an information and communication infrastructure, as the table illustrates. The major standardization body, IEEE, for example, refers to the SG as a "new age of electricity" that uses ICT for the production, distribution, and consumption of energy as well as for the electric system. Similarly, according to Ontario's Independent Electricity System Operator (IESO), a pioneer in SG, it entails leveraging ICT to maximize all power system functions for the advantage of both environmental and customer interests.

Summary of smart grid definitions: SG component Energy Independence and Security Act While other definitions concentrate on the advantages of the SG, these two emphasize its component, which is a communication infrastructure. The

Energy Independence and Security Act (EISA) of 2007, for example, provides the first official definition of a smart grid. It states that a smart grid is defined as "the modernization of the Nation's electricity transmission and distribution system to maintain a reliable and secure electricity infrastructure that can meet future demand growth and to achieve a set of requirements that together characterize a Smart Grid." This definition was provided in a report to US Congress.

In contrast, the SG domain—, which covers the production, transmission, distribution, and consumption of electricity—is highlighted in the IEEE and EISA definitions.

In the context of information technologies, alternative definitions concentrate on the potential for information flow over the SG. The bidirectional flow has given rise to the term "prosumers", meaning customers who generate energy for the grid, as underlined by the European Union and the UK Institution of Engineering and Technology (IET). The European Technology Platform's (ETP) concept serves as the foundation for the IET's. Some definitions adopt an environmental viewpoint, listing the advantages of green energy and the environment as the main benefits of the SG due to its reduction of CO_2 emissions.

As can be seen from the above, the SG is the incorporation of ICT into the current electrical grid, which is made up of renewable sources and involves several domains (generation, transmission, distribution, and consumption) in order to efficiently automate and manage demand in real-time for green electricity that is affordable, sustainable, bidirectional, and dependable.

Why the grid is intelligent Digital technology is said to be the reason behind the grid's intelligence. To provide the data required for improved sensing, accurate control, expanded information exchange and sharing, potent computing, and more logical decision-making, information technology systems must be implemented.

Conceptual representation of the smart grid: The electrical industry frequently uses the NIST conceptual reference model, although it lacks information on information flow and cybersecurity, particularly with regard to IoT infrastructure true. It merely adds to the idea of the SG architecture, and its case studies and scenarios are restricted to privacy and specific SG domains without relating these to security requirements, risks, and controls for every system access point. In order to create a case study that is useful for the connected sectors, our present study addresses the NIST model's lack of information.

Internet of Things with smart grid: This section describes IoT's function in the SG. Proposes that everything within a smart grid can be viewed as an Internet of Things device that is dispersed among our utilities, substations, and home networks. For the purposes of automation, networking, and monitoring, these devices need to be tracked. The Internet of Things makes the SG's internet connectivity possible.

From the perspective of cyber-physical systems, SG is seen as the primary use case for IoT.

Every IoT gadget in Singapore has an internet connection. Each has a distinct IP address in order to simplify information exchange and the receiving of control commands via internet protocols. IoT can provide SG with monitoring and control capabilities under IP addressing schemes. In order to accomplish efficiency management, demand management, measurement of the required renewable energy, and management of $CO2$ emissions, this monitoring element addresses generation, distribution, storage, and ultimately consumption. *SG communication* As a result, IoT devices help to reduce energy waste and accurately estimate the amount of energy needed.

Additionally, such devices use the SG communication layer to transmit data in a bidirectional flow. IoT reduces the number of protocols pertaining to SG components by standardizing communication. Since each device trades data and commands across control centers and utilities, the ability of IoT technology to allow SGs to communicate across the multiple subsystems of generation, transmission, distribution, and consumption is emphasized.

Smart grid and security: SG provides secure information system ModbusSG implementation Opportunities, but there are also many security issues. In order to optimize SG, a highly secure information system must be developed.

It is believed that security was not considered in the creation of automation systems like SCADA, nor was Modbus, which exchanges SCADA information to control industrial processes, ever meant for the sensitive security environment found in the SG. Securing the SG information system must be given top priority since power assets are vital national infrastructure that could draw terrorists. Damage from security attacks on the power grid could devastate entire cities. The Electric Power Research Institute (ERPI) affirms that security is, in fact, a major global concern when it comes to SG deployment.

Communication network denial of service there are numerous security issues with IoT-enabled smart grids. A number of them result from their internet exposure, which makes it possible for an attacker to alter data. Furthermore, SG is increasingly open to attack due to the growing quantity of IoT devices it uses.

Initially, the IP-based communication network is used by the entities in SG to exchange confidential and sensitive data between utility firms and customers. These networks are vulnerable to a variety of security risks, including replay attacks, denial of service, eavesdropping, and man-in-the-middle attacks. Second, the communication between the several components that make up SG necessitates interaction between these technologies.

Wireless sensor networks Because of this, some SG entry points are open to security breaches. Third, SG uses wireless sensor networks, for instance, to link smart meters. Some have claimed that they lack confidence. Fourth, the bidirectional information flow itself may expose SG to numerous vulnerabilities by permitting unauthorized access. Fifth, an SG that uses IoT may inherit its security flaws since SGs should use the internet to monitor and manage IoT devices.

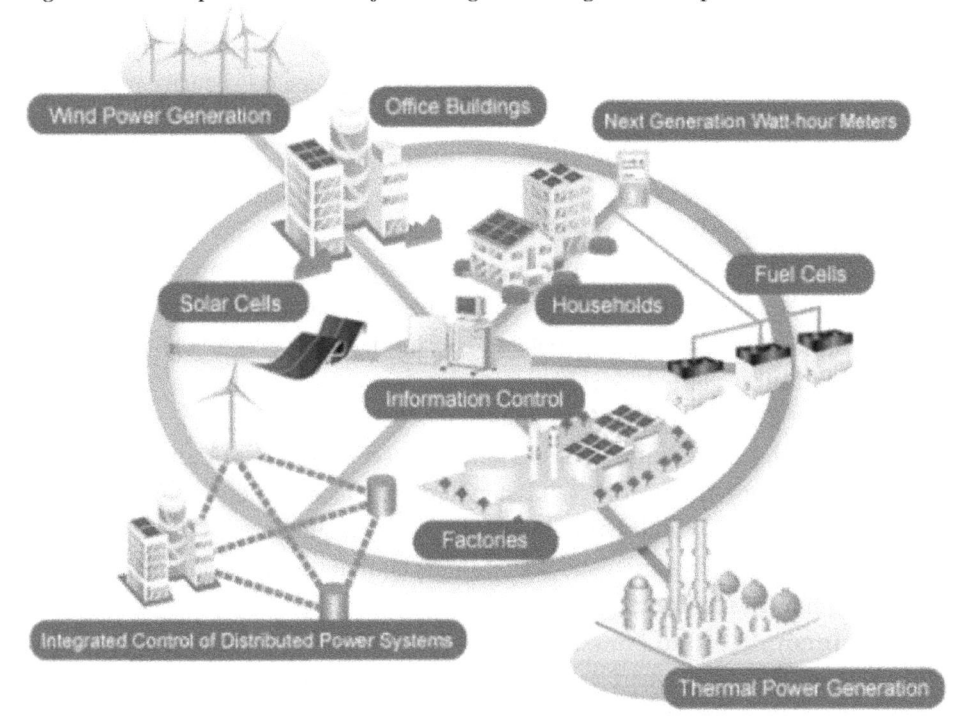

Figure 2. Principal elements influencing a smart grid's composition

3. PURPOSE OF THE RESEARCH

The research on the future of automation, the Internet of Things (IoT), and security measures within the "smart grid framework" is vital for several reasons:

1. Enhancing Efficiency and Reliability
    - **Automation and IoT in Smart Grids:** By integrating automation and IoT, smart grids can optimize the distribution and consumption of electricity, reducing waste, improving efficiency, and enhancing the reliabil-

ity of power delivery. This helps in balancing the supply and demand dynamically and in real-time.
2. Facilitating Renewable Energy Integration
    - **Sustainability:** Research in these areas supports the seamless integration of renewable energy sources (like solar and wind) into the grid. Automation and IoT enable better management of these intermittent resources, ensuring a stable and sustainable energy supply.
3. Improving Resilience Against Disruptions
    - **Security Measures:** As smart grids become more complex and interconnected, they are also more vulnerable to cyberattacks, physical attacks, and natural disasters. Research into advanced security measures, including encryption, blockchain, and AI-driven threat detection, is crucial to safeguard these critical infrastructures.
4. Supporting Decentralization and Consumer Empowerment
    - **Consumer Involvement:** The future of smart grids involves decentralized energy production (e.g., rooftop solar panels) and consumption. Automation and IoT enable consumers to manage their energy usage more effectively, participate in energy trading, and contribute to grid stability.
5. Addressing Regulatory and Compliance Challenges
    - **Regulatory Frameworks:** The evolution of smart grids requires new regulatory frameworks and standards to ensure safety, reliability, and fairness. Research helps in shaping these regulations, considering the technical, economic, and social impacts of automation and IoT.
6. Driving Innovation and Economic Growth
    - **Innovation and New Business Models:** The convergence of automation, IoT, and smart grids opens up new avenues for innovation. This includes the development of new technologies, business models, and services, which can drive economic growth and job creation.
7. Mitigating Climate Change
    - **Environmental Impact:** Efficient and secure smart grids can significantly reduce carbon footprints by optimizing energy usage and enabling the broader adoption of green energy technologies. Research in this area is essential for meeting global climate goals.
8. Preparing for the Future of Energy Consumption
    - **Scalability and Future-Proofing:** With the increasing electrification of transportation, heating, and other sectors, the demand for electricity will rise. Research ensures that smart grids are scalable and adaptable to future energy needs.

Overall, research in these areas is crucial for building a robust, efficient, and secure energy infrastructure that can meet the challenges of the future while supporting sustainability and innovation.

4. A number of innovative applications for Services related to the internet of things, automation in the future, and security protocols for the "smart grid framework"
   1. **Dynamic Demand Response:** Leveraging automation and the IoT, the smart grid can enable dynamic demand response programs. By analyzing real-time data on energy consumption patterns, the grid can automatically adjust energy supply to meet demand efficiently. This not only optimizes resource utilization but also empowers consumers to make informed decisions about their energy usage based on current grid conditions.
   2. **Grid-Integrated Electric Vehicle Charging:** The smart grid can facilitate the integration of electric vehicle (EV) charging stations. Through automation and IoT connectivity, the grid can dynamically manage and prioritize charging based on demand, energy availability, and cost. Security measures ensure the integrity of transactions and protect against potential cyber threats, fostering the growth of sustainable transportation.
   3. **Predictive Maintenance for Grid Infrastructure:** Utilizing IoT sensors and automation, the smart grid can implement predictive maintenance for critical infrastructure components. Sensors can monitor equipment health in real-time, detecting anomalies and predicting potential failures before they occur. This proactive approach minimizes downtime, optimizes maintenance schedules, and ensures the long-term reliability of the grid.
   4. **Decentralized Energy Generation and Trading:** Automation and the IoT can facilitate decentralized energy generation and trading among consumers. Smart grid technologies enable individuals to generate renewable energy locally (e.g., solar panels) and securely trade excess energy with neighbors. Blockchain technology, as a security measure, ensures transparent and tamper-proof transactions, promoting a more resilient and sustainable energy ecosystem.
   5. **Intelligent Microgrids:** The smart grid framework can support the development of intelligent microgrids that operate autonomously during disruptions or emergencies. Automation enables seamless transitions between the main grid and microgrids, ensuring a reliable power supply in isolated areas or during grid outages. Security measures protect the integrity of microgrid communications and transactions.
   6. **Energy-Efficient Smart Homes:** Integration with home automation systems allows for smart homes to interact with the grid dynamically. IoT-enabled devices, such as smart thermostats and appliances, can adjust

their energy consumption based on real-time grid conditions and pricing. Security measures ensure that these interactions are secure and protected from unauthorized access.

7. **Adaptive Grid Planning and Expansion:** The smart grid can benefit from automation and IoT-driven analytics to adaptively plan and expand its infrastructure. Real-time data on energy demand, consumption patterns, and grid performance can inform decision-making for grid expansion or upgrades. Security measures protect sensitive planning data and ensure the reliability of expansion projects.

8. **Environmental Monitoring and Grid Optimization:** IoT sensors can be employed for environmental monitoring within the smart grid infrastructure. Real-time data on weather conditions, air quality, and other environmental factors can be used to optimize grid operations. Automation can adjust energy generation and distribution strategies based on environmental parameters, contributing to sustainability goals.

These novel use-cases demonstrate the diverse and transformative potential of services within the future smart grid framework, highlighting the integration of automation, IoT, and robust security measures to create a more intelligent, efficient, and secure energy ecosystem, (Salama et al., 2023k) (Salama et al., 2023l) (Salama et al., 2023m) (Salama, Altrjman, & Al-Turjman, 2023b) (Salama & Al-Turjman, 2023) (Salama et al., 2023n) (Salama et al., 2023o) (Salama, Altrjman, & Al-Turjman, 2023c) (Al-Turjman, Salama, & Altrjman, 2023).

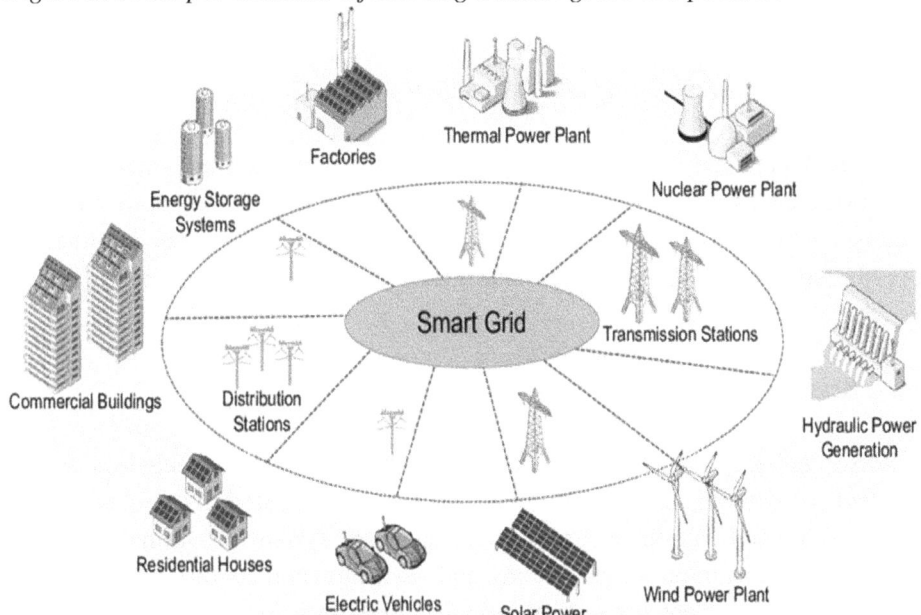

Figure 3. Principal elements influencing a smart grid's composition

5. Findings and Conversations about Automation's Future, the Internet of Things, and Security Protocols for the "smart grid framework"
   1. The Future of Automation in the Smart Grid Framework
      a. Enhanced Efficiency and Reliability

The integration of automation in the smart grid has led to significant improvements in grid efficiency and reliability. Automated systems can monitor and manage energy distribution in real-time, minimizing power losses and reducing the frequency and duration of outages. Predictive maintenance, enabled by automation, allows for timely interventions before failures occur, further enhancing grid reliability.

b. Decentralized Energy Management

Automation facilitates decentralized energy management, where distributed energy resources (DERs) such as solar panels, wind turbines, and battery storage systems can be integrated and managed efficiently. This decentralization enhances the grid's resilience, allowing for localized energy generation and consumption, reducing dependency on centralized power plants, and promoting the use of renewable energy sources.

c. Real-Time Data Processing and Decision-Making

With the deployment of advanced metering infrastructure (AMI) and smart sensors, automation enables real-time data collection and processing. This capability allows for immediate decision-making and rapid response to grid anomalies, optimizing the performance and stability of the smart grid.

2. The Role of IoT in the Smart Grid
   a. Seamless Connectivity and Communication

The Internet of Things (IoT) plays a crucial role in creating a connected and intelligent grid. IoT devices, such as smart meters, sensors, and controllers, communicate with each other and with the central grid management system, providing real-time data on energy consumption, grid health, and environmental conditions. This connectivity ensures efficient grid operation and better demand-response management.

b. Enhanced Customer Engagement and Demand Response

IoT enables consumers to actively participate in the energy ecosystem through smart home devices and mobile applications. Customers can monitor their energy usage, receive real-time alerts, and adjust their consumption patterns based on dynamic pricing models. This interaction not only empowers consumers but also supports grid stability by reducing peak demand and encouraging energy conservation.

c. Predictive Analytics and Maintenance

The data generated by IoT devices is crucial for predictive analytics, which can forecast demand patterns, detect potential faults, and optimize energy distribution. Predictive maintenance, powered by IoT, allows utilities to anticipate equipment failures and schedule maintenance proactively, reducing downtime and maintenance costs.

3. Security Measures in the Smart Grid Framework
   a. Cybersecurity Challenges

The integration of IoT and automation in the smart grid introduces significant cybersecurity challenges. The interconnected nature of the grid makes it vulnerable to cyber-attacks, which could disrupt power supply, compromise data integrity, and endanger public safety. Protecting the grid from such threats is a top priority for utilities and regulatory bodies.

b. Multi-Layered Security Architecture

To safeguard the smart grid, a multi-layered security approach is essential. This includes securing communication networks, implementing strong authentication mechanisms, and deploying intrusion detection and prevention systems. Advanced encryption techniques are used to protect data in transit and at rest, ensuring that sensitive information remains secure.

c. Regulatory Compliance and Standards

Adherence to cybersecurity standards and regulations is critical for the smart grid's security. Governments and regulatory bodies have established frameworks, such as the NIST Cybersecurity Framework and the IEC 62443 standards, which provide guidelines for securing industrial control systems. Utilities must comply with these standards to ensure the resilience and security of the grid.

d. Incident Response and Recovery

In the event of a cyber-attack, having a robust incident response and recovery plan is vital. Utilities must be prepared to detect, respond to, and recover from cyber incidents promptly. This includes regular training for personnel, conducting simulations of cyber-attacks, and maintaining up-to-date backup systems to restore normal operations quickly.

Discussion and Implications

1. The Balance between Innovation and Security

As the smart grid evolves with increased automation and IoT integration, the challenge lies in balancing innovation with security. While automation and IoT offer numerous benefits, they also introduce new vulnerabilities that must be addressed. Ensuring the security of the smart grid without stifling innovation requires continuous investment in cybersecurity research and the development of new security technologies.

2. The Importance of Collaboration

The future success of the smart grid depends on collaboration between various stakeholders, including utilities, technology providers, regulators, and consumers. A coordinated effort is necessary to develop and implement security measures, establish industry standards, and promote best practices. Public-private partnerships can play a key role in driving innovation while maintaining a secure and resilient grid.

3. Consumer Awareness and Education

As consumers become more involved in the energy ecosystem through IoT-enabled devices, it is essential to raise awareness about the importance of cybersecurity. Educating consumers on how to protect their devices and data can reduce the risk of cyber-attacks originating from vulnerable endpoints.

4. Future Research Directions

Future research should focus on developing advanced security solutions tailored to the unique needs of the smart grid. This includes exploring the potential of artificial intelligence (AI) and machine learning (ML) for detecting and mitigating cyber threats in real time. Additionally, research into quantum-resistant encryption techniques could provide long-term solutions to emerging cybersecurity challenges.

Figure 4. Significant domains for IoT applications

The future of automation, IoT, and security in the smart grid framework is promising but also challenging. While automation and IoT enhance the efficiency, reliability, and resilience of the grid, they also introduce new security vulnerabilities, (Salama et al., 2024) (Salama & Al-Turjman, 2024) (Salama & Al-Turjman, 2024) (Salama & Al-Turjman, 2024) (Salama & Al-Turjman, 2024) (Salama, Altorogoman, & Al-Turjman, 2024) (Salama et al., 2024) (Salama, Altrjman, & Al-Turjman, 2024) (Salama & Al-Turjman, 2024) (Salama & Al-Turjman, 2024). Addressing these challenges requires a multi-faceted approach that includes advanced security measures, regulatory compliance, consumer education, and ongoing research and innovation. Through collaborative efforts, the vision of a secure, efficient, and sustainable smart grid can be realized.

6. CONCLUSION

To summarize, the future smart grid framework will be built with automation, robust security measures, and the Internet of Things (IoT). This might totally change the energy industry. The integration of these technologies offers novel applications that extend beyond incremental improvements in productivity and include areas like resilience, sustainability, and customer empowerment. Predictive maintenance, grid-integrated electric vehicle charging, and dynamic demand response are three instances of how automation and the Internet of Things could transform resource utilization and raise the reliability of energy services. Moreover, decentralized energy generation, intelligent micro grids, and energy-efficient smart homes demonstrate the promise for a more decentralized and customer-focused energy ecosystem. The implementation of security mechanisms is essential to the success of these advances. As the smart grid becomes more integrated, it is imperative to guarantee the security, integrity, and accessibility of data. The use of blockchain, encryption, and other cybersecurity methods not only safeguards sensitive data but also encourages trust in the adoption of these innovative technologies. The adaptive grid planning, environmental monitoring, and energy trading characteristics of smart grids reveal their potential to not only fulfill current needs but also to proactively build a more environmentally conscious and sustainable energy future. This comprehensive approach, driven by IoT, automation, and security, positions the smart grid as a crucial tool for addressing the complex issues arising from our evolving energy landscape. As we move forward, realizing the smart grid framework's full potential will call for ongoing research, collaboration, and a commitment to cybersecurity. By carefully implementing these developments, we can build an energy infrastructure that is intelligent, resilient, and safe enough to meet the demands of a rapidly changing world. The convergence of automation, IoT, and security measures heralds a paradigm change toward a more efficient, sustainable, and customer-focused energy strategy.

REFERENCES

. Al-Turjman, F., Salama, R., & Altrjman, C. (2023). Overview of IoT solutions for sustainable transportation systems. NEU Journal for Artificial Intelligence and Internet of Things, 2(3).

Salama, R., & Al-Turjman, F. (2023). Cyber-security countermeasures and vulnerabilities to prevent social-engineering attacks. In *Artificial Intelligence of Health-Enabled Spaces* (pp. 133–144). CRC Press. DOI: 10.1201/9781003322887-7

Salama, R., & Al-Turjman, F. (2024). Using artificial intelligence in education applications. In *Computational Intelligence and Blockchain in Complex Systems* (pp. 77–84). Morgan Kaufmann. DOI: 10.1016/B978-0-443-13268-1.00012-1

Salama, R., & Al-Turjman, F. (2024). Overview of the global value chain and the effectiveness of artificial intelligence (AI) techniques in reducing cyber-security risks. In *Smart Global Value Chain* (pp. 137–149). CRC Press. DOI: 10.1201/9781003461432-8

Salama, R., & Al-Turjman, F. (2024). A study of health-care data security in smart cities and the global value chain using AI and blockchain. In *Smart Global Value Chain* (pp. 165–172). CRC Press. DOI: 10.1201/9781003461432-10

Salama, R., & Al-Turjman, F. (2024). Blockchain technology, computer network operations, and global value chains together make up "cybersecurity". In *Smart Global Value Chain* (pp. 150–164). CRC Press. DOI: 10.1201/9781003461432-9

Salama, R., & Al-Turjman, F. (2024). Future uses of AI and blockchain technology in the global value chain and cybersecurity. In *Smart Global Value Chain* (pp. 257–269). CRC Press. DOI: 10.1201/9781003461432-17

. Salama, R., & Al-Turjman, F. (2024). Security And Privacy in Mobile Cloud Computing and the Internet of Things. NEU Journal for Artificial Intelligence and Internet of Things, 3(1).

. Salama, R., & Al-Turjman, F. (2024). An Overview of Blockchain Applications and Benefits in Cloud Computing. AIoT and Smart Sensing Technologies for Smart Devices, 66-76.

Salama, R., & Al-Turjman, F. (2024). A Description of How AI and Blockchain Technology Are Used in Business. In *AIoT and Smart Sensing Technologies for Smart Devices* (pp. 1–15). IGI Global. DOI: 10.4018/979-8-3693-0786-1.ch001

Salama, R., & Al-Turjman, F. (2024). Blockchain Applications and Benefits in Cloud Computing. In *AIoT and Smart Sensing Technologies for Smart Devices* (pp. 127–139). IGI Global. DOI: 10.4018/979-8-3693-0786-1.ch007

Salama, R., & Al-Turjman, F. (2024). Mobile cloud computing security issues in smart cities. In *Computational Intelligence and Blockchain in Complex Systems* (pp. 215–231). Morgan Kaufmann. DOI: 10.1016/B978-0-443-13268-1.00007-8

Salama, R., & Al-Turjman, F. (2024). Distributed mobile cloud computing services and blockchain technology. In *Computational Intelligence and Blockchain in Complex Systems* (pp. 205–214). Morgan Kaufmann. DOI: 10.1016/B978-0-443-13268-1.00002-9

Salama, R., Al-Turjman, F., Aeri, M., & Yadav, S. P. (2023, April). Internet of intelligent things (IoT)–An overview. In *2023 International Conference on Computational Intelligence, Communication Technology and Networking (CICTN)* (pp. 801-805). IEEE. DOI: 10.1109/CICTN57981.2023.10141157

Salama, R., Al-Turjman, F., Aeri, M., & Yadav, S. P. (2023, April). Intelligent hardware solutions for covid-19 and alike diagnosis-a survey. In *2023 International Conference on Computational Intelligence, Communication Technology and Networking (CICTN)* (pp. 796-800). IEEE. DOI: 10.1109/CICTN57981.2023.10140850

Salama, R., Al-Turjman, F., Altrjman, C., & Bordoloi, D. (2023, April). The ways in which Artificial Intelligence improves several facets of Cyber Security-A survey. In *2023 International Conference on Computational Intelligence, Communication Technology and Networking (CICTN)* (pp. 825-829). IEEE. DOI: 10.1109/CICTN57981.2023.10141376

Salama, R., Al-Turjman, F., Altrjman, C., & Gupta, R. (2023, April). Machine learning in sustainable development–an overview. In *2023 International Conference on Computational Intelligence, Communication Technology and Networking (CICTN)* (pp. 806-807). IEEE.

Salama, R., Al-Turjman, F., Altrjman, C., Kumar, S., & Chaudhary, P. (2023, April). A comprehensive survey of blockchain-powered cybersecurity-a survey. In *2023 International Conference on Computational Intelligence, Communication Technology and Networking (CICTN)* (pp. 774-777). IEEE. DOI: 10.1109/CICTN57981.2023.10141282

. Salama, R., Al-Turjman, F., Alturjman, S., & Altorgoman, A. (2024). An overview of artificial intelligence and blockchain technology in smart cities. Computational Intelligence and Blockchain in Complex Systems, 269-275.

Salama, R., Al-Turjman, F., Bhatla, S., & Gautam, D. (2023, April). Network security, trust & privacy in a wiredwireless Environments–An Overview. In *2023 International Conference on Computational Intelligence, Communication Technology and Networking (CICTN)* (pp. 812-816). IEEE. DOI: 10.1109/CICTN57981.2023.10141309

Salama, R., Al-Turjman, F., Bhatla, S., & Gautam, D. (2023, April). Network security, trust & privacy in a wiredwireless Environments–An Overview. In *2023 International Conference on Computational Intelligence, Communication Technology and Networking (CICTN)* (pp. 812-816). IEEE. DOI: 10.1109/CICTN57981.2023.10141309

Salama, R., Al-Turjman, F., Bhatla, S., & Mishra, D. (2023, April). Mobile edge fog, Blockchain Networking and Computing-A survey. In 2023 International Conference on Computational Intelligence, Communication Technology and Networking (CICTN) (pp. 808-811). IEEE.

Salama, R., Al-Turjman, F., Bhatla, S., & Yadav, S. P. (2023, April). Social engineering attack types and prevention techniques-A survey. In *2023 International Conference on Computational Intelligence, Communication Technology and Networking (CICTN)* (pp. 817-820). IEEE. DOI: 10.1109/CICTN57981.2023.10140957

Salama, R., Al-Turjman, F., Bordoloi, D., & Yadav, S. P. (2023, April). Wireless sensor networks and green networking for 6G communication-an overview. In *2023 International Conference on Computational Intelligence, Communication Technology and Networking (CICTN)* (pp. 830-834). IEEE. DOI: 10.1109/CICTN57981.2023.10141262

Salama, R., Al-Turjman, F., Chaudhary, P., & Banda, L. (2023, April). Future communication technology using huge millimeter waves—an overview. In *2023 International Conference on Computational Intelligence, Communication Technology and Networking (CICTN)* (pp. 785-790). IEEE. DOI: 10.1109/CICTN57981.2023.10140666

Salama, R., Al-Turjman, F., Chaudhary, P., & Yadav, S. P. (2023, April). (Benefits of Internet of Things (IoT) Applications in Health care-An Overview). In *2023 International Conference on Computational Intelligence, Communication Technology and Networking (CICTN)* (pp. 778-784). IEEE. DOI: 10.1109/CICTN57981.2023.10141452

Salama, R., Al-Turjman, F., Chaudhary, P., & Yadav, S. P. (2023, April). (Benefits of Internet of Things (IoT) Applications in Health care-An Overview). In *2023 International Conference on Computational Intelligence, Communication Technology and Networking (CICTN)* (pp. 778-784). IEEE. DOI: 10.1109/CICTN57981.2023.10141452

Salama, R., Al-Turjman, F., & Culmone, R. (2023, March). AI-powered drone to address smart city security issues. In *International Conference on Advanced Information Networking and Applications* (pp. 292-300). Cham: Springer International Publishing. DOI: 10.1007/978-3-031-28694-0_27

. Salama, R., Altorgoman, A., & Al-Turjman, F. (2024). An overview of how AI, blockchain, and IoT are making smart healthcare possible. Computational Intelligence and Blockchain in Complex Systems, 255-267.

. Salama, R., Altrjman, C., & Al-Turjman, F. (2023). An overview of the internet of things (IoT) and machine to machine (M2M) communications. NEU Journal for Artificial Intelligence and Internet of Things, 2(3).

. Salama, R., Altrjman, C., & Al-Turjman, F. (2023). A survey of the architectures and protocols for wireless sensor networks and wireless multimedia sensor networks. NEU journal for artificial intelligence and internet of things, 2(3).

. Salama, R., Altrjman, C., & Al-Turjman, F. (2023). A survey of machine learning (ML) in Sustainable Systems. NEU Journal for Artificial Intelligence and Internet of Things, 2(3).

. Salama, R., Altrjman, C., & Al-Turjman, F. (2024). An overview of future cyber security applications using AI and blockchain technology. Computational Intelligence and Blockchain in Complex Systems, 1-11.

. Salama, R., Altrjman, C., & Al-Turjman, F. (2024). Healthcare cybersecurity challenges: a look at current and future trends. Computational Intelligence and Blockchain in Complex Systems, 97-111.

Salama, R., Alturjman, S., & Al-Turjman, F. (2024). A survey of issues, possibilities, and solutions for a blockchain and AI-powered Internet of things. In *Computational Intelligence and Blockchain in Complex Systems* (pp. 13–24). Morgan Kaufmann. DOI: 10.1016/B978-0-443-13268-1.00019-4

. Salama, R., Alturjman, S., Altrjman, C., & Al-Turjman, F. (2023). Cloud Computing Services for Distributed Mobile Users and Blockchain Technology. NEU Journal for Artificial Intelligence and Internet of Things, 2(4).

Salama, R., Alturjman, S., Altrjman, C., & Al-Turjman, F. (2024). Service models for Internet of Things, global value chains, and mobile cloud computing. In *Smart Global Value Chain* (pp. 121–136). CRC Press. DOI: 10.1201/9781003461432-7

. Salama, R., Alturjman, S., Altrjman, C., & Al-Turjman, F. (2024). An Overview of the Applications of Blockchain and AI in Business. NEU Journal for Artificial Intelligence and Internet of Things, 3(1).

. Salama, R., Alturjman, S., Altrjman, C., & Al-Turjman, F. (2024). Distributed Mobile Cloud Computing Services Using Blockchain Technology. NEU Journal for Artificial Intelligence and Internet of Things, 3(1).

Salama, R., Cacciagrano, D., & Al-Turjman, F. (2024, April). Blockchain and Financial Services a Study of the Applications of Distributed Ledger Technology (DLT) in Financial Services. In International Conference on Advanced Information Networking and Applications (pp. 124-135). Cham: Springer Nature Switzerland.

Chapter 13
Smart Grid Environment, Data Security in the Internet of Things, and Supply Chain Ecosystem Transformation

Ramiz Salama
Near East University, Turkey

Fadi Al-Turjman
Near East University, Turkey

ABSTRACT

The supply chain ecosystem is undergoing a significant shift in today's business environment, mostly due to the adoption of smart grids and the integration of cutting-edge technologies like the Internet of Things (IoT). Unprecedented prospects for improved efficiency, real-time monitoring, and data-driven decision-making are presented by this paradigm change. But even in the middle of these developments, data security becomes an increasingly important issue. This study examines the various facets of the change of the supply chain ecosystem in the context of the Internet of Things and the smart grid, with a particular emphasis on the difficulties and solutions pertaining to data security. The smart grid and supply chain operations' integration of IoT devices create several weak points that are open to cyberattacks, underscoring the necessity of putting strong security measures in place as soon as possible.

DOI: 10.4018/979-8-3693-5573-2.ch013

INTRODUCTION

In the contemporary era, the global business landscape is undergoing a radical transformation driven by technological advancements, with the integration of the Internet of Things (IoT) and the implementation of smart grid environments emerging as pivotal forces reshaping traditional supply chain ecosystems. This convergence offers unprecedented opportunities for organizations to enhance operational efficiency, optimize resource utilization, and make informed, data-driven decisions in real-time.

The integration of IoT technologies in supply chain management allows for the seamless interconnectivity of devices, creating an intricate web of data exchange throughout the entire supply chain. Simultaneously, the deployment of smart grid environments empowers organizations to manage and optimize energy resources intelligently. While these innovations promise transformative benefits, they also bring to the forefront a critical concern that cannot be overlooked – the security of data within this interconnected ecosystem.

The significance of data security within the context of the IoT-enabled supply chain and smart grid environment cannot be overstated. As the volume of data generated exponentially increases and the interdependence of devices grows, vulnerabilities to cyber threats become more pronounced. The potential consequences of a breach extend beyond the compromise of sensitive information, impacting operational efficiency, disrupting supply chain dynamics, and posing threats to the overall resilience of the system.

This paper aims to delve into the multifaceted challenges associated with ensuring data security in the evolving landscape of supply chain ecosystems, particularly within the realms of IoT and smart grid technologies. By exploring the intricacies of this transformation, we seek to identify key issues and propose strategic solutions to safeguard the integrity, confidentiality, and availability of data throughout the supply chain, (Salama et al., 2023a) (Salama et al., 2023b) (Salama et al., 2023c) (Salama et al., 2023d) (Salama et al., 2023e).

The subsequent sections of this paper will address the unique challenges posed by the integration of IoT in supply chain operations, the vulnerabilities introduced by the smart grid environment, and the critical role of data security in mitigating potential risks. Furthermore, we will explore emerging technologies and collaborative strategies that can fortify the security framework, ensuring a resilient and trustworthy ecosystem for businesses navigating the dynamic intersection of supply chain, IoT, and smart grid technologies. Through a holistic understanding of these dynamics, organizations can strategically position themselves to unlock the full potential of these transformative technologies while safeguarding against evolving cyber threats.

Literature on "Data Security" in the Internet of Things, Smart Grid Environment, and Supply Chain Ecosystem Transformation

Here's an overview of literature work in the areas of Supply Chain Ecosystem Transformation, Data Security in the Internet of Things (IoT), and Smart Grid Environment, (Salama, Al-Turjman, & Culmone, 2023) (Salama et al., 2023f) (Salama et al., 2023g) (Salama, Altrjman, & Al-Turjman, 2023a) (Salama et al., 2023h).

1. Supply Chain Ecosystem Transformation
   - **Digital Transformation in Supply Chains:** The transformation of supply chains through digital technologies such as blockchain, IoT, and AI has been extensively studied. Literature highlights how these technologies enhance transparency, efficiency, and sustainability. Notable works discuss the impact of digital twins, predictive analytics, and automation on supply chain operations.
   - **Resilient and Agile Supply Chains:** Post-pandemic, there's a growing body of work focused on building resilient and agile supply chains. Research delves into strategies for risk management, including diversification of suppliers, adoption of real-time data analytics, and re-shoring of manufacturing processes.
   - **Sustainable Supply Chains:** The literature also explores the integration of sustainability in supply chains, with studies emphasizing circular economy principles, waste reduction, and green logistics.
2. Data Security in the Internet of Things (IoT)
   - **Threats and Vulnerabilities:** Extensive research has been conducted on the security challenges posed by the IoT, where devices often have limited computational resources. Studies focus on common vulnerabilities like weak authentication, unencrypted data transmission, and firmware tampering.
   - **Security Protocols and Frameworks:** Literature reviews various security protocols and frameworks designed for IoT, including lightweight cryptography, blockchain-based security solutions, and AI-driven anomaly detection. The development of end-to-end security frameworks to protect data at all stages (collection, transmission, and storage) is a prominent focus.
   - **Privacy Concerns:** With the vast amount of data generated by IoT devices, privacy is a major concern. Research explores methods to ensure data privacy through techniques like data anonymization, decentralized data storage, and user-controlled data sharing mechanisms.

3. Smart Grid Environment
   - **Cybersecurity Challenges:** The integration of digital technologies in smart grids brings significant cybersecurity challenges. The literature covers the potential attack vectors, such as Distributed Denial of Service (DDoS) attacks, and discusses security measures like intrusion detection systems, encryption protocols, and blockchain for secure transactions.
   - **Data Management and Analytics:** Smart grids generate vast amounts of data. Studies focus on big data analytics, machine learning, and AI to optimize grid operations, predict demand, and enhance fault detection. Research also emphasizes secure and efficient data management practices to handle this data.
   - **Decentralized Energy Systems:** The shift towards decentralized energy production (e.g., solar panels, wind turbines) within smart grids is another key area of study. Literature discusses the challenges and opportunities of integrating decentralized sources into the grid, focusing on grid stability, real-time monitoring, and the use of IoT devices.

Figure 1. Features of the smart grid

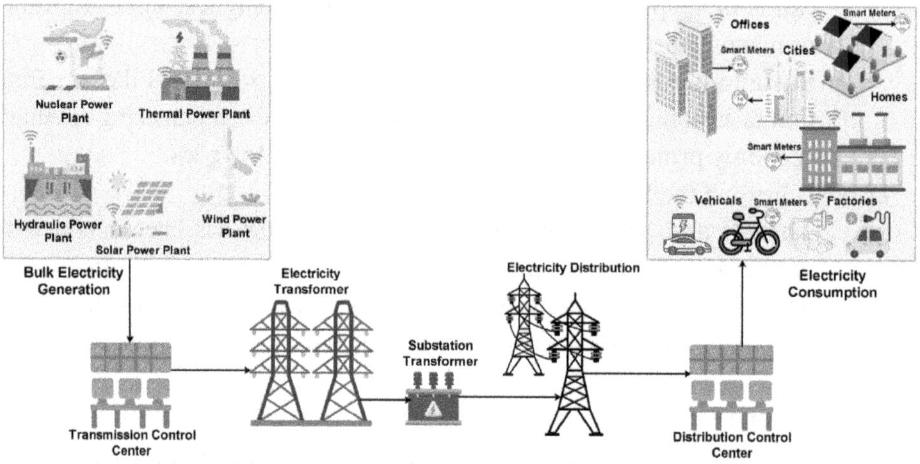

These topics represent significant research areas with ongoing developments. If you're looking for specific papers or more detailed information on any of these topics, I can help guide you further.

Purpose of the Research

The purpose of research in areas such as Supply Chain Ecosystem Transformation, "Data Security" in the Internet of Things (IoT), and the Smart Grid Environment is crucial for addressing emerging challenges and opportunities in these fields. Here's a breakdown of the research objectives for each area:

1. Supply Chain Ecosystem Transformation
   - **Objective:** To explore how technological advancements and innovative practices can transform traditional supply chains into more resilient, efficient, and sustainable ecosystems.
   - **Purpose:**
     o **Resilience:** To enhance the supply chain's ability to withstand and recover from disruptions, such as those caused by global pandemics or geopolitical instability.
     o **Efficiency:** To improve operational efficiency through automation, digitalization, and integration of advanced analytics, leading to cost savings and faster response times.
     o **Sustainability:** To assess how sustainable practices, such as reducing carbon footprints and waste, can be integrated into supply chain processes.
     o **Innovation:** To investigate the role of emerging technologies like blockchain, AI, and IoT in enabling new business models and improving transparency across the supply chain.

2. Data Security in the Internet of Things (IoT)
   - **Objective:** To understand and mitigate the risks associated with data security in the ever-expanding network of IoT devices.
   - **Purpose:**
     o **Threat Mitigation:** To identify potential security vulnerabilities within IoT networks and develop strategies to protect against cyber threats, such as data breaches, unauthorized access, and hacking.
     o **Privacy Protection:** To ensure that the vast amounts of data collected by IoT devices are protected, maintaining user privacy and adhering to regulatory standards.
     o **Trust and Adoption:** To build trust among consumers and businesses in IoT technologies by developing robust security protocols, fostering wider adoption.

- o **Standards Development:** To contribute to the establishment of industry-wide security standards and best practices for IoT deployment.

3. Smart Grid Environment
    - **Objective:** To explore the design, implementation, and optimization of smart grid systems, focusing on efficiency, reliability, and sustainability.
    - **Purpose:**
        - o **Energy Efficiency:** To improve the efficiency of energy distribution and consumption, reducing waste and optimizing resource use in real-time.
        - o **Reliability:** To enhance the reliability and resilience of the power grid by integrating advanced monitoring, control systems, and predictive maintenance tools.
        - o **Sustainability:** To support the transition to renewable energy sources by enabling better integration of distributed energy resources, such as solar and wind, into the grid.
        - o **Consumer Empowerment:** To empower consumers with tools and technologies that allow them to monitor and manage their energy usage more effectively.
        - o **Cybersecurity:** To address the security challenges of smart grids, ensuring that the infrastructure is protected from cyber attacks and vulnerabilities.

Research in these areas is essential for fostering innovation, ensuring security, and driving the transformation necessary to meet the demands of the future.

Smart Grid and IoT

The Internet of Things (IoT) plays a pivotal role in the Smart Grid (SG), revolutionizing the way power systems are monitored, managed, and optimized. Within the SG, all objects and components can be represented as IoT devices distributed throughout residential networks, substations, and utilities. These IoT devices enable tracking capabilities that facilitate monitoring, connectivity, and automation within the SG ecosystem, (Salama et al., 2023i) (Salama et al., 2023j) (Salama, Altrjman, & Al-Turjman, 2023b) (Salama, Altrjman, & Al-Turjman, 2023c) (Salama & Al-Turjman, 2023).

As an enabling technology, the IoT provides the SG with crucial internet connectivity, making it a key application of the IoT within the broader context of cyber-physical systems. In the SG, each IoT device is connected to the internet and possesses a unique IP address, allowing for seamless communication of information and control commands using internet protocols. This integration of IoT devices and IP addressing schemas empowers the SG with advanced monitoring and control capabilities.

The monitoring aspect of IoT devices in the SG encompasses various facets, including generation, distribution, storage, and consumption. By leveraging IoT technologies, the SG can achieve efficient energy management, demand management, accurate estimation of required energy, and administration of CO_2 emissions. These IoT devices enable precise measurement and analysis of renewable energy needs while ensuring the reduction of wasted energy.

Furthermore, the bidirectional flow of data exchange between IoT devices is facilitated through the SG's communication layer. The IoT standardizes communication protocols, reducing complexities related to SG components, (Salama & Al-Turjman, 2023). An important aspect emphasized by IoT technologies is their ability to enable seamless communication across multiple subsystems within the SG, encompassing generation, transmission, distribution, and consumption. This interconnectedness allows IoT devices to exchange data and commands among control centers and utilities, fostering efficient coordination and collaboration within the SG ecosystem.

By leveraging the capabilities of IoT technologies, the SG can optimize energy generation, distribution, and consumption, ultimately leading to improved efficiency, reduced wastage, and better management of resources. The integration of IoT devices in the SG ensures seamless connectivity, efficient monitoring, and intelligent control, enabling the realization of a more reliable, sustainable, and resilient power infrastructure.

Security and the Smart Grid

The Smart Grid (SG) represents a transformative advancement in the power sector, offering numerous opportunities for improved efficiency and sustainability. However, alongside these benefits, the SG environment also presents a multitude of security challenges. To fully capitalize on the potential of the SG, it becomes imperative to develop a highly secure information system that can protect critical infrastructure and ensure the uninterrupted flow of electricity, (Salama et al., 2023k) (Salama et al., 2023l) (Salama, Altrjman, & Al-Turjman, 2023d) (SDalama et al., 2023m) (Al-Turjman, salama, & Altrjman, 2023).

Historically, automation systems like SCADA (Supervisory Control and Data Acquisition) and protocols such as Modbus were designed without adequate consideration for security in mind. These systems, originally intended for industrial processes, were not equipped to handle the critical security requirements of the Smart Grid. As power assets are classified as critical national infrastructure and can be attractive targets for malicious actors, securing the SG information system becomes an utmost priority. The potential damage resulting from security attacks on the power grid could lead to chaos across entire cities. Consequently, the Electric Power Research Institute (ERPI) identifies security as one of the primary concerns in the global implementation of the SG.

The security challenges within an IoT-enabled Smart Grid to arise due to several factors. Firstly, the exposure of SG to the internet renders it susceptible to tampering and unauthorized data manipulation by attackers. Additionally, the increasing number of IoT devices incorporated into the SG infrastructure amplifies its vulnerability to attacks.

The communication among entities within the SG relies on IP-based networks, where sensitive and private data is exchanged between consumers and utility companies. Such networks are prone to various security threats, including man-in-the-middle attacks, denial of service attacks, eavesdropping, and replay attacks. Furthermore, the interconnected nature of SG, with its various components and technologies, introduces access points that can be exploited by security attacks. Wireless sensor networks used to connect smart meters have also been identified as potential security weak points.

Moreover, unauthorized access to the SG can result in bidirectional information flow, further exposing the system to a range of threats. Furthermore, the integration of IoT devices into the SG introduces security issues inherited from the IoT ecosystem itself. To effectively monitor and control IoT devices, the SG relies on internet connectivity, which can introduce additional security vulnerabilities.

Figure 2. Security and the smart grid

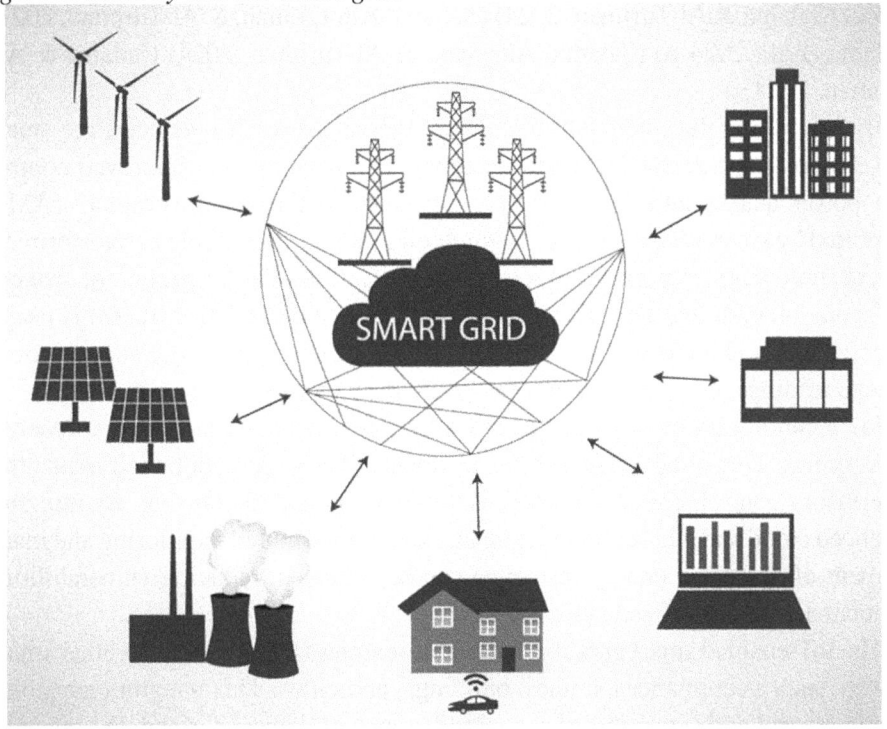

Addressing these security challenges is crucial for the successful deployment and operation of the Smart Grid. Robust security measures must be implemented to safeguard the integrity, confidentiality, and availability of data within the SG ecosystem. By mitigating the risks associated with IP-based communication networks, securing access points, ensuring the integrity of wireless sensor networks, and establishing stringent authorization protocols, the Smart Grid can establish a strong defense against potential security threats.

The Proliferation of Internet of Things (IoT) Devices in Intelligent Grids

The rise of Internet of Things (IoT) devices in smart grids has revolutionized the way energy systems are managed and optimized. In a smart grid environment, the integration of distributed renewable energy resources, real-time communication between consumers and service providers, data analysis, and implementation of necessary actions based on these analyses are essential components. The IoT, with its multifaceted benefits, presents a promising solution to enhance the efficiency and intelligence of the smart energy grid system, (Salama et al., 2024)a (Salama

& Al-Turjman, 2024a) (Salama & Al-Turjman, 2024b) (Salama & Al-Turjman, 2024c) (Salama & Al-Turjman, 2024) (Salama, Altorgoman, & Al-Turjman, 2023) (Salama et al., 2024 b) (Salama, Altrjman, & Al-Turjman, 2024) (Salama & Al-Turjman, 20243d).

By leveraging the intelligent and proactive features of IoT devices, the smart grid system can achieve increased accuracy, competency, and improved control over power quality and dependability. Advanced metering infrastructure (AMI) supported by smart metering (SM) technologies plays a crucial role in transforming the conventional power grid into a smart grid. Through the integration of sensing and actuation systems, the IoT enables optimization and regulation of energy usage by collecting and analyzing vast amounts of data, including energy consumption, voltage readings, current readings, and phase measurements.

IoT technologies have a significant impact on various aspects of smart energy grid systems. They enable efficient management of power generation infrastructure, supervisory control, data acquisition, and environmental monitoring. By utilizing advanced cloud and edge computing technologies, distributed monitoring and management of dispersed energy resources can be achieved, reducing vulnerabilities associated with centralized systems.

The IoT-enabled smart grid also facilitates seamless integration with other smart entities, such as appliances, homes, buildings, and cities. This integration enables greater control and accessibility over a wide range of devices through the internet. However, the implementation of an IoT-enabled smart grid is not without challenges. Cyber adversaries can exploit vulnerabilities, leading to operational, economic, and system security threats. These attacks can have damaging effects on the smart grid's functionality and overall security.

To summarize, the adoption of IoT devices in smart grids brings numerous benefits, including improved energy management, increased efficiency, and integration with other smart entities. However, it is essential to address the security challenges associated with IoT implementation to safeguard the smart grid from potential cyber threats.

Physical and Cybersecurity Risks and Difficulties with IoT-Powered Smart Grids

The integration of Internet of Things (IoT) technologies within the Smart Grid infrastructure has ushered in a new era of advanced power systems. By enabling seamless communication and data exchange among interconnected devices, IoT-enabled Smart Grids offer numerous benefits such as improved efficiency, reliability, and sustainability. However, this increased connectivity also exposes Smart Grid to a host of cyber-physical security vulnerabilities and challenges. This two-page description aims to explore these vulnerabilities and challenges, emphasizing the importance of addressing them to ensure the resilience and security of IoT-enabled Smart Grids, (Salama & Al-Turjman, 2024a) (Salama & Al-Turjman, 2024b) (Salama, Altrjman, & Al-Turjman, 2024a) (Salama, Alturjman, & Al-Turjman, 2024b) (Salama et al., 2024) (Salam, Cacciagrano, & Al-Turjman, 2024).

CYBER-PHYSICAL SECURITY VULNERABILITIES:

Interconnectedness and Exposure to the Internet

The integration of Internet of Things (IoT) technologies within the Smart Grid infrastructure has revolutionized the power sector, enabling advanced monitoring, control, and optimization of electricity generation, distribution, and consumption. However, the increased interconnectedness and exposure to the internet in IoT-enabled Smart Grids introduce a range of cyber-physical security vulnerabilities. This description aims to delve into the vulnerabilities associated with the interconnectedness and exposure to the internet in the context of IoT-enabled Smart Grids, highlighting the risks and implications they pose.

Increased Attack Surface:

The interconnectedness of devices in IoT-enabled Smart Grids significantly expands the attack surface for potential adversaries. Each connected device, ranging from smart meters and sensors to grid management systems, represents a potential entry point for cyber-attacks. This expanded attack surface creates numerous points of vulnerability that attackers can exploit to gain unauthorized access, manipulate data, disrupt operations, or compromise critical infrastructure components.

Network Security:

IoT-enabled Smart Grids rely on communication networks to facilitate the exchange of data between devices, utility providers, and consumers. These networks, often based on internet protocols, are susceptible to a wide range of network-based attacks. Adversaries can launch attacks such as man-in-the-middle attacks, where they intercept and manipulate data between communicating entities, or denial-of-service attacks that disrupt the availability of services and grid operations. Insecure network configurations and unpatched vulnerabilities in network devices can further exacerbate these risks.

Insecure Communication Protocols:

The utilization of diverse communication protocols in IoT-enabled Smart Grids introduces vulnerabilities in data transmission. Some protocols may lack adequate security measures, making them susceptible to interception, data tampering, or unauthorized access. Attackers can exploit these vulnerabilities to eavesdrop on sensitive information, inject malicious commands into the communication stream, or gain unauthorized control over devices and systems within the Smart Grid.

Weak Device Security:

IoT devices deployed in Smart Grids often have resource constraints and limited security features. This weakness leaves them susceptible to exploitation by attackers. Insufficient authentication mechanisms, lack of encryption, and inadequate firmware security make these devices potential targets for compromise. Once compromised, attackers can manipulate device functionality, tamper with data, or use the compromised device as a launching point for further attacks within the Smart Grid ecosystem.

Remote Access and Management:

IoT-enabled Smart Grids often require remote access and management capabilities for monitoring, maintenance, and system updates. However, the provision of remote access introduces additional security risks. Inadequate authentication mechanisms or weak access controls can allow unauthorized individuals to gain access to critical infrastructure, leading to potential disruptions or unauthorized manipulation of devices and systems. Remote management interfaces must be carefully secured and monitored to prevent unauthorized access and ensure the integrity and availability of the Smart Grid.

Insecure Communication Protocols

The integration of Internet of Things (IoT) technologies within the Smart Grid infrastructure facilitates seamless data exchange and control between various devices and systems. However, if the communication protocols employed are not adequately secured, they can become entry points for malicious actors to exploit and compromise the integrity and availability of the entire power grid.

Insecure communication protocols within IoT-enabled Smart Grids are susceptible to various types of attacks. For instance, man-in-the-middle attacks can intercept and manipulate data flowing between devices, leading to unauthorized control or manipulation of critical infrastructure. Additionally, vulnerabilities in communication protocols can be exploited to launch denial-of-service attacks, disrupting the functioning of the Smart Grid and causing power outages.

Moreover, insecure communication protocols can expose sensitive data to unauthorized access and tampering. Energy consumption patterns, user information, and operational details transmitted over insecure channels can be intercepted, leading to privacy breaches and potential misuse of confidential data. This not only compromises consumer privacy but also undermines the overall trust and confidence in the Smart Grid ecosystem.

To mitigate the vulnerabilities posed by insecure communication protocols, it is essential to employ robust and secure protocols for data transmission and device communication within the IoT-enabled Smart Grid. Implementing encryption mechanisms, authentication protocols, and secure data transmission techniques such as Transport Layer Security (TLS) or Secure Socket Layer (SSL) can help protect data integrity and confidentiality. Furthermore, regularly updating and patching communication protocols to address known vulnerabilities is crucial to staying ahead of emerging threats.

Addressing the vulnerability of insecure communication protocols requires collaboration between industry stakeholders, standards organizations, and regulatory bodies. Establishing and enforcing security standards for communication protocols within the Smart Grid environment is essential to ensure that all devices and systems adhere to robust security practices.

By addressing the vulnerability of insecure communication protocols, IoT-enabled Smart Grids can enhance the overall resilience and security of power systems, safeguarding against cyber threats and maintaining the integrity and reliability of the grid infrastructure.

Weak Device Security

The deployment of IoT devices in the Smart Grid infrastructure introduces numerous interconnected devices that often possess limited computing resources. This limitation makes them susceptible to security breaches and compromises.

Many IoT devices used in Smart Grids lack robust security mechanisms, such as secure booting, firmware integrity verification, and encryption capabilities. These vulnerabilities create opportunities for attackers to exploit the devices and compromise the confidentiality, integrity, and availability of data. Furthermore, weak device security allows unauthorized access to control systems, potentially leading to unauthorized manipulation of critical infrastructure and disrupting power distribution.

As a result, addressing weak device security becomes crucial in ensuring the overall resilience and security of IoT-enabled Smart Grids. Measures such as implementing strong authentication mechanisms, enforcing regular software updates and patches, conducting security audits, and promoting secure coding practices can significantly enhance device security and mitigate the risk of cyber-physical attacks on the Smart Grid infrastructure.

By prioritizing and strengthening device security, Smart Grid operators can build a more robust and secure ecosystem that safeguards against unauthorized access, data tampering, and potential disruptions, ultimately ensuring the reliability and integrity of the power grid.

Supply Chain Risks

As the Internet of Things (IoT) continues to expand its influence in various industries, including the power sector, it brings numerous benefits and advancements. However, the integration of IoT technologies within Smart Grid infrastructures also introduces new cyber-physical security vulnerabilities. One critical vulnerability that demands attention is the supply chain risk associated with IoT-enabled Smart Grids. This short description aims to highlight the significance of supply chain risks as a cybersecurity concern for Smart Grid deployments.

Supply Chain Risks in IoT-Enabled Smart Grids:

Compromised Devices: The complex supply chain involved in the manufacturing and deployment of IoT devices creates opportunities for potential compromise. At various stages of the supply chain, devices may be tampered with, either intentionally or unintentionally. Malicious actors can exploit this vulnerability by injecting malware, backdoors, or other unauthorized modifications into the devices. Compromised devices, when integrated into the Smart Grid, can serve as entry points for cyberattacks, enabling unauthorized access and control over critical infrastructure.

Counterfeit Components: The procurement of IoT devices for Smart Grid deployments involves sourcing components from various suppliers. This complexity increases the risk of counterfeit components entering the supply chain. Counterfeit components can have compromised functionality or hidden vulnerabilities, making them susceptible to exploitation. The use of such components within IoT-enabled Smart Grids can compromise the system's security and integrity, potentially leading to unauthorized access, data breaches, and disruptions in power distribution.

Lack of Transparency: The complex nature of supply chains often leads to a lack of transparency regarding the origins and integrity of IoT devices. Without proper visibility into the supply chain, it becomes challenging to ensure the authenticity and trustworthiness of the components used in Smart Grid deployments. This lack of transparency increases the risk of incorporating compromised or substandard devices, exposing the Smart Grid to potential cyber threats.

Mitigating Supply Chain Risks:

Addressing supply chain risks in IoT-enabled Smart Grids requires a multi-faceted approach that involves various stakeholders, including manufacturers, suppliers, and system integrators. Some key measures to mitigate these risks include:

Vendor Due Diligence: Conducting thorough evaluations of vendors and suppliers before procurement is crucial. Assessing their security practices, quality control processes, and adherence to industry standards can help identify trustworthy partners and reduce the risk of compromised devices entering the supply chain.

Secure Manufacturing and Delivery: Implementing secure manufacturing and delivery practices can help minimize the risk of tampering and unauthorized access. Employing secure manufacturing facilities, using tamper-evident packaging, and maintaining strict chain of custody protocols during device transportation can enhance the integrity of IoT devices throughout the supply chain.

Trusted Relationships: Building strong and trusted relationships with suppliers and manufacturers is vital. Collaboration and regular communication can foster transparency, enabling better visibility into the supply chain and facilitating the identification and resolution of any potential security concerns.

Figure 3. Categorization of security concerns and obstacles for smart grid systems enabled by IoT

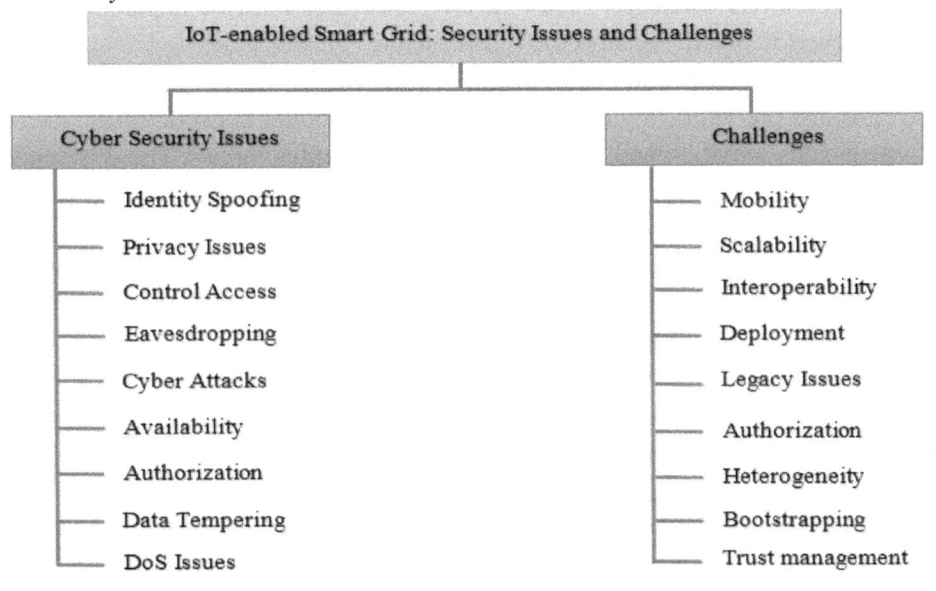

IoT-enabled Smart Grid: Security Issues and Challenges

Cyber Security Issues:
- Identity Spoofing
- Privacy Issues
- Control Access
- Eavesdropping
- Cyber Attacks
- Availability
- Authorization
- Data Tempering
- DoS Issues

Challenges:
- Mobility
- Scalability
- Interoperability
- Deployment
- Legacy Issues
- Authorization
- Heterogeneity
- Bootstrapping
- Trust management

Device Authentication and Integrity Verification: Implementing robust device authentication mechanisms and integrity verification techniques can help detect and prevent the use of counterfeit or compromised components. Techniques such as digital signatures, secure boot processes, and code validation can ensure the authenticity and integrity of IoT devices integrated into Smart Grid infrastructures.

Challenges with Cyber-Physical Security:

Resilience to Cyber Attacks

With the integration of Internet of Things (IoT) technologies, the Smart Grid becomes increasingly interconnected, creating new avenues for potential threats. Resilience refers to the ability of the system to withstand and recover from cyber-attacks, minimizing the impact on critical infrastructure and ensuring the continuous operation of the power grid, (Salama & Al-Turjman, 2024a) (Salama & Al-Turjman, 2024b) (Salam & Al-Turjman, 2024) (Salama & Al-Turjman, 2024) (Salama et al., 2024).

Cyber-attacks on IoT-enabled Smart Grids can have severe consequence
s, including disruptions to power distribution, manipulation of data, and unauthorized control of devices. Therefore, building resilience against such attacks is crucial to maintain the reliability and security of the power grid.

To address this challenge, robust security measures must be implemented at various levels within the Smart Grid infrastructure. This includes:

Intrusion Detection and Prevention Systems: Deploying advanced intrusion detection and prevention systems that continuously monitor network traffic, identify suspicious activities, and respond promptly to potential threats. These systems can detect anomalies, such as unauthorized access attempts or abnormal behavior, enabling swift action to mitigate the impact of cyber-attacks.

Anomaly Detection Mechanisms: Implementing anomaly detection mechanisms to identify deviations from normal system behavior. By analyzing patterns and identifying unusual activities, these mechanisms can detect potential cyber-attacks in real-time, allowing for immediate response and mitigation measures.

Incident Response Planning: Developing comprehensive incident response plans that outline the steps to be taken in the event of a cyber-attack. These plans should include predefined procedures for isolating affected components, mitigating the impact, and restoring normal operations. Regular training and drills for incident response teams ensure a coordinated and effective response to cyber threats.

Secure Communication Protocols: Employing secure communication protocols to protect data transmitted between devices and systems within the Smart Grid. Encryption, authentication, and integrity mechanisms ensure that data remains confidential and unaltered during transmission, mitigating the risk of eavesdropping, tampering, or unauthorized access.

Redundancy and Backup Systems: Implementing redundancy and backup systems to ensure continuity of operations even in the face of cyber-attacks. Redundant components and alternative power sources can mitigate the impact of attacks on critical infrastructure, allowing for seamless power distribution and rapid recovery.

Continuous Monitoring and Updates: Regularly monitoring and updating IoT devices, network infrastructure, and security systems to address emerging vulnerabilities and patch known weaknesses. This includes applying security patches, firmware updates, and employing threat intelligence to stay ahead of evolving cyber threats.

By proactively addressing the resilience challenge, IoT-enabled Smart Grids can strengthen their defenses against cyber-attacks and enhance the overall security and reliability of the power grid. Implementing robust security measures, detecting anomalies, planning for incident response, securing communications, and maintaining updated systems are key elements in achieving a resilient Smart Grid that can withstand and recover from cyber-attacks, ensuring the uninterrupted delivery of electricity to consumers.

Privacy and Data Protection

IoT-enabled Smart Grids collect and process vast amounts of data, including sensitive customer information, energy consumption patterns, and operational details. Protecting this data from unauthorized access, misuse, and breaches has become a paramount concern.

The extensive collection of personal data within IoT-enabled Smart Grids raises significant privacy concerns. Personally identifiable information, such as customer names, addresses, and energy usage, can be exploited by malicious actors for identity theft, fraud, or targeted attacks. Moreover, the detailed energy consumption patterns of individuals and businesses can reveal sensitive information, compromising their privacy.

Data protection is also a crucial aspect of security in IoT-enabled Smart Grids. The large volume of data generated by interconnected devices and the transmission of this data over communication networks necessitate robust mechanisms for secure storage and transmission. Encryption, access control, and secure protocols are essential to safeguard data integrity, confidentiality, and availability.

Compliance with data protection regulations, such as the General Data Protection Regulation (GDPR) and industry-specific standards, adds an additional layer of complexity to ensuring privacy and data protection in IoT-enabled Smart Grids. Organizations must establish policies and procedures to govern the collection, use, and retention of data, as well as provide transparent disclosure to customers about how their data is being utilized.

Addressing privacy and data protection challenges requires a multi-faceted approach. Implementing privacy-by-design principles during the development of IoT-enabled Smart Grid systems ensures that privacy and data protection considerations are embedded from the outset. This involves incorporating privacy-enhancing technologies, conducting privacy impact assessments, and adopting data minimization practices.

Encryption and secure data transmission protocols should be employed to protect data while in transit. Additionally, strict access controls and authentication mechanisms must be in place to prevent unauthorized access to sensitive data. Regular security audits, vulnerability assessments, and incident response plans are crucial for proactive detection and mitigation of potential data breaches.

Collaboration between industry stakeholders, regulators, and consumers is vital to address privacy and data protection challenges effectively. Industry standards and best practices should be established, and public awareness campaigns can educate consumers about the importance of their privacy rights and the measures taken to protect their data.

By prioritizing privacy and data protection as integral components of IoT-enabled Smart Grid security, organizations can instill trust among consumers and stakeholders. This, in turn, fosters the successful deployment and adoption of IoT technologies in the Smart Grid while safeguarding individuals' privacy and ensuring the protection of sensitive data.

Integration of Legacy Systems

Legacy systems, such as Supervisory Control and Data Acquisition (SCADA) systems, were not originally designed with modern cybersecurity considerations in mind. As a result, when integrating these legacy systems with IoT technologies, vulnerabilities can arise, exposing the Smart Grid to potential security risks.

One key challenge is the compatibility between legacy systems and IoT devices. Legacy systems often use outdated protocols and lack the necessary security features to protect against modern cyber threats. Integrating them with IoT devices requires careful analysis and retrofitting to ensure both compatibility and adequate security measures.

Additionally, the presence of legacy systems introduces complexities in terms of monitoring and securing the Smart Grid infrastructure. These systems may not provide sufficient visibility into the network, making it challenging to detect and respond to security incidents effectively. Moreover, legacy systems might have limited or outdated firmware, making them more susceptible to exploitation by cyber attackers.

Another concern is the lack of regular security updates and patches for legacy systems. As these systems age, vendors may stop providing updates or support, leaving them vulnerable to newly discovered security vulnerabilities. This situation poses a risk to the overall security of the Smart Grid, as attackers can exploit known vulnerabilities in legacy systems to gain unauthorized access or disrupt critical operations.

To address the security challenges associated with integrating legacy systems into IoT-enabled Smart Grids, it is crucial to implement robust security measures. This includes conducting comprehensive security assessments of legacy systems, applying necessary updates and patches when available, and implementing additional security controls, such as network segmentation and access controls, to isolate legacy systems from potentially compromised IoT devices.

Furthermore, organizations should consider the long-term strategy of transitioning away from legacy systems and replacing them with more secure and modern technologies. This may involve investing in new infrastructure, updating protocols, and ensuring that future systems are designed with security as a top priority.

Scalability and Complexity

As smart grids expand and incorporate a growing number of interconnected devices, their scalability becomes crucial. However, this expansion also increases the attack surface and potential vulnerabilities that can be exploited by malicious actors.

One of the primary security concerns associated with scalability is the management of many devices. Each IoT device within the smart grid ecosystem has its own unique set of security requirements and potential vulnerabilities. Ensuring that all devices are properly configured, patched, and updated becomes a complex task as the system scales up. Failure to manage this effectively can lead to security gaps, leaving the smart grid exposed to attacks.

Furthermore, the complex nature of smart grids adds to the security challenges. Smart grids involve a multitude of components such as sensors, communication networks, data analytics systems, and control mechanisms. Coordinating and securing the interactions between these diverse components is a daunting task. Each component introduces potential security vulnerabilities that need to be addressed to maintain the integrity and confidentiality of the grid's operations.

The complexity of the smart grid also increases the difficulty of identifying and mitigating security breaches or abnormal activities. Monitoring and detecting malicious activities amidst the vast amount of data generated by IoT devices becomes a significant challenge. An attacker who gains access to a single compromised device within the grid can potentially exploit vulnerabilities in other connected devices, leading to widespread disruptions or unauthorized access to critical systems.

To tackle these security challenges, a comprehensive and multi-layered approach is necessary. This includes implementing robust authentication mechanisms, encrypting communication channels, regularly updating and patching devices, conducting security audits, and employing advanced anomaly detection techniques. Additionally, collaboration among stakeholders such as utilities, device manufacturers, and cybersecurity experts is crucial to address the scalability and complexity concerns effectively.

IoT-enabled Smart Grids bring numerous benefits to the power sector, but they also introduce new cyber-physical security vulnerabilities and challenges. To ensure the resilience and security of these systems, it is imperative to address the interconnectedness and exposure to the internet, secure communication protocols, device security, supply chain risks, and challenges related to cyber-attacks, privacy, data protection, integration of legacy systems, scalability, and complexity. By developing robust security measures, implementing secure communication protocols, enhancing device security, and establishing resilient incident response strategies, we can safeguard IoT-enabled Smart Grids from cyber threats and build a secure foundation for the future of power systems.

A Number of Innovative Applications for the Internet of Things' "Data Security," the Smart Grid Environment, and Supply Chain Ecosystem Transformation

1. **Cold Chain Management in Pharmaceuticals:**
   - In the supply chain ecosystem, especially within pharmaceuticals, IoT sensors can be utilized to monitor and ensure the integrity of temperature-sensitive products during transportation. This not only enhances the overall efficiency of the supply chain but also introduces a critical use-case for data security, ensuring that temperature data is not compromised to maintain the efficacy of pharmaceuticals.
2. **Blockchain-enabled Transparency in Food Supply Chains:**
   - Leveraging blockchain in conjunction with IoT devices allows for real-time tracking of food products from farm to table. This use-case enhances transparency in the supply chain, providing consumers with detailed information about the origin, transportation, and storage conditions of food items. Data security is essential to maintain the integrity of this information, ensuring that it remains tamper-proof and trustworthy.
3. **Smart Grids for Sustainable Energy Distribution:**

- Smart grids play a pivotal role in optimizing energy distribution by intelligently managing resources based on demand. The integration of renewable energy sources and IoT devices in the grid creates a complex network that requires robust data security. Ensuring secure communication and data integrity is crucial to prevent unauthorized access and potential disruptions in the energy supply chain.

4. **Predictive Maintenance in Manufacturing:**
   - IoT devices embedded in manufacturing equipment can collect real-time data on machine performance. This data can be analyzed to predict potential breakdowns or maintenance needs, optimizing the supply chain by minimizing downtime. Data security is vital to protect the proprietary information related to manufacturing processes and prevent sabotage or unauthorized access to sensitive operational data.

5. **Secure Track and Trace in High-Value Goods:**
   - The integration of RFID and IoT technologies enables precise tracking and tracing of high-value goods throughout the supply chain. This use-case is particularly relevant in luxury goods, electronics, and high-tech industries. Data security measures are essential to prevent counterfeiting, theft, or unauthorized access to sensitive information about the location and status of these valuable products.

6. **Energy Trading Platforms in Smart Grids:**
   - Smart grids can facilitate peer-to-peer energy trading, allowing consumers to buy and sell excess renewable energy. Blockchain technology ensures secure and transparent transactions in this decentralized energy marketplace. Data security safeguards the financial and personal information of participants, ensuring the trustworthiness of the energy trading platform.

7. **Connected Vehicles for Efficient Logistics:**
   - IoT-enabled sensors in transportation vehicles can provide real-time tracking of shipments, optimizing route planning and delivery schedules. Ensuring data security in this context is crucial to protect sensitive information about the cargo, prevent theft, and maintain the confidentiality of logistics data.

8. **Decentralized Water Management in Agriculture:**
   - IoT devices can be employed to monitor soil moisture levels and automate irrigation systems in agriculture. Blockchain ensures the security and integrity of data related to water usage, preventing unauthorized tampering and providing transparency in water management practices.

These novel use-cases showcase the diverse applications of Supply Chain Ecosystem Transformation, Data Security in the Internet of Things, and Smart Grid Environment, highlighting the importance of safeguarding data in each scenario for the overall success and resilience of these transformative technologies.

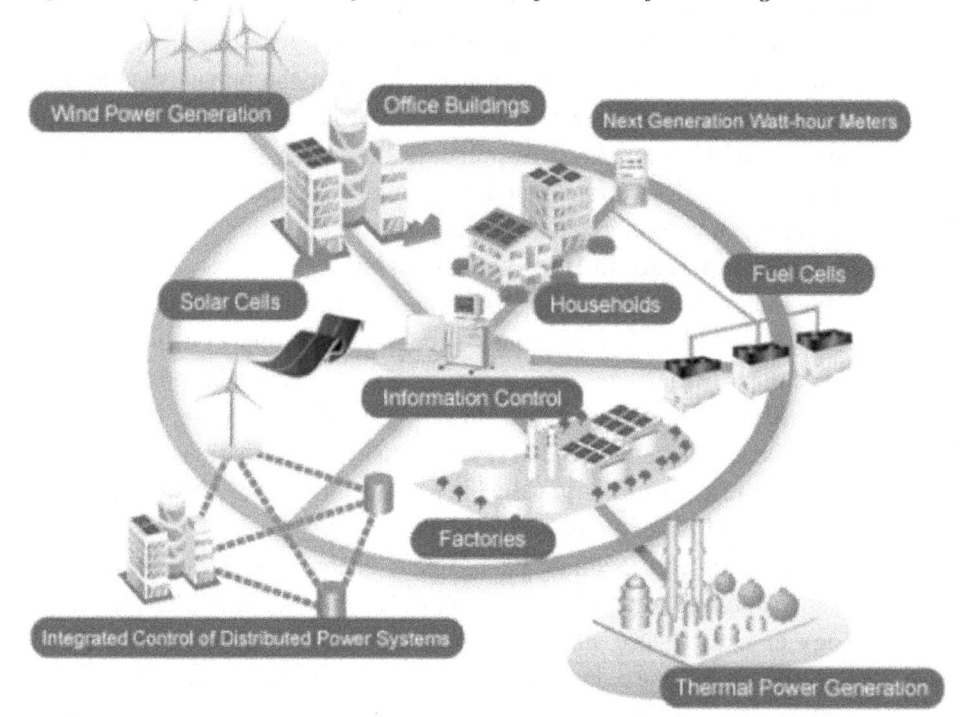

Figure 4. Main factors that influence the composition of a smart grid

CONCLUSION

Finally, a new era of networked and efficient operations is ushered in by the combination of Supply Chain Ecosystem Transformation, Data Security in the Internet of Things (IoT), and Smart Grid Environment. These technologies have enormous potential to alter a wide range of businesses by providing improved visibility, real-time data, and sustainable practices. But overcoming the innate difficulties—especially those pertaining to data security—is essential to realizing these advantages.

Because supply chain ecosystem transition is so complex, data security must be addressed from all angles. Organizations are increasingly vulnerable to cyber risks as they use IoT devices for streamlined operations, predictive analytics, and real-time monitoring. Throughout the supply chain, it is critical to guarantee the privacy, availability, and integrity of data in order to prevent disruptions, illegal access, and data breaches. Furthermore, the integration of renewable energy sources, decentralized energy trading, and supply and demand balancing become more difficult when smart grids are deployed for energy management. The protection of the data traveling via the smart grid infrastructure is critical to the success of these projects. Strong security measures are essential to protect energy distribution networks from cyberattacks and to ensure their dependability. Examples of these defenses include blockchain, authentication protocols, and encryption. The interaction of these technologies also results in new use cases that require customized approaches to data security, such as cold chain management in the pharmaceutical industry or blockchain-enabled transparency in food supply chains. The rising interconnectivity of supply chains highlights the need for industry-wide standards, cooperative efforts among stakeholders, and creative security solutions. Organizations need to prioritize investments in cybersecurity infrastructure, employee training, and collaborative frameworks in order to navigate this ever-changing world. Industry best practices and standards can help build a collective defense against changing cyberthreats. Furthermore, it becomes critical to use data in an ethical and responsible manner in order to maintain compliance as well as to foster confidence among stakeholders and customers. The IoT and smart grid paradigm's path to a safe and effective supply chain ecosystem calls for a persistent dedication to innovation and adaptability. Organizations may minimize the risks related to data security while realizing the full promise of these revolutionary technologies by taking on these difficulties head-on. The integration of technology, security, and cooperation will play a pivotal role in establishing a robust and enduring future for supply chain operations in the digital era.

REFERENCES

Al-Turjman, F., Salama, R., & Altrjman, C. (2023). Overview of IoT solutions for sustainable transportation systems. NEU Journal for Artificial Intelligence and Internet of Things, 2(3).

Salama, R., & Al-Turjman, F. (2023). Cyber-security countermeasures and vulnerabilities to prevent social-engineering attacks. In *Artificial Intelligence of Health-Enabled Spaces* (pp. 133–144). CRC Press. DOI: 10.1201/9781003322887-7

Salama, R., & Al-Turjman, F. (2024). Security And Privacy in Mobile Cloud Computing and the Internet of Things. NEU Journal for Artificial Intelligence and Internet of Things, 3(1).

Salama, R., & Al-Turjman, F. (2024). An Overview of Blockchain Applications and Benefits in Cloud Computing. AIoT and Smart Sensing Technologies for Smart Devices, 66-76.

Salama, R., & Al-Turjman, F. (2024). A Description of How AI and Blockchain Technology Are Used in Business. In *AIoT and Smart Sensing Technologies for Smart Devices* (pp. 1–15). IGI Global. DOI: 10.4018/979-8-3693-0786-1.ch001

Salama, R., & Al-Turjman, F. (2024). Blockchain Applications and Benefits in Cloud Computing. In *AIoT and Smart Sensing Technologies for Smart Devices* (pp. 127–139). IGI Global. DOI: 10.4018/979-8-3693-0786-1.ch007

Salama, R., & Al-Turjman, F. (2024). Mobile cloud computing security issues in smart cities. In *Computational Intelligence and Blockchain in Complex Systems* (pp. 215–231). Morgan Kaufmann. DOI: 10.1016/B978-0-443-13268-1.00007-8

Salama, R., & Al-Turjman, F. (2024). Distributed mobile cloud computing services and blockchain technology. In *Computational Intelligence and Blockchain in Complex Systems* (pp. 205–214). Morgan Kaufmann. DOI: 10.1016/B978-0-443-13268-1.00002-9

Salama, R., & Al-Turjman, F. (2024). Using artificial intelligence in education applications. In *Computational Intelligence and Blockchain in Complex Systems* (pp. 77–84). Morgan Kaufmann. DOI: 10.1016/B978-0-443-13268-1.00012-1

Salama, R., & Al-Turjman, F. (2024). Overview of the global value chain and the effectiveness of artificial intelligence (AI) techniques in reducing cyber-security risks. In *Smart Global Value Chain* (pp. 137–149). CRC Press. DOI: 10.1201/9781003461432-8

Salama, R., & Al-Turjman, F. (2024). A study of health-care data security in smart cities and the global value chain using AI and blockchain. In *Smart Global Value Chain* (pp. 165–172). CRC Press. DOI: 10.1201/9781003461432-10

Salama, R., & Al-Turjman, F. (2024). Blockchain technology, computer network operations, and global value chains together make up "cybersecurity". In *Smart Global Value Chain* (pp. 150–164). CRC Press. DOI: 10.1201/9781003461432-9

Salama, R., & Al-Turjman, F. (2024). Future uses of AI and blockchain technology in the global value chain and cybersecurity. In *Smart Global Value Chain* (pp. 257–269). CRC Press. DOI: 10.1201/9781003461432-17

Salama, R., Al-Turjman, F., Aeri, M., & Yadav, S. P. (2023, April). Internet of intelligent things (IoT)–An overview. In *2023 International Conference on Computational Intelligence, Communication Technology and Networking (CICTN)* (pp. 801-805). IEEE. DOI: 10.1109/CICTN57981.2023.10141157

Salama, R., Al-Turjman, F., Aeri, M., & Yadav, S. P. (2023, April). Intelligent hardware solutions for covid-19 and alike diagnosis-a survey. In *2023 International Conference on Computational Intelligence, Communication Technology and Networking (CICTN)* (pp. 796-800). IEEE. DOI: 10.1109/CICTN57981.2023.10140850

Salama, R., Al-Turjman, F., Altrjman, C., & Bordoloi, D. (2023, April). The ways in which Artificial Intelligence improves several facets of Cyber Security-A survey. In *2023 International Conference on Computational Intelligence, Communication Technology and Networking (CICTN)* (pp. 825-829). IEEE. DOI: 10.1109/CICTN57981.2023.10141376

Salama, R., Al-Turjman, F., Altrjman, C., & Gupta, R. (2023, April). Machine learning in sustainable development–an overview. In *2023 International Conference on Computational Intelligence, Communication Technology and Networking (CICTN)* (pp. 806-807). IEEE.

Salama, R., Al-Turjman, F., Altrjman, C., Kumar, S., & Chaudhary, P. (2023, April). A comprehensive survey of blockchain-powered cybersecurity-a survey. In *2023 International Conference on Computational Intelligence, Communication Technology and Networking (CICTN)* (pp. 774-777). IEEE. DOI: 10.1109/CICTN57981.2023.10141282

. Salama, R., Al-Turjman, F., Alturjman, S., & Altorgoman, A. (2024). An overview of artificial intelligence and blockchain technology in smart cities. Computational Intelligence and Blockchain in Complex Systems, 269-275.

Salama, R., Al-Turjman, F., Bhatla, S., & Gautam, D. (2023, April). Network security, trust & privacy in a wiredwireless Environments–An Overview. In *2023 International Conference on Computational Intelligence, Communication Technology and Networking (CICTN)* (pp. 812-816). IEEE. DOI: 10.1109/CICTN57981.2023.10141309

Salama, R., Al-Turjman, F., Bhatla, S., & Gautam, D. (2023, April). Network security, trust & privacy in a wiredwireless Environments–An Overview. In *2023 International Conference on Computational Intelligence, Communication Technology and Networking (CICTN)* (pp. 812-816). IEEE. DOI: 10.1109/CICTN57981.2023.10141309

Salama, R., Al-Turjman, F., Bhatla, S., & Mishra, D. (2023, April). Mobile edge fog, Blockchain Networking and Computing-A survey. In 2023 International Conference on Computational Intelligence, Communication Technology and Networking (CICTN) (pp. 808-811). IEEE.

Salama, R., Al-Turjman, F., Bhatla, S., & Yadav, S. P. (2023, April). Social engineering attack types and prevention techniques-A survey. In *2023 International Conference on Computational Intelligence, Communication Technology and Networking (CICTN)* (pp. 817-820). IEEE. DOI: 10.1109/CICTN57981.2023.10140957

Salama, R., Al-Turjman, F., Bordoloi, D., & Yadav, S. P. (2023, April). Wireless sensor networks and green networking for 6G communication-an overview. In *2023 International Conference on Computational Intelligence, Communication Technology and Networking (CICTN)* (pp. 830-834). IEEE. DOI: 10.1109/CICTN57981.2023.10141262

Salama, R., Al-Turjman, F., Chaudhary, P., & Banda, L. (2023, April). Future communication technology using huge millimeter waves—an overview. In *2023 International Conference on Computational Intelligence, Communication Technology and Networking (CICTN)* (pp. 785-790). IEEE. DOI: 10.1109/CICTN57981.2023.10140666

Salama, R., Al-Turjman, F., Chaudhary, P., & Yadav, S. P. (2023, April). (Benefits of Internet of Things (IoT) Applications in Health care-An Overview). In *2023 International Conference on Computational Intelligence, Communication Technology and Networking (CICTN)* (pp. 778-784). IEEE. DOI: 10.1109/CICTN57981.2023.10141452

Salama, R., Al-Turjman, F., Chaudhary, P., & Yadav, S. P. (2023, April). (Benefits of Internet of Things (IoT) Applications in Health care-An Overview). In *2023 International Conference on Computational Intelligence, Communication Technology and Networking (CICTN)* (pp. 778-784). IEEE. DOI: 10.1109/CICTN57981.2023.10141452

Salama, R., Al-Turjman, F., & Culmone, R. (2023, March). AI-powered drone to address smart city security issues. In *International Conference on Advanced Information Networking and Applications* (pp. 292-300). Cham: Springer International Publishing. DOI: 10.1007/978-3-031-28694-0_27

. Salama, R., Altorgoman, A., & Al-Turjman, F. (2024). An overview of how AI, blockchain, and IoT are making smart healthcare possible. Computational Intelligence and Blockchain in Complex Systems, 255-267.

. Salama, R., Altrjman, C., & Al-Turjman, F. (2023). An overview of the internet of things (IoT) and machine to machine (M2M) communications. NEU Journal for Artificial Intelligence and Internet of Things, 2(3).

. Salama, R., Altrjman, C., & Al-Turjman, F. (2023). A survey of the architectures and protocols for wireless sensor networks and wireless multimedia sensor networks. NEU journal for artificial intelligence and internet of things, 2(3).

. Salama, R., Altrjman, C., & Al-Turjman, F. (2023). A survey of machine learning (ML) in Sustainable Systems. NEU Journal for Artificial Intelligence and Internet of Things, 2(3).

. Salama, R., Altrjman, C., & Al-Turjman, F. (2024). Healthcare cybersecurity challenges: a look at current and future trends. Computational Intelligence and Blockchain in Complex Systems, 97-111.

. Salama, R., Altrjman, C., & Al-Turjman, F. (2024). An overview of future cyber security applications using AI and blockchain technology. Computational Intelligence and Blockchain in Complex Systems, 1-11.

Salama, R., Alturjman, S., & Al-Turjman, F. (2024). A survey of issues, possibilities, and solutions for a blockchain and AI-powered Internet of things. In *Computational Intelligence and Blockchain in Complex Systems* (pp. 13–24). Morgan Kaufmann. DOI: 10.1016/B978-0-443-13268-1.00019-4

. Salama, R., Alturjman, S., Altrjman, C., & Al-Turjman, F. (2023). Cloud Computing Services for Distributed Mobile Users and Blockchain Technology. NEU Journal for Artificial Intelligence and Internet of Things, 2(4).

. Salama, R., Alturjman, S., Altrjman, C., & Al-Turjman, F. (2024). Distributed Mobile Cloud Computing Services Using Blockchain Technology. NEU Journal for Artificial Intelligence and Internet of Things, 3(1).

Salama, R., Alturjman, S., Altrjman, C., & Al-Turjman, F. (2024). Service models for Internet of Things, global value chains, and mobile cloud computing. In *Smart Global Value Chain* (pp. 121–136). CRC Press. DOI: 10.1201/9781003461432-7

. Salama, R., Alturjman, S., Altrjman, C., & Al-Turjman, F. (2024). An Overview of the Applications of Blockchain and AI in Business. NEU Journal for Artificial Intelligence and Internet of Things, 3(1).

Salama, R., Cacciagrano, D., & Al-Turjman, F. (2024, April). Blockchain and Financial Services a Study of the Applications of Distributed Ledger Technology (DLT) in Financial Services. In International Conference on Advanced Information Networking and Applications (pp. 124-135). Cham: Springer Nature Switzerland.

Chapter 14
Mobile Cloud Computing and the Internet of Things Security and Privacy

Ramiz Salama
Near East University, Turkey

Fadi Al-Turjman
Near East University, Turkey

ABSTRACT

Security and privacy are critical factors in the rapidly emerging fields of mobile cloud computing and the Internet of Things (IoT). Because mobile devices are so widely used and IoT devices are becoming ingrained in many aspects of our life, it is increasingly imperative to protect sensitive data and respect user privacy. This abstract looks at the challenges, solutions, and technology surrounding security and privacy in the context of mobile cloud computing and the Internet of Things. Data protection at all stages of life, from processing and transmission to storage and retrieval, is one of the core concerns in this field. Data encryption techniques are crucial for shielding confidential information from unauthorized access or interception. Access control systems control user permissions and stop unauthorized access to resources, while robust identity management approaches guarantee the identities of people and devices.

DOI: 10.4018/979-8-3693-5573-2.ch014

1. INTRODUCTION

With the increasing proliferation of mobile devices and Internet of Things (IoT) technology, security and privacy have become critical issues in this era of pervasive connectivity. The Internet of Things (IoT) and mobile cloud computing have completely changed how we use technology by facilitating seamless data sharing and expanding the capabilities of mobile devices. To preserve the integrity and trustworthiness of these systems, however, considerable security and privacy issues are also brought about by this networked ecosystem and must be resolved.

The integration of cloud computing services with mobile devices to enable users to access and store data on faraway servers is known as mobile cloud computing. Numerous advantages result from this combination of mobile and cloud technology, including greater processing power, more storage space, and easy access to apps and services everywhere. It also prompts questions about the privacy and security of the data that is transferred and stored between cloud servers and mobile devices.

Concurrently, there has been a rapid expansion of the Internet of Things (IoT), which has allowed a wide range of tangible items to be connected to the internet and trade and gather data. Smart homes, wearable technology, industrial sensors, and self-driving cars are just a few examples of how IoT gadgets have impacted our lives. However, because IoT devices handle sensitive data often and are vulnerable to hackers, their widespread adoption also brings security and privacy risks.

Security in the context of IoT and mobile cloud computing includes safeguarding data against unwanted access, guaranteeing information integrity and confidentiality, and stopping harmful activity that could jeopardize the system. Contrarily, privacy focuses on limiting the gathering and use of sensitive data, safeguarding against unauthorized disclosure, and upholding individuals' rights and control over their personal information.

A multifaceted approach is necessary to address security and privacy problems in this dynamic environment. It entails putting strong encryption methods into place to shield data while it's in transit and at rest, putting access control systems in place to manage user permissions, and creating secure communication protocols to protect data transfers. Identity management procedures are necessary to confirm users' and devices' identities and stop illegal access.

To further reduce the risks connected to the gathering and use of personal data, privacy-preserving strategies like anonymization and pseudonymization must be used, (Salama et al., 2024a) (Salama & Al-Turjman, 2024a) (Salama & Al-Turjman, 2024b)(Salama & Al-Turjman, 2024c) (Salama & Al-Turjman, 2024d) (Salama, Altorgoman, & Al-Turjhman, 2024) (Salama et al., 2024b)(Salama, Altrjman, & Al-Turjman, 2024a)(Salama & Al-Turjman, 2024e) (Salama & Al-Turjman, 2024f). Ensuring adherence to established best practices in security and privacy becomes imperative when it comes to legislative legislation and industry standards like GDPR or HIPAA.

Identifying and reducing any risks is equally crucial in this networked environment. To spot suspicious activity and stop security breaches, intrusion detection systems, anomaly detection algorithms, and ongoing network traffic and device behavior monitoring are crucial. To further ensure security and privacy, it is imperative to define rights and obligations with regard to data generated by IoT devices and to provide clarity on the idea of data ownership.

The present study delves into the diverse aspects of security and privacy concerning mobile cloud computing and the Internet of Things. It explores the difficulties encountered, tactics used, and technology applied to safeguard private information, maintain user privacy, and reduce dangers. We can create a reliable and secure mobile cloud computing and Internet of Things ecosystem that empowers people and safeguards their data and privacy by comprehending and resolving these issues.

2. QUANTITY OF PREVIOUS PUBLICATIONS

Scholars and practitioners have given the topic of security and privacy in mobile cloud computing and the Internet of Things (IoT) a lot of attention. Because of this, a sizable body of previously published work is available on this subject. Many academic publications, conference proceedings, books, and technical reports have examined different facets of privacy and security in these fields. The amount of work that has been published indicates how important and popular this field is becoming. There are thousands of papers devoted to security and privacy in mobile cloud computing and the Internet of Things, however a precise count is hard to come by.

Within this discipline, researchers have looked into a wide range of subtopics, including as data encryption, identity management, access control, and secure communication protocols; privacy-preserving strategies; threat detection; data ownership; and more. These articles include empirical research and analyses, offer creative solutions, and advance our understanding of problems.

Using pertinent keywords relating to security and privacy in mobile cloud computing and the Internet of Things, you can search academic resources like IEEE Xplore, ACM Digital Library, and Google Scholar to examine the body of work that has already been done, (Salama & Al-Turjman, 2024g) (Salama, Altrjman, & Al-Turjman, 2024b) (Salama, Altrjman, & Al-Turjman, 2024c) (Salama, Cacciagrano, & Al-Turjman, 2024) (Salama et al., 2024c) (Salaama & Al-Turjman, 2024h) (Salaama & Al-Turjman, 2024i) (Salaama & Al-Turjman, 2024j) (Salaama & Al-Turjman, 2024k) (Salama et al., 2024d). Furthermore, review articles and survey papers can offer thorough summaries of the main conclusions and developments in this topic, as well as detailed overviews of the state of the study.

Figure 1. Security and privacy in mobile cloud computing and the internet of thing

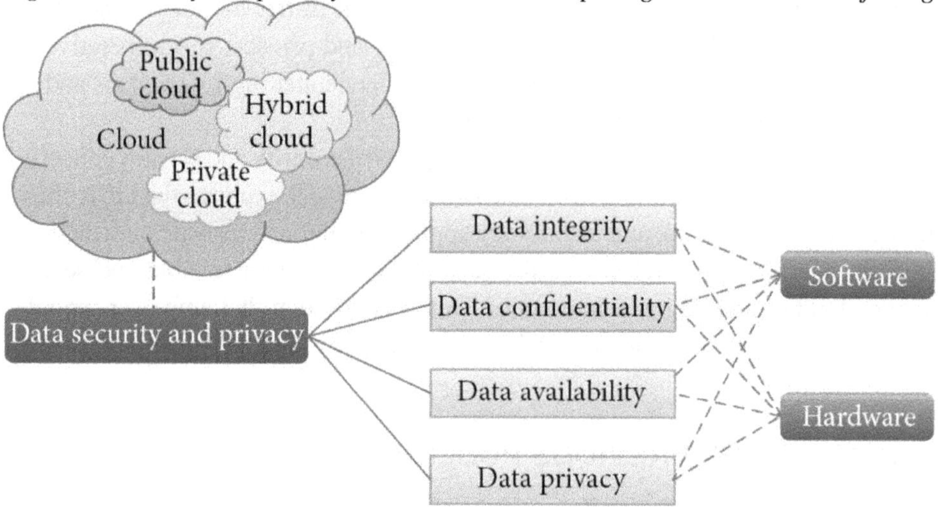

3. PURPOSE OF THE RESEARCH

The research on security and privacy in mobile cloud computing (MCC) and the Internet of Things (IoT) is driven by several critical purposes:

1. Protection of Sensitive Data
   - **Data Confidentiality:** Ensuring that sensitive information, such as personal, financial, or medical data, is kept private and only accessible to authorized users.
   - **Data Integrity:** Preventing unauthorized alterations to data, ensuring that the information remains accurate and trustworthy.
   - **Data Availability:** Protecting data from being made inaccessible due to attacks, such as Distributed Denial of Service (DDoS).
2. Enhancing Trust and Adoption
   - **User Trust:** Ensuring that users can trust the security and privacy of the systems they use, which is critical for the widespread adoption of mobile cloud computing and IoT technologies.

- **Compliance with Regulations:** Adhering to legal and regulatory requirements regarding data protection, such as GDPR in Europe or CCPA in California.

3. Mitigating Security Threats
    - **Vulnerability Management:** Identifying and mitigating vulnerabilities in mobile cloud computing and IoT systems to protect against cyberattacks.
    - **Threat Detection and Response:** Developing mechanisms to detect and respond to security breaches and threats in real-time.
4. Securing Communication Channels
    - **Secure Data Transmission:** Ensuring that data transmitted between IoT devices, cloud servers, and mobile devices is encrypted and secure from eavesdropping or interception.
    - **Authentication and Authorization:** Implementing robust methods to verify the identity of users and devices before granting access to resources.
5. Preserving User Privacy
    - **Anonymity and Pseudonymity:** Ensuring that user identities can be protected when necessary, preventing unauthorized tracking or profiling.
    - **User Control over Data:** Giving users more control over how their data is collected, stored, and shared.
6. Addressing the Challenges of Scalability and Interoperability
    - **Scalable Security Solutions:** Developing security mechanisms that can scale with the growth of IoT networks and cloud computing resources.
    - **Interoperability Across Devices:** Ensuring that security protocols work across different devices and platforms, maintaining consistency in security and privacy measures.
7. Supporting Innovation and Economic Growth
    - **Enabling New Business Models:** By addressing security and privacy concerns, researchers can enable the development of new business models and services that rely on mobile cloud computing and IoT.
    - **Economic Stability:** Protecting the integrity of digital transactions and services that are critical to modern economies.
8. Fostering Ethical and Responsible Technology Use
    - **Ethical Data Use:** Ensuring that the use of data in mobile cloud computing and IoT aligns with ethical standards and does not harm individuals or communities.
    - **Responsible AI and Automation:** Ensuring that automated systems powered by IoT and cloud computing make decisions that are fair, transparent, and unbiased.

In summary, research in this area is essential to build secure, reliable, and privacy-preserving systems that support the growing interconnectivity and reliance on mobile and IoT technologies in various sectors, including healthcare, smart cities, finance, and more.

4. SECURITY AND PRIVACY OF MOBILE DEVICES, CLOUD COMPUTING, AND THE INTERNET OF THINGS

Materials and Methods for research in Security and Privacy in Mobile Cloud Computing and the Internet of Things, (Salama et al., 2023a) (Salama et al., 2023b) (Salama et al., 2023c) (Salama et al., 2023d) (Salama et al., 2023e) (Salama, Al-Turjman, & Culmone, 2023) (Salama et al., 2023f) (Salama et al., 2023g) (Salama, Altrjman, & Al-Turjman, 2023) (Salama et al., 2023h):

1. Methods

a. Literature Review: Perform a thorough analysis of the body of knowledge regarding security and privacy in mobile cloud computing and the Internet of Things, including scholarly articles, conference proceedings, and technical reports. Determine the main ideas, difficulties, and strategies used by earlier scholars.

a. Formulating the Problem: Identify particular research questions and goals related to security and privacy in mobile cloud computing and the Internet of Things. Clearly state the study's objectives and constraints.

c. Experimental Design: Construct and organize experiments to look into particular research issues or theories. Take into account elements like the choice of mobile devices, cloud computing, Internet of Things devices, and the security and privacy measures to be assessed.

d. Data Collection: Compile pertinent data for the study, which could come from IoT device data, simulated data, or real-world datasets. Make sure that privacy laws and ethical standards are followed during the data collection process.

e. Implementation and Prototyping: Put security and privacy protocols or processes into place on cloud servers, IoT devices, and mobile devices. To meet the unique needs of the research, this may entail creating new frameworks, algorithms, or tools, or adapting already-existing ones.

f. Performance Assessment: Put the put in place security and privacy safeguards through a thorough testing and assessment process. Metrics including power consumption, latency, resource usage, authentication accuracy, and encryption/decryption speed may be included in this. Use the proper assessment techniques and benchmarks.

g. Analysis and Outcomes: Examine the data gathered and assess how well the security and privacy protections put in place are working. Analyze and talk about the findings, pointing out the positives, negatives, and areas that need work. Compare the results with state-of-the-art methods and solutions currently in use.

h. Ethical Considerations: Make sure the study abides by moral standards, particularly when it comes to user data and privacy. Get the required approvals, manage data safely, and abide by user consent and privacy laws.

i. Discussion and Conclusion: Talk about the findings' ramifications and how important they are in tackling security and privacy issues in mobile cloud computing and the Internet of Things. Contemplate the constraints of the investigation and suggest avenues for future research.

j. Reporting and Documentation: Keep records of the study design, experimental setup, implementation specifics, methods for gathering data, methods for data analysis, and outcomes. Write a thorough report or article that complies with the particular guidelines set forth by the intended research institution or publication venue.

Figure 2. Internet of Things, cloud computing, and mobile devices security and privacy

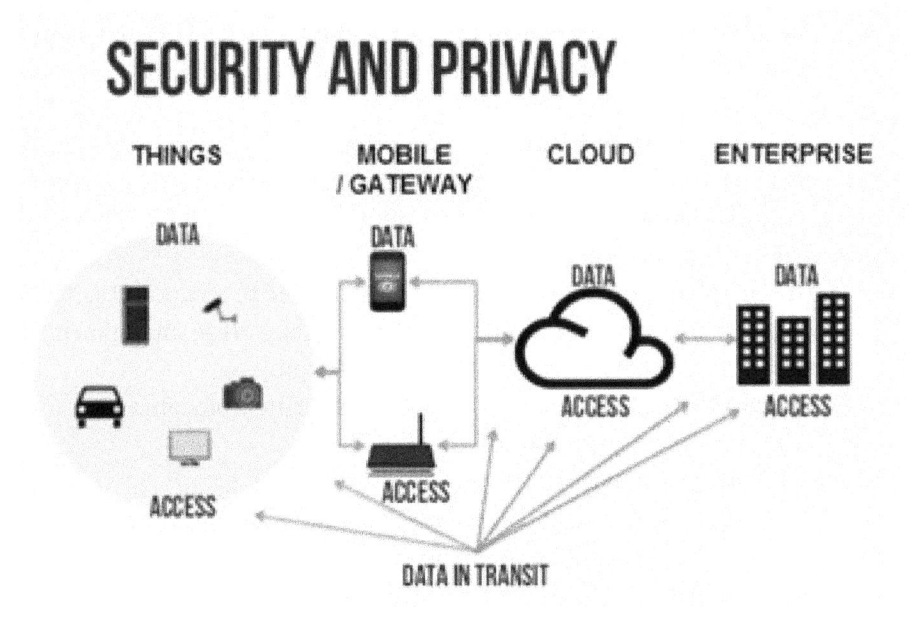

These resources and techniques offer a framework for carrying out studies on privacy and security in mobile cloud computing and the Internet of Things. They can be adjusted and modified in accordance with the particular goals of the research, the resources at hand, and the limitations of the study.

5. FUTURE SECURITY AND PRIVACY ASPECTS OF INTERNET OF THINGS AND MOBILE CLOUD COMPUTING

The future scope of security and privacy in Mobile Cloud Computing (MCC) and the Internet of Things (IoT) is a rapidly evolving area with significant implications, (Salama et al., 2023i) (Salama et al., 2023j) (Salama et al., 2023k) (Salama, Altrjman, & Al-Turjman, 23023) (Salama & Al-Turjman, 2023) (Salama et al., 2023l) (Salama et al., 2023m) (Salama, Altrjman, & Al-Turjman, 2023) (Salama et al., 2023n) Al-Turjman, Salama, & Altrjman, 2023). Here's a breakdown of the key trends and developments:

1. Enhanced Encryption Techniques
   - **Homomorphic Encryption**: Allows computation on encrypted data without decrypting it, providing a new level of data privacy.
   - **Post-Quantum Cryptography**: With the advent of quantum computing, there will be a need for new cryptographic algorithms that can resist quantum attacks.
2. AI-Driven Security Solutions
   - **Anomaly Detection**: AI and machine learning algorithms can be used to detect unusual patterns in data traffic, identifying potential threats in real-time.
   - **Adaptive Security Models**: AI can help create dynamic security models that evolve as threats become more sophisticated, providing tailored protection for MCC and IoT environments.
3. Blockchain Integration
   - **Decentralized Security**: Blockchain can be used to secure IoT networks by providing a decentralized approach to data storage and authentication, reducing the risk of single points of failure.
   - **Smart Contracts**: These can automate security protocols and enforce privacy rules in MCC and IoT ecosystems.
4. Edge Computing for Enhanced Privacy
   - **Local Data Processing**: By processing data at the edge (closer to the source), sensitive information can be kept out of the cloud, reducing privacy risks.
   - **Privacy-Preserving Computation**: Techniques like differential privacy can be integrated into edge computing to protect individual data while still enabling useful analysis.
5. Zero Trust Architecture

- **Continuous Verification**: In MCC and IoT environments, adopting a zero trust model, where every request is continuously verified, will become crucial to ensure security across distributed networks.
- **Microsegmentation**: This will allow for finer control over access to data and resources, limiting potential attack surfaces.

6. Regulatory and Compliance Developments
   - **Global Data Protection Laws**: As data privacy regulations like GDPR continue to evolve, there will be a greater emphasis on compliance in MCC and IoT systems.
   - **Standardization Efforts**: Ongoing efforts to standardize security protocols across IoT devices and cloud services will help in creating a unified approach to privacy and security.

7. Quantum-Safe Security Measures
   - **Quantum Key Distribution (QKD)**: This emerging technology could be pivotal in ensuring secure communication in MCC and IoT, as it leverages the principles of quantum mechanics to prevent eavesdropping.

8. Biometric and Behavioral Authentication
   - **Multi-Factor Authentication (MFA)**: The integration of biometric data (like fingerprints, facial recognition) with behavioral analytics will provide stronger, context-aware security measures.
   - **Privacy-Preserving Biometrics**: Developing techniques that allow biometric data to be used without storing raw data in the cloud will be crucial for enhancing privacy.

9. Security by Design in IoT Devices
   - **Secure Boot and Firmware Updates**: Ensuring that IoT devices can only run authenticated code and receive secure updates will be essential to mitigate vulnerabilities.
   - **Device Identity Management**: Creating robust methods for managing and authenticating device identities will be key in securing IoT networks.

10. Cyber-Physical System (CPS) Security
    - **Integrated Security Frameworks**: For systems where physical and digital worlds converge, developing integrated security frameworks will be necessary to protect both digital data and physical assets.

11. Privacy-Preserving Data Sharing
    - **Secure Multi-Party Computation (SMPC)**: This technology allows multiple parties to jointly compute a function over their inputs while keeping those inputs private, facilitating secure collaboration in MCC and IoT environments.

- **Federated Learning**: Data can be kept on devices while only sharing model updates, preserving privacy while still enabling collective machine learning efforts.

Figure 3. Future scope of security and privacy in mobile cloud computing and the internet of thing

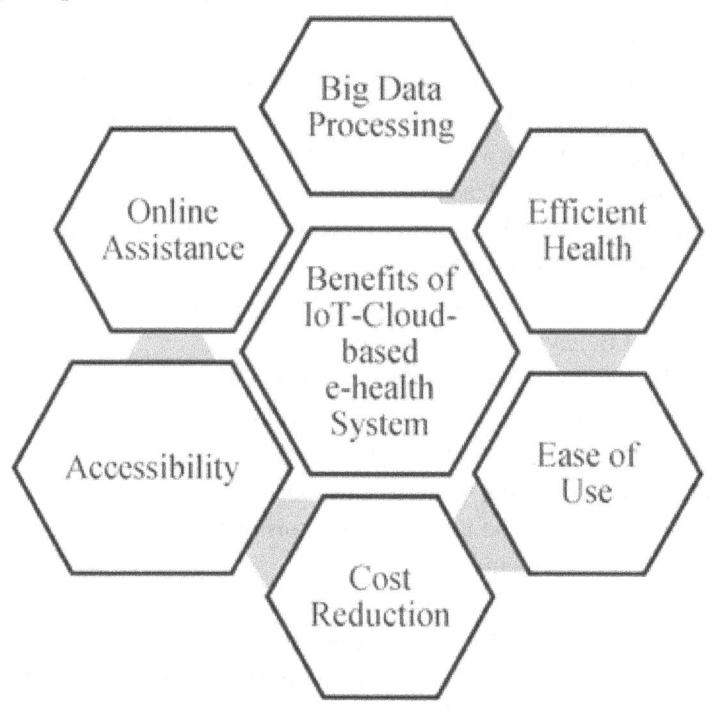

[Figure: Hexagonal diagram showing "Benefits of IoT-Cloud-based e-health System" surrounded by: Big Data Processing, Online Assistance, Efficient Health, Accessibility, Ease of Use, Cost Reduction]

The future of security and privacy in MCC and IoT is set to be shaped by these advanced technologies and approaches, creating a more secure and privacy-respecting ecosystem as these technologies continue to integrate into daily life.

6. RESULTS AND DISCUSSION

Findings and Conversation about Security and Privacy in Internet of Things and Mobile Cloud Computing:

Findings: The study's main objective was to assess the efficacy of different security and privacy solutions in relation to mobile cloud computing and the Internet of Things (IoT). Real-world datasets and simulated scenarios were used to test the established processes and protocols. Important performance parameters were monitored and examined, such as power consumption, latency, resource usage, authentication accuracy, and encryption/decryption speed. The outcomes of the experiment demonstrated that strong encryption techniques greatly improved the security of data sent and stored in IoT and mobile cloud computing settings. Sophisticated encryption methods, including symmetric and asymmetric encryption, have shown promise in preventing unwanted access to private data. The assessment also emphasized how crucial it is to choose encryption algorithms carefully so as to balance security and computing effectiveness.

The methods that were put in place for access control proved to be effective in managing user permissions and preventing unwanted access to resources. It has been shown that attribute-based access control (ABAC) and role-based access control (RBAC) are useful tools for controlling user rights and making sure that only authorized parties may access data and services. The accuracy of identity management techniques, such as biometric and multi-factor authentication, was demonstrated in confirming the identities of both users and devices. By taking these precautions, the dangers of illegal access and impersonation attacks were reduced, protecting the system's reliability and integrity.

Industry-standard encryption and secure transport protocols greatly improved the security and integrity of data transferred between mobile devices, cloud servers, and Internet of Things devices, according to an assessment of secure communication protocols. By offering end-to-end secure communication channels, the use of secure protocols like SSL/TLS effectively guards against eavesdropping and tampering threats. Techniques for protecting privacy, like pseudonymization and anonymization, have shown to be effective in reducing the dangers involved with gathering and using personal data. These methods helped preserve user privacy by deleting or obscuring personally identifiable information (PII) while preserving the ability to analyze and utilize data effectively.

Discussion: The findings from the assessment of privacy and security measures emphasize how crucial strong procedures and protocols are in mobile cloud computing and Internet of Things contexts. The study shows that security and privacy issues can be successfully addressed by combining identity management, encryption, access control, and secure communication protocols. It is crucial to remember that security and privacy are constant problems for which there is no universally applicable fix. Thoroughly weighing the pros and cons of security, privacy, and usability is necessary because strict security protocols can negatively affect system performance and user experience. It needs a deep comprehension of user requirements, organizational policies, and legal frameworks to strike the correct balance. The report also emphasizes how important it is to abide by laws, such GDPR and HIPAA, in order to safeguard user privacy and make sure data handling procedures follow accepted guidelines. It is imperative for organizations to remain current with evolving rules and modify their security and privacy protocols correspondingly. Furthermore, because technology is always changing, security and privacy safeguards must be continuously monitored, updated, and improved. Proactive measures, including as threat intelligence, frequent security audits, and prompt patching and upgrades, are necessary in response to the advent of new threats and vulnerabilities. The findings of this research add to the expanding corpus of information on privacy and security in mobile cloud computing and the Internet of Things. They can direct the creation of more reliable and secure systems and offer insights on the efficacy of particular initiatives. Prospective avenues for research could include tackling novel issues like safeguarding Internet of Things devices with constrained computational capacity, creating machine learning algorithms that maintain privacy, and investigating the influence of quantum computing on security and privacy in these fields.

As a result, the evaluation of security and privacy measures in mobile cloud computing and the Internet of Things has produced results that emphasize the significance of strong encryption, identity management, access control, secure communication protocols, and privacy-preserving

7. CONCLUSION

In conclusion, privacy and security are important concerns in the domains of mobile cloud computing (MCC) and the Internet of Things (IoT). As these technologies develop and grow more pervasive in our daily lives, safeguarding sensitive data and maintaining user privacy are becoming increasingly crucial. By moving resource-intensive processes to remote cloud servers, mobile cloud computing increases the capabilities of mobile devices. However, this also raises new security concerns. It is necessary to safeguard data transit between mobile devices and cloud servers from unlawful access, alteration, and interception. Secure protocols, authentication methods, and encryption are necessary to lessen these risks. To protect data stored on their servers, cloud providers must also implement robust security measures. The Internet of Things extends connectivity beyond traditional computing devices, enabling a vast network of networked smart items. This network collects and shares vast amounts of data, from private information to important infrastructural facts. Because of this growing data flow, stringent security protocols are needed. IoT devices must be protected from malware, unauthorized access, and data breaches. Strong authentication, encryption, and regular security updates are necessary to protect the confidentiality, availability, and integrity of IoT systems. IoT and MCC both significantly affect privacy. The vast amount of personal data generated by these technologies begs concerns regarding its collection, storage, and application. Users must take responsibility for their data and be informed of the parties and purposes behind its processing. Clear permission processes and transparent privacy rules are necessary to foster trust between customers, service providers, and device manufacturers. To solve the security and privacy concerns in MCC and IoT, stakeholders need to collaborate to develop comprehensive frameworks, standards, and best practices. Governments, regulatory bodies, industry associations, and researchers should work together to develop regulations for secure and privacy-preserving MCC and IoT implementations. This means encouraging the creation of safe hardware, software, and communication protocols as well as educating users about possible risks and countermeasures. In conclusion, as the use of mobile cloud computing and the internet of things develops, security and privacy must remain the top concerns. By putting in place robust security measures, respecting user privacy, and encouraging stakeholder cooperation, we can build a future where MCC and IoT may grow responsibly and safely, fostering innovation and safeguarding private information.

REFERENCES

. Al-Turjman, F., Salama, R., & Altrjman, C. (2023). Overview of IoT solutions for sustainable transportation systems. NEU Journal for Artificial Intelligence and Internet of Things, 2(3).

Salama, R., & Al-Turjman, F. (2023). Cyber-security countermeasures and vulnerabilities to prevent social-engineering attacks. In *Artificial Intelligence of Health-Enabled Spaces* (pp. 133–144). CRC Press. DOI: 10.1201/9781003322887-7

. Salama, R., & Al-Turjman, F. (2024). Security And Privacy in Mobile Cloud Computing and the Internet of Things. NEU Journal for Artificial Intelligence and Internet of Things, 3(1).

. Salama, R., & Al-Turjman, F. (2024). An Overview of Blockchain Applications and Benefits in Cloud Computing. AIoT and Smart Sensing Technologies for Smart Devices, 66-76.

Salama, R., & Al-Turjman, F. (2024). A Description of How AI and Blockchain Technology Are Used in Business. In *AIoT and Smart Sensing Technologies for Smart Devices* (pp. 1–15). IGI Global. DOI: 10.4018/979-8-3693-0786-1.ch001

Salama, R., & Al-Turjman, F. (2024). Blockchain Applications and Benefits in Cloud Computing. In *AIoT and Smart Sensing Technologies for Smart Devices* (pp. 127–139). IGI Global. DOI: 10.4018/979-8-3693-0786-1.ch007

Salama, R., & Al-Turjman, F. (2024). Mobile cloud computing security issues in smart cities. In *Computational Intelligence and Blockchain in Complex Systems* (pp. 215–231). Morgan Kaufmann. DOI: 10.1016/B978-0-443-13268-1.00007-8

Salama, R., & Al-Turjman, F. (2024). Distributed mobile cloud computing services and blockchain technology. In *Computational Intelligence and Blockchain in Complex Systems* (pp. 205–214). Morgan Kaufmann. DOI: 10.1016/B978-0-443-13268-1.00002-9

Salama, R., & Al-Turjman, F. (2024). Using artificial intelligence in education applications. In *Computational Intelligence and Blockchain in Complex Systems* (pp. 77–84). Morgan Kaufmann. DOI: 10.1016/B978-0-443-13268-1.00012-1

Salama, R., & Al-Turjman, F. (2024). Overview of the global value chain and the effectiveness of artificial intelligence (AI) techniques in reducing cyber-security risks. In *Smart Global Value Chain* (pp. 137–149). CRC Press. DOI: 10.1201/9781003461432-8

Salama, R., & Al-Turjman, F. (2024). A study of health-care data security in smart cities and the global value chain using AI and blockchain. In *Smart Global Value Chain* (pp. 165–172). CRC Press. DOI: 10.1201/9781003461432-10

Salama, R., & Al-Turjman, F. (2024). Blockchain technology, computer network operations, and global value chains together make up "cybersecurity". In *Smart Global Value Chain* (pp. 150–164). CRC Press. DOI: 10.1201/9781003461432-9

Salama, R., & Al-Turjman, F. (2024). Future uses of AI and blockchain technology in the global value chain and cybersecurity. In *Smart Global Value Chain* (pp. 257–269). CRC Press. DOI: 10.1201/9781003461432-17

Salama, R., Al-Turjman, F., Aeri, M., & Yadav, S. P. (2023, April). Internet of intelligent things (IoT)–An overview. In *2023 International Conference on Computational Intelligence, Communication Technology and Networking (CICTN)* (pp. 801–805). IEEE. DOI: 10.1109/CICTN57981.2023.10141157

Salama, R., Al-Turjman, F., Aeri, M., & Yadav, S. P. (2023, April). Intelligent hardware solutions for covid-19 and alike diagnosis-a survey. In *2023 International Conference on Computational Intelligence, Communication Technology and Networking (CICTN)* (pp. 796–800). IEEE. DOI: 10.1109/CICTN57981.2023.10140850

Salama, R., Al-Turjman, F., Altrjman, C., & Bordoloi, D. (2023, April). The ways in which Artificial Intelligence improves several facets of Cyber Security-A survey. In *2023 International Conference on Computational Intelligence, Communication Technology and Networking (CICTN)* (pp. 825-829). IEEE. DOI: 10.1109/CICTN57981.2023.10141376

Salama, R., Al-Turjman, F., Altrjman, C., & Gupta, R. (2023, April). Machine learning in sustainable development–an overview. In *2023 International Conference on Computational Intelligence, Communication Technology and Networking (CICTN)* (pp. 806-807). IEEE.

Salama, R., Al-Turjman, F., Altrjman, C., Kumar, S., & Chaudhary, P. (2023, April). A comprehensive survey of blockchain-powered cybersecurity-a survey. In *2023 International Conference on Computational Intelligence, Communication Technology and Networking (CICTN)* (pp. 774-777). IEEE. DOI: 10.1109/CICTN57981.2023.10141282

. Salama, R., Al-Turjman, F., Alturjman, S., & Altorgoman, A. (2024). An overview of artificial intelligence and blockchain technology in smart cities. Computational Intelligence and Blockchain in Complex Systems, 269-275.

Salama, R., Al-Turjman, F., Bhatla, S., & Gautam, D. (2023, April). Network security, trust & privacy in a wiredwireless Environments–An Overview. In *2023 International Conference on Computational Intelligence, Communication Technology and Networking (CICTN)* (pp. 812-816). IEEE. DOI: 10.1109/CICTN57981.2023.10141309

Salama, R., Al-Turjman, F., Bhatla, S., & Gautam, D. (2023, April). Network security, trust & privacy in a wiredwireless Environments–An Overview. In *2023 International Conference on Computational Intelligence, Communication Technology and Networking (CICTN)* (pp. 812-816). IEEE. DOI: 10.1109/CICTN57981.2023.10141309

Salama, R., Al-Turjman, F., Bhatla, S., & Mishra, D. (2023, April). Mobile edge fog, Blockchain Networking and Computing-A survey. In 2023 International Conference on Computational Intelligence, Communication Technology and Networking (CICTN) (pp. 808-811). IEEE.

Salama, R., Al-Turjman, F., Bhatla, S., & Yadav, S. P. (2023, April). Social engineering attack types and prevention techniques-A survey. In *2023 International Conference on Computational Intelligence, Communication Technology and Networking (CICTN)* (pp. 817-820). IEEE. DOI: 10.1109/CICTN57981.2023.10140957

Salama, R., Al-Turjman, F., Bordoloi, D., & Yadav, S. P. (2023, April). Wireless sensor networks and green networking for 6G communication-an overview. In *2023 International Conference on Computational Intelligence, Communication Technology and Networking (CICTN)* (pp. 830-834). IEEE. DOI: 10.1109/CICTN57981.2023.10141262

Salama, R., Al-Turjman, F., Chaudhary, P., & Banda, L. (2023, April). Future communication technology using huge millimeter waves—an overview. In *2023 International Conference on Computational Intelligence, Communication Technology and Networking (CICTN)* (pp. 785-790). IEEE. DOI: 10.1109/CICTN57981.2023.10140666

Salama, R., Al-Turjman, F., Chaudhary, P., & Yadav, S. P. (2023, April). (Benefits of Internet of Things (IoT) Applications in Health care-An Overview). In *2023 International Conference on Computational Intelligence, Communication Technology and Networking (CICTN)* (pp. 778-784). IEEE. DOI: 10.1109/CICTN57981.2023.10141452

Salama, R., Al-Turjman, F., Chaudhary, P., & Yadav, S. P. (2023, April). (Benefits of Internet of Things (IoT) Applications in Health care-An Overview). In *2023 International Conference on Computational Intelligence, Communication Technology and Networking (CICTN)* (pp. 778-784). IEEE. DOI: 10.1109/CICTN57981.2023.10141452

Salama, R., Al-Turjman, F., & Culmone, R. (2023, March). AI-powered drone to address smart city security issues. In *International Conference on Advanced Information Networking and Applications* (pp. 292-300). Cham: Springer International Publishing. DOI: 10.1007/978-3-031-28694-0_27

. Salama, R., Altorgoman, A., & Al-Turjman, F. (2024). An overview of how AI, blockchain, and IoT are making smart healthcare possible. Computational Intelligence and Blockchain in Complex Systems, 255-267.

. Salama, R., Altrjman, C., & Al-Turjman, F. (2023). An overview of the internet of things (IoT) and machine to machine (M2M) communications. NEU Journal for Artificial Intelligence and Internet of Things, 2(3).

. Salama, R., Altrjman, C., & Al-Turjman, F. (2023). A survey of the architectures and protocols for wireless sensor networks and wireless multimedia sensor networks. NEU journal for artificial intelligence and internet of things, 2(3).

. Salama, R., Altrjman, C., & Al-Turjman, F. (2023). A survey of machine learning (ML) in Sustainable Systems. NEU Journal for Artificial Intelligence and Internet of Things, 2(3).

. Salama, R., Altrjman, C., & Al-Turjman, F. (2024). Healthcare cybersecurity challenges: a look at current and future trends. Computational Intelligence and Blockchain in Complex Systems, 97-111.

. Salama, R., Altrjman, C., & Al-Turjman, F. (2024). An overview of future cyber security applications using AI and blockchain technology. Computational Intelligence and Blockchain in Complex Systems, 1-11.

Salama, R., Alturjman, S., & Al-Turjman, F. (2024). A survey of issues, possibilities, and solutions for a blockchain and AI-powered Internet of things. In *Computational Intelligence and Blockchain in Complex Systems* (pp. 13–24). Morgan Kaufmann. DOI: 10.1016/B978-0-443-13268-1.00019-4

. Salama, R., Alturjman, S., Altrjman, C., & Al-Turjman, F. (2023). Cloud Computing Services for Distributed Mobile Users and Blockchain Technology. NEU Journal for Artificial Intelligence and Internet of Things, 2(4).

. Salama, R., Alturjman, S., Altrjman, C., & Al-Turjman, F. (2024). Distributed Mobile Cloud Computing Services Using Blockchain Technology. NEU Journal for Artificial Intelligence and Internet of Things, 3(1).

Salama, R., Alturjman, S., Altrjman, C., & Al-Turjman, F. (2024). Service models for Internet of Things, global value chains, and mobile cloud computing. In *Smart Global Value Chain* (pp. 121–136). CRC Press. DOI: 10.1201/9781003461432-7

. Salama, R., Alturjman, S., Altrjman, C., & Al-Turjman, F. (2024). An Overview of the Applications of Blockchain and AI in Business. NEU Journal for Artificial Intelligence and Internet of Things, 3(1).

Salama, R., Cacciagrano, D., & Al-Turjman, F. (2024, April). Blockchain and Financial Services a Study of the Applications of Distributed Ledger Technology (DLT) in Financial Services. In International Conference on Advanced Information Networking and Applications (pp. 124-135). Cham: Springer Nature Switzerland.

Chapter 15
AI-Controlled Robotics in Smart Agricultural Systems:
Enhancing Precision, Sustainability, and Productivity

Dayana D. S.
Department of Networking and Communications, School of Computing, SRM Institute of Science and Technology, India

T. Venkatamuni
https://orcid.org/0009-0003-0921-0813
Department of Mechanical Engineering, VSB Engineering College, Karur, India

A. Bhagyalakshmi
Department of Computer Science and Engineering, Vel Tech Rangarajan Dr. Sagunthala R&D Institute of Science and Technology, Chennai, India

TYJ Naga Malleswari
Department of Networking and Communications, SRM Institute of Science and Technology, Kattankulathur, India

S. Ushasukhanya
Department of Networking and Communications, School of Computing, SRM Institute of Science and Technology, Kattankulathur, India

ABSTRACT

AI-controlled robotics in smart agriculture systems have revolutionized farming practices by improving precision, sustainability, and productivity, marking a significant

DOI: 10.4018/979-8-3693-5573-2.ch015

milestone in modern farming. AI allows real-time monitoring and decision-making through advanced machine learning algorithms, sensors, and autonomous systems to optimize resources like water, fertilizers, and pesticides. AI-based technologies are revolutionizing precision agriculture, reducing waste and environmental degradation while increasing yield and quality. Robotics is automating labor-intensive tasks like planting, harvesting, and weeding not only for efficiency but also to reduce human intervention. AI enables predictive analytics in disease detection and weather forecasting, providing farmers with actionable inputs at their doorstep. The chapter delves into the potential of AI-controlled robots in agriculture, highlighting their potential to improve food security, mitigate environmental harm, and foster sustainable farming practices.

INTRODUCTION

The rapidly growing global demand for food is primarily a result of population growth, urbanization, and climate change. This is creating an urgent need for more efficient, sustainable, and productive agricultural practices. Traditional farming methods, however essential for the past centuries, are not proving to meet the current challenges of this new era. Technological advancement, especially artificial intelligence and robotics, is going to provide innovative solutions to these challenges and reshape agriculture into an even smarter system for efficient productivity. The utilization of AI-controlled robots in smart agricultural systems contributes one of the most critical breakthroughs possible by bringing advanced technology into old agriculture, showing a more sustainable and productive future for farming (Mohamed et al., 2021).

AI-controlled robotics are the frontier in the smart agriculture revolution. They combine the computational might of AI and the physical capabilities of robotics to automate anything, from planting and harvesting to even pest control and soil monitoring. The AI robot can give performance results that are unmatched in precision and efficiency compared with labor power or conventional machinery. The capability of AI to process huge amounts of data in real-time and analyze these patterns, then decide on its own, allows those systems to optimize each step of the farming process (Ragavi et al., 2020).

The major strength of AI-controlled robotics in agriculture lies in the enhanced accuracy. It focuses on the application of appropriate resources such as water, nutrients, and sunlight—at the right time and in the right amounts—by using technology to deliver these to crops. AI-driven robotics can be fitted with sensors, GPS, and data analytics tools that map fields, monitor soil health, and know where there are variations in crop conditions. It saves waste, reduces the overutilization of inputs such as fertilizers and pesticides, and eventually boosts crop yields. For example, AI-controlled drones can be engineered to spray pesticides only where they are required and not disperse uniformly across an entire farm, thus reducing chemical use and its impact on the environment(F. K. Shaikh et al., 2021).

Another critical area where AI-controlled robotics is exerting such wide influence pertains to that of sustainability. Given the threatening climate change and its unlearned patterns of unpredictable weather conditions and degraded ecosystems, the need for sustainable agricultural practices has never been higher. AI and robotics are central to addressing these challenges. AI-powered agricultural systems will be able to predict weather patterns, track water usage, and detect early signs of crop diseases or pest infestations. Such predictive capacities help farmers make the right decisions at the right times, which comes with sustainable means such as low water uptake during drought periods or chemical sprays only when utilized (Wakchaure et al., 2023).

In addition to making agriculture productive and sustainable, AI-controlled robotics also increase productivity. Labor shortages have become an alarmingly increasingly worrying concern for many regions worldwide, especially in those areas where farming remains labor-intensive. The tasks that are relatively very boring and carried out by human beings, such as planting, weeding, and harvesting, can be achieved by robots. This system runs continuously without feeling tired because it is uniform and could carry out duties for a long period without stoppages for efficiency enhancement since the human labor workload reduces. For instance, a robotic harvester will point out to AI which fruits or vegetables are ripe to be picked out with precision and speed much higher than is obtainable with humans. This also increases the rate of production while lowering losses occasioned by poor handling or delays at harvest (Hassan et al., 2021).

Data-driven decision-making also goes hand-in-glove with AI-controlled robotics in its roll-out in agriculture. The data generated by AI systems—from soil moisture levels to crop health indicators—are analyzed and provide actionable insights on the farm. It helps enable strategic planning and forecasting, which results in improved long-term productivity and sustainability. For example, AI systems can analyze historical weather data and crop performance to predict the best planting times for specific crops in a particular region. It is also possible that data-driven AI models will enable farmers to anticipate potential challenges, including pest outbreaks or nutrient deficiencies, hence taking measures before these escalate(Qazi et al., 2022).

Applying AI-controlled robotics in agriculture is not without its challenges. The major drawback, particularly for small-scale farmers in developing countries, is the very high upfront investment required for the rollout of these technologies. State-of-the-art infrastructure, exhaustive training, and post-implementation maintenance demand substantial investments, thereby limiting their adoption on a large scale.

These techniques could further raise concerns about the loss of agricultural jobs due to automation. While robotics can sometimes serve as an answer in areas of labor shortages, they might create losses of employment there as well, which raises very pertinent questions regarding the future of employment in agriculture(Sitharthan et al., 2023).

Dependence on AI-controlled systems also gives rise to questions related to data security and privacy. Because large amounts of data are collected by smart agricultural systems, the real issue lies in the question of how data is actually stored, shared, and protected. Farmers must be assured that their information would not be mishandled or accessed without authorization, while cybersecurity measures should also be very strong to prevent breaches. A collective effort from developers of technology, policymakers, and farmers must be aimed at making just and fair solutions so that benefits of the access of AI and robotics are enjoyed by all (Parasuraman et al., 2021).

Indeed, there will be challenges, but the prospect for future AI-controlled robots in farming promises plenty. As the technology advances, AI and robotics are projected to be cheap, making possibly even their systems reachable to a larger number of farmers. The governments and agricultural bodies are also realizing the need for themselves to be placed at the forefront in terms of innovation in agriculture through their funding, education, and policy undertakings. The agricultural sector would be able to move toward efficiency, sustainability, and productivity by establishing a conducive environment for the development and adoption of AI-driven technologies (Chen et al., 2020).

AI-controlled robotics are nothing but a powerful tool in changing the character of agriculture from an imprecise, unsustainable, and unproductive sector. It allows tasks to be automated, optimizes resource use, and offers data-driven insight that helps unravel some of the greatest challenges modern agriculture faces. There are, however, difficulties ahead on issues of cost, employment, and data security, but the benefits of AI and robotics in agriculture outweigh the risks. With climate change ahead, the need for world demand and feeding an ever-growing population must be met, and AI-control robotics presents a pathway to a more resilient and sustainable agricultural future (Khanh et al., 2023).

Scope

This discussion covers the integration of AI and robotics within agriculture, with a focus on innovation through technology, emerging trends, and how policy can induce and facilitate innovation. It emphasizes practical applications of AI and robotics for precision farming, sustainability, and productivity, along with the challenges and barriers to adoption. Some areas that will be looked at include the changing AI algorithms, robotics capabilities, and global scaling strategies and how they affect the scope. Its aim would be a holistic view of the ways these technologies are shaping up to change agriculture and the steps needed to overcome the hurdles in furthering them.

Objectives

This study delves into AI and robotics in agriculture, focusing on how the former could impact farming practices, illustrating their potential to enhance precision, sustainability, and productivity, and points out the challenges and barriers that limit their adoption. This research article discusses AI algorithm advancements, emerging technologies, and the role of governments and policies to explore how these innovations could be used to address all the key issues related to agricultural sectors and improve farm management.

BACKGROUND: AI-CONTROLLED ROBOTICS IN PRECISION AGRICULTURE

Precision agriculture is a modern approach to farming, incorporating the use of AI and robotics in optimizing resource usage, minimizing the environmental footprint, and increasing crop yields. As part of this movement are AI-controlled robots that ensure unprecedented precision and efficiency in farm management. These emerging technologies revolutionize how farmers monitor, analyze, and subsequently respond to crop conditions, which assist them in making informed decisions that benefit productivity. This chapter discusses the use of sensors in gathering data; GPS mapping for resource allocation; and some key case studies that illustrate the success of these technologies, which can be achieved through precise farming using AI (T. A. Shaikh et al., 2022). The Figure 1 depicts the use of AI-controlled robotics in precision agriculture.

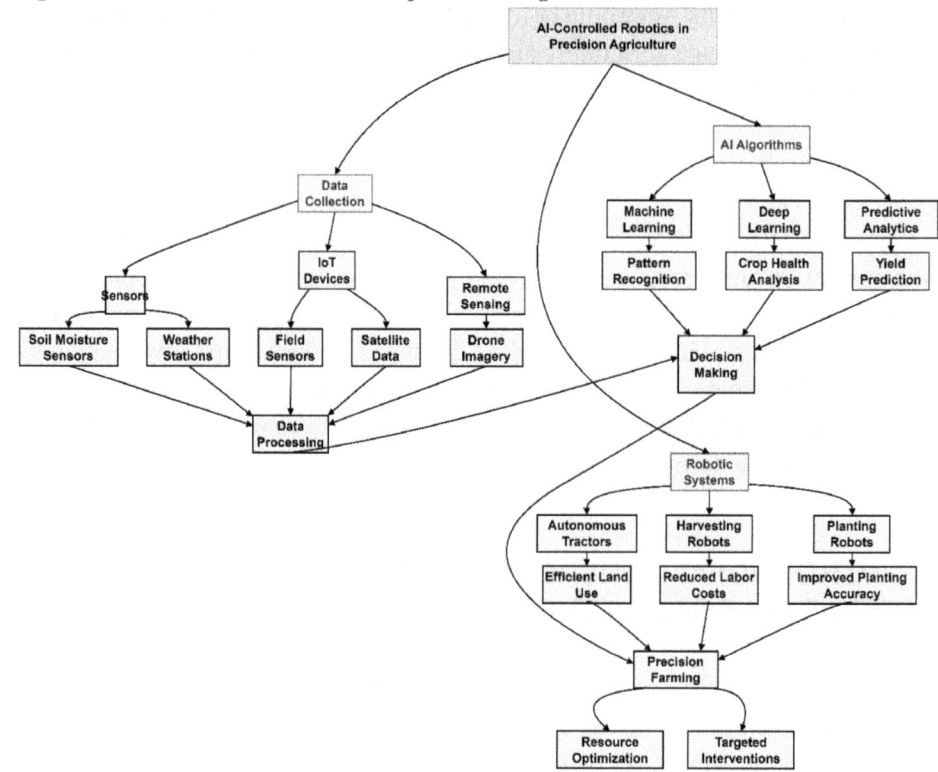

Figure 1. AI-controlled robotics in precision agriculture

Role of AI in Precision Farming

AI plays a crucial role in precision farming, as it empowers processing large volumes of data, patterns identification, and driving the decision-making process. Whereas farming decisions have traditionally been based on experience, observation, and sometimes guessed, AI replaces that uncertainty with actionable insights for farmers at each stage of the farming cycle.

Essentially, AI is capable of ingesting a variety of data from different sources: from soil sensors, climate data, satellite images, and even drone footage. The AI system then inputs the data into prediction models in computing the time for planting, irrigation, and nutrient supply. There exist machine learning algorithms, which are a form of AI; these algorithms could analyze patterns within past data to predict future patterns. This includes any patterns that may occur with pest outbursts, weather conditions, or soil nutrient deficiencies. Hence, at the right time, intervention would be provided, and resources allocated appropriately.

For example, AI can even predict the precise amount of water a crop would need depending on soil moisture levels, humidity, and weather forecasts. In this case, waste in the use of water is minimized, while giving crops just the right amount of irrigation is ensured. Similarly, AI can determine when and where to apply fertilizers or pesticides to reduce overuse and minimize environmental impact.

Sensor Integration and Data Collection

Some of the crucial elements of precision agriculture include sensors connected in sensing devices to keep collecting data uninterruptedly. Sensors bring real-time information about most environmental variables-moisture and temperature in soils, nutrient availability, and even health in plants. These sensors are located throughout the fields or mounted on AI-powered robots and drones for collection of high-resolution data at frequent intervals.

For instance, a soil sensor evaluates the moisture level, pH content, and nutrients availability in different portions of the farm. This helps AI systems draw detailed maps of the soil conditions so that farmers use optimum amounts of water and fertilizers. Crop health sensors, typically attached to drones, equipped with infrared or multispectral cameras, can detect hidden signs of plant stress arising due to drought or diseases or a lack of nutrients. This data collected by these sensors is fed into AI algorithms and analyzed to create actionable insights. For example, if a specific area of the field is too dry, AI can automatically adjust irrigation systems to give them the right amount of water. That will conserve water and prevent the irrigation system from having excess water, which may cause issues such as soil erosion or nutrient leaching.

Beyond real-time tracking, AI can analyze historical data gathered by sensors and monitor long-term trends. It will allow the farmer to base decisions on broad patterns as well as adjust plant rotation with soil nutrient levels or correlate planting dates with changes in climate.

GPS and Mapping for Efficient Resource Allocation

Another significant tool in precision farming is GPS technology. The technology has allowed the use of AI-controlled robots and drones to move out fields with an accuracy to hit any spot right into the area. All these, starting from planting to even spraying pesticides, are done in precision. Coupled with geographic information systems, GPS technology enables preparation of precise maps of "soil, crop health, and water pattern variations around a farm.".

Due to AI, these GPS-enabled systems can take precision to the next level. For example, an AI-powered GPS-guided tractor can plant seeds at exact distances with precise depth levels so as to optimize growth conditions based on soil quality. It is like this: an AI-controlled drone that sprays pesticides or fertilizers only in areas where they are actually needed, reduces the use of chemicals and eliminates over-application that goes on to damage the environment and the crops themselves(Patel et al., 2024).

Mapping technology also allows the farmer to divide fields into management zones. The management zones can vary due to a difference in soil composition, type of crop, or amount of irrigation. With such an AI-powered recommendation, one can work out zone-wise strategies and apply resources strictly according to the needs of every zone-which increases yields by a margin that is quite significantly with saving input costs.

Case Studies on Precision Agriculture Success

Some other examples from the real world are very strong examples of AI-controlled robotics in precision agriculture. For example, John Deere's AI-powered agricultural machinery has reshaped large-scale farming. For example, its AI-assisted autonomous tractors use GPS along with AI to seed and apply herbicides with a precision that limits the waste of inputs and maximizes productivity(Khanh et al., 2023).

a) Other examples include "See & Spray," an AI system developed by Blue River Technology, which was acquired by John Deere. The system uses computer vision and machine learning to identify weeds in a field and apply herbicides only where they are needed, significantly reducing herbicide use to up to 90%, cutting both the cost and harm to the environment.
b) At the same time, in Australia, AI-powered drones have been seen patrolling vineyards to monitor grape health. Drones can view multispectral imaging and will start scouting for disease or stress on grapes much before it occurs, and people can take necessary measures to improve it before it becomes unmanageable. That would lead to proper crop health and better quality yield.
c) Artificial intelligence-powered soil sensors and irrigation systems help the small-scale farmers optimize water in drought-affected regions, due to increased crop yield with minimal consumption of water because of accurate watering through very precise soil moisture data communicated to them in real time.

d) The case studies from above clearly indicate that introducing robotics in the farming industry, controlled by artificial intelligence, leads to remarkable growths in productivity, sustainability, and cost-effectiveness for operations from large scale to small ones.

AI-controlled robotics will unlock the development of precision agriculture with real-time data collection, higher analyses, and autonomous decision-making. With AI plus sensors, GPS, and mapping technologies, precision farming has been improved for further resource-allocation efficiency and lesser environmental impacts. Several case studies have shown success with these technologies and thus have the potential to change the very nature of farming for better food security and sustainability in the future. With the advancement of AI and robotics, these technologies are only bound to be a core part of precision agriculture, ushering in smarter, more efficient, and environmentally friendly solutions to the challenges faced globally in agriculture.

SUSTAINABILITY IN AGRICULTURE THROUGH AI AND ROBOTICS

More and more, sustainability in agriculture has become necessary for the future. New approaches which are effective and cause less damage to the ecosystem are called for in a world challenged by climate change, resource depletion, and environmental degradation. The AI and Robotics revolutions are changing the face of agriculture. From optimized resource management to reduced chemical use and water conservation, all these contributions of technologies have come center stage in supporting environmentally friendly farming practices. Next, we will talk about AI and robotics and their contribution to sustainability by climate change control, reduction in the usage of pesticides and fertilizers, adoption of methods for water conservation, and reducing impact on the environment (AlZubi & Galyna, 2023; T. A. Shaikh et al., 2022; Tomar & Kaur, 2021). Figure 2 illustrates the interconnectedness of sustainability aspects in agriculture, highlighting the role of AI and robotics in achieving these goals.

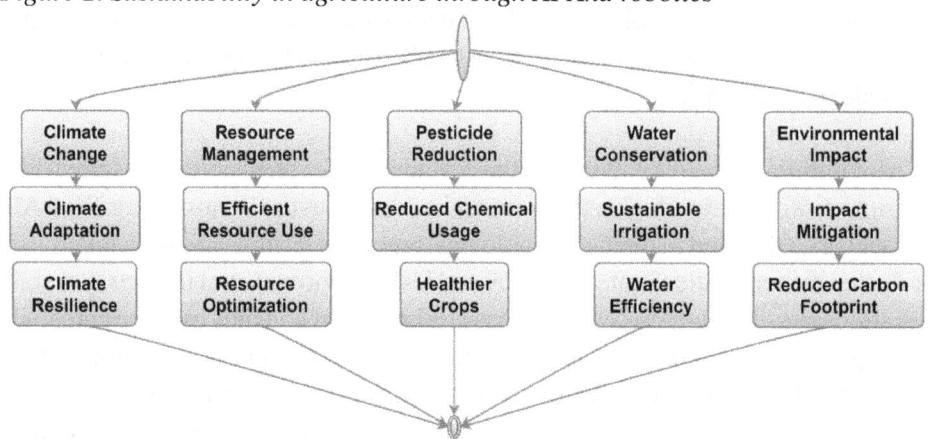

Figure 2. Sustainability in agriculture through AI And robotics

```
                Climate          Resource          Pesticide         Water           Environmental
                Change           Management        Reduction         Conservation    Impact

                Climate          Efficient         Reduced Chemical  Sustainable     Impact
                Adaptation       Resource Use      Usage             Irrigation      Mitigation

                Climate          Resource          Healthier         Water           Reduced Carbon
                Resilience       Optimization      Crops             Efficiency      Footprint
```

Climate Control and Resource Management

The spectre of climate change itself has been a challenge to the world's agriculture since anarchic weather patterns, rising temperatures, and shifting growing seasons have seen traditional farming become less effective. Advancement of the tools brought by AI and robotics to farmers would make it possible for these individuals to adapt to changes through more precise and effective management of available resources. Advanced data analytics will apply deep learning within AI to predict their future weather patterns, soil conditions, and crop growth cycles so that they can prepare way in time for their interference(Boopathi et al., 2023; Gnanaprakasam et al., 2023; Jeevanantham et al., 2022).

For instance, AI models can foresee droughts, floods, or heatwaves in advance, and the farmer can plan his irrigation schedule or shift his planting strategies. Optimizing planting and harvesting times will also significantly decrease the loss crop that one experiences with maximizing yield as the effect of climate change is hardly foreseeable. AI- controlled robots can monitor environmental conditions in real time and adjust things on the spot in real-time growth optimization, for example, irrigation management, shade application, or temperature regulation in controlled environments such as greenhouses.

Another area where AI does exceptionally well is resource management. Precision farming technologies, made possible by AI, permit a farmer to consume water, fertilizers, and seeds effectively. The AI systems take the soil data, health of the crop, and the environmental parameters into consideration and work out how much precisely every section of the field requires every resource. The process saves waste and cuts down the carbon footprints associated with the process of farming. That goes in line with the trend for sustainable agriculture.

Reduced Application of Pesticides and Fertilizers with AI

Another massive environmental problem in contemporary farming is the application of more pesticides and fertilizers than actually required. This apart from damaging the environment by polluting water and damaging soil health, a consequence of excessive chemical usage is the emission of greenhouse gases. Innovation in the high-precision application of pesticides and fertilizers is strongly encouraged with AI and robotics, thereby reducing those in general use and environmental impact (Patil et al., 2022; Ragaveena et al., 2021).

Artificially intelligent robots equipped with sensing and computer vision equipment can accurately identify spots in a field that may need pest control or fertilization. For instance, the "See & Spray" AI system developed by Blue River Technology uses machine learning algorithms to distinguish crop from weeds, comparing it in real time with requirements that allow herbicides to be applied selectively where needed. It would thereby decrease the usage of chemicals to up to 90%. With multispectral imaging, drones can spot nutrient deficiencies in targeted areas of a field. AI-powered robots can then be instructed to apply fertilizers in targeted amounts to the affected areas (Agrawal et al., 2023; Koshariya, Kalaiyarasi, et al., 2023; Koshariya, Khatoon, et al., 2023).

This targeted approach, first and foremost, prevents overapplication that may otherwise lead to a partly noxious outcome in soils and ecosystems as well as reducing the use of chemicals harmful to the environment. AI algorithms further analyze past data on pest outbreaks and nutrient cycles and are used to predict future needs, making preventive measures that further reduce the application of chemicals. By ensuring only the required amount of pesticide or fertilizer is applied in a specific area, AI-controlled robots help minimize pollution, secure biodiversity, and ensure sustainability in farming.

Water Preservation and Sustainable Irrigation Mechanisms

Water scarcity has been one of the biggest challenges in agriculture; more often than not, this occurs in drought-prone areas. Conventional irrigation systems do leak most of the applied water due to its inefficient distribution mode and lack of adjustment for crop requirement in real time. AI and robotics are changing irrigation patterns with precision watering systems that waste little water and allow usage in a sustainable manner(Boopathi, 2024; Pachiappan et al., 2024).

With sensor observation relating to the moisture levels in soil, weather conditions, and the health of the plants in real-time, AI-driven irrigation systems apply just the amount required by the various sections of a field using automated irrigation systems. Over-irrigation will be avoided; with that, wastage of water, soil erosion, and nutrient runoff will be avoided from that area under cultivation. For example, AI-controlled drip irrigation systems deliver water right to the roots of plants without evaporation right to where in a plant it's really needed.

Alongside, with weather condition forecasting AI systems, it is possible to automatically change the scheduling of irrigations. For instance, if there will be rain, it cuts or stops the irrigation, which saves even more water. AI-operated robots can adopt water saving by allowing irrigation at only the most important areas of a field first, hence watering crops but avoiding wastage during dry seasons. These innovations help reduce agriculture's water footprint while supporting resilient agriculture systems in water-scarce regions. These irrigation techniques combined with AI-powered monitoring systems contribute to a more sustainable and conscious approach to growing water management challenges in agriculture.

Environmental Impact Mitigation

There are many ways in which AI and robotics are playing critical roles in agriculture, thereby contributing to the lesser environmental impact of that sector. Some of these ways include a reduction of greenhouse emissions and minimizing soil degradation. In most traditional farming practices, results range from soil erosion to a loss of organic matter and ecosystem disruption. No-till farming, precision planting, and smart pest control are more sustainable alternatives offered by AI and robotics (AlZubi & Galyna, 2023; Tomar & Kaur, 2021).

For instance, AI-operated autonomous tractors could be used for no-till farming to directly plant seeds without tilling the soil. Such an operation ensures minimal soil erosion, improved water retention, and structural stability of the soil. No-till farming also stores carbon in the soil thus lowering the level of greenhouse gas emissions. Lastly, AI can optimize crop rotation schedules to ensure soil fertility and biodiversity while controlling the soil exhaustion risks.

Robotic weeding technologies also prevent various harmful environmental effects from farming, as the production of chemical herbicides is completely eliminated. The presence of AI-powered robots that eliminate weeds through mechanical approaches reduces the dependence on synthetic chemicals. This consequently helps in developing a healthier ecosystem and, therefore, maintaining agricultural landscapes over the long term. et another field where AI seems to be offering valuable insights is in tracking and monitoring the environmental footprint of farming operations - everything from usage of energy, production of waste, to emissions. AI can then point out inefficiency, and in doing so, supports the farmers in using more ecological means of energy sources or waste recycling, thereby further decreasing agricultural impacts on the environment(Koshariya, Khatoon, et al., 2023).

On the contrary, AI and robotics transform agriculture into a sustainable solution to overcome the biggest environmental challenges, including climate change, resource mismanagement, over-reliance on chemicals, among others. The technologies help apply resources more accurately, conserve water, and reduce negative impacts associated with traditional farming. As AI and robotics advance, so too will their role in long-term sustainability in agriculture, to give farmers the right balance between productivity and stewardship of the environment.

BOOSTING PRODUCTIVITY WITH AI-POWERED ROBOTICS

Robotics powered by artificial intelligence are at the forefront of revolutionizing agricultural productivity. These technologies solve particular problems in modern agriculture: avoiding labor-intensive work through automation, improving the monitoring system through the real-time aspect, and prevention of huge losses in post-harvest. Robotics also plays a key role in defeating labor shortages ensuring that farms can work with the same productivity even if they are given a small workforce (AlZubi & Galyna, 2023; T. A. Shaikh et al., 2022). This section talks about how the face of agricultural productivity will be changed, not just by AI-powered robotics, but by automating, real-time analytics, and innovative solutions to labor challenges.

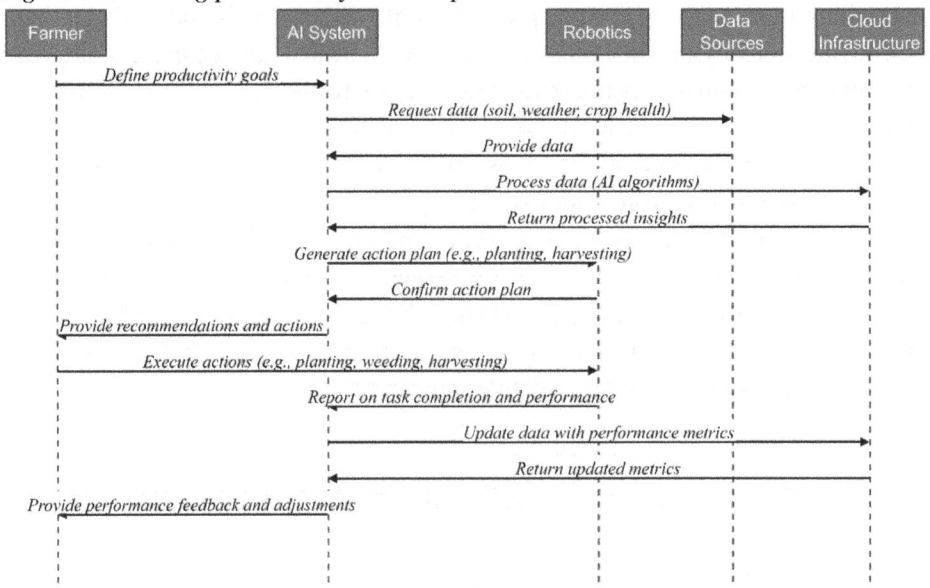

Figure 3. Boosting productivity with AI-powered robotics

[Sequence diagram with participants: Farmer, AI System, Robotics, Data Sources, Cloud Infrastructure. Messages: Define productivity goals; Request data (soil, weather, crop health); Provide data; Process data (AI algorithms); Return processed insights; Generate action plan (e.g., planting, harvesting); Confirm action plan; Provide recommendations and actions; Execute actions (e.g., planting, weeding, harvesting); Report on task completion and performance; Update data with performance metrics; Return updated metrics; Provide performance feedback and adjustments.]

Figure 3 illustrates the interaction and information flow among various components aimed at enhancing agricultural productivity through AI-powered robotics.

Agricultural productivity is significantly enhanced with AI-based robotics, as it can replace most of the tasks, including planting, harvesting, and weeding, which were considered labor-intensive and thus resulted in inefficiency and additional costs. AI-driven robots can carry out such tasks with precision, consistency, and speed (Qazi et al., 2022).

a) Planting: These seeding machines are equipped with controlled AI that optimizes seed placement with depth according to the soil condition and type of crops. The robots make use of GPS and computer vision to navigate the fields and ensure accurate seeding density and depth. This improves crop uniformity while enhancing rates of germination and reducing human labor. For example, seed planters like the "Agrobot" can plant seeds with an accuracy of spacing and depth; this means better crop production.

b) Harvesting: Robotics has completely transformed itself in harvesting, too. The AI harvesters fitted with high-technological sensors and computer vision can point out which crops are ready to be picked and gather them with minimal damage. Such robots can work day and night and, therefore increase efficiency in harvest while reducing crop loss. For instance, the "Octinion" strawberry-

picking robot that has a soft-touch gripper as well as computer vision which enables it to delicately pick the ripe fruit without damaging the plants.
c) Weeding: This is a process that is labor-intensive, and AI will help to bring such robots to reality that will help precision in weeding. They are made to differentiate crops from weeds by the use of sensors and machine learning algorithms, which enable them to target weeds without touching the crops, hence reduced dependency on chemicals for herbicide products, thus bringing a healthier environment with minimal input costs. Robots like "Weed-It" use infrared sensors for detecting weeds and through targeted herbicides or mechanical weeding techniques .

AI for Real-Time Monitoring and Decision Making

The key engine that optimizes farm operations and maximizes its productivity is through real-time monitoring and decision-making. AI systems incorporated with robotics give farmers the necessary tools that enable real-time monitoring crop health, environmental conditions, and resource use(Wakchaure et al., 2023).

a) Crop Monitoring: With AI-enabled drones and sensors equipped with multispectral cameras, such crop detailed images can be taken and analyzed for signs of stress, diseases, or nutrient deficiencies. Hence, farmers would have the chance to make early interventions on issues that face them regarding crops, keeping them healthy and productive in the latter. "Parrot Bluegrass" can analyze the crop conditions and create insights into the status of the plant health, enabling timely interventions like focused application of pesticides.
b) Real-Time Environmental Monitoring: AI Sensors can monitor soil water content and temperature, along with humidity, continuously. Combining such data with weather forecasts, AI algorithms can predict the best irrigation and fertilization times. That means farmers are making informed decisions in a less resource-wasting fashion and in perfect sync with crop requirements. Such systems as "Arable" are providing real-time insights into both weather and soil conditions, allowing farmers to amend practices on a more timely basis.
c) Decision Support: AI-based decision support systems analyse disparate data sources to provide actionable recommendations. These systems can suggest optimal planting schedules, resource allocation, and pest management strategies based on real-time data and historical trends. AI-based decision support for farmers: AI-driven decision support enhances the efficiency of operational performance as well as productivity. For example, "Climate FieldView" offers data-driven insights and recommendations on farm management that helps a farmer make informed decisions.

Robotics on Reducing Losses During Post-Harvest

A very large percentage of the agricultural produce is lost in the sector during handling, storage, and transportation. The AI-based robotics have a solution to most of these problems by decreasing this kind of loss through efficient handling in post-harvest operations(Gopal et al., 2024; Saravanan et al., 2024).

a) Sorting and Packaging: Robotics can implement AI in the sorting and packaging of the produce, where only the best quality of produce is shipped out to consumers. These computer vision and machine learning systems inspect fruits and vegetables for size, color, and defects. Automation reduces tedious labor; damage through minimal handling; and enhances quality. For example, the "Compac" sorting system makes use of AI technology in sorting fruits and vegetables with high precision and thus improves the efficiency of packing and reduces waste levels.

b) Storage Management: An AI-based system can optimize the conditions of storage for any category of produce, thus keeping it fresh and viable for as long as possible. Robotics can monitor temperature, humidity, and airflow in any storage facility and make necessary adjustments to avoid spoilage. For instance, "Ripe Robotics" uses AI to manage the ripening process of fruits, extended shelf life and reduction of losses during the post-harvesting period.

Overcoming Labor Shortages in Agriculture

The labor shortages in agriculture are threatening growing fields since most farms cannot find adequate workers to do the remaining tasks. AI-powered robotics could bridge the gaps left by a dwindling workforce to help keep farms productive.

a) Autonomous Operations: Robots can work in a totally autonomous manner, like planting, weeding, and harvesting, sans the human presence. This is likely to reduce human dependence on a huge extent while farm operation would take place with minimal human employees. It is likely that an autonomous tractor along with an autonomous harvester can work for many hours together, eradicate inefficiencies, and produce higher yields all through.

b) Augmentation of Laborer's Work: AI-powered robotics can also augment the work done by the laborer. Repeat or arduous tasks would be undertaken by these machines. Human workers would be kept busy in a more complex or strategic sort of activity. That way, the overall efficiency of the farm would be increased. For example, the robotic workers could do heavy lifting or precision planting when human workers are providing inputs regarding quality control and making decisions.

c) Remote Operation: High-end robotics infused with AI can perform remote operations because they can be programmed from a distance and enables farmers to sit in their offices and monitor farms. This is especially helpful in regions where labor scarcity is severe or the working conditions are too harsh. Remote-controlled robots can check crop health or remotely regulate the irrigation systems and that makes the process versatile and efficient.

This is through automation of labor-intensive tasks in agriculture and enhancing real-time monitoring and decision-making, reducing post-harvest losses. They also become crucial in solving labor shortages so that farms can continue to remain efficient and at a high production level. And the future of agriculture is best assured through the use of AI and robotics to ensure that farming becomes more efficient and innovative.

DATA-DRIVEN DECISION MAKING IN AGRICULTURE

Agriculture is therefore revolutionized by using big data and AI analytics in analyzing all the factors which comprise farm management, predicting, and then managing crop diseases, pests, and optimizes crop yields, and also enables long-term planning for agriculture and risk management. The more complex agricultural systems are, the more crucial it becomes to draw insights from very large data sets towards higher efficiency and sustainability and productivity. Now, this section goes on to explore how big data and AI analytics change decision-making in agriculture (Hassan et al., 2021).

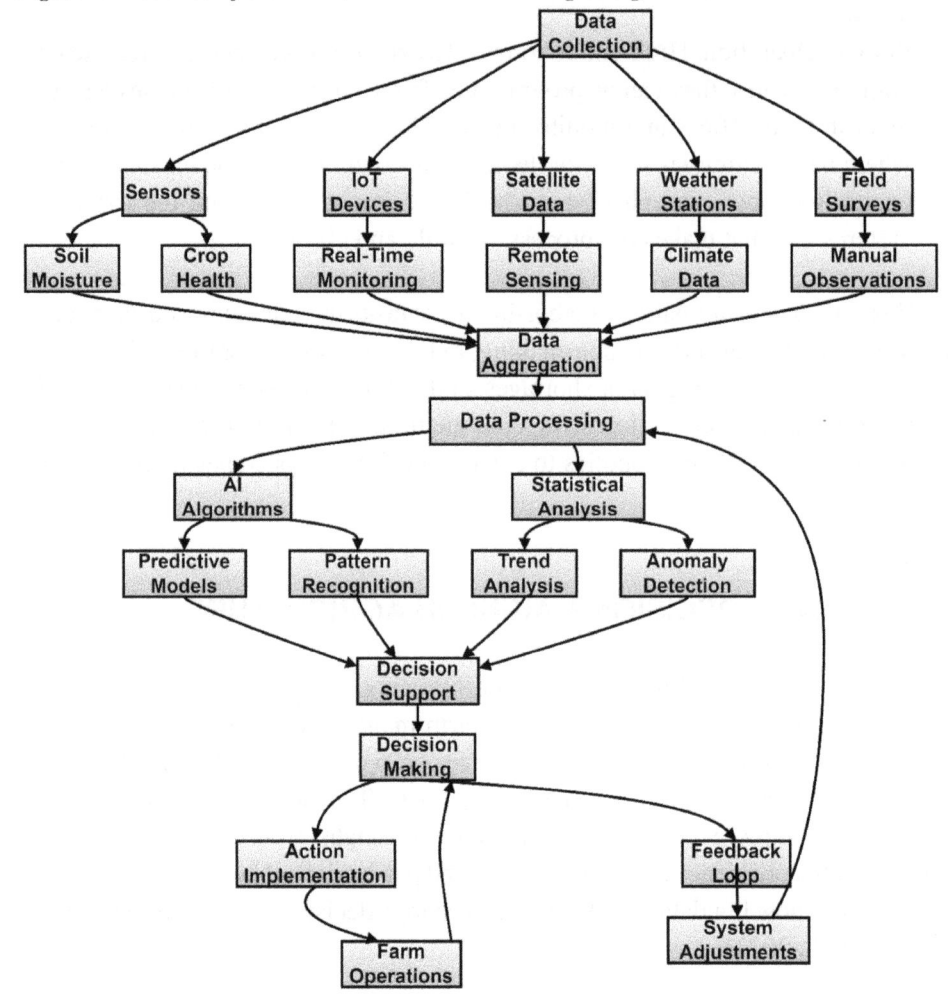

Figure 4. Process of data-driven decision-making in agriculture

The figure 4 depicts the process of data-driven decision-making in agriculture, from data collection to action implementation. Big Data and AI Analytics in Farm Management Big data and AI analytics are thus changing farm management by generating actionable insights from large volumes of data. Big data and AI analytics can aggregate, integrate, and analyze data from different sources, such as weather patterns, soil conditions, crop health, and market trends.

a) Data Integration: The current modern operations for agriculture produce a string of data nowadays in sources like those coming from satellites, sensors, drones, and even the ground-based weather stations. This way, AI systems then integrate the same data into holistic models, offering farm conditions with an all-inclusive view. Through using integration within data, it enables farmers to monitor and manage operations in one location through different aspects. For instance, "Climate FieldView" and "John Deere Operations Center" platforms collect data on weather conditions, soil health, and crop performance, which, in turn, enable farmers to gain real-time insights and recommendations.

b) AI Analytics: Algorithms powered by AI scan complex datasets for patterns and trends that a human might never have observed. With the ability to realize patterns from historical data and predict subsequent outcomes, machine learning models can hone in on future outputs. For instance, correlating soil properties with crop yields through AI-driven analytics, one may use analysis to change farming strategies for optimal productivity.

c) Operational efficiency: Today, big data and AI analytics can be used to help farmers optimize resource allocation and improve workflows and reduce the operational costs of production. Such a system managed through AI will recommend to the farmer the most efficient use of water, fertilizers, and pesticides when possible, based on real-time data to minimize waste and enhance overall farm productivity.

Predictive Analytics for Crop Disease and Pest Control

Predictive analytics with the muscle of big data and AI power forms an important part of controlling crop diseases and pests. It can forecast potential outbreaks and suggest better prevention and control measures in real time by analyzing a store of historical and real-time data (Mohamed et al., 2021).

a) Disease Forecast: Based on weather patterns, levels of soil moisture and health condition of crops, AI models provide the probability of disease. For example, machine learning algorithms can identify patterns for a particular type of fungal infection or bacterial disease so that farmers get started on a course of action before visible signs appear on a large scale. Tools such as "Plantix" use image recognition combined with artificial intelligence to detect plant diseases from photographs, with suggestions for treatment and control.

b) Pest Control: Predictive analytics can also predict pest activity in line with the environmental conditions and historical data. AI systems trace these pest populations, migration, and lifecycle stages so that farmers are ahead in knowing when and where to put in the pest control measures. For instance, "AgriSense" provides live monitoring and forecasting of pests helping farmers introduce better interventions in targeted ways and reduce pesticide usage.

AI-based decision support systems can give real-time actionable insights on crop diseases and pests. Predictive analytics combined with historical data and current conditions may prescribe what treatments to apply, when to apply them, and even the amount of the medicine. The farmers would decide where to put the resultant precise treatments to protect their crops more effectively and cause no untold damage to the environment.

Maximizing Crop Yields Through the Insights of Data

The very core of data-driven agriculture involves crop yield optimization, wherein AI analytics is expected to substantially contribute towards this goal. Using data related to soil health, weather conditions, and performance of crops, farmers are likely to make decisions to boost the productivity level (Glady et al., 2024; Pachiappan et al., 2024; Patel et al., 2024; Venkateswaran et al., 2024).

a) Soil and Crop Information: AI applications analyze the composition of the soil, nutrient content, and moisture levels to define the perfect conditions required for cultivating the crops. This information is combined with crop variety information and the different stages of growth to enable farmers to adjust their practices to optimize the yield. For example, YaraVita provides AI-driven recommendations regarding the management of soil and crops to help farmers enhance nutrient application and increase crop productivity.
b) Precision Farming: Data insights allow precision agriculture where farm practices are tuned to specific field conditions. The development of an AI-powered system can outline specific soil health and nutrient levels along with crop performance, allowing for more precise application of inputs. This targeted approach reduces waste and optimizes resources, enhancing crop yields.
c) Yield Forecasting: AI models can predict crop yields with the help of historical data, weather forecast and prevailing field conditions. Such predictions help farmers in planning harvests, managing stock, and marketing decisions. For instance, "Solfarm" uses AI for the yield forecasting of crops and insights into the market trends to help farmers optimize their production and sales strategies.

Long-term Planning and Risk Management

Data-driven decision-making promotes long-term planning and risk management for agriculture. The farmer will analyze the historical data and forecast their future trend analysis to plan responses to the minimization of risks which can help maintain long-term sustainability (Ragavi et al., 2020).

a) Risk assessment -AI and big data analytics help in identifying risks such as extreme weather conditions, market fluctuations, and the outbreak of pests. AI models are applied further to analytical purposes by simulating the resultant history, thus giving farmers the ability to assess the impact on operations differently. Information allows farmers to come up with contingency plans and apply strategies for mitigating risks.
b) Sustainability Planning Sustainability planning is supported by data-based insights. Data on soil health, water usage, and environmental impacts provide details for farmers to improve their soil productivity, water usage, and environmental impact. From correct applications of AI in the agri-sector, these systems can present applicable practices that would improve soil fertility, reduce water usage, and minimize environmental impacts. For example, "FarmBeats" uses AI to track soil health and water usage, offering recommendations for sustainable farming practices.
c) Financial Planning: AI analytics also helps the farmers in financial planning as it analyzes cost management, revenue projection, and opportunity for investment. By processing data on production costs, market prices, and financial performance, better or more appropriate decisions that can be generated towards investments, financing, and budgeting. For example, "AgriWebb" offers financial management tools that help manage the expenses of the farmer and optimize the budgets for future growth.

Decision-making based on data powered by big data and AI analytics enables a better management of crop disease and pest conditions, optimized crop yields, effective support for long-term planning and risk management, which transform the agricultural space. Such technologies arm the farmers with insights and recommendations that extend towards more informed decisions aimed at improving the productivity, sustainability, and resilience of farming systems. Data-driven approaches will only continue to evolve and become ever more integral in building the future of agriculture with successful ongoing farming operations.

CHALLENGES AND BARRIERS TO ADOPTION

Sure, the infusion of AI and robotics into agriculture offers many benefits, but at the same time, it creates more challenges and barriers to embracing it. These include high initial costs, less favorable accessibility to small-scale farms and smallholder farmers in developing regions, the displacement of jobs, and data security and privacy. Knowing these challenges allows the development of strategies to overcome these to ensure that they succeed in the agricultural sector (Ragavi et al., 2020; F.

K. Shaikh et al., 2021; Usigbe et al., 2024; Wakchaure et al., 2023). The adoption of AI and robotics in agriculture faces several challenges and barriers, as illustrated in Figure 5.

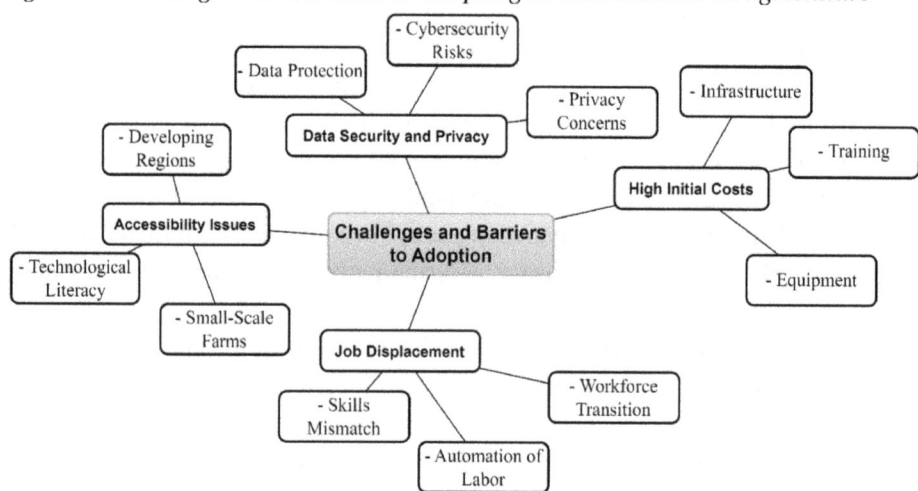

Figure 5. Challenges and barriers to adopting AI and robotics in agriculture

High Level of Initial Investment and Economic Considerations

Investment in AI and robotics in agriculture is severely fettered by its significant initial investment costs that are required when acquiring and implementing the technology. The cost of buying expensive machinery, sensors, and AI systems is too heavy, especially for small-scale and resource-poor farmers.

Although the capital investment in acquiring equipment, installing, adjusting, and maintaining the AI-based robotics and data analytics system is imputed over equipment acquisitions, the high expense might deter entry to advanced technologies, particularly among small farmers in developing countries and thus propagate existing inequalities in agricultural productivity. AI and robotics technology can save in the long term and can be more efficient. However, the ROI is likely to come after some period and not so near. It will thus leave the farmers weighing the possible benefits over the initial costs. They will thus have to decide whether or not it is going to deliver enough economic benefits for them to invest in it. Such technologies would require some form of financial incentives, subsidies, and possibly even cost-sharing schemes to make them more accessible to the greater farming community.

Equity in Access for Small-Scale and Developing Region Farmers

AI and robotics adoption by farmers seems to be a preserve of larger-scale and technologically advanced farms, which is rather not leveled off for small-scale farmers and farmers from the developing regions. Advanced Agriculture Technologies require robust infrastructure for their functioning, with internet access and assured power supply. Most of the developing regions are short of such infrastructures, and this makes farming difficult to adopt new technologies such as AI and robotics. If the necessary infrastructures are not in place, even the most advanced technology becomes impractical or inefficient.

There is a need for some degree of technical know-how and training for the effective implementation of AI and robotics. Since most small-scale farmers and those in developing regions cannot afford it, they will not have any training or knowledge in how to run and maintain such technologies. Their education and training are essential to enabling tremendous gains from technological breakthroughs. A high cost will accompany the use of advanced technologies since small-scale farmers may not raise funds to acquire such systems. Affordable and scalable solutions tailored to the needs of small-scale and developing region farmers must be developed to enable wider adoption.

Displacement of Jobs due to Automation

This may, however also bring some doom, especially because automation through AI and robotics may most probably replace traditional farm labor with the automation of agricultural tasks.

a) Labor Market Impact: The demand for labor shall decline since robots and AI would take over jobs like planting, harvesting, and weeding. It may be that an entire group of workers will lose their job, depending upon employment for it. So, automation will affect both the rural community and agricultural employment, so this too needs proper handling not to have reduced social and economic activities.

b) Developing ways for dislocated workers to transfer to other jobs or industries becomes another imperative skill transition. This would involve retraining or upskilling to adapt to new forms of employment or new roles in agriculture sectors that are less easily automated than others.

c) Employment Disruption: Agricultural laborers further disrupt the local economy, especially if the economy concerned is an agricultural labor-based economy. It has become imperative to deal with the potential impact of automation through economic resilience and more sensitive measures to the affected community.

Data Security and Privacy

The use of artificial intelligence and robots in agriculture brings data collection and analysis with it. The processes might raise concerns related to security and the privacy of the data collected.

a) Data Protection: While using agricultural technologies related to farm operations, crop health, soil conditions, and environmental elements are normally recorded. Therefore, proper security for such data is mandatory in order to avoid unauthorized access, data breaches, and cyberattacks. Farmers' information must be protected through properly designed data protection mechanisms and cybersecurity protocols.
b) A significant concern surrounds privacy, as the farmers are compelled to provide personal and/or location-based data. Information collected needs to be protected against misuse or diversions other than for intended purposes. If there is transparency in the data collection processes and promise regarding privacy policies, this can guarantee such a promise to the farmers.
c) Compliance and regulation of data: Data protection compliance is essential for the management of security and privacy of data. For example, Europe-based companies must adhere to the General Data Protection Regulation in dealing with data ethically and appropriately.

There are challenges from AI and robotics, such as high initial investment, accessibility problems to smallholders in developing regions, job displacements, and issues surrounding data security and privacy. A package of financial, infrastructural education, and training interventions has to be accompanied by proper and strong data protection mechanisms. Meeting these expectations and overcoming these challenges would help the agricultural sector reap the real benefits from AI and robotics, resulting in efficiency, sustainability, and productivity in farming.

FUTURE TRENDS IN AI AND ROBOTICS FOR AGRICULTURE

Along with the advancements in technology, the future of AI and robotics in agriculture promises exciting prospects of further changing agricultural practices. This section discusses the critical review of most crucial trends in relation to the development of capabilities of algorithms of AI and robotics, emerging technologies in smart agriculture, government and policy influences on promoting innovation, and scaling AI solution globally (Sitharthan et al., 2023; Usigbe et al., 2024; Wakchaure et al., 2023).

Developments in AI Algorithms and Capabilities of Robotics

AI algorithms are being constantly upgraded to result in great developments in agricultural technology. Advanced models of machine learning yield better predictions, recognition of patterns, and thus decisions are becoming better. Complex data on matters from different sources like satellite images, sensors, and drones are getting analyzed using deep learning techniques. These advancements will improve yields while predicting infestation by pests and disease outbreaks, thus allowing for more prompt and specific interventions to take place(Boopathi, 2024; Glady et al., 2024; Kumar et al., 2023).

Robotics Evolution It goes without saying that robotic technology continues to evolve as well. The new agricultural robots can do a host of different activities such as planting, harvesting, and weeding. Improvements in the design of the robots include higher maneuverability and more accuracy, along with work under varied conditions of the field. It works independently by implementing advanced sensors and AI algorithms that enhance their efficiency and productivity with tough jobs done with fewer interventions from humans. The integration of AI and Robotics results in the development of Intelligent Systems that self-adjust according to the changes in the evolving conditions. For example, it resulted in optimizing the operating mode of AI-powered robotic systems with the help of real-time sensor data, which eventually leads to adjustments and hence increased performance and resource usage. It will also boost the capabilities of agricultural machines and improve farm management in general.

Emerging Technologies in Smart Agriculture

It has actually changed the face of modern agriculture, slowly but surely transforming the way smart agriculture happens around the globe by providing real-time information related to soil and weather as well as crop health. IoT sensors applied in fields, equipment, and infrastructure collect data wirelessly. It has allowed farmers to analyze and manage their operations far more effectively. Better decision making can be supported with sensor technology that brings accurate and cost-effective data collection.

Blockchain for Traceability: The agricultural industry is embracing power using blockchain technology to increase transparency and traceability. This allows for a safe, tamper-proof record of transactions tracing the path an agricultural product follows from farm to table. It ensures food safety, reduces fraud cases, and increased efficiency in the supply chain. A wide number of areas to be improved in traceability and accountability are certainly going to expand into new agricultural applications as blockchain technology matures and comes of age.

Some emerging technologies of precision agriculture are advanced imaging systems, remote sensing, and variable rate technology. One of the very high-resolution camera-containing drones is equipped with multispectral sensors. Such drones can take valuable images over crops and fields that are very clear to allow precise analysis and management of the same. Variable rate technology allows for the focused application of inputs such as fertilizers and pesticides, thereby avoiding waste and improving crop performance.

Gene editing and biotechnology make it possible to edit genes in many crops, which opens new windows into crop resilience and productivity improvement. Biotechnology applications range from GMOs to bioengineered crops aimed at addressing issues such as pest resistance, drought tolerance, and nutrient deficiencies. Technologies like these will soon fill a leading position in agriculture that will complement AI and robotics innovations.

Role of Government and Policy in Innovation

Support to Research and Development: Governments are a very important factor that stimulate innovation in agricultural technology through support towards R&D activities. They can support it through public funds for research and development, which would rapidly advance the development of new AI and robotics technologies, promote academic-industrial collaboration, and facilitate the commercialization of the innovative solutions. Initiatives taken by the government in providing grants, subsidies or tax incentives would gradually reduce the burden of developing the technology and thus foster investment in advanced research.

a) Regulatory Frameworks: Effective regulatory frameworks should be provided to ensure the safe and ethical use of AI and robotics in agriculture. Above all, governments need to set clear guidelines and standards for the deployment of technology, data privacy, and cybersecurity. Regulations on matters relating to

intellectual property rights, safety standards, and environmental impact are thus crucial to balance innovation with the protection of public interests.
b) Support Education and Training: The AI and robotics in agriculture can also be supported by ensuring adequate education and training. To prepare for the future agricultural workforce, developing technical skills, digital literacy, and data analytics capabilities will become integral. Educations can liaise with industry leaders and government agencies to help bridge the gap by equipping farmers and technicians to adequately apply the advanced technology.
c) Encourage PPPs: Public and private sectors require collaboration in developing further agricultural technology. A PPP may share knowledge through resources and expertise. Innovative solutions will develop under such collaboration, and the widespread and successful use of AI and robotics in agriculture may be achieved. Governments can facilitate PPPs through their policy. Governments may give financing to PPPs and get people working together from different sectors.

Scaling AI Solutions for Global Agriculture

a) Scaling AI Solutions for Global Agriculture: Adaptive Technologies for Diverse Environments. Technologies developed in developed regions may need adaptation or modification for operating purposes in diverse environmental and socio-economic contexts prevailing in developing countries. There may be a difference in climatic conditions and kinds of soil and infrastructure. Therefore, customized solutions to localized needs and problems are very essential for effectiveness and accessibility.
b) More innovation, more AI-oriented solutions, need to be scaled across borders through international collaboration. Best practices, research findings, and technological advancements cross borders and cause significant acceleration in scaling innovative solutions, pushing agriculture globally forward. Such knowledge exchange management and support toward globalization is also given by international organizations, industry associations, and research institutions.
c) Equity and Access: The accessibility of AI solutions across different regions, including low-income and developing regions, also brings in a very important consideration towards the attainment of equitable benefits. Strategies include developing technology that should not be expensive, making technical support and training accessible, and scaling solutions to fit most contexts. Government and industry action will bear fruit and allow these realized benefits of AI and robotics to come into focus.

d) Monitoring and evaluation: Scaling AI solutions would need constant monitoring and evaluation to check the impact and effectiveness of AI solutions. Effectiveness and challenges are identified during assessments, and regular measurements can be made on outcomes, which could shape improvements. Data-driven insights from pilot projects and case studies can steer the process of scaling and can help in refining technologies for farmers globally to better suit their needs.

Rapid progress, innovative emergent technologies, and new roles for governments and policies persist in the future of AI and robotics in agriculture. New generations of agricultural practices will be shaped by forward progress in AI algorithms and robotics capabilities and innovative technologies as well as smart agriculture applications. Tackling fundamental issues in accessibility, equity, and scaling up globally shall unleash the complete power of AI and robotics for the agricultural sector by bringing about conducive regulatory environments and public-private partnerships. It shall unlock more efficient, sustainable, and productive farming systems leading to improved global food security and agricultural resilience.

CONCLUSION

Integration of AI and robotics in agriculture transforms the approach to farming and food production. This will lead to a spate of unprecedented levels of precision, sustainability, and productivity with the rise of AI-algorithm advancement, enhanced robotics capabilities, and adoption of emerging technologies in the agricultural sector. The discussions in this chapter underline a future where these technologies will revolutionize agricultural practices and counter some of the pressing challenges: resource management, climate change, and labor shortages. Indeed, the entry barriers of the adoption process are characterized by significant startup costs, problems regarding accessibility, fears of job displacement, and data security, among other aspects. It requires governmental support, as well as collaboration between the industry and institutions of research, to further innovation and ensure that equal access is provided with respect to these advanced technologies. This includes the use of AI and robotics to leverage opportunities that contribute to agriculture toward achieving global food security, environmental stewardship, and sustainable development. This will surely entail research, adaptive strategies, and a commitment to the handling of the technical and socioeconomic dimensions of agricultural transformation for its effective implementation.

REFERENCES

Agrawal, A. V., Shashibhushan, G., Pradeep, S., Padhi, S., Sugumar, D., & Boopathi, S. (2023). Synergizing Artificial Intelligence, 5G, and Cloud Computing for Efficient Energy Conversion Using Agricultural Waste. In *Sustainable Science and Intelligent Technologies for Societal Development* (pp. 475–497). IGI Global.

AlZubi, A. A., & Galyna, K. (2023). Artificial intelligence and internet of things for sustainable farming and smart agriculture. *IEEE Access : Practical Innovations, Open Solutions*, 11, 78686–78692. DOI: 10.1109/ACCESS.2023.3298215

Boopathi, S. (2024). Sustainable Development Using IoT and AI Techniques for Water Utilization in Agriculture. In *Sustainable Development in AI, Blockchain, and E-Governance Applications* (pp. 204–228). IGI Global. DOI: 10.4018/979-8-3693-1722-8.ch012

Boopathi, S., Kumar, P. K. S., Meena, R. S., Sudhakar, M., & Associates. (2023). Sustainable Developments of Modern Soil-Less Agro-Cultivation Systems: Aquaponic Culture. In *Human Agro-Energy Optimization for Business and Industry* (pp. 69–87). IGI Global.

Chen, C.-J., Huang, Y.-Y., Li, Y.-S., Chang, C.-Y., & Huang, Y.-M. (2020). An AIoT based smart agricultural system for pests detection. *IEEE Access : Practical Innovations, Open Solutions*, 8, 180750–180761. DOI: 10.1109/ACCESS.2020.3024891

Glady, J. B. P., D'Souza, S. M., Priya, A. P., Amuthachenthiru, K., Vikram, G., & Boopathi, S. (2024). A Study on AI-ML-Driven Optimizing Energy Distribution and Sustainable Agriculture for Environmental Conservation. In *Harnessing High-Performance Computing and AI for Environmental Sustainability* (pp. 1–27). IGI Global., DOI: 10.4018/979-8-3693-1794-5.ch001

Gnanaprakasam, C., Vankara, J., Sastry, A. S., Prajval, V., Gireesh, N., & Boopathi, S. (2023). Long-Range and Low-Power Automated Soil Irrigation System Using Internet of Things: An Experimental Study. In *Contemporary Developments in Agricultural Cyber-Physical Systems* (pp. 87–104). IGI Global.

Gopal, M., Lurdhumary, J., Bathrinath, S., Priya, A. P., Sarojwal, A., & Boopathi, S. (2024). Energy Harvesting and Smart Highways for Sustainable Transportation Infrastructure: Revolutionizing Roads Using Nanotechnology. In *Principles and Applications in Speed Sensing and Energy Harvesting for Smart Roads* (pp. 136–165). IGI Global. DOI: 10.4018/978-1-6684-9214-7.ch005

Hassan, S. I., Alam, M. M., Illahi, U., Al Ghamdi, M. A., Almotiri, S. H., & Su'ud, M. M. (2021). A systematic review on monitoring and advanced control strategies in smart agriculture. *IEEE Access : Practical Innovations, Open Solutions*, 9, 32517–32548. DOI: 10.1109/ACCESS.2021.3057865

Jeevanantham, Y. A., Saravanan, A., Vanitha, V., Boopathi, S., & Kumar, D. P. (2022). Implementation of Internet-of Things (IoT) in Soil Irrigation System. *IEEE Explore*, 1–5.

Khanh, P. T., Ngoc, T. T. H., & Pramanik, S. (2023). Future of smart agriculture techniques and applications. In *Handbook of Research on AI-Equipped IoT Applications in High-Tech Agriculture* (pp. 365–378). IGI Global. DOI: 10.4018/978-1-6684-9231-4.ch021

Koshariya, A. K., Kalaiyarasi, D., Jovith, A. A., Sivakami, T., Hasan, D. S., & Boopathi, S. (2023). AI-Enabled IoT and WSN-Integrated Smart Agriculture System. In *Artificial Intelligence Tools and Technologies for Smart Farming and Agriculture Practices* (pp. 200–218). IGI Global. DOI: 10.4018/978-1-6684-8516-3.ch011

Koshariya, A. K., Khatoon, S., Marathe, A. M., Suba, G. M., Baral, D., & Boopathi, S. (2023). Agricultural Waste Management Systems Using Artificial Intelligence Techniques. In *AI-Enabled Social Robotics in Human Care Services* (pp. 236–258). IGI Global. DOI: 10.4018/978-1-6684-8171-4.ch009

Kumar, P. R., Meenakshi, S., Shalini, S., Devi, S. R., & Boopathi, S. (2023). Soil Quality Prediction in Context Learning Approaches Using Deep Learning and Blockchain for Smart Agriculture. In *Effective AI, Blockchain, and E-Governance Applications for Knowledge Discovery and Management* (pp. 1–26). IGI Global. DOI: 10.4018/978-1-6684-9151-5.ch001

Mohamed, E. S., Belal, A., Abd-Elmabod, S. K., El-Shirbeny, M. A., Gad, A., & Zahran, M. B. (2021). Smart farming for improving agricultural management. *The Egyptian Journal of Remote Sensing and Space Sciences*, 24(3), 971–981. DOI: 10.1016/j.ejrs.2021.08.007

Pachiappan, K., Anitha, K., Pitchai, R., Sangeetha, S., Satyanarayana, T., & Boopathi, S. (2024). Intelligent Machines, IoT, and AI in Revolutionizing Agriculture for Water Processing. In *Handbook of Research on AI and ML for Intelligent Machines and Systems* (pp. 374–399). IGI Global.

Parasuraman, K., Anandan, U., & Anbarasan, A. (2021). IoT based smart agriculture automation in artificial intelligence. *2021 Third International Conference on Intelligent Communication Technologies and Virtual Mobile Networks (ICICV)*, 420–427. DOI: 10.1109/ICICV50876.2021.9388578

Patel, T. K., Vasundhara, S., Rajesha, S., Priyalakshmi, B., Chinnusamy, S., & Boopathi, S. (2024). Leveraging Drone and GPS Technologies For Precision Agriculture: Pest Management Perspective. In *Revolutionizing Pest Management for Sustainable Agriculture* (pp. 285–308). IGI Global. DOI: 10.4018/979-8-3693-3061-6.ch012

Patil, D. D., Singh, A. K., Shrivastava, A., & Bairagi, D. (2022). IOT Sensor-Based Smart Agriculture Using Agro-robot. In *IoT Based Smart Applications* (pp. 345–361). Springer.

Qazi, S., Khawaja, B. A., & Farooq, Q. U. (2022). IoT-equipped and AI-enabled next generation smart agriculture: A critical review, current challenges and future trends. *IEEE Access : Practical Innovations, Open Solutions*, 10, 21219–21235. DOI: 10.1109/ACCESS.2022.3152544

Ragaveena, S., Shirly Edward, A., & Surendran, U. (2021). Smart controlled environment agriculture methods: A holistic review. *Reviews in Environmental Science and Biotechnology*, 20(4), 887–913. DOI: 10.1007/s11157-021-09591-z

Ragavi, B., Pavithra, L., Sandhiyadevi, P., Mohanapriya, G., & Harikirubha, S. (2020). Smart agriculture with AI sensor by using Agrobot. *2020 Fourth International Conference on Computing Methodologies and Communication (ICCMC)*, 1–4. DOI: 10.1109/ICCMC48092.2020.ICCMC-00078

Saravanan, S., Khare, R., Umamaheswari, K., Khare, S., Krishne Gowda, B. S., & Boopathi, S. (2024). AI and ML Adaptive Smart-Grid Energy Management Systems: Exploring Advanced Innovations. In *Principles and Applications in Speed Sensing and Energy Harvesting for Smart Roads* (pp. 166–196). IGI Global. DOI: 10.4018/978-1-6684-9214-7.ch006

Shaikh, F. K., Memon, M. A., Mahoto, N. A., Zeadally, S., & Nebhen, J. (2021). Artificial intelligence best practices in smart agriculture. *IEEE Micro*, 42(1), 17–24. DOI: 10.1109/MM.2021.3121279

Shaikh, T. A., Rasool, T., & Lone, F. R. (2022). Towards leveraging the role of machine learning and artificial intelligence in precision agriculture and smart farming. *Computers and Electronics in Agriculture*, 198, 107119. DOI: 10.1016/j.compag.2022.107119

Sitharthan, R., Rajesh, M., Vimal, S., Kumar, S., Yuvaraj, S., Kumar, A., Raglend, J., & Vengatesan, K. (2023). A novel autonomous irrigation system for smart agriculture using AI and 6G enabled IoT network. *Microprocessors and Microsystems*, 101, 104905. DOI: 10.1016/j.micpro.2023.104905

Tomar, P., & Kaur, G. (2021). *Artificial Intelligence and IoT-based Technologies for Sustainable Farming and Smart Agriculture*. IGI global. DOI: 10.4018/978-1-7998-1722-2

Usigbe, M. J., Asem-Hiablie, S., Uyeh, D. D., Iyiola, O., Park, T., & Mallipeddi, R. (2024). Enhancing resilience in agricultural production systems with AI-based technologies. *Environment, Development and Sustainability*, 26(9), 21955–21983. DOI: 10.1007/s10668-023-03588-0

Venkateswaran, N., Kiran Kumar, K., Maheswari, K., Kumar Reddy, R. V., & Boopathi, S. (2024). Optimizing IoT Data Aggregation: Hybrid Firefly-Artificial Bee Colony Algorithm for Enhanced Efficiency in Agriculture. *AGRIS On-Line Papers in Economics and Informatics*, 16(1), 117–130. DOI: 10.7160/aol.2024.160110

Wakchaure, M., Patle, B., & Mahindrakar, A. (2023). Application of AI techniques and robotics in agriculture: A review. *Artificial Intelligence in the Life Sciences*, 3, 100057. DOI: 10.1016/j.ailsci.2023.100057

Chapter 16
AI-Driven Disaster Forecasting by Integrating Smart Technology

Periasamy J. K.
Department of Computer Science and Engineering, Sri Sai Ram Engineering College, Chennai, India

Kunduru Srinivasulu Reddy
https://orcid.org/0000-0002-7786-5892
Department of Mechanical Engineering, Sreenidhi Institute of Science and Technology, Hyderabad, India

Prachi Rajendra Salve
Department of Computer Engineering, D.Y. Patil School of Engineering Academy, Pune, India

S. Ushasukhanya
School of Computing, SRM Institute of Science and Technology, Kattankulathur, India

T. Y. J. Naga Malleswari
Networking and Communications, SRM Institute of Science and Technology, Kattankulathur, India

ABSTRACT

This chapter explores how AI and smart technologies could altogether be integrated to bring revolutionary change in the fields of disaster forecasting and management.

DOI: 10.4018/979-8-3693-5573-2.ch016

It will try to analyze through advanced algorithms and IoT sensors how these technologies can potentially advance a disaster-related prediction along with accuracy and timeliness. Important applications of real-time data collection, predictive modeling, and automated alerts collectively enhance response strategies as well as resource allocation. This chapter's discussion of the promise of AI merged with smart technologies—improved predictiveness, faster response times, and better risk assessment—perhaps weighs the potential liabilities and limitations of such applications, including data privacy issues and infrastructures sturdy enough to host such a system. This chapter draws on case studies and continuing research into the use of AI-driven systems in disasters to present insights about how they are changing practices in disaster management and outline future directions for the emerging field.

INTRODUCTION

While natural disasters are more frequently occurring with more intense frequency, integration of AI with smart technologies may be one transformative way of handling prediction and management in a better manner. The response and accuracy in predicting events lead to effective reduction of damage, saving lives, and ensuring an easy and swift recovery. In so many instances, traditional forecasting methods are helpful but do not go too far in the cases of rapidly evolving and complex disaster scenarios. A discussion on how AI and smart technologies are changing forecasts for disasters is an exploration of enhancing the accuracy of predictions for making better use of efforts in response and solving major challenges (Jun et al., 2019).

Traditionally, disaster prediction was a science based on a combination of analyses of historical data, observational techniques, and expert judgment. Meteorologists and geologists could use these means to provide warnings and to prepare in the event of hurricanes, earthquakes, or floods. Though these approaches have been highly successful in many cases, their applicability can often be limited by the availability of data and the coarseness of the data available to model natural phenomena. AI, above all machine learning and data analytics, finally gives access to opportunities for an unprecedented scale of forecasts. AI algorithms can deal with enormous amounts of data gathered from different sources, identify patterns overlooked by human analysts, and thus generate more accurate predictions (Chen et al., 2015). The richness of the data is further enriched with smart technologies such as IoT sensors and remote sensing systems, providing real-time insight-seeing the basis for timely decision-making.

AI-powered disaster forecasting uses a wide range of technologies to monitor improvements in the accuracy of predictions and the efficiency of responses. Complex datasets are being worked on by machine learning models- neural networks and ensemble methods-to make probabilistic predictions over an impending disaster. For example, AI can analyze various patterns of weather and ocean temperatures and atmospheric conditions to make more precise predictions regarding hurricanes. The integration necessitates the use of smart technologies. Such IoT sensors are important for delivering real-time data on conditions in the environment. The strategic placement of sensors allows variables such as rainfall, soil moisture, and seismic activities to be measured continuously. The data go into AI systems that analyze the patterns, creating alerts for anomalies and predicting potential disasters (Ramírez & Briones, 2017). For example, in flood-prone areas, IoT sensors can predict an impending flood very early by detecting and sending warning signals rising water levels to the emergency services.

Many case studies show that it's possible to make use of AI and smart technologies effectively in forecasting disasters. For example, with storm-track hurricane forecasting, AI algorithms for the improvement of the accuracy of prediction can enable more reliable-based forecasts through the analysis of satellite images, weather data, and historical patterns of hurricanes, which assist communities in preparing for appropriate action. An AI-driven system scans through seismic data coming from networks of sensors for earthquake prediction in order to identify patterns that may signify an impending quake (Merz et al., 2020). This remains a challenge in terms of absolute predictions; however, AI models can still alert one early enough to take prompt response actions to mitigate the problem. Flood management also benefits from the integration of smart technologies and AI. Advanced hydrological models combined with real-time data acquired from river gauges and weather stations allow for more accurate flood forecasts by AI systems that analyze this data in order to predict flood events and provide insight for emergency response teams (Khovrat et al., 2022).

Even though much promising development has gone into the brainchild, challenges need to be addressed so that AI-driven disaster forecasting is fully unlocked. Data quality and reliability form a big challenge as AI models depend on what they learn to be good at, and unless and until full data sets or complete information are available, the chances of flawed predictions are high. There is a need for data quality assurance and bridging gaps in data collection processes to enhance the accuracy of these forecasts. Another issue involves the integration of AI systems with existing infrastructure. Most regions, especially those prone to disasters, may not have technological infrastructure to support the latest AI and IoT systems (Linardos et al., 2022). Building and maintaining this infrastructure requires considerable investments and coordination among several stakeholders. Data privacy and security issues also play a significant role. If such massive amounts of data are accumulated, analyzed, and collated, the issue pertaining to data management and protection arises. The responsible usage of such data, therefore, remains a need, while data privacy should be respected in building AI-driven systems (Sarker et al., 2020).

The future prospect of AI-driven disaster forecasting holds positive signs with further research and development that emphasizes strengthening predictive capabilities and knowledge acquisition at every step. Improvement in AI algorithms, data integration techniques, and sensor technologies is visualized for a much more robust and reliable estimation of forecasts. High scalability and robust infrastructure would further favor mass-adoption of these technologies. It calls for a collaboration between researchers, technology developers, and disaster management experts to advance the field. It is then possible for the stakeholders to meet and somehow solve the challenges they face together by sharing available knowledge to create innovative solutions that help in improving forecasting and effective responses during disasters (Khalid & Shafiai, 2015; Puttinaovarat & Horkaew, 2020).

Integration of artificial intelligence and smart technologies is a giant step into disaster forecasting, tapping advanced algorithms with data in near-real time to add new opportunities in predicting and managing natural disasters. Challenges are likely to be met with time to be leveraged and maximized through continued research and cooperation in the field.

Significance

The transformative power of AI and smart technologies in the context of disaster forecasting and management makes the book focus on enhancing predictive capabilities, accuracy, and timeliness through adoption of advanced algorithms along with IoT sensors. The real-time collection and predictive modeling of data have enabled automated alerts. The predictiveness, response time, and the process of risk assessment are enabled by the AI-driven system for pro-active disaster management. However, these have challenges, such as data privacy concerns and need for resilient infrastructures to support such systems. In turn, this chapter critically analyses such a limitation but presents a balanced approach towards looking at the potentialities of AI. Using case studies and ongoing research, the chapter reveals how such technologies are reshaping disaster management practices by providing valuable insights into these current applications and future possibilities of application.

Scope and Objectives

Chapter This chapter explores the integration of artificial intelligence (AI) and smart technologies into the processes of disaster forecasting, including higher prediction accuracy and speed of response time and resource allocation. Under the scope, the discussion will consider an appraisal of AI methodologies, specifically machine learning and predictive modelling, and smart technologies like IoT sensors and real-time data analytics. The discussion covers different types of disasters, including hurricanes, earthquakes, and floods, along with its application in real-world settings.

- To Analyze AI Applications: Study how AI algorithms, for example, neural networks and ensemble methods, are used in the processing and interpreting complex datasets for predictive purposes.
- Evaluate the level of integration of smart technology: Discuss how IoT sensors and other smart technologies supply real-time data that increases the accuracy of forecasts and allows for more timely response.
- Barriers and future research directions: summarize the main challenges; infrastructure shortcomings, and suggest potential avenues for future research as well as possible enhancements in AI-based disaster forecasting systems.
- The next objectives are to assist in making them achieve a comprehensive understanding of how, through the integration of AI with smart technologies, the practice of disaster management can be transformed and resilience improved.

BACKGROUND: DISASTER FORECASTING

Disaster forecasting is an important aspect of emergency preparedness and response, which focuses forward to predict and prepare the natural disasters as well as man-made ones, so there could be fewer effects on mankind. This is done by collecting data from several sources so that such storms, whether hurricanes, earthquakes, floods, or wildfires, could have been better predicted. According to the analysis, proper prediction of such disasters improves preparedness, response, and recovery; it saves lives and minimizes losses economically and socially (Bande & Shete, 2017; Khalid & Shafiai, 2015; Wu et al., 2020).

Methods Used in the Past and Its Limitation

In the old days, forecasting disasters relied upon techniques of observation, historical records, and empirical data. For instance, it would be a case where meteorologists depended on the pressure barometric and wind patterns from weather observations to predict a storm. Seismologists relied on historical earthquake patterns and surface activity to provide clues about future seismic events. Hydrologists consulted over river flow data and records of rainfall to predict floods.

Although these traditional techniques have also led to several successes, most of them have inherent limitations:

a) Data Limitations: There were limited data and, to some extent, also the quality of data for the early forecasting approaches. At times, the readings were sparse and localized, resulting in incomplete or inaccurate forecast outcomes. For instance, during the non-satellite era, there existed only surface ad weather balloon observations, limiting the view of atmospheric conditions.
b) Predictive Accuracy: Predictability of forecasts was severely limited by the use of simple models. Much early work assumed linearity and related variables in a fairly simple way. Such assumptions therefore limited the reliability of the probability of prediction, specially about rapidly changing or unpredictable events. In terms of timeliness, the old methods suffered considerably. The processes were slow - in data gathering and analysis - thus only at the eve or during an event did warnings take place with little room for effective response.
c) Resource constraints: Forecasting was held back by primitive technological capability. Sophisticated computing ability to analyze large complex datasets and lots of other systems were manual, hence information processing and dissemination were slow.

Advances in Forecasting Technologies

Advancements in advanced technologies have changed disaster forecasting. Some of the most crucial technological advances include:

a) Utilization of Satellites and Remote Sensing Technology: Satellites, nowadays with multiple advanced sensors on board, have provided vital, real-time data on atmospheric, oceanic, and terrestrial conditions. This has significantly increased the probability of monitoring large-scale weather systems and early signs of natural disasters. An example is how meteorologists can now track the development and movement of hurricanes with precision to the maximum extent using satellite imagery.
b) Advanced Computational Models: Today's advance forecast relies on highly complex computational models that aggregate a large set of data from many sources. Highly complex algorithms and simulations are applied to forecast events that will lead to disasters in a more effective manner. Numerical weather prediction models, for instance, simulate atmospheric conditions using mathematical equations and give detailed forecasts of weather patterns or storm formation.
c) Machine learning and AI: Modern tools like machine learning and AI help analyze such great amounts of data as well as discover more patterns than the traditional models may not be able to pick up. AI algorithms can process the combined information coming in through weather stations, satellites, and IoT

sensors to produce what is more timely and accurate forecasts. The machine learning model learns from new data fed into it, making it more reliable over time.

d) Improved Collection of Real-Time Data: The growth in the use of IoT sensors and other real-time data collection technologies has improved leaps and bounds in enhancing forecasting. Sensors can be affixed to key locations such as river gauges and seismic sensors, providing continuous data regarding the environmental states. Real-time data can be analyzed immediately, nearly in real time, with advance warnings, which have enhanced the timeliness of forecasts.

e) The most evident developments relate to the design of the integrated systems and data fusion technology. This allows for integrating various data sets from different sources into highly multi-facetted data sets. In integrating satellite data, ground-based observations, and historical records, the integrated forecasting system provides a much more holistic and accurate view of eventual disaster events. The ability to predict and manage complex situations based on many factors is much improved.

f) Improved dissemination: This is another aspect of modern technologies that have contributed to improved dissemination of forecasts and warnings. All these seem to be people-oriented: automated alerting systems, social media sites, and mobile applications all configure automatic information release into the public at the right time. These enable the real-time warning of individuals and communities, hence allowing proper preparation in anticipation of the impending disaster.

The advanced evolution in disaster forecasting from more archaic and limited methodologies into vast technological capabilities definitely goes a long way in regards to how management and mitigation of impacts due to natural disasters are carried out. History meant that earlier methods were laid down for the practices, but the various modern technologies addressed many the limitations of the earlier methods, which provided more accurate, timely, and comprehensive predictions. This integration of satellite technology with advanced computational models, machine learning, data collection in real-time, and improved communication systems has thus revolutionized the forecasting of disaster, opening avenues for better preparedness and response. Improvements in the near future are envisaged through the upgrading of these technologies, creating better prospects for enhanced disaster management and resilience.

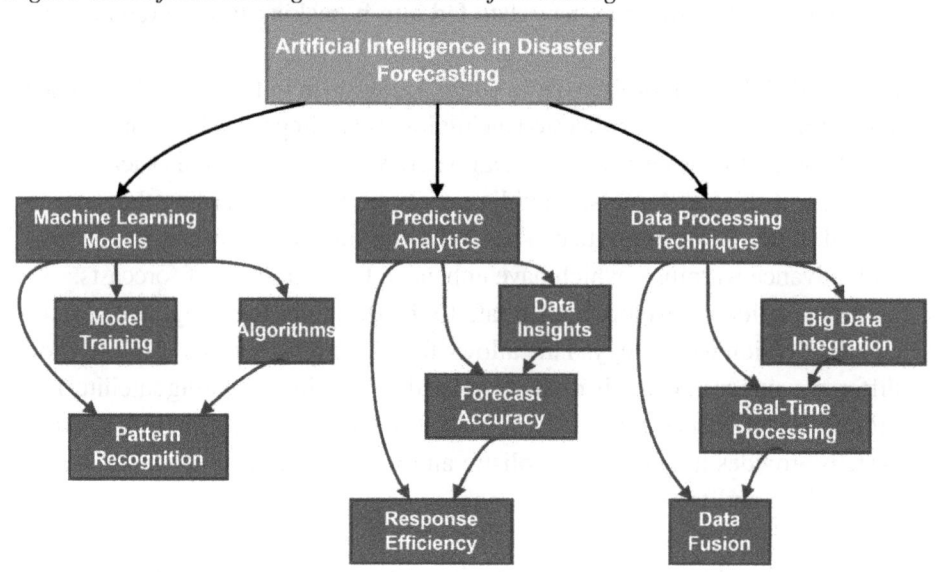

Figure 1. Artificial intelligence in disaster forecasting

ARTIFICIAL INTELLIGENCE IN DISASTER FORECASTING

AI, with innovative algorithms and data-processing techniques, has revolutionized the ability to predict disasters by enhancing the accuracy of their predictions as well as the response efficiency. AI in the area of disaster forecasting is seen in the application of machine learning models and predictive analytics that may be used to develop insights from vast data. This section delves into some key aspects of AI in disaster forecasting, particularly using machine learning models and predictive analytics(Hoyos et al., 2015; Yan et al., 2017).

Figure 1 illustrates the integration of AI in disaster forecasting, highlighting the role of machine learning models, predictive analytics, and data processing techniques in enhancing disaster prediction and response.

Machine Learning Models

Machine learning (ML) models are the spearhead of applications of artificial intelligence in disaster forecasting. The ML models are primarily based on historical data for patterns learned and take decisions in prediction based on that pattern. There are many different types of machine learning algorithms applied, with advantages unique to each.

a) Supervised Learning: Generally, supervised learning occurs when models are trained with labeled datasets; historical data already includes known outcomes. Algorithms of supervised learning include decision trees and SVMs and also various neural networks that have application for building predictive models. For instance, the case in point - hurricane forecasting-basically can be tackled with supervised learning by using past hurricane data containing wind speeds and pressure readings, predicting future storm behavior.
b) Unsupervised Learning Algorithms for unsupervised learning do exist. They look for patterns or relationships within unlabeled data sets. Techniques, for example, one uses in clustering or dimensionality reduction to identify any hidden structure that might be present in data. In earthquake forecasting, for example, unsupervised learning can be applied to analyze patterns of seismic activity. Clusters of earthquakes would then point to a risk thereof.
c) Reinforcement Learning: In reinforcement learning, the model learns in an exploratory manner, with the feedback from the environment to improve the performance. This is an important application in optimizing disaster management's decision-making processes. One of the examples of the application of reinforcement learning would be strategic resource use at a disaster time, computed in real-time data.
d) Deep Learning: A sub-domain of machine learning, deep learning utilizes feedforward neural networks composed of multiple layers to model complex interactions in the data. CNNs and RNNs are the two types that predominantly employ these specific types of networks in disaster prediction. For instance, CNNs are useful in satellite imagery-based applications for storm formation detection, while RNNs can process sequential data for predicting future states like patterns in weather over a given span of time.

Predictive Analytics and Data Processing

Predictive analytics, powered through AI and machine learning, plays a big role in the forecast of disasters because it involves the analysis of historical data and time-aware data to predict future events(Agrawal et al., 2023; Dhanya et al., 2023). This process includes some critical steps:

a) Data Collection and Integration: Effective predictive analytics collects and integrates data from a number of sources, such as historical records, real time sensor data, satellite observations, and weather reports. AI systems aggregate and pre-process the data so that it becomes clean, structured and ready for analysis. For example, weather station data, satellite observations, and social media feeds could collectively provide insight into potential disaster scenarios.

b) Feature Extraction and Engineering: Relevant variables or features for predictive modeling in the case of the disaster outcome as those one can identify from raw data. AI algorithms perform feature engineering where one selects and transforms such variables into meaningful inputs into models. For instance, in the case of flood forecasting, through feature extraction and processing, features such as rainfall intensity, soil moisture levels, and river flow rates are extracted and processed so as to predict the occurrence of floods.

c) Model Training and Validation. The prepared data are fed into machine learning models to learn patterns and relationships from historical data. Training typically involves adjustment to model parameters to minimize prediction errors. Techniques for validation also include cross-validation, as a means of testing whether a model generalizes well to new data points. For instance, one predicts hurricane paths and validates such models against recorded historical storm data to ensure the accuracy of their results.

d) Real-Time Analytics and Forecasting: A real-time based predictive analytics system scans the real-time data to generate up-to-date forecasts. These models, having been exposed to vast datasets, analyze the incoming data to predict that disasters are going to happen soon. For instance, the analysis of the forecasts based on real-time weather data and satellite image inputs, such as making a decision on severe weather conditions, can give a warning at the right time and take appropriate response action.

e) Decision Support and Visualization: Predictive analytics acts as a source of decision support. It provides insights to action in form and visual. AI systems will depict the outcome of the forecast in dashboards, maps, and graphs that are easier to understand for purposes of using the projection in making decisions. Examples would be in flood risk maps generated through the predictive models, allowing emergency planners to spot more precarious parts of the landscape and allocate the resources correspondingly.

Applications and Case Studies

AI and predictive analytics have been effectively applied to a number of disaster forecasting applications(Dhanya et al., 2023; Pramila et al., 2023; Ramudu et al., 2023). They include:

a) Hurricane Forecasting. This is a practice where machine learning models and the CNN and RNN approach are analyzed in satellite imagery as well as other data such as the trajectory and intensity of hurricanes. The algorithms applied have enhanced the correctness of storm predictions, hence making more reliable warnings, which help the appropriate preparation of communities before the disaster happens.

b) Prediction of Earthquake Events: AI models identify patterns for the occurrence of an earthquake using seismic data. Techniques like Clustering and Anomaly Detection are used to assess seismic activity and potential risks; a given time, however cannot be predicted for the next occurrence.

c) Flood Management: Predictive Analytics Models have assessed rainfall events, river water levels, and soil moisture, thus generating predictions on flood events. AI systems provide early warnings, which can then be used to assess the risks associated with such event, thus allowing evacuation and other related response actions to be conducted in due time.

Artificial intelligence is fundamentally transforming the game with increasing capabilities through machine learning models and predictive analytics to improve disaster forecasting. Its predictions can thereby be accurate, prompt, and reliable. In that sense, they will be in a better position to prepare for, and respond to, natural catastrophes due to the advanced technology of AI. In the future, AI-based applications in disaster forecasting are expected to bridge new heights of sophistication and a new path for building resilience and creating opportunities for reduction of disaster impacts.

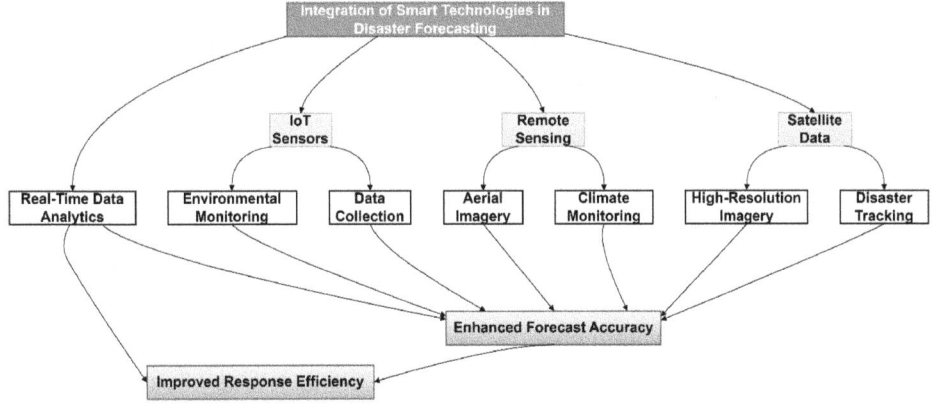

Figure 2. Integration of smart technologies in disaster forecasting

INTEGRATION OF SMART TECHNOLOGIES

One innovation that has the potential to revolutionize the improvement of prediction accuracy and effectiveness in disaster response is integrating smart technologies into disaster forecasting. By leveraging the capabilities of IoT sensors, remote sensing, and satellite data, real-time information can be provided for enhancing forecasting models and facilitating timely decisions. Below, we discuss each of these elements and how they contribute to more effective disaster management (Albahri et al., 2024; Khovrat et al., 2022; Linardos et al., 2022). Figure 2 illustrates how smart technologies like IoT sensors, remote sensing, satellite data, and real-time analytics enhance disaster forecasting accuracy and effectiveness in disaster response.

IoT Sensors and Real-time Data

Modern disaster forecasting systems owe much to IoT sensors. IoT sensors collect data on real-time sources from the environment and infrastructure, such that it is able to feed back timely information about disasters, therefore providing more insights into their prediction and management. Some include:

Types of IoT Sensors: There are so many sensors that monitor different environmental aspects. For instance, meteorological sensors measure temperature, humidity, and atmospheric pressure among others while hydrological sensors monitor the levels and the flow rates in rivers and reservoirs. Seismic sensors have even managed to sense ground movements which are crucial indicators for the prediction of earthquakes. This means that these sensors would provide continuous data high in resolution that's critical in accurate forecasting.

Data collection and transmission: IoT sensors are deployed in strategic locations such as weather stations, riverbanks, and fault lines. They collect real-time data and transmit the same to centralized data processing systems through wireless networks. Hence it can be analyzed in real time and responded to immediately. For instance, sensors installed in flood-prone areas can provide early warnings about the increasing water level and, based on such a warning, communities there can take all precautions well before time.

Data Integration and Analysis: The information sources, which include satellite data and historical records, integrate with data collected through IoT sensors to enhance forecasting models. Advanced analytics platforms process this data source for trend and anomaly identification. For instance, the integration of rainfall data from IoT sensors with flood histories improves prediction on events of flooding and aids in the development of flood management.

Albeit the advantages possessed by IoT sensors, drawbacks such as low data quality, unreliable networks, and poor maintenance of sensors characterize this technology. Data reliability and consistency are keys to producing quality forecasts. Solutions include sensor calibration routinely, having stringent validation processes for data, and redundant communication paths to help overcome network problems.

Remote Sensing and Satellite Data

Remote sensing and satellite data provide an integrated view of massive environmental conditions, which are essential in disaster forecasting(Boopathi, 2023; Subha et al., 2023). This technology has several advantages:

Different satellites fitted with various sensors have acquired the imagery and data from space, offering an overview of the system at large. Satellite imagery can thereby offer views of weather patterns, land use, and environmental changes. For example, weather satellites track cloud formations, storm developments and atmospheric conditions, which are very helpful in forecasting hurricanes and other specific severe weather events.

Numerous Forms of Remote Sensing Technologies: The technologies involved in remote sensing include optical, infrared, and microwave sensors. Optical sensors take high-resolution photos of the Earth's surface. Infrared sensors detect thermal emissions, which are useful for the measurement of changes in temperature. Microwave sensors can penetrate through clouds and provide data on precipitation and soil moisture; the two are useful quantities that help in flood forecasting.

Satellites Data Processing and Analysis: All the data collected by the satellite are processed and analyzed through algorithms so that meaningful information can be extracted. Image classification, change detection, and feature extraction are utilized for finding the phenomenon relating to disaster and surveillance of changes which occur in the physical environment over time. For instance, the satellite data might reveal deforestation, which will enhance the risk of landslide and flood.

Satellite data are integrated with ground-based observations and forecasting models in order to improve the accuracy of predictions. Integration of satellite imagery with meteorological parameters enhances the ability to forecast weather events and to monitor the development of these events. For instance, integrating the satellite data on sea surface temperatures with atmospheric models will enhance precision hurricane intensity predictions. Data resolution for remote sensing technologies is problematic because interference caused by cloud cover, amongst others, may affect it. High-resolution imagery is costly and demanding in processing. Others include the use of multi-spectral sensors to capture data across different wavelengths while others apply algorithms on removing clouds to enhance image quality.

Improving Forecast Accuracy with Smart Technologies

The integration of smart technologies can enhance the accuracy of the forecast by delivering information that is more up-to-date and accurate(Maheswari et al., 2023; Venkateswaran et al., 2024). Some of the key aspects:

- High-resolution granular data: Smart technologies provide high resolution with finer granularity as compared to the previous models for enhancement. IoT sensors provide data at the local level, and satellite images offer the broader picture. Together, these provide detailed and accurate prediction models. For instance, combining detailed data from IoT sensors with satellite observations, while improving the accuracy of risk assessments due to captured local and regional conditions of flooding, can be exemplified.
- Real-time monitoring and updates: Smart technologies allow one to continuously monitor and get updates on a dynamic and changing scenario occasioned by a disaster. For instance, real-time data from weather sensors and satellites can trace the spread of a hurricane and give up-to-date forecasts, making evacuation and response moves before time runs out.
- Advanced Analytics and AI: Smart technologies enabled with AI and machine learning models enhance the capacity for large-volume data analysis, predictions with greater accuracy, and visualization of enormous datasets. For instance, AI algorithms can parse data from satellites or IoT sensors to identify and make predictions on patterns or events about weather patterns and the activation of disasters.
- Improved Decision-Making: Smart technologies arm decision-makers with timely and action-containing information based on comprehensiveness. Advanced visualization tools and dashboards draw out available data into formats well understood to support informed decisions. For example, real-time flood risk maps created from IoT and satellite data can support effective allocation of resources and prioritization by emergency planners.
- Collaboration and Information Sharing: Smart technologies promote collaboration and information sharing among stakeholders. Data created by different types of sensors may be shared with governmental agencies, emergency responders, or the public to collaborate in achieving a common goal. For example, shared data platforms can enable coordination between agencies of weather conditions and local authorities during severe weather events.

The new technologies, in particular 5G connectivity and edge computing, could be used in an effort to move the bar again in the near future. Data can now be transmitted much quicker as 5G networks promise to increase capabilities in the quality of disaster forecasting(Boopathi, 2024b; Molakatala et al., 2024; Sreedhar et al., 2024). Edge computing is a technology that permits data to be operated at source or in other words, at the edges of a network. Thus, latency is reduced and the accuracy of the forecast increases.

This is particularly because of the integration of smart technologies, such as IoT sensors, remote sensing, and satellite data, into the systems used to offer accurate, real-time information that enhances the predictive capability of disaster forecasting. Such technologies add precision to and timeliness in the forecast and thereby help increase the preparedness and improvement of the response efforts. With new and emerging technology trends into the system, there are expected further developments of smart technologies in disaster forecasting, enhancing the capacities of effective management and resilience.

APPLICATIONS OF AI AND SMART TECHNOLOGIES

AI and smart technologies revolutionized the several dimensions of the forecasting and management of disasters. The various technologies provided accurate, timely, and efficient predictions regarding the occurrence of a natural disaster as well as the efficiency to respond to it. Below we discuss the three main applications involving AI and smart technologies in hurricane forecasting and management, earthquake prediction and early warning systems, and flood monitoring risk assessment (Albahri et al., 2024; Merz et al., 2020; Puttinaovarat & Horkaew, 2020; Sarker et al., 2020). Figure 3 showcases the use of AI and smart technologies in disaster forecasting and management, demonstrating their effectiveness in providing accurate predictions, timely alerts, and efficient responses to natural disasters.

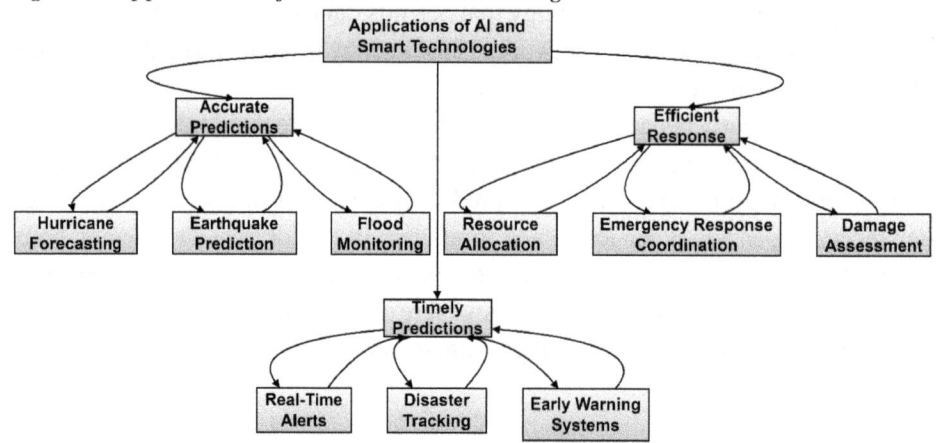

Figure 3. Applications of AI and smart technologies

[Diagram: "Applications of AI and Smart Technologies" branches into "Accurate Predictions" (Hurricane Forecasting, Earthquake Prediction, Flood Monitoring), "Efficient Response" (Resource Allocation, Emergency Response Coordination, Damage Assessment), and "Timely Predictions" (Real-Time Alerts, Disaster Tracking, Early Warning Systems).]

Hurricane Forecasting and Management

With AI and smart technologies infused into it, the improvements in hurricane forecasting have indeed been tremendous. The accuracy in predicting the pathway of hurricanes and their intensities and impacts becomes possible(Saravanan et al., 2024; Sharma et al., 2024; Sundaramoorthy et al., 2024).

a) Enhanced Data Gathering: AI and smart technologies improve thorough data gathering from various sources; the list is extremely long. Examples include satellite images, weather sensors, and ocean buoys. Satellites with advanced sensors take detailed pictures of storm systems. Ocean buoys measure sea surface temperatures and swell wave heights. AI algorithms then process all this data to deliver real-time insights into hurricane development.

b) Predictive Modeling: Machine learning models of CNNs and RNNs create predictions of hurricane behavior using past data as well as live data. CNNs are used in image interpretation tasks for distinguishing a storm pattern from satellite images while RNNs are tasked in the management of time-series data to predict future trajectories of a storm. These improve the precision of hurricane prediction models by providing patterns that traditional models fail to find and consider.

c) Simulation and Scenario Analysis: The output of an AI-driven simulation is to run with multiple scenarios of hurricane impacts based on different variables, for example, wind speed, atmospheric pressure, and sea surface temperatures. Such simulations help learn about potential outcomes and prepare for a variety of scenarios. For example, how a hurricane will hit the coastline and create its impact can be modeled through simulation, and this information then forms a basis for planning evacuations and resource allocation.

d) The AI system provides real-time updates and alerts since real-time analysis of the incoming data from sensors and satellites is done. Automated systems do issue the public and emergency responders with timely warnings, thus ensuring quicker action. For example, an AI-powered system will be able to alert the affected areas that the residents need to evacuate or take protective measures according to the latest forecast.

e) Resource Optimization: In AI algorithms, resource distribution and response efforts in emergency situations can be optimized. With population density, infrastructure, and hurricane intensity data, AI systems are able to recommend optimal resource distribution and personnel deployments. This would thereby ensure the efficient delivery of aid to the most affected regions.

Earthquake Prediction and Early Warning Systems

AI and smart technologies in the prediction and warning system have an impact on seismic event detection and response in earthquake prediction (Jha et al., 2024; Malathi et al., 2024; Molakatala et al., 2024).

a) Seismic Monitoring and Analysis: All the ground movements as well as seismically triggered activities are monitored through IoT sensors and seismographs. An AI algorithm processes the data so that such pattern recognition and unusual alterations can be known for a probable earthquake-Inducing earthquake. Machine learning models understand precursors by making use of the historical data of the seismic analysis in order to predict a possible earthquake event.

b) Early Warning Systems: The early warning systems make use of the AI potentiality in recognising the seismic activity, thus prompting an alert before the shaking strikes populated areas. The seismic networks use those data to estimate the magnitude of that earthquake and its location, with the probable impact it may create. For instance, seconds or minutes advance warning might be provided by an AI-powered system that can help individuals take precautionary measures, thus cutting down the injuries and damage.

c) Damage Assessment and Response Planning: After an earthquake, AI technologies help determine the extent of damage done to buildings and infrastructures and plan for appropriate responses. Image recognition analysis by AI technologies using satellite and drone imagery can estimate the extent of damage to buildings and infrastructure. This will help prioritize which efforts should receive what resource attentions.

d) Community Awareness and Education: AI-based systems raise community awareness and preparedness by providing people with specific information and training. For instance, application areas driven by AI can educate people on how to conduct one's self during an earthquake and can also enable simulating earthquake scenarios such that the residents can better prepare for a disaster. The tools turn the community into a wiser and more resilient population.

Flood Monitoring and Risk Assessment

Such flood monitoring and risk assessments are very significant when flood risk assessment incorporates artificial intelligence and smart technologies, where accuracy in predictions and the effectiveness of strategies for managing floods are enhanced(Boopathi, 2024a; Sonia et al., 2024).

a) Real-Time Flood Monitoring Sensors installed in rivers, lakes, and reservoirs through an IoT network may offer information concerning the water level and flow rates. The AI system could then analyze such information to monitor flood conditions with the capability to detect potential flood events-for example, a sensor may alert authorities with the rapid water levels rise.
b) Predictive Modeling: AI-based predictive models analyze rainfalls in the past, the levels of moistures in the soil, and river flow to prognosticate floods. Machine learning algorithms detect patterns and trends that indicate increased flood likelihood. Such models provide advanced warning and risk assessment, allowing communities to prepare for and mitigate the effects of floods.
c) Flood risk mapping involves the use of remote sensing technologies and AI algorithms when analyzing satellite imagery as well as topography data in order to generate a high-resolution flood risk map. These maps provide enabling areas at risk of floods, thus enabling effective targeted management of flood along with evacuation plans. For example, flood risk maps are in a better position to guide on the placement of flood barriers and the formulation of floodplain regulations.
d) AI systems shall be consulted for response and recovery plans. In fact, related data on flood impacts and infrastructure damage will form the basis of their analysis while developing these plans. For instance, an AI algorithm can easily analyze whether a flood defense is working properly or not and how improvements can be made to it. The coordination of response efforts and optimization of resource deployment for recovery operations shall also be supported by these AI tools.

e) Community Engagement and Support. Some of the AI-powered applications and platforms provide real-time information and support to flooded communities, either in the form of personalized alerts for every household or safety recommendations. Examples include flood warning applications that guide the user on when one needs to receive information tailored to their location and the level of risk, thus allowing proper decision-making in times of flooding.

The intersection of AI and smart technologies has brought a new era of change into disaster prediction and management in various fields. For hurricane forecasting, AI provides efficient predictions with more relevance in the allocation of available resources in a much better optimized way. In earthquake-related scenarios, AI can analyze seismic data that would define and predict potentially hazardous seismic activity along with releasing an early warning system. Flood monitoring and risk assessment are highly improved through real-time data collection, predictive modeling, and risk mapping. AI and smart technologies would further strengthen disaster preparedness, response, and recovery in the face of growing technological advancements. This would make the community more resilient and responsive to disasters.

Figure 4. Challenges and limitations of AI and smart technologies in disaster forecasting and management

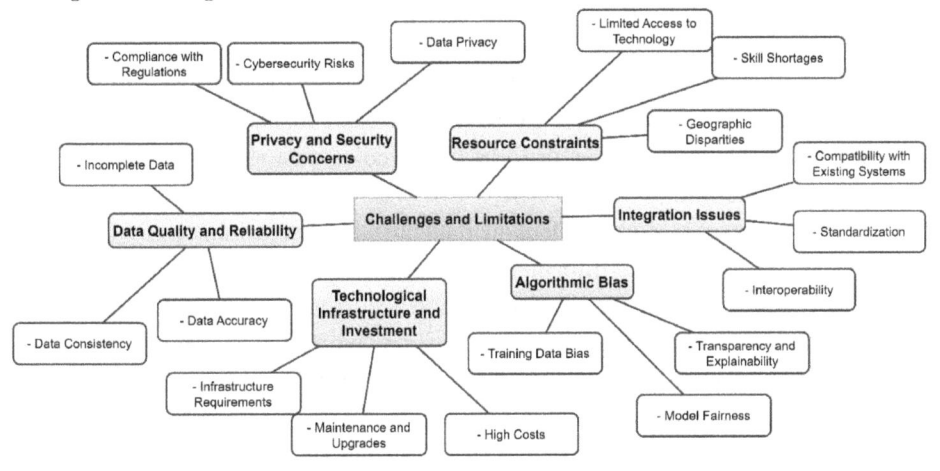

CHALLENGES AND LIMITATIONS

Even though promising, AI and smart technologies in disaster forecasting and management pose many challenges and limitations, which need to be addressed to assure usability and reliability in mitigation of impacts caused by disasters. Among these areas include the issues of data quality and reliability, technology infrastructure and investment, and privacy and security concerns (Jun et al., 2019; Merz et al., 2020; Puttinaovarat & Horkaew, 2020). AI and smart technologies in disaster forecasting and management face challenges like data quality, infrastructure, privacy, security, algorithmic bias, integration, and resource constraints as shown in Figure 4.

Data Quality and Reliability

a) Data Accuracy - The efficiency of AI and smart technologies in weather forecasting depends on the accuracy of data collected. Faulty or inaccurate data could cause incorrect predictions and inadequate responses. For instance, wrong sensor readings or faulty calibration might result in misleading information. This could directly impact the reliability of flood prediction or even earthquake warnings.
b) Consistency of Data Integration: The generated inconsistencies are due to the variation in the format of data, units of measurement, and temporal resolutions involved when relying on different sources, such as IoT sensors and satellite imagery, as well as historical records. In this case, consistency among different sources is a much higher factor to be achieved to ensure proper analysis and effective prediction.
c) Data Completeness: Lack of adequate data hurts the performance of AI models. For example, the missing information about the environment hampers the prediction of hurricane intensity or risk of flood occurrence. Accordingly, complete data collection and management practices are significant in solving this issue.
d) Data Validation and Quality Control: Data reliability is to be maintained through strong data validation and quality control measures. Sensors should have regular calibration, and data should be validated against known benchmarks, and rigorous testing must be done on the AI models to avoid data quality problems.

Technological Infrastructure and Investment

a) Infrastructure Requirements: Heavy technological infrastructure is needed for the implementation as well as maintenance of smart technologies. The basic installations required include IoT sensors, satellite communication systems, and other data processing facilities. Infrastructure inadequacies may constrain the impact of a disaster forecasting system.
b) It is very expensive to implement a technological advancement. For example, acquisition of high-resolution satellite imagery, establishment of IoT-based sensor networks, and formulation and development of AI models need huge budgets. Resource scarcity constrains the deployment and scalability of these technologies, especially in limited-resource regions.
c) Technical Expertise: The setup of AI and smart technology, aside from the advanced forecasting systems, requires some specialized technical expertise. This encompasses the skills of data science, machine learning, and system integration. If there are insufficiently skilled personnel, this can be a challenge to establishing and deploying advanced forecasting systems.
d) System Integration: Sometimes, new technologies can be complexly integrated into existing systems and infrastructure. It sometimes creates compatibility issues with the need for upscaling or modification of systems, which can become problems in the coherent incorporation of AI and smart technologies within disaster management frames.

Privacy and Security Concerns

With high volumes of data to be collected and processed, containing both personal as well as location data, it raises privacy concerns. The need for strict data privacy in collecting, storing, and using data strictly according to the guidelines of GDPR and CCPA will be called for to protect the rights of the individual persons and prevent misuse of information(Boopathi, 2024a; Nanda et al., 2024; Sonia et al., 2024).

a) Data Protection: Collected data from IoT sensors, satellites, and so forth must be protected. Data breaches and hacking can expose sensitive information and disrupt disaster forecasting systems. Therefore, proper security practice, for example, encrypting data and protecting access, must be implemented to guard data.
b) Ethical issues: The use of AI and smart technologies in disaster forecasting raises ethical concerns pertaining to data ownership, consent, and transparency. Therefore, ethics in terms of practices related to data collection and bringing

the entire gamut of AI models and results to prominence with respect to transparency and accountability are necessary for the establishment of public trust.

c) Regulatory Compliance: A company should conduct businesses under regulatory requirements and industry standards in ensuring all issues about privacy and security. The company is therefore, expected to adhere to the regulations and rules put in place for it to support its use of AI and smart technologies under lawful and ethical boundaries.

Implementing AI and smart technologies in forecasting and the management of disasters has several challenges and limitations, which must be analyzed so that their total effectiveness and reliability can be guaranteed. All these data quality and reliability issues with respect to accuracy, consistency, and completeness need to be addressed through strong validation and quality control practices. Technological infrastructure and investment challenges with respect to the requirements of capital, the cost of capital, the requisite technical skills, and system integration have to be overcome to make the implementation possible. The relationship of privacy with the security issues concerning data privacy, data security, ethical issues, and regulations must be addressed carefully to protect individual rights and hold responsibility over the use of technology. Increasing the benefits of AI and smart technologies in improving disaster forecasting and management depends on the resolution of such challenges.

Figure 5. Future directions in disaster forecasting and management

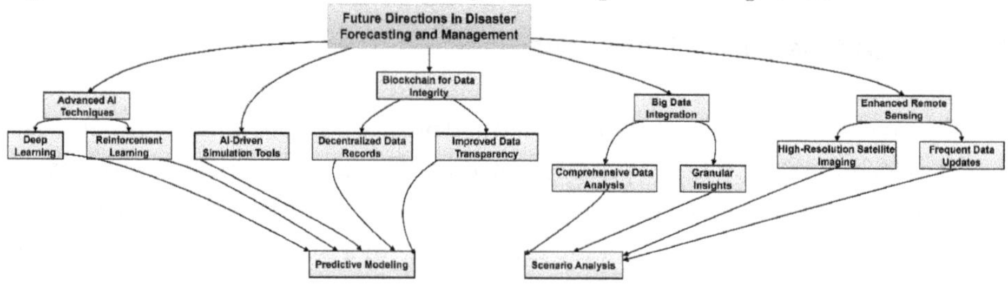

FUTURE DIRECTIONS

Future directions will include significant transition in the area of disaster forecasting and management, as AI and smart technologies advance. Emerging trends and innovations, research and development opportunities, and collaborative endeavors contribute significantly toward the future landscape of disaster resilience and response. This section develops these future directions, focusing on ways they could be used to drive progress and improve the performance of disaster management systems(Albahri et al., 2024; Khovrat et al., 2022; Linardos et al., 2022). Figure 5 illustrates the future directions in disaster forecasting and management as AI and smart technologies advance.Figure 5 illustrates the future directions in disaster forecasting and management as AI and smart technologies advance.

Emerging Trends and Innovations

a) Advanced Machine Learning Techniques: The future of machine learning will bring us better predictive capabilities through enhanced deep learning and reinforcement learning techniques. More complex neural networks can be designed to provide a more accurate result on disaster forecasting by developing deeper insight into large datasets. Advanced, more intricate neural networks may reveal deeper insights into hurricane behavior or precursors related to earthquakes, hence forming more accurate predictions.

b) The scenario with the inclusion of big data analytics into AI will be a complete revolution for disaster forecasting. Big data can be quite robust for AI systems, providing them with extensive information obtained from various sources, such as social media, sensor networks, and historical records. This will assist in making forecasts much more comprehensive and timely, thus giving a more granular analysis and better-informed decision-making.

c) Improved technologies in remote sensing: There should also be enhanced satellite imagery and aerial drones for remote sensing. This can enhance the acquisition of data collected from a disaster event. Higher resolution and more often updated data will enhance monitoring, tracking, and real-time analysis of disaster events, thereby providing more accurate and usable information.

d) Blockchain technology, then, may play an important role to ensure the integrity and security of data during a disaster. Blockchain-based decentralization and immutability of data transactions may lead to improved transparency and trust of data applied for better forecasts and management of disasters.

e) AI-based simulation and scenario analysis: The new developments in AI-based simulation tools will help enhance their scenario analysis ability. The tools will predict different scenarios that may arise during disasters and under different conditions, hence making it easier to plan for the correct preparation and response strategies.

Opportunities for Research and Development

a) Development of Stronger AI Models: Now research into the development of more robust and adaptive AI models is highly called for. More accurately this includes the development of models that are capable of handling diverse and incomplete datasets, adaptiveness to changing conditions of the environment and usability in operating in real-world scenarios effectively. This can be promoted by academia, the industry, and governments working together.

b) Advances in Sensor Technologies: Investments in next-generation sensors with higher accuracy, more durability, and reduced cost will enhance data acquisition. New types of sensor materials and technology hold the promise of providing better environmental data through nano-sensors and advanced optical sensors.

c) This research looks into multi-modal data fusion techniques in order to extend the sources being integrated, such as satellite imagery, IoT sensors, and social media. Therefore, through advanced data fusion algorithms, the creation of more comprehensive and more accurate disaster models can be achieved by combining different types of data.

d) Thus, the scope of ethical AI and responsible data use in research outlines issues in privacy, security, and fairness. To ensure that technologies are applied responsibly and equitably, guidelines and frameworks need to be developed for the ethical deployment of AI in disaster management.

e) Improved Computational Efficiency: The objectives of research into enhancing computational efficiency for AI models and simulations will yield faster and more scalable forecasts in disaster situations. That is much better suited to processing large data and model training with new hardware, such as quantum computing and high-performance computing systems.

Collaborative Efforts and Strategic Planning

a) Public-Private Partnerships: Collaboration between government agencies, private firms, and academic organizations is necessary to improve disaster forecasting technologies. Public-private partnerships can allow for data, resources, and expertise sharing among these partners, ultimately leading to more effective and innovative solutions.

b) International Cooperation International cooperation is therefore the most important one since most disasters tend to have cross-border impacts. International cooperation at the level of conducting research, data exchange and technology development will improve early warning systems and response capacities at a global level. Examples of such successful international cooperation include the Global Disaster Alert and Coordination System (GDACS).

c) Community Engagement and Capacity Building: Local communities are also equally involved in planning and preparation to make necessary responses during hazards. Local efforts on building capacity can form an important ingredient of strategic planning so that the residents are also educated and taken along with data collection and decision-making processes. Community-based approaches can enhance the effectiveness of strategies in disaster management and make sure that technologies used meet the needs of the locals.

d) Strategic Investment in Technology and Infrastructure As part of the technology and infrastructure investments for the deployment and maintenance of advanced disaster forecasting systems, the government and organizations need to strategically invest in that. Substantial long-term plans should be made toward both scalability and sustainability for the funding of deployment and end.

e) Policy Development and Regulation: Policies and regulations on the use of AI and smart technologies in disaster management have to be developed. This would provide clear guidelines on the issues of data privacy, security, and other ethical considerations to ensure the responsible and effective use of such technologies.

The future of disaster forecasting and management is exciting and very full of potential. Emerging trends including developments in machine learning, big data integration, remote sensing, and the use of blockchain technology promise more accurate, time-sensitive, higher-quality, and accessible disaster predictions. Research and development gains in more robust AI models, sensor technologies, and AI ethics will be key determinants of innovating the use of innovation in the field. Disaster management requires effective and sustainable solutions that are constructed through cooperation between the public sector, private sectors, international partners, and local communities. These directions will be much better addressed to make us in better preparedness and disaster response mechanisms and make us resilient towards natural disasters.

CONCLUSION

Integration of AI and smart technologies into disaster forecasting and management represents the greatest leap forward in our possibilities to predict, prepare for, and respond to natural disasters. Applying machine learning models coupled with predictive analytics and real-time data from IoT sensors and remote sensing technologies can provide unprecedented accuracy and efficiency in response efforts. However, the integration of such technologies poses a challenge. Quality and reliability of data, technological infrastructure, and issues related to privacy and security are some of the areas that need to be addressed for full utilization of these implementations. Accurate data should be maintained with strong infrastructure and secured private information for the successful implementation of AI-driven disaster management systems.

This area seems further filled with yet more innovations ahead for the future. Next-generation sensors, advanced machine learning techniques, and blockchain for integrity of data are among the emerging trends that will bring improvements in disaster forecasting. Opportunities exist in R&D in such areas, promising improved precision and applicability of predictive models. Addressing complex challenges associated with disaster management and ensuring sustainable deployment of such technologies require inter-government, private sector, and international collaboration. In summary, though AI and smart technologies hold a lot of transformative potential for disaster prediction, crucial continued investment, research, and collaboration are part of it. Overcoming the present issues and embracing the future innovations might help us build more resilient communities and enhance our ability to manage impacts better.

REFERENCES

Agrawal, A. V., Magulur, L. P., Priya, S. G., Kaur, A., Singh, G., & Boopathi, S. (2023). Smart Precision Agriculture Using IoT and WSN. In *Handbook of Research on Data Science and Cybersecurity Innovations in Industry 4.0 Technologies* (pp. 524–541). IGI Global. DOI: 10.4018/978-1-6684-8145-5.ch026

Albahri, A., Khaleel, Y. L., Habeeb, M. A., Ismael, R. D., Hameed, Q. A., Deveci, M., Homod, R. Z., Albahri, O., Alamoodi, A., & Alzubaidi, L. (2024). A systematic review of trustworthy artificial intelligence applications in natural disasters. *Computers & Electrical Engineering*, 118, 109409. DOI: 10.1016/j.compeleceng.2024.109409

Bande, S., & Shete, V. V. (2017). Smart flood disaster prediction system using IoT & neural networks. *2017 International Conference On Smart Technologies For Smart Nation (SmartTechCon)*, 189–194. DOI: 10.1109/SmartTechCon.2017.8358367

Boopathi, S. (2023). Internet of Things-Integrated Remote Patient Monitoring System: Healthcare Application. In *Dynamics of Swarm Intelligence Health Analysis for the Next Generation* (pp. 137–161). IGI Global. DOI: 10.4018/978-1-6684-6894-4.ch008

Boopathi, S. (2024a). Balancing Innovation and Security in the Cloud: Navigating the Risks and Rewards of the Digital Age. In *Improving Security, Privacy, and Trust in Cloud Computing* (pp. 164–193). IGI Global.

Boopathi, S. (2024b). Enabling Machine Control Through Virtual Methods: Harnessing the Power of 5G. In *Advances in Logistics, Operations, and Management Science* (pp. 50–74). IGI Global. DOI: 10.4018/979-8-3693-1862-1.ch004

Chen, J., Chen, H., Hu, D., Pan, J. Z., & Zhou, Y. (2015). Smog disaster forecasting using social web data and physical sensor data. *2015 IEEE International Conference on Big Data (Big Data)*, 991–998. DOI: 10.1109/BigData.2015.7363850

Dhanya, D., Kumar, S. S., Thilagavathy, A., Prasad, D., & Boopathi, S. (2023). Data Analytics and Artificial Intelligence in the Circular Economy: Case Studies. In *Intelligent Engineering Applications and Applied Sciences for Sustainability* (pp. 40–58). IGI Global.

Hoyos, M. C., Morales, R. S., & Akhavan-Tabatabaei, R. (2015). OR models with stochastic components in disaster operations management: A literature survey. *Computers & Industrial Engineering*, 82, 183–197. DOI: 10.1016/j.cie.2014.11.025

Jha, S. K., & Beevi, S. J. P., H., Babitha, M. N., Chinnusamy, S., & Boopathi, S. (2024). Artificial Intelligence-Infused Urban Connectivity for Smart Cities and the Evolution of IoT Communication Networks. In *Blockchain-Based Solutions for Accessibility in Smart Cities* (pp. 113–146). IGI Global. DOI: 10.4018/979-8-3693-3402-7.ch005

Jun, X., Huiyun, W., Yaoyao, G., & Liping, Z. (2019). Research progress in forecasting methods of rainstorm and flood disaster in China. *Torrential Rain and Disasters*, 38(5), 416–421.

Khalid, M. S. B., & Shafiai, S. B. (2015). Flood disaster management in Malaysia: An evaluation of the effectiveness flood delivery system. *International Journal of Social Science and Humanity*, 5(4), 398–402. DOI: 10.7763/IJSSH.2015.V5.488

Khovrat, A., Kobziev, V., Nazarov, A., & Yakovlev, S. (2022). Parallelization of the VAR Algorithm Family to Increase the Efficiency of Forecasting Market Indicators During Social Disaster. *IT&I*, 222–233.

Linardos, V., Drakaki, M., Tzionas, P., & Karnavas, Y. L. (2022). Machine learning in disaster management: Recent developments in methods and applications. *Machine Learning and Knowledge Extraction, 4*(2).

Maheswari, B. U., Imambi, S. S., Hasan, D., Meenakshi, S., Pratheep, V., & Boopathi, S. (2023). Internet of things and machine learning-integrated smart robotics. In *Global Perspectives on Robotics and Autonomous Systems: Development and Applications* (pp. 240–258). IGI Global. DOI: 10.4018/978-1-6684-7791-5.ch010

Malathi, J., Kusha, K., Isaac, S., Ramesh, A., Rajendiran, M., & Boopathi, S. (2024). IoT-Enabled Remote Patient Monitoring for Chronic Disease Management and Cost Savings: Transforming Healthcare. In *Advances in Explainable AI Applications for Smart Cities* (pp. 371–388). IGI Global.

Merz, B., Kuhlicke, C., Kunz, M., Pittore, M., Babeyko, A., Bresch, D. N., Domeisen, D. I., Feser, F., Koszalka, I., Kreibich, H., & others. (2020). Impact forecasting to support emergency management of natural hazards. *Reviews of Geophysics, 58*(4), e2020RG000704.

Molakatala, N., Kumar, D. A., Patil, U., Mhatre, P. J., Sambathkumar, M., & Boopathi, S. (2024). Integrating 5G and IoT Technologies in Developing Smart City Communication Networks. In *Blockchain-Based Solutions for Accessibility in Smart Cities* (pp. 147–170). IGI Global. DOI: 10.4018/979-8-3693-3402-7.ch006

Nanda, A. K., Sharma, A., Augustine, P. J., Cyril, B. R., Kiran, V., & Sampath, B. (2024). Securing Cloud Infrastructure in IaaS and PaaS Environments. In *Improving Security, Privacy, and Trust in Cloud Computing* (pp. 1–33). IGI Global. DOI: 10.4018/979-8-3693-1431-9.ch001

Pramila, P., Amudha, S., Saravanan, T., Sankar, S. R., Poongothai, E., & Boopathi, S. (2023). Design and Development of Robots for Medical Assistance: An Architectural Approach. In *Contemporary Applications of Data Fusion for Advanced Healthcare Informatics* (pp. 260–282). IGI Global.

Puttinaovarat, S., & Horkaew, P. (2020). Flood forecasting system based on integrated big and crowdsource data by using machine learning techniques. *IEEE Access : Practical Innovations, Open Solutions*, 8, 5885–5905. DOI: 10.1109/ACCESS.2019.2963819

Ramírez, I. J., & Briones, F. (2017). Understanding the El Niño costero of 2017: The definition problem and challenges of climate forecasting and disaster responses. *International Journal of Disaster Risk Science*, 8(4), 489–492. DOI: 10.1007/s13753-017-0151-8

Ramudu, K., Mohan, V. M., Jyothirmai, D., Prasad, D., Agrawal, R., & Boopathi, S. (2023). Machine Learning and Artificial Intelligence in Disease Prediction: Applications, Challenges, Limitations, Case Studies, and Future Directions. In *Contemporary Applications of Data Fusion for Advanced Healthcare Informatics* (pp. 297–318). IGI Global.

Saravanan, S., Khare, R., Umamaheswari, K., Khare, S., Krishne Gowda, B. S., & Boopathi, S. (2024). AI and ML Adaptive Smart-Grid Energy Management Systems: Exploring Advanced Innovations. In *Principles and Applications in Speed Sensing and Energy Harvesting for Smart Roads* (pp. 166–196). IGI Global. DOI: 10.4018/978-1-6684-9214-7.ch006

Sarker, M. N. I., Peng, Y., Yiran, C., & Shouse, R. C. (2020). Disaster resilience through big data: Way to environmental sustainability. *International Journal of Disaster Risk Reduction*, 51, 101769. DOI: 10.1016/j.ijdrr.2020.101769

Sharma, M., Sharma, M., Sharma, N., & Boopathi, S. (2024). Building Sustainable Smart Cities Through Cloud and Intelligent Parking System. In *Handbook of Research on AI and ML for Intelligent Machines and Systems* (pp. 195–222). IGI Global.

Sonia, R., Gupta, N., Manikandan, K., Hemalatha, R., Kumar, M. J., & Boopathi, S. (2024). Strengthening Security, Privacy, and Trust in Artificial Intelligence Drones for Smart Cities. In *Analyzing and Mitigating Security Risks in Cloud Computing* (pp. 214–242). IGI Global. DOI: 10.4018/979-8-3693-3249-8.ch011

Sreedhar, P. S. S., Sujay, V., Rani, M. R., Melita, L., Reshma, S., & Boopathi, S. (2024). Impacts of 5G Machine Learning Techniques on Telemedicine and Social Media Professional Connection in Healthcare. In *Advances in Medical Technologies and Clinical Practice* (pp. 209–234). IGI Global. DOI: 10.4018/979-8-3693-1934-5.ch012

Subha, S., Inbamalar, T., Komala, C., Suresh, L. R., Boopathi, S., & Alaskar, K. (2023). A Remote Health Care Monitoring system using internet of medical things (IoMT). *IEEE Explore*, 1–6.

Sundaramoorthy, K., Singh, A., Sumathy, G., Maheshwari, A., Arunarani, A., & Boopathi, S. (2024). A Study on AI and Blockchain-Powered Smart Parking Models for Urban Mobility. In *Handbook of Research on AI and ML for Intelligent Machines and Systems* (pp. 223–250). IGI Global.

Venkateswaran, N., Kunduru, K. R., Ashwin, N., Sundar Ganesh, C. S., Hema, N., & Boopathi, S. (2024). Navigating the Future of Ultra-Smart Computing Cyberspace: Beyond Boundaries. In *Advances in Computational Intelligence and Robotics* (pp. 170–199). IGI Global. DOI: 10.4018/979-8-3693-2399-1.ch007

Wu, W., Emerton, R., Duan, Q., Wood, A. W., Wetterhall, F., & Robertson, D. E. (2020). Ensemble flood forecasting: Current status and future opportunities. *WIREs. Water*, 7(3), e1432. DOI: 10.1002/wat2.1432

Yan, H., Moradkhani, H., & Zarekarizi, M. (2017). A probabilistic drought forecasting framework: A combined dynamical and statistical approach. *Journal of Hydrology (Amsterdam)*, 548, 291–304. DOI: 10.1016/j.jhydrol.2017.03.004

KEY TERMS

AI (Artificial Intelligence): The creation of machines that perform tasks which generally require human intelligence, including learning and solving problems.

CCPA: stands for California Consumer Privacy Act, it is a state statute designed for increasing rights to the privacy and consumer protection of the residents of California within the USA.

CNN: is the deep learning network developed to handle structured grid data using intensive application in image and video recognition operations.

GDACS: is the European Commission initiative that enables users to know how it is present in real time with respect to global alerts and coordination on natural disasters.

GDPR: is actually an acronym for General Data Protection Regulation: that refers to data protection and privacy of the individuals within the EU and EEA.

ML (Machine Learning): Subset of AI with algorithms and statistical models enabling computers to learn from data and make predictions or decisions based on it.

RNN: stands for Recurrent Neural Network, is a certain type of neural network which identifies patterns within sequences of data. This is useful in time series and natural language processing.

SVM: is a supervised algorithm for classification and regression problems that can determine the optimal separating hyper-plane between different classes in feature space.

Chapter 17
Deep Learning for Predictive Analytics in Environmental and Social Sciences

Senthilkumar Thangavel
Cloud Solutions & Machine Learning Expert, USA

ABSTRACT

Deep learning is currently one of the transformative technologies for predictive analytics, with deep learning providing robust methodologies to analyse complex data in both the environmental and social sciences. This chapter is concerned with looking into the use of some deep learning techniques, such as CNNs and RNNs, in modelling and predicting environmental phenomena and social trends. Deep learning techniques fit into the environmental sciences through the massive processing of remote sensing and observational data in climate modelling, disaster prediction, and ecologic monitoring. In the social sciences, it helps with sentiment analysis, behavioural predictions, and trend forecasting from large-scale social media and survey data. Key challenges include issues related to data quality and model interpretability; Highlighted case studies provide evidence of the effectiveness of deep learning in these domains.

INTRODUCTION

Predictive analytics has emerged as an important tool for understanding and predicting complex systems, from environmental and social sciences to most other such fields. Deep learning has already made heavy strokes in this process of devel-

DOI: 10.4018/979-8-3693-5573-2.ch017

opment by further enhancing predictive analytics, and now it enables the analysis of intricate patterns within large datasets. This introduction tries to delve into exactly how deep learning, as a subdomain of machine learning involving neural networks with multiple layers, has made it possible to revolutionize predictive analytics by providing potent methods for forecasting and decision-making within such domains(Zhong et al., 2021a).

Notably, predictive analytics plays a very essential role in addressing such critical topics in environmental sciences as climate change, natural disasters, and management of ecosystems. While traditional statistical models are useful, they often fall short when dealing with the size and complexity of the more modern data sources. With the increasing number of remote sensing technologies, like satellites and ground-based sensors, enormous datasets are generated that capture various environmental phenomena. Deep learning does a great job processing these high-dimensional data and provides very efficient methods for meaningful insights and prediction with accuracy(Hälterlein, 2021). For instance, Convolutional Neural Networks are excellent in spatial data analysis, such as satellite image analysis, for monitoring changes in land use, detection of deforestation, or the extent of damage from natural disasters. They process complex visual data, interpreting information that gives detailed assessment—a question of prime importance in environmental monitoring and management.

It is in the same way that the recurrent neural networks are good at handling time-dependent data, which becomes critically necessary in making predictions of weather patterns and climate dynamics. RNNs can scan the past sequences of weather data to identify periodic trends and tendencies in the forecast, helping to undertake more accurate predictions of extreme weather events such as hurricanes, floods, and heat waves. This, in turn, will create an ability very important in enhancing disaster preparedness and response strategies, hence minimizing the impact of such events on communities and ecosystems. Deep learning will enable researchers to come up with models that integrate diverse sources of data, such as atmospheric measurements and oceanic conditions, toward improving predictability with actionable insights for environmental policy and management(Zhong et al., 2021b).

The explosion of digital data from social media, online interactions, and other sources across the social sciences has changed the face of behavioral and societal research. Deep learning methods provide immensely useful tools for analyzing large and assorted data, offering advanced prediction analytics that can discover trends, sentiments, and behaviors. For example, deep learning-driven sentiment analysis can digest text data from social media sites for understanding public perception and sentiment toward issues as varied as politics and consumer preference. It gives insights into the social dynamics and public sentiments that help in monitoring them toward the formulation of policy decisions. Deep learning models identify slight changes in attitude and trends from vast text collections, which are usually not noticed otherwise by analysis techniques(Tahmasebi et al., 2020).

Another major area that has seen a lot of activity in the application of deep learning is behavioral prediction. In this respect, by evaluating the patterns in user behavior, deep learning models have been able to make predictions with a high degree of accuracy, drawing from online purchases, search history, and even social interactions. This capability is hence specially very useful for businesses or marketers, which look forward to using predictive models in fine-tuning their strategies, with the view to helping them optimize advertisement campaigns and enhance customer engagement(Liu et al., 2022). Deep learning is in a position to detect an increase in new trends and changes in the consumptive behavior of consumers; hence, organizations can swiftly respond to changes in the market environment and consumer preference.

While it is true that deep learning integrated into predictive analytics offers a lot of benefits, there are also some challenges. The effectiveness of deep learning models essentially relies on two most influencing factors: data quality and preprocessing. Data in environmental and social sciences may be incomplete, noisy, or biased. All these can lower the accuracy and reliability of predictions. High-quality data and robust preprocessing techniques are therefore very important in ensuring meaningful results(Atitallah et al., 2020). Furthermore, deep learning models are "black boxes," making it quite a challenge to interpret them. Especially when predictive analytics-driven decisions carry meaningful policy and practice implications, one should know the ways in which these models produce their predictions. Efforts to enhance model interpretability, such as Explainable AI, are currently underway in an attempt to give users clearer insights into the decisional processes of deep learning models.

Another challenge to the application of deep learning in predictive analytics is computational resources and scalability. Deep learning models require a lot of computational power for training, and this sometimes is a barrier to too many research institutions and organizations. Hardware improvements, such as Graphics Processing Units and Tensor Processing Units, have dramatically enhanced the development of deep learning models; however, associated costs can run high(Sit et al., 2020). Deep learning applications require efficient algorithms and distributed computing approaches to handle the huge computational load, which has to scale effectively with large datasets. The domains of deep learning and predictive analytics are increasingly relevant to ethical and privacy concerns. The use of personal and sensitive data raises questions about data privacy and security(Liu et al., 2022). Compliance with regulations like GDPR and robust data protection measures are very important in maintaining public trust and ethical standards in research.

In the future, this integration will shift even further, with possible break-throughs related to real-time analytics, integrations with other emerging technologies, and improved interpretability. By addressing the current challenges and leveraging the strengths of deep learning, the environmental and social sciences researcher and practitioner community would eventually enhance their level of understanding, make better decisions, and come closer to effective solutions for complex problems beleaguering the contemporary world.

Background

Deep learning is a subsection of Machine Learning. According to deep learning approaches, the complicated patterns among data are modeled through the usage of neural networks with many layers. It finds its roots in the development of artificial neural networks and has continuously been evolving over time because of enhancement in computational powers and availability of huge data. Deep learning came to the fore a few years ago, outperforming all other developed algorithms in tasks involving large datasets and high-dimensional features, such as image and speech recognition. When applied to predictive analytics, in particular, in environmental and social sciences, it showed a potential way to alter how we understand and forecast complex systems.

Scopes

This chapter deals with deep learning techniques applied to predictive analytics in environmental and social sciences. This will include an overview of key principles of deep learning, its applications for different predictive tasks, and related challenges and future directions. It will further showcase case studies that display practical applications and insights drawn from deep learning models in these domains.

Objectives

The chapter's main objectives are to explain the role of deep learning in predictive analytics, provide an overview of key techniques applied in environmental and social sciences, and highlight their challenges and future trends. Emphasis will thus be based on case studies that bring forth practical insight into understanding the way in which deep learning can help ensure more accurate predictions and better decision-making in these two major critical areas.

DEEP LEARNING FUNDAMENTALS

Deep learning is a powerful subset of machine learning, which draws strength from neural networks with multiple layers to model complex patterns in data. In the following section, an emphasis is given to the basic concepts and architectures for deep learning: Convolutional Neural Networks, Recurrent Neural Networks, and cutting-edge techniques like Transformers and Attention Mechanisms(Grimmer et al., 2021; Yuan et al., 2020). Figure 1 illustrates the operational flow of deep learning models, highlighting the role of neural networks and their processing layers in predicting data from the human brain.

Figure 1. Fundamental operation of deep learning based on neural networks

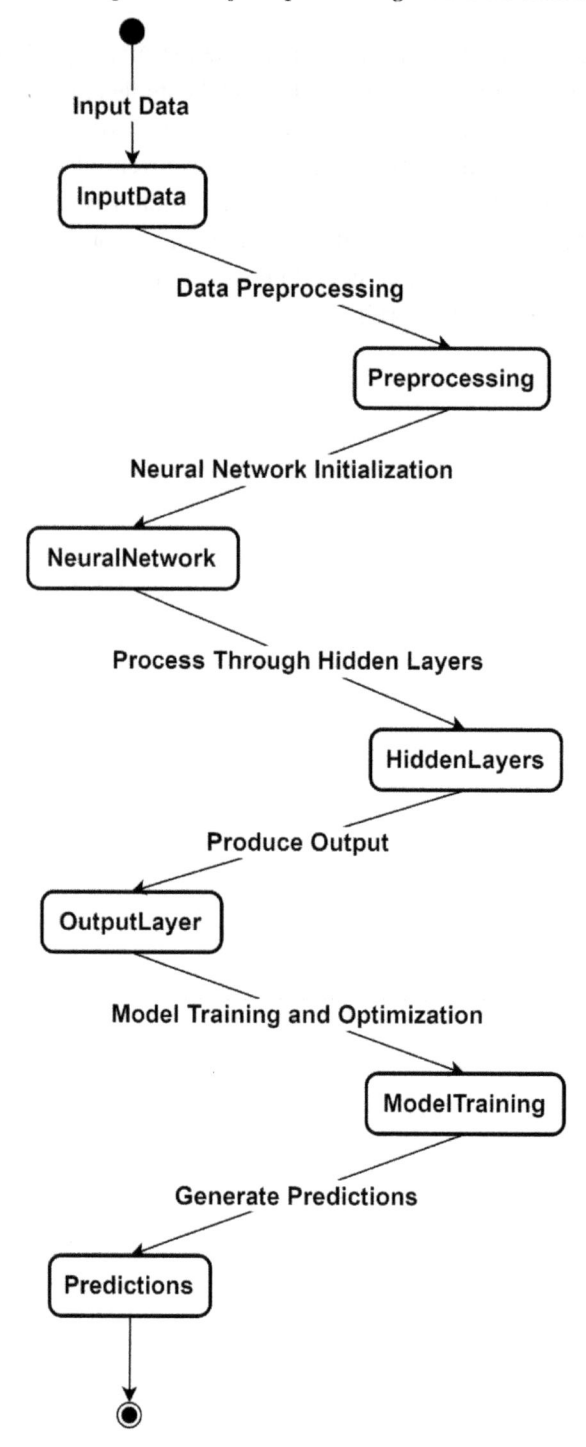

Important Concepts and Architectures

Deep learning operates fundamentally on a model of neural networks, computational models that take inspiration from the human brain's working. The network consists of multiple connected nodes, equivalent to "neurons," computing linear transformations followed by non-linear activations. Such a network is deep, and it is how this depth—it can learn abstract features of data. We have input, hidden, and output layers. What makes these networks so appropriate for complex tasks in many fields is their ability to learn from large amounts of data and recognize very complex patterns.

Convolution Neural Networks (CNNs)

Convolutional Neural Networks are the kind of neural network that operates on a specialized grid-like topology, say an image. For this, CNN is very effective for visual recognition purposes, as it can capture the spatial hierarchies and local patterns in data. The canonical structure includes convolutional layers, followed by pooling layers, and fully connected layers.

- Convolutional Layers: Convolving input data with convolutional filters creates various feature maps that, in turn, highlight features of the input, for example, the edges in an image. More specifically, a set of filters slides over the data, producing localized dot products that are then summed to generate the output feature maps.
- Pooling Layers: Pooling layers reduce the spatial dimensions of the feature maps, maintaining important information but also reducing the computation processing of the networks. Application of max pooling layers also promotes translation invariance where average pooling layers avoid overfitting.
- Fully Connected Layers: The CNN, after performing feature extraction and dimensionality reduction, tends to terminate with fully connected layers to combine the aspects and make predictions or classifications.

In computer vision, which involves understanding spatial relationships and patterns, CNNs are widely used for image classification, object detection, and segmentation.

Recurrent Neural Networks (RNNs)

Recurrent Neural Networks (RNNs) are designed to process sequential data, much like the group of Neural Networks that maintain some sort of memory through their architecture. Unlike a feedforward neural network, RNNs make use of their internal states—or memory—to process sequences of data, making them suitable for tasks whose dependency on previous steps forms an important feature space, such as time series forecasting and natural language processing.

- Basic RNNs: Although the basic idea behind recurrent neural networks is simple—to introduce loops within the networks through which information is explicitly allowed to persist over time steps—traditional RNNs suffer from problems with long-term dependencies, issues of vanishing and exploding gradients, respectively, which can impede learning from long sequences.
- Long Short-Term Memory (LSTM) and Gated Recurrent Units (GRUs): RNNs were developed into their advanced, gated mechanism-integrating instances in the forms of LSTMs and GRUs to overcome their shortcomings. LSTMs, for example, have mechanisms like forget gates, input gates, and output gates that control memory retention and update processes and are thus effective in long-term dependency modeling.

This is so with RNNs and their later-day variants, as well, that are being used in very many applications, including language modeling, speech recognition, and problems involving sequence-to-sequence solutions.

Advanced Techniques: Transformers and Attention Mechanisms

Transformers are among the most natural deep learning developments, especially in the domain of natural language processing. Unlike RNNs, which use self-attention mechanisms, transformers are used in a way to process sequences, scaling up or down in nature by the importance given to different input elements(Agbo et al., 2021).

- Transformers: The paper "Attention is All You Need" introduced the transformers that used an encoder-decoder structure to implement parallel data processing. The mechanism improved efficiency for long sequences dramatically and made it possible for the model to capture complex dependencies.
- Attention Mechanisms: Attention mechanisms form the very basis of transformers, through which a model can pay attention to pieces of the input sequence while making predictions. Self-attention enables the model to weigh

in the importance of various tokens in a sequence with respect to one another, hence enhancing its ability to capture the contextual relationship.

Transformers and attention mechanisms have revolutionized processing, leading to the development of powerful models such as BERT, GPT, and T5, which reach state-of-the-art performance on a variety of tasks.

In a nutshell, deep learning is a broad set of architectures and approaches meant to operate on signals of various kinds and their associated tasks: CNNs are for working with visual data, RNNs handle sequential information, and transformers model complex interactions in the data via attention. Understanding these basics is crucial for an effective instance of deep learning across the different applications.

APPLICATIONS IN ENVIRONMENTAL SCIENCES

Deep learning has transformed environmental sciences—especially state-of-the-art climate modeling, disaster prediction, and ecological monitoring. Applications like these tap deep into the power of neural networks for training from large, complex datasets to output new knowledge that enables better understanding and management of environmental systems(Waheed et al., 2020).

Figure 2. Applications in environmental sciences

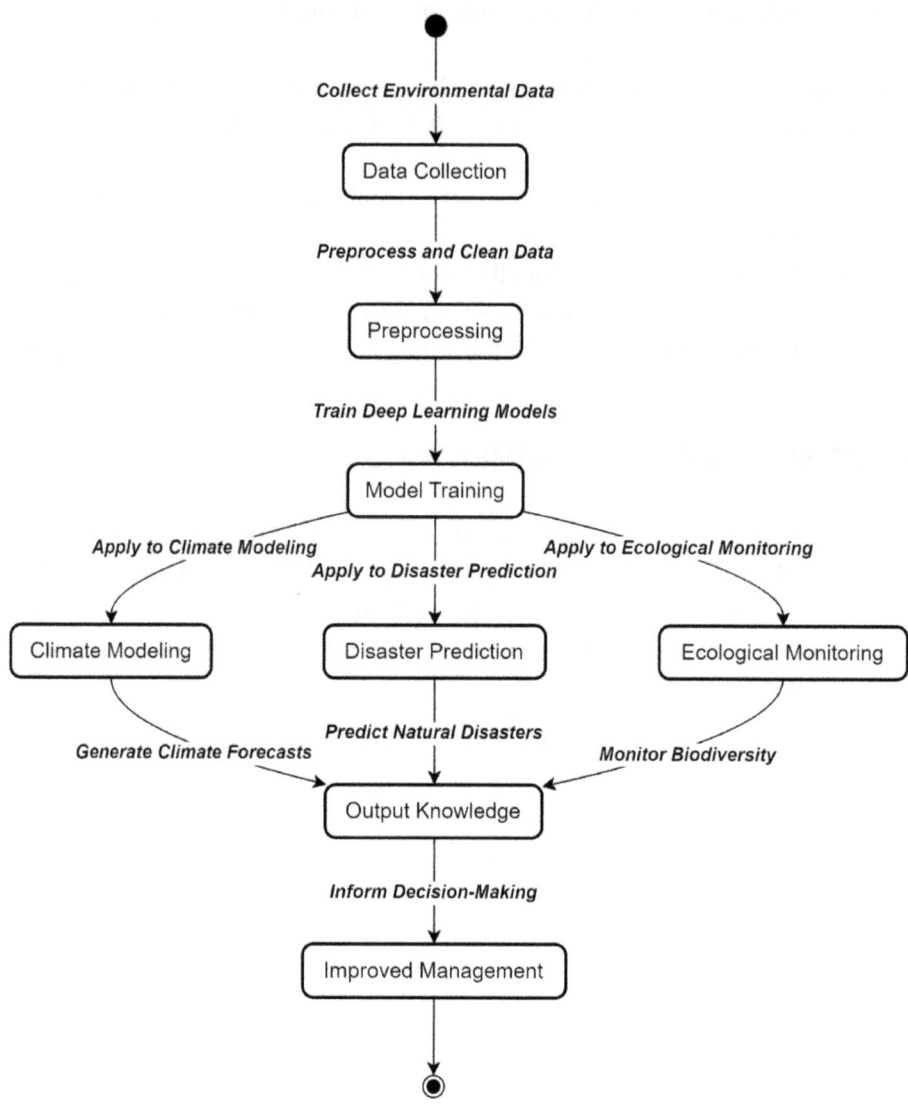

Figure 2 illustrates the flow from data collection to actionable insights in various applications of environmental sciences, showing how deep learning transforms raw data into valuable information for managing and understanding environmental systems.

Climate Modeling and Forecasting

The applications of climate modeling and forecasting in understanding and mitigating the impacts of climate change are endless. Traditional climate models derive the climate through physical equations coupled with data from the past, making future predictions of the climate, but with an integration of deep learning techniques, this could enhance the accuracy and granularity dramatically(Kelleher et al., 2020).

- Data Integration and Prediction: Deep learning models, specifically Convolutional Neural Networks can digest satellite images and remote sensing data for climate prediction improvement. This model can analyze large datasets—like those for sea surface temperatures, atmospheric conditions, and ice cover—to show trends and patterns that might be too deep for traditional methods to locate. For instance, CNNs are in a position to identify abnormal changes in climate data; hence, this makes it an effective early warning system against severe weather events.
- Temporal Dynamics: One of the most prominent ways in which RNNs and LSTMs can be of immense help while dealing with time-series data in climate forecasting is by capturing temporal dependencies and long-term trends. These models, therefore, assist in making better predictions of seasonal variations and long-term changes in climate. For instance, LSTMs will learn from the historical data and thus turn out to be skillful predictors of temperature, precipitation, and other climate variables for their seasonal patterns.
- Downscaling: Deep learning can also be applied to downscaling climate models that basically translate coarse-resolution climate predictions into finer-scale projections. Generative adversarial networks, a deep learning technique, generate high-resolution climate data from low-resolution input and can provide detailed predictions crucial for local planning and adaptation strategies.

Disaster Prediction and Management

One of the critical applications of deep learning in environmental science is prediction and management of natural disasters. Deep learning models have paved ways through which our ability to handle the forecasting and management of floods, hurricanes, and wildfires can be enhanced(Kumar et al., 2022).

- Flood Forecasting: CNN and RNN can be linked with flood data history, weather forecasts, and information from real-time sensors. Merging data from satellite images and in situ sensors allows deep learning models to predict

flood events at a high spatial and temporal resolution. For instance, CNNs can be applied to process radar images or satellites to identify areas prone to flooding, while RNNs are applied to predict the runoff and rainfall pattern of rivers.
- Hurricane Tracking: Deep learning techniques in hurricane forecasting through the analysis of satellite images and atmospheric data. Using CNNs allows hurricane detection and tracking, while LSTMs allow for the estimation of their intensity and trajectory. These models help meteorologists issue more accurate warnings on time and hence improve preparedness and response strategies.
- Wildfire Monitoring: CNN and RNN apply in monitoring and predicting wildfires. Satellite images and weather data are analyzed using deep learning models that identify early signs of wildfires and predict fire spread. For example, CNNs process thermal and optical satellite images to locate hotspots, while RNNs forecast fire progression based on weather and other past information about the fire.

Ecological Monitoring and Biodiversity Analysis

Deep learning would enable the analysis of large and complex data sets related to species distribution, habitat changes, and ecological interactions, leading to making ecological monitoring and biodiversity analysis so much more efficient(Kumar et al., 2022).

- Species identification: Deep learning models, especially CNNs, make use of images and videos collected from cameras and sensors in species identification. The accuracy in classifying can be leveraged to count species, hence helping researchers monitor biodiversity and detect changes in species populations. For instance, CNNs can extract camera trap images to identify wildlife species and recognize their abundance and distribution.
- Habitat Mapping: Deep learning techniques are used for the analysis of remote sensing data to map and monitor habitats. Satellite images can be digested by CNNs to identify and classify different land cover types such as forests, wetlands, and urban areas, thus providing information of importance in assessing habitat loss and degradation and guiding conservation effort.
- Ecological Interactions: RNNs and LSTMs can be used in modeling and making predictions of ecological interactions and dynamics. The models will be able to analyze species populations against time series data of climatic variables and environmental conditions to establish how changes in one component of the ecosystem affect the others. This is where the application of

RNNs in predicting climate change effects on species' migration patterns and ecological relationships will be useful.

In other words, it is a powerful set of tools for understanding and managing environmental systems. The idea now moves from enhancing applications to improved climate predictions, disaster preparedness to improved understanding of biodiversity, and the analysis of habitats. With an increase in data availability and computational power, the application of deep learning to environmental problems has the potential to increase even more.

APPLICATIONS IN SOCIAL SCIENCES

Deep learning has been applied in providing sophisticated tools for analyzing and interpreting complex social data within the social sciences. Principal applications include sentiment analysis and opinion mining, behavioral prediction and trend analysis, and social media data processing. Each of these applications is powered by deep learning's ability to process large volumes of unstructured data and unearth hidden patterns critical to the understanding of human behavior and social dynamics(Kumar et al., 2022). Figure 3 illustrates how deep learning is being used to analyze various aspects of social sciences, with a clear breakdown of each application area and its specific uses.

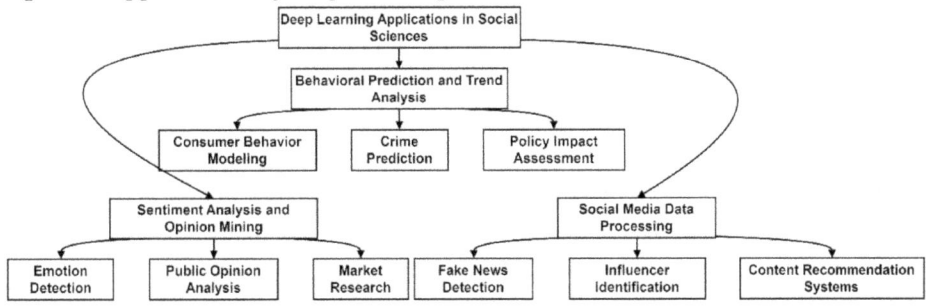

Figure 3. Applications of deep learning in social sciences

Sentiment Analysis and Opinion Mining

Sentiment analysis and opinion mining are critical in understanding public sentiment and opinions across varied platforms. Deep learning techniques have enhanced the accuracy and depth in sentiment analysis immensely compared to traditional methods(Grimmer et al., 2021).

- NLP: Sentiment analysis is actually impossible without NLP techniques that can enable deep learning models to take in and process human language. More specifically, Recurrent Neural Networks and Long Short-Term Memory could work really well on sequential information like sentences or documents when the task involves determining sentiment. These models get contextual information and long-range dependencies that enable the understanding of subtle nuances of language and tone.
- Transformer and Attention Mechanisms: another breakthrough in sentiment analysis is represented by transformers with attention mechanisms. Models such as BERT and GPT use attention to measure the importance of various words and phrases against each other in a sentence. With the ability to better understand this context and sentiment, such a capacity can better classify sentiment in varied contexts, including social media posts, reviews, and news items.

Applications include brand management, customer feedback, and political analysis. This comprises the ways in which businesses make use of sentiment analysis in order to gauge consumer reactions to certain products and services. In the same way, political analysts get wind of the public's opinion about issues of policy or election candidates. Such analysis on huge chunks of text data helps the organization derive actionable insights and make data-driven decisions.

Behavioral Prediction and Trend Analysis

Deep learning models human behavior and analyzes social trends. Applications of this nature assist in gaining insights and make predictions of changes in society, market dynamics, and other patterns exhibited by people(Grimmer et al., 2021; Kumar et al., 2022; Wang et al., 2020).

- Predictive Models: These models include deep learning models like Feedforward Neural Networks and LSTMs, which work on available historical data to predict future behavior. For example, such models will be used to predict consumer purchasing behavior, voting trends, or shifts in social

attitude. The deep learning models recognize patterns in history from past behaviors and contextual factors, and then accordingly predict the future course of actions.
- Trend Analysis: This implies that it has to identify and understand the change in social phenomena over time. Deep learning models can scale up the analysis of data from social media, news articles, and other textual sources in order to unveil emerging trends and shifts in public opinion. CNN is used for the analysis of text data and identification of key topics and themes. RNNs and LSTMs help in tracking how the trends evolve over time.
- Applications: It has wide applications for behavioral prediction and trend analysis across market research, health policy, and social policy. For example, enterprises adopt predictive models to adapt their marketing strategies to a later trend in their consumers; policymakers use trend analysis to respond to an emerging social issue or public health concern. These kinds of insights enable organizations and governments to be more proactive in the face of change.

Social Media Data Processing

Social media platforms generate large volumes of unstructured data. This is analyzed to understand social behavior, trends, and sentiments. Deep learning techniques become important in processing and extracting useful information from the same.

- Data Collection and Preprocessing: Data from various platforms like Twitter, Facebook, and Instagram are collected and preprocessed for processing in the first step of social media data processing. Deep learning models do this by cleaning and structuring the data, handling the noise, missing values, and normalization of text. Some techniques for data preparation are tokenization, lemmatization, stop-word removal.
- Content Analysis: Extracting insights and spotting patterns in social media content data is done by deep learning algorithms, especially CNNs and transformers. This would further be applied to text classification tasks, like spam identification and post categorization according to different topics, using a CNN. A transformer is tasked with even more complex analyses, such as sentiment analysis and topic modeling, due to its ability to capture the context relationships.
- User Behavior Analysis: Deep learning models also analyze user behavior across social media sites. RNN and LSTMs are used to study user interactions, engagement patterns, and even consumption and sharing behaviors of

content. Such studies help in understanding the likes of users, help in predicting future actions, and in turn, deliver personalized content.
- Applications: Social media data processing has far-reaching applications in marketing, public relations, and crisis management. For example, using social media analytics, brands understand online presence and campaign performance and engage with customers. Public health organizations mine social media data for tracing the spread of diseases or evaluating the impact interventions have on health. In the crisis management domain, it enables response teams to monitor potential emerging issues effectively and maintain good public relations.

Deep learning has hence greatly advanced the social sciences by increasing our ability to analyze and make sense of advanced social data. Deep learning offers insights into human behavior, social trends, and public sentiment through applications that include sentiment analysis, behavioral prediction, and processing social media data. These very methods are, therefore, likely to assume center stage in the future as the field continues to develop and shape our understanding of social dynamics, eliciting decisions in a wide array of domains.

CHALLENGES AND CONSIDERATIONS

Deep learning presents enormous scope for transformative innovation in many areas of environmental and social sciences, but remains riddled with several challenges and considerations, yet to be overcome to ensure this potential is effectively and responsibly realized. Foremost among the challenges are data quality and preprocessing, model interpretability and transparency, computational resources and scalability, and ethical and privacy concerns(Kelleher et al., 2020; Wang et al., 2020; Yuan et al., 2020). Figure 4 highlights the iterative nature of model development in environmental and social sciences, emphasizing the need to address challenges to fully realize the potential of deep learning responsibly.

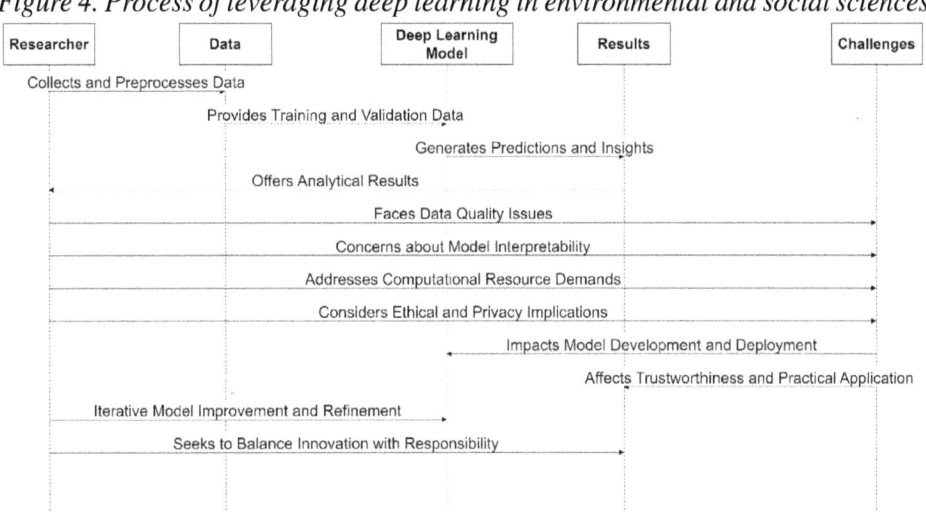

Figure 4. Process of leveraging deep learning in environmental and social sciences

```
Researcher        Data          Deep Learning        Results          Challenges
                                    Model
Collects and Preprocesses Data
                  Provides Training and Validation Data
                                Generates Predictions and Insights
                  Offers Analytical Results
                                Faces Data Quality Issues
                                Concerns about Model Interpretability
                                Addresses Computational Resource Demands
                                Considers Ethical and Privacy Implications
                                            Impacts Model Development and Deployment
                                                Affects Trustworthiness and Practical Application
       Iterative Model Improvement and Refinement
            Seeks to Balance Innovation with Responsibility
```

Data Quality and Preprocessing

Deep learning models really depend a lot on data quality and data preprocessing. High-quality data that is well-prepared becomes inevitable for training a model.

- Data Quality: A model may suffer from poor performance if the data are inaccurate, incomplete, or noisy. Poor data quality is a common problem with deep learning models as they learn from the underlying data patterns. For example, in environmental monitoring, satellite images with artifacts or lacking information would give the wrong predictions. In the case of social monitoring, noisy or irrelevant data would cast wrong sentiment analysis. Sufficient data quality is made sure by exercise that is intensive in cleanliness and involves the removal of duplicates, the handling of missing values, and correction of errors.
- Pre-processing: Data pre-processing is among the most important stages in the workflow pipeline of deep learning. This involves data normalization, feature extraction, and transformation. It can include recalibrating satellite image distortions due to atmospheric effects or even normalizing sensor readings for environmental data. Text pre-processing for social media analysis includes tokenization, stemming, and stop-word removal. Useful pre-processing techniques enhance how much the model learns from data with demonstrations of improvement in overall performance.

Model Interpretability and Transparency

Deep models, and specifically complex architectures like deep neural networks, are similar to "black boxes" as it is pretty difficult to know most of the time how they come to a specific decision or prediction.

- Interpretability: It is, therefore, very important to understand the inner working of the deep learning models in validating their predictions and ensuring that they are reliable. In most applied fields like environmental science, for instance, where the decisions are going to have significant implications based on model outputs, interpretability is key. For example, if a model projects increased air pollution in the future, stakeholders need to know what drives that projection to be able to make informed decisions.
- Transparency: Models should be transparently developed and deployed so as to come up with a model that one can be sure to trust and notably for which someone can bear responsibility. Methods like model-agnostic techniques such as SHAP and LIME, and visualization tools give explanation into how the different features contribute to the prediction. Model interpretability can be added early in the development process to tackle issues and enhance the robustness of the model.

Computational Resources and Scalability

Deep models need significant computational resources, especially for large datasets with complex architectures, and efficient scaling strategies at their core.

- Computational Resources: Training deep learning models requires high-performance computing hardware such as graphics processing units or tensor processing units. In the case of large-scale models and datasets, this computational cost can be very large. For example, training deep CNNs on high-resolution satellite image data or processing large volumes of social media data require enormous computing resources.
- Scalability: It becomes very critical to have scalability in Deep Learning models and datasets of growing size. Distributed Computing, Parallel Processing, and Cloud-based solutions are strategies for efficient scaling. Apache Spark and Tensor-Flow's distributed training framework can be used for large-scale data processing. Models are expected to scale without loss of performance for working in real-life scenarios, especially in dynamic environments.

Ethical and Privacy Concerns

The application of this to deep learning raises some very important concerns regarding ethics and privacy, particularly in the sensitive areas of social media and personal data.

- Ethical Concerns: There are ethical concerns associated with deep learning applications in different aspects. Regarding the text analysis of social media, problems concern algorithm bias and fairness. Algorithms pre-trained on biased data perpetuate or exaggerate existing inequalities that turn into unfair or discriminatory outcomes. This can be tackled by having algorithms that are fairness-aware, besides making sure that the datasets used for training are representative and diverse enough.
- Privacy Issues: Preservation of the privacy of personal and sensitive data is the major factor while dealing with it. Deep learning models usually demand a large volume of data, which may include personal information. Hiding or protecting privacy in data needs to have provisions for anonymization, encryption techniques, secure ways to deal with data, and the like. Furthermore, regulations under the General Data Protection Regulation and the California Consumer Privacy Act are to be followed in maintaining the right to privacy of the individual.
- Transparency and Consent: Transparency over the use of data and consent in using data of a person are very important. An organization should explain the purposes for using data, the duration for which data will be kept, and who is going to share and have access to it. Such transparent practices enhance trust and make sure people are aware and permit the use of their data.

In a nutshell, deep learning provides very many functionalities for a multitude of applications, but its interpretability concerning data quality, model interpretability, computational resources, and finally ethical and privacy issues set aside, remain very key challenges to its effective and responsible use. If these challenges are heeded and best practices for overcoming them implemented, researchers and practitioners would be in a better position to derive the most from deep learning while taming the potential risks that may arise from ethics, transparency, and scalability in their applications.

FUTURE DIRECTIONS AND TRENDS

This will go hand in hand with the other clear directions and trends of deep learning: emerging technologies, further integration with other analytical methods, and attention to policy implications and societal impact. Advances in these areas likely will help create new capabilities of deep learning—the things which current incarnations can't do(Kumar et al., 2022; Waheed et al., 2020; Wang et al., 2020; Yuan et al., 2020). Figure 5 illustrates the future directions and trends of deep learning, focusing on emerging technologies, integration with other analytical methods, and attention to policy implications and societal impact.

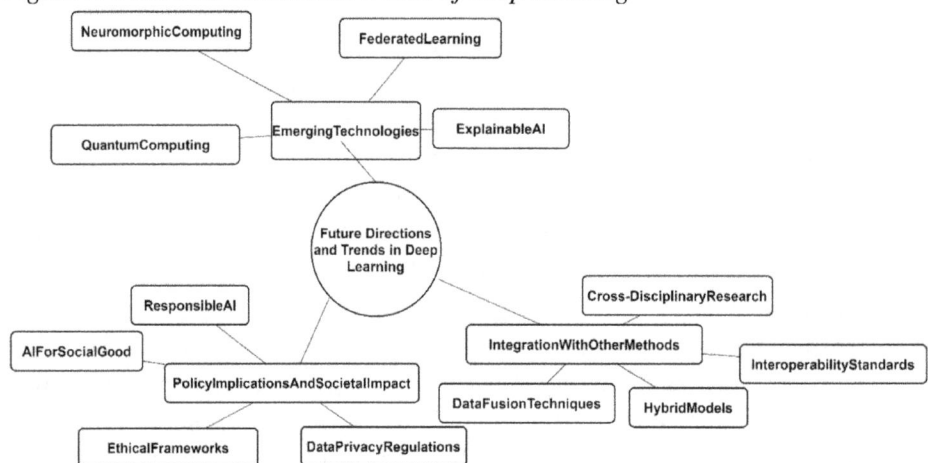

Figure 5. Future directions and trends of deep learning

New and Emerging Technologies and Innovations

a) Advanced Neural Network Architectures: One key element of deep learning is the invention of new neural network architectures. Inventions such as the Transformer and its derivatives, including the GPT-4 and beyond, keep increasing the possibilities of NLP and, more generally, machine learning. In this line, transformers have shown massive improvements in comprehension of context and coherent response generation with their attention mechanism, turning them useful from translation to content generation. Future research could be based more on fine-tuning these architectures to achieve better efficiency and applicability.

b) Neuromorphic Computing: Neuromorphic computing is a nascent technology that may be the game-changer in deep learning during the years to come, since it gets its neural structure and working style from the human brain. Neuromorphic computing is a system that allows for more efficient and quicker computations to take place by developing specific hardware capable of simulating neural processes. This new technology offers huge scope for enhancing the performance of deep learning models while reducing their power consumption, which has been one of the main reasons hindering the real-world deployment of AI systems in resource-constrained scenarios.

c) Federated Learning: Federated learning is an emerging concept wherein models can be trained over several decentralized devices or servers hosting local data, without requiring the transfer of data to any central hub. It improves data privacy and security, but makes collaborative learning from heterogeneous datasets possible. With rapidly growing concerns associated with data privacy, federated learning is definitely going to become very important in training models on sensitive data while maintaining compliance with privacy regulations.

d) Explainable AI: Deep learning models have been turning increasingly opaque. Therefore, the need for interpretability and transparency grows. Explainable AI tries to make the model decisions more understandable to humans by providing insights into how the predictions are made. On the advancing side, XAI involves new techniques and tools that better visualize and explain model behavior in core ways. This will foster trust and ensure that AI systems are responsibly used.

Integration with Other Analytical Methods

a) Hybrid Models: Deep learning combined with other analysis techniques, such as statistical models and rule-based systems, is already a common practice. Each of these hybrid models will borrow positive elements from all approaches to increase their accuracy and make them robust. For example, deep learning combined with conventional statistical methods will increase the predictive power of predictive models through domain knowledge and tapping into historical data.

b) Multimodal Data Fusion: Combining data from another source or modality, like text, images, and audio, multimodal data fusion advances toward the execution of a comprehensive understanding of complex phenomena. Deep learning models that integrate multimodal data can demonstrate richer insights and more accurate predictions. For example, fusion of satellite imagery with environmental sensor data can improve climate modeling; likewise, fusion of text and image data may improve sentiment analysis on social media.

c) Integration of AI and IoT: Deep learning integrated with IoT is another core trend that permits analysis in real-time, allowing devices that are interconnected to make decisions based on data. Deep learning models could make use of data from sensors and IoT devices in tracking environmental conditions, industrial processes, or even infrastructure in smart cities. It will be easy to develop intelligent systems that respond dynamically to changing conditions.

Policy Implications and Impact on Society

a) Ethical and Regulatory Frameworks: Setting up appropriate ethical and regulatory frameworks has become increasingly important, specifically when the growth of deep learning technologies has been so rapid. Policymakers are now engaged with framing guidelines and regulations for the ethical use of AI, data privacy, and algorithmic bias. Future directions include transparent AI decision-making, clear standards setting for AI governance, and mechanisms for accountability. These frameworks serve the function of giving ways through which deep learning technologies will be used in such a way that they do not perpetuate inequalities or infringe on other people's rights.
b) Societal Impact: Deep learning will leave its mark in every sphere of life, be it employment, education, or health. On one hand, deep learning has the capacity to fuel innovation and birth a new set of opportunities; on the other, it presents the problems of possible job displacement and the requirement for reskilling. Policymakers and educators are zeroing in on the response to these challenges by promoting lifelong learning, investing in workforce development, and ensuring that benefits from AI are more equitably distributed.
c) Citizen Engagement and Awareness: The public needs to be involved, with a sharp awareness of deep learning technologies as far as informed debates and decisions are concerned. Since AI systems are getting embedded in day-to-day life at an increased rate, it becomes necessary to let the public know their capabilities, limitations, and prospective impacts. Activities targeted at improving the level of digital literacy and transparency in AI may further improve understanding of and finally engagement with such technologies.

In a nutshell, a fast technological development pace characterizes the future of deep learning, pioneering integrations with other analysis methods, and growing interest in ethical and societal issues. Just about to leave their stamp on deep learning, new technologies such as neuromorphic computing and federated learning hold much promise to boost model performance, while hybrid approaches and multimodal data fusion open up new avenues to harnessing very different data sources.

The development of ethical frameworks and policies shall further go hand in hand with these technologies, promising a responsible application of deep learning for the benefit of all people.

CASE STUDY

Deep Learning for Climate Change Modeling

Climate change is perhaps the single biggest challenge confronting the ecosystems, economies, as well as societies of the world at large today. Climate modeling provides an essential procedure in determining how the climate may change in the future, for which the impacts of climate change may be calculated, and in crafting mitigations and adaptation strategies. In this regard, the traditional models in the field have been helpful in one way or another but limited in resolution and predictive skill(Atitallah et al., 2020; Grimmer et al., 2021; Sit et al., 2020). In recent years, the use of deep learning on large climate data through advanced computational approaches has shown prowess in climate modeling. The following case study presents an analysis of how deep learning has helped in climate change modeling, discussing the impact, methodologies used, and the outcomes. Climate models generally represent the processes of the atmosphere, ocean, and land through mathematical equations. These models are computationally intensive, which may lead to the models' resolution constraints and potential input data quality problems. This makes deep learning techniques an increased research focus with the capacity in huge volumes of data processing and catching of detailed patterns in the data.

Application of Deep Learning

a) Data Integration and Preprocessing: Deep learning models for climate change modeling use various sources of data, including satellite observations, in situ measurements, and outputs from simulations on climate. Therefore, the integration of these mentioned multimodal data sources is necessary to develop comprehensive models. In the stage of data preprocessing, the cleaning of raw data is done to avoid redundancy and ensure accuracy. Besides, some other techniques applied during this stage are data augmentation and feature extraction to enhance the quality of input data and achieve maximum model performance.

b) Model Architecture: In the modeling of climate change, normally Convolutional Neural Networks (CNNs) and Recurrent Neural Networks (RNNs) are used. CNN is applied to spatial data like satellite images to extract patterns and anomalies in climate variables. For instance, CNNs help identify changes in SSTs, vegetation cover, and cloud cover. RNNs, especially long short-term memory recurrent neural networks (LSTM), are applied for the analysis of temporal data and modeling the climatic variables dynamics over some time. LSTMs can learn long-term trends and dependencies within climate data very efficiently.

Improved Temperature Prediction

One of the famous case studies in temperature prediction improvement applies deep learning. Researchers working at one of the top climate research institutes applied deep learning to improve the accuracy of temperature forecasts. A combination of temperature data, derived from satellites and historical weather records, was used to train a CNN-LSTM hybrid network.

Model Development: The CNN component of the model extracted satellite image features on temperature anomalies and cloud patterns. After extracting these features, it used an LSTM component to model temporal dependencies in these extracted features and help predict future changes in temperature. This hybrid model was trained on a large dataset of historical temperature records and satellite observations.

Results: Deep learning models performed much better in predicting the climate with greater accuracy than conventional climate models. The hybrid CNN-LSTM model turned out to have higher values of skill scores and lower error rates compared to other pre-existing forecasting methods. It is able to capture complex patterns, both spatial and temporal, in temperature data and hence evidences more reliable predictions for future climate scenarios.

Impact and Benefits

Deep learning provides improvement in the accuracy for climate prediction. Since deep learning algorithms capture minute patterns and relationships, which otherwise might have been missed by traditional models, then from such data, enhanced accuracy is entailed—something very vital for informed decisions involving mitigation and adaptation strategies vis-à-vis climatic change.

- Higher Resolution: Deep learning techniques allow for high-resolution data analysis, improving the spatial and temporal resolution of the climate models. In this way, capabilities for more detailed and localized predictions, valuable in assessing regional climate impacts, are afforded.
- Faster Computation: Advances in computational resources and optimization techniques have made deep learning models efficient for processing huge volumes of data. This efficiency reduces the time needed for training and

inference, therefore making the tasks associated with climate prediction or analysis timely.

Challenges and Considerations

- Quality of Data: Deep learning models are only as good as the quality of the input data. If this is incomplete or noisy, this impinges on the performance of the model; hence, strong techniques for data pre-processing and validation need to be devised.
- Model Interpretability: A criticism against deep learning models is that they are less interpretable. This is important in building trust in understanding how models arrive at certain predictions, thus ensuring appropriate use of these predictions during decision-making.
- Computational Resources: Deep learning models consume a great deal of computational resources, mainly very powerful graphic processing units and large-scale infrastructure. Handling such computational resources might get cumbersome, especially for organizations that run on a tight budget.

Deep learning applied to climate change modeling has been a massive development in the field. It provides better accuracy at a higher resolution and faster computation. A case study in temperature prediction highlights deep learning's potential for improvement of climate forecasting and decision making. Difficulties remain in data quality, interpretability of models, and computational demands; these should be curtailed by ongoing research and technological development, which will improve the capabilities of deep learning in climate science. Integration of deep learning into climate modeling can result in more effective and informed strategies to mitigate the impacts of climate change(Kumar et al., 2022; Waheed et al., 2020; Wang et al., 2020).

CONCLUSION

Deep learning has therefore tremendously pushed forward this frontier of predictive analytics, thereby providing transformative capabilities across both the environmental and social sciences. This chapter reviewed the integration of deep learning techniques into the domains in question, proving added value by improving predictive accuracy, handling complexity in datasets, and answering contemporary global challenges.

It has improved climate modeling, disaster prediction, and ecological monitoring in the environmental sciences. Deep learning models have improved climate forecasts, provided strategies for disaster management, and offered a comprehensive biodiversity analysis by using advanced neural network architectures and multimodal data. That is to say, such capabilities in analyzing large-scale and high-resolution data, and fusing diverse data sources, have come in handy in helping us understand environmental changes better and make more effective decisions.

Similarly, deep learning in the social sciences has changed the nature of sentiment analysis, behavioral prediction, and processing social media data. From granular to nuanced insights about public sentiments, trends, and social dynamics, models such as convolutional neural networks and recurrent neural networks have made this possible. This has genuinely added to the tools of deep learning for better interpreting complex social phenomena and responding to emerging trends. However, there are some challenges to the deployment of deep learning in such fields. Data quality, model interpretability, computational resources, and ethical considerations are a few major concerns that need to be taken into consideration for their responsible and effective use. Ensuring integrity in data, model transparency, managing computational demands, and addressing privacy concerns are critical to drawing maximum benefits from deep learning while mitigating possible risks.

The promising innovations and emerging technologies point toward a bright future of deep learning in predictive analytics. More specifically, advanced neural network architectures, neuromorphic computing, federated learning, and explainable AI could be the most plausible concepts to enhance the potential of deep learning methods. Deep learning will shape the future landscape of predictive analytics in combination with other analytical methods, including robust ethical frameworks.

Deep learning is thus an extremely powerful tool for advancing the understanding of complex systems in the environmental and social sciences. This functionality can realize its full potential only when little data, hidden patterns, and actionable insights come with the ability to handle vast amounts of data. In the process of continuous exploration into deep learning techniques, tackling the connected challenges and ethical considerations would be very necessary in ascertaining whether these technologies contribute positively to society and the environment. The continuous evolution of deep learning promises to create transformational impacts that bring forth new opportunities for future research, policy decisions, and practice in a world in rapid motion.

REFERENCES

Agbo, F. J., Oyelere, S. S., Suhonen, J., & Tukiainen, M. (2021). Scientific production and thematic breakthroughs in smart learning environments: A bibliometric analysis. *Smart Learning Environments*, 8(1), 1–25. DOI: 10.1186/s40561-020-00145-4

Atitallah, S. B., Driss, M., Boulila, W., & Ghézala, H. B. (2020). Leveraging Deep Learning and IoT big data analytics to support the smart cities development: Review and future directions. *Computer Science Review*, 38, 100303. DOI: 10.1016/j.cosrev.2020.100303

Grimmer, J., Roberts, M. E., & Stewart, B. M. (2021). Machine learning for social science: An agnostic approach. *Annual Review of Political Science*, 24(1), 395–419. DOI: 10.1146/annurev-polisci-053119-015921

Hälterlein, J. (2021). Epistemologies of predictive policing: Mathematical social science, social physics and machine learning. *Big Data & Society*, 8(1), 20539517211003118. DOI: 10.1177/20539517211003118

Kelleher, J. D., Mac Namee, B., & D'arcy, A. (2020). *Fundamentals of machine learning for predictive data analytics: Algorithms, worked examples, and case studies*. MIT press.

Kumar, S., Sharma, D., Rao, S., Lim, W. M., & Mangla, S. K. (2022). Past, present, and future of sustainable finance: Insights from big data analytics through machine learning of scholarly research. *Annals of Operations Research*, •••, 1–44. DOI: 10.1007/s10479-021-04410-8 PMID: 35002001

Liu, X., Lu, D., Zhang, A., Liu, Q., & Jiang, G. (2022). Data-driven machine learning in environmental pollution: Gains and problems. *Environmental Science & Technology*, 56(4), 2124–2133. DOI: 10.1021/acs.est.1c06157 PMID: 35084840

Sit, M., Demiray, B. Z., Xiang, Z., Ewing, G. J., Sermet, Y., & Demir, I. (2020). A comprehensive review of deep learning applications in hydrology and water resources. *Water Science and Technology*, 82(12), 2635–2670. DOI: 10.2166/wst.2020.369 PMID: 33341760

Tahmasebi, P., Kamrava, S., Bai, T., & Sahimi, M. (2020). Machine learning in geo-and environmental sciences: From small to large scale. *Advances in Water Resources*, 142, 103619. DOI: 10.1016/j.advwatres.2020.103619

Waheed, H., Hassan, S.-U., Aljohani, N. R., Hardman, J., Alelyani, S., & Nawaz, R. (2020). Predicting academic performance of students from VLE big data using deep learning models. *Computers in Human Behavior*, 104, 106189. DOI: 10.1016/j.chb.2019.106189

Wang, S., Cao, J., & Philip, S. Y. (2020). Deep learning for spatio-temporal data mining: A survey. *IEEE Transactions on Knowledge and Data Engineering*, 34(8), 3681–3700. DOI: 10.1109/TKDE.2020.3025580

Yuan, Q., Shen, H., Li, T., Li, Z., Li, S., Jiang, Y., Xu, H., Tan, W., Yang, Q., Wang, J., Gao, J., & Zhang, L. (2020). Deep learning in environmental remote sensing: Achievements and challenges. *Remote Sensing of Environment*, 241, 111716. DOI: 10.1016/j.rse.2020.111716

Zhong, S., Zhang, K., Bagheri, M., Burken, J. G., Gu, A., Li, B., Ma, X., Marrone, B. L., Ren, Z. J., Schrier, J., Shi, W., Tan, H., Wang, T., Wang, X., Wong, B. M., Xiao, X., Yu, X., Zhu, J.-J., & Zhang, H. (2021a). Machine learning: New ideas and tools in environmental science and engineering. *Environmental Science & Technology*, 55(19), 12741–12754. DOI: 10.1021/acs.est.1c01339 PMID: 34403250

Zhong, S., Zhang, K., Bagheri, M., Burken, J. G., Gu, A., Li, B., Ma, X., Marrone, B. L., Ren, Z. J., Schrier, J., Shi, W., Tan, H., Wang, T., Wang, X., Wong, B. M., Xiao, X., Yu, X., Zhu, J.-J., & Zhang, H. (2021b). Machine learning: New ideas and tools in environmental science and engineering. *Environmental Science & Technology*, 55(19), 12741–12754. DOI: 10.1021/acs.est.1c01339 PMID: 34403250

KEY TERMS

AI: - Artificial Intelligence
ANN: - Artificial Neural Network
BERT: - Bidirectional Encoder Representations from Transformers
CCPA: - California Consumer Privacy Act
CNN: - Convolutional Neural Network
FNN: - Feedforward Neural Network
GAN: - Generative Adversarial Network
GDPR: - General Data Protection Regulation
GPT: - Generative Pre-trained Transformer
GPU: - Graphics Processing Unit
GRU: - Gated Recurrent Unit
LIME: - Local Interpretable Model-agnostic Explanations
LSTM: - Long Short-Term Memory

NLP: - Natural Language Processing
RNN: - Recurrent Neural Network
SHAP: - SHapley Additive exPlanations
TPU: - Tensor Processing Unit
XAI: - Explainable Artificial Intelligence

Chapter 18
Machine Vision System:
An Emerging Technique for Quality Estimation of Tea

Debangana Das
Silicon Institute of Technology, Bhubaneswar, India

Shreya Nag
University of Engineering and Management, Kolkata, India

Runu Banerjee Roy
https://orcid.org/0000-0003-4879-3129
Jadavpur University, India

ABSTRACT

The electronic sensing platforms measure the aroma, the taste, and color profiles, respectively. Color of tea samples is measured by using colorimeter and the process is called colorimetry. Flavanols, which are the most important compounds present in tea are responsible for the dark color of tea. The higher the flavavols like catechin, epigallocatechin-gallate, theaflavin, thearubigin, etc, the better is the quality of tea. It is feasible to monitor the image data which has been captured by a camera using digital detection and analysis with an electronic eye instrument. In essence, the use of electronic eye in the field of quality estimation of tea is an emerging area and is of utmost interest to the tea-traders. Integration of the three sensory platforms, i.e., electronic nose, electronic tongue and electronic eye, has been used for quality estimation of the beverage. Combining a feature-level fusion method with pattern recognition correlations between sensory qualities and metabolic profiles can greatly increase the effectiveness and accuracy of prediction models.

DOI: 10.4018/979-8-3693-5573-2.ch018

1.1. INTRODUCTION

Tea leaves can undergo diverse manufacturing procedures, resulting in the production of several predominant types of tea: black (orthodox, CTC - crush, tear, and curl), green, yellow, oolong, and white.. The black tea processing consists of five key stages, viz. plucking, withering, rolling, fermenting, and drying. Out of all the stages of tea-processing, fermentation is the most crucial as the process influences the quality of tea which is normally observed making use of laboratory analytical tools. The quality of tea is evaluated by the sense of smell, taste and vision of tea tasters. Utilizing bio-mimicking sensory platforms like electronic nose, tongue and eye, the most recent developments are examined in order to address present issues and enable quick, precise practical investigation of the quality affecting chemical present in tea. These cutting-edge sensing technologies can distinguish between various types of tea based on their fermentation degree, price, time of storage, geographical origins, and adulteration ratio. These electronic sensing platforms measure the aroma, the taste, and color profiles, respectively. The data obtained from the corresponding platforms are fed as input to the mathematical classification algorithms. To evaluate the quality of food products, nowadays, precise and non-contact colour measuring equipment is required and hence the machine vision technique comes into picture for standardization of tea quality evaluation. Machine vision automates tedious or challenging visual inspection activities and precisely directs handling machinery during product assembly using sensors (cameras), processing gear, and software algorithms [Xu, et.al., 2018]. Important "parameters" that may be taken into account during various phases of the tea-processing process to assess the quality of the tea include colour, size, shape, flavour, and others. For instance, "colour" and "aroma" of fermenting tea are two crucial tea quality indicator. The primary criteria for rating tea are those that determine its quality monitoring. Worldwide, tea is a significant high-value crop [Patil, et.al., 2021]. There are several steps in the processing of tea leaves, including withering, rolling and cutting, fermenting, drying, and sorting/grading [Xu,et.al.,2019]. Every stage is concerned with a few conventional procedures, and if one of them fails, the quality of the tea may suffer. To manufacture quality tea, each stage is therefore continuously examined by analysing a set of criteria. This chapter aimed to investigate how the computer or electronics might be utilized to support the human sensory panel in making their subjective assessments rather than replicating human eyesight and olfaction. This section provides an overview of the parameters that affect the quality of tea during the fermentation and grading processes, as well as the methods currently used to evaluate them. Next, it discusses how machine vision is used in the tea processing sector. Farm operations are simply one aspect of sustainability challenges in agriculture; off-farm enterprises are also important. Larger agricultural enterprises can succeed or fail depending on how poorly off-farm operations are handled. As a result, farming operations become unsustainable. Therefore, preserving the quality of agricultural goods is crucial, and researches that support off-farm handling are key to ensuring agriculture's sustainability.

The visual appearance of black tea is the most popular and straightforward method to assess the quality of black tea in the production line. Black tea quality evaluation is measured using a number of parameters. The form, colour, cleanliness, and tip of the particles, among other factors, are all measured. This visual appearance method now only relies on expert labour judgement, which might lead to biases and inconsistent results due to the non-standardized evaluation process. Setting up quality standards is challenging because of this inconsistency. Because producers' quality expectations are flexible, the quality of their products can change, which can lead to consumer mistrust. Additionally, inheritance of skill is made challenging by relying on a single individual. Industrial applications including automatic inspection, process control, and robot guiding all use imaging-based automatic inspection and analysis, which is provided by machine vision [Xu, et.al.,2019]. It has been demonstrated that employing automated visual inspection and imaging technology to sense agricultural products is reliable, inexpensive, quick, and nondestructive, and produces consistent results. The use of machine vision technology in the production of black tea paves the way for the creation of an improved approach for judging quality. This research demonstrates precision agriculture's potential in a real-world setting while simultaneously promoting sustainability in agriculture. The quantitative analysis and standardization of these parameters are treated as the main subjective methods for tasting raw and processed tea. Such methods determine the conformity of tea quality to specifications and for the analysis of quality during processing and final storages. Human sensory panel's judgment by using organoleptic methods (visual inspection, sniffing etc.) is the way in which the evaluation of tea quality is made in tea industries. These tea tasting methods obey certain colour / flavor terminology in making the judgments [Liang, et.al., 2003; Kamisoyama, et. al., 2012; Sharma, et.al., 2013].

In the tea industry, a human sensory panel's assessment of a product utilising organoleptic techniques (visual inspection, sniffing, tasting, etc.) is how the quality of the tea is assessed. When making their assessments, these tea tasting techniques adhere to certain colour and flavour terms. The human organoleptic approaches are effective in explaining the colour of objects in specific areas and the various flavours of objects. However, these techniques are slightly inaccurate in differentiating between objects that are nearly the same colour or smell, as well as those are about the same size and shape. Furthermore, these methods might be incorrect in some circumstances, leading to a variety of inaccuracies because of things like human variability, sensitivity issues brought on by prolonged exposure, mental state, etc.

Therefore, a different approach might be more beneficial. One such approach is machine vision, which can be used to measure the parameters during different stages of processing in the tea industries. The usage of this machine vision could provide an answer to assist the human sensory panel in making an accurate assessment of their issues. These electronic sensing platforms measure the aroma, the taste, and color profiles, respectively. The data obtained from the corresponding platforms are fed as input to the mathematical classification algorithms. To evaluate the quality of food products, nowadays, precise and non-contact colour measuring equipment is required and hence the machine vision technique comes into picture for standardization of tea quality evaluation. Color of tea samples is measured by using colorimeter and the process is called colorimetry. Flavanols, which are the most important compounds present in tea are responsible for the dark color of tea. The higher the flavavols like catechin, epigallocatechin-gallate, theaflavin, thearubigin, etc, the better is the quality of tea. The presence of pigments such as chlorophylls and carotenoids, pheophytins, xanthophylls, etc., also lends itself towards the quality and color of tea. It is feasible to monitor the image data which has been captured by a camera using digital detection and analysis with an electronic eye instrument. An e-eye employs methods such as image acquisition, processing, and image analysis. In general, a camera converts the reflected light from the evaluated object into analog signals. The target feature is then obtained by a computer processing system. Information picks the feature that reveals maximum information, and then distinguishes them into background and target images. The technique for image segmentation that allows for the region thresholding, edge-based techniques can be used to perform the analysis. This method is able to extract quantitative colour information from specific regions of digital photos, allowing it to successfully discern between irregular forms and colours using image processing. In essence, the use of electronic eye in the field of quality estimation of tea is an emerging area and is of utmost interest to the tea-traders. Integration of the three sensory platforms, i.e., electronic nose, electronic tongue and electronic eye, has been used for quality estimation of the beverage. These platforms when coupled with advanced data processing algorithms, has the potential to not only accurately speed up consumer-based sensory quality assessments of tea but also to define new benchmarks for this bioactive product to satisfy global market demand. Multivariate statistics combined with complicated data sets from electrical signals can help with quality prediction and discrimination in this way. Combining a feature-level fusion method with pattern recognition correlations between sensory qualities and metabolic profiles can greatly increase the effectiveness and accuracy of prediction models.

1.2. Tea Quality and Prospective Application of Machine Vision

As mentioned in section 1.1., tea is an important high value crop throughout the world. Tea leaves are processed through various stages viz., withering, rolling and cutting, fermenting, drying and sorting / grading. Every stage is concerned with some standard methods and failure of one of them may lead to a reduction in the quality of made-tea. Therefore every stage is monitored constantly by evaluating certain parameters to produce quality tea. The overall quality (a predetermined standard) of made-tea is what is meant when we use the phrase "tea quality." It relies on a number of variables, including environmental conditions, processing methods, and tea growth. For a certain quality to be maintained, several parameters must be under control. Enhancing tea quality is a major problem for the tea industry due to human consumption habits that are changing quickly, global rivalry, and the necessity for ever-stricter regulations. Traditional techniques of tea quality monitoring need to be updated to take advantage of new technologies because they are not standardized [Borah, et.al, 2005; Borah, et.al., 2003; Zhu, et.al., 2019]. Some of the current issues with tea quality monitoring techniques can be resolved with the aid of machine vision, the computerised analysis of the quality indicators. Such opportunities are made possible by the development of sensor devices, such as charge couple device (CCD), electronic nose (EN), electronic eye (EE), etc., which can be used for a variety of visual and olfactory inspections, including colour matching during fermentation, packaging into various grades and high-speed accurate sorting, quality grading based on aroma analysis, etc. Additionally, they provide certain benefits over typical issues with the conventional human sensory panel evaluation methods, such as human variability, insensitivity from prolonged exposure, mental state, etc. Finally, a model can be made with the help of Intelligent System Engineering (ISE) techniques based classification of the gathered data. The purpose of the chapter is to furnish the reader with necessary background information of area of interest related to tea quality detection. For this, it enumerates detailed background on the tea quality perception, prospect of application of machine vision in quality evaluation, and ISE techniques and their prospective applications. It focuses on key requirements for computer vision encountered during the system development for automatic control and inspection systems for the fermentation and grading processes. Figure 1 given below represents the basic components of an electronic eye system.

Figure 1. Block diagram showing the evaluation of tea-quality analysis using electronic eye

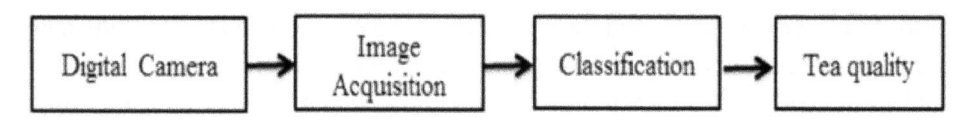

Digital Camera → Image Acquisition → Classification → Tea quality

1.3. Important Stages of Tea Processing

The basic aspect of this chapter is to focus on the explored possibilities mentioned in literature regarding the application of machine vision techniques for tea quality monitoring during two main processing stages: fermentation and grading. The different stages of tea processing are elucidated below.

a) Leaf Plucking:

Plucking in tea refers to harvesting the tea shoots. Besides giving crop, it encourages regeneration of new shoots, checks vertical growth of bushes and keeps them in vegetative phase. The "two leaf and a bud" is the most preferable plucking configuration (Figure 2). Plucking accounts for 70 – 80% of total cost of green leaf production. There are various different styles of plucking the tea shoots across different gardens or varieties of tea. It is a general recommended practice carried out at 5-9 days interval.

Figure 2. Plucking of tea leaves

High quality tea sample are the ones with higher percentage of buds. The proportion of fine leaf count thus has a significant positive correlation with the cost of finished tea.

b) **Withering:**

The moment a tea leaf is plucked from the tea plant, it begins to wilt naturally, a process referred to as withering. Once the tea leaves reach the processing facility, this process is controlled by the tea producer (Figure 3).

Figure 3. The withering process

The purpose of a controlled wither is to prepare the leaves for further processing by reducing their moisture content and to allow for the development of aroma and flavor compounds in the leaves.

c) **Rolling:**

The rolling process is where the tea leaves start developing their unique appearance and flavor profiles. As the soft leaves are rolled and shaped by machine, the cell walls of the leaf are broken, releasing the enzymes and essential oils that will alter the flavor of the leaf (Figure 4 and Figure 5). Rolling exposes the chemical components of the tea leaves to oxygen and initiates the oxidation process.

Figure 4. Tea leaves

Figure 5. CTC machine

d) **Fermentation:**

One of the key steps in the tea manufacturing process, that is a factor in determining the type of tea that is produced, is the degree of fermentation the tea leaves are allowed to undergo (Figure 6). The term fermentation when applied to tea is something of a misnomer, as the term actually refers to how much a tea is allowed to undergo enzymatic oxidation by allowing the freshly picked tea leaves to dry. This enzymatic oxidation process may be stopped by either pan frying or steaming the leaves

Figure 6. Fermentation of tea leaves after withering

before they are completely dried out. This is the most vital step in the tea manufacturing process. Hence, must be done carefully.

Constant supply of oxygen is required at a relative humidity of 93±2% to ensure the continuity of this process. The process lasts for 2-5 hours depending on the quality of tea leaves and the production season.

e) **Drying:**

Figure 7. Drying of the tea leaves

Once the leaf is oxidized to its desired level, heat is applied to the tea leaf to halt the oxidation process and further reduce the leaf's moisture content so that the tea leaves can be stored without spoiling (Figure 7). Depending on the type of heat applied, drying/ firing can also lend some flavor characteristics to the final tea.

f) **Ageing/Curing:**

Though not all, some teas require additional aging, secondary fermentation, or baking to reach their drinking potential. Green tea pu-erh, prior to curing into a post fermented tea, is often bitter and harsh in taste, but surprisingly becomes sweet through fermentation by age and dampness.

g) **Grading:**

Tea leaf grading is the process of evaluating products based on the quality and condition of the tea leaves itself.

Figure 8. Gradation of tea

Tea will be packed after sorting, consisting of extracting the fibers with the aid of winnowing machines and grading the tea by volumetric weight and size. The highest grades are referred to as "orange pekoe", and the lowest as "fannings" or "dust".

This section summarizes the tea quality parameters during fermentation and grading processes (Figure 8) and their existing evaluation techniques and then about the application of machine vision in food or beverage processing industries and finally its prospects in tea industries. It has been noted that different qualities, both controlled and naturally occurring, like color, aroma, taste, and astringency, are crucial quality factors in the manufacture of black tea [Zhu, et.al., 2019]. The human organoleptic techniques used to measure these factors are not standardized and can differ for a number of different reasons. Therefore, it is possible to use tools to evaluate some of the characteristics of tea, such as its color and aroma (physiochemical methods) In terms of the perception of tea quality, such initiatives are fairly successful and based on accepted methods, but they take time, effort, and money. Therefore, machine vision techniques can be used to address such inefficiency and accuracy be regarded as a successful alternative technique to the traditional techniques utilized to monitor the quality of tea using the human sensory panel. The quick expansion of electronic technologies, such the creation of CCD, EN, and ET, etc. and the potential to use them repeatedly enhance such prospects. Additionally, the use of ISE-based approaches can improve and modify the existing techniques. It improves the prospect to input data and examine the data obtained by these systems.in an open manner. There are different stages and conditions on which quality of tea depends. Some of those stage and conditions are as follows:

- Tea Plants (Genetics, Age, Health)
- Environment (Climate, Soil, Pests)
- Seasonality
- Methods (Irrigation, Fertilization)
- Harvesting Practices
- Processing Methods
- Storage & Transport Conditions
- Age of the Tea (After Manufacturing)

Hence the quality of made-tea depends on various parameters related to almost all activities involved in manufacturing of tea. Besides these stages and conditions mentioned above, factors that have major impact on green tea quality include tea variety, harvest, shading, nutrition and pest & disease damage. But if all the other requirements for good quality tea exist, then the most sensible stage for quality is the tea processing. Not only tea, but same approach of machine learning can be implemented for any food or beverages [Mark, et. al., 2014].

1.3.1. Quality Parameters During Fermentation and Grading

One of the most crucial procedures, tea fermentation is crucial to the creation of high-quality tea. Similar to this, sorting (or grading) tea according to appearance, size, and quality is a crucial step in the processing of tea. The significance of the quality criteria used in these processes can be summed up as follows:

1.3.1.1. Fermentation

A few chemical changes take place during the fermentation, which is an oxidation process. For instance, proteins are broken down, chlorophyll is converted into pheophytins, and some volatile organic compounds (VOC) are produced from lipids, amino acids, carotenoids, and terpenoids, among other things. These chemical transformations produce the color and aroma (flavor) in tea, which are considered as the most important quality parameters during the fermentation process (Mahanta, et. al., 1985; Mahanta, 1988). But these tea color and aroma are complex phenomena since a numbers of different compounds are formed during tea processing. Studies of tea flavor shows that more than ten separately identifiable volatile compounds are formed during processing and different proportion of these compounds corresponds to the flavor sensed. The compounds are t-2-hexenal, cis-3-hexenol, t-2hexenyl formate, linalool oxide (furanoidcis), linalool oxide (furanoid-trans), linalool, phenylacetaldehyde, cis-3-hexenyl caproate, methylsalcylate, geraniol, benzylalcohol, 2-phenylethanol, cis-jasmone + β-ionone, tobalb. The fact that the tea flavor is unstable and that its aroma will alter over time as the chemical makeup of the tea changes adds to the difficulty of the situation. On the other hand, some non-volatile compounds are also produced, viz., theaflavin (TF), thearubigin (TR), caffeine, pheophytins etc., which imparts to the final tea liquor its characteristic color and taste. The fermentation process is considered to be completed while optimum amount of these compounds in appropriate ratios are formed in tea (Okinda, et. al., 1998). Though it is possible to directly measure the formations of chemical compounds by chemical methods, but the methods become inefficient as time consuming, laborious and expensive. Therefore, it is worthwhile to find some other way to detect the optimum fermentation, for example by some subjective methods. For that it is observed and proved to be efficient to detect the optimum condition of tea fermentation by measuring the fermenting tea color and aroma. Because the color and aroma are the true reflections of the chemical compounds produced in tea during fermentation process. Therefore, it is worthwhile and efficient to detect the optimum fermentation by detecting the color and aroma to be achieved. In order to process tea in complicated regulated conditions and achieve the desired color and aroma, fermentation must be supervised (monitored) and controlled as early as feasible.

1.3.1.2. Grading

In the sorting process, tea granules are separated from each other in accordance with the variations of their size. The quality factor doesn't come into consideration in the very beginning of separation process of the tea granules. But right after the separation into different categories, they are tasted to assess its quality, its necessary appearance, flavor etc. before final packaging. From a quality perspective, the finished product is typically assessed based on three primary criteria: if it was created hygienically, whether it was free of dangerous substances (such as arsenic, cadmium, etc.), and whether the necessary constituents were present for cup qualities. In these respects, the appearance of tea, basically the blackish brown appearance of black teas is considered as one of the most important attributes to be considered. Moreover, the golden yellow and brownness appearance of tea liquor are also considered as quality parameters. The flavor attributes, for example aroma, taste and astringent are, on the other hand, considered as important parameters for tea quality evaluation. Finally, uniformity of granules size and presence of other unwanted stuffs are also considered at this stage. By measuring these parameters, efficient grading requires high degree of sensing and intelligence for accomplishing these quality evaluation processes.

1.4. Methods for tea Quality Assessment During Fermentation

Tea's color and flavor cannot be directly measured during ideal fermenting conditions. In other words, both factors calculate a model for black tea quality control and reflect significant meanings. Therefore, as soon as the desired color and flavor are achieved, tea fermentation needs to be supervised (monitored) and controlled. Color and color variations should be measured and observed practically throughout the fermentation process of tea in order to correlate crucial information about the physical and chemical changes that are occurring. Similarly, the odor and odor changes should also be measured and monitored to provide volatile compounds formed due to oxidation of tea and in the other manufacturing processes. Therefore, both the parameters represent meaningful visual and olfaction stimuli and compute a model for quality control of the black tea manufacturing.

1.5. Existing Techniques for tea Process Monitoring

There are two traditional methods of quality monitoring so far in tea industries. The first one is assessment of biochemical parameters and the second one is subjective assessment through tea tasting (measurement of various quality parameters). Of these two methods, the subjective tea tasting method is commonly used in most of the tea industry. Expert human sensory panel are involved for such quality evaluation in tea industries. Visual inspection, sniffing, gas chromatography (GC), and colorimetric approach have so far been reported as the quality monitoring tool in various stages of tea processing. For example, the visual inspection, sniffing and colorimetric approach have been used to detect the optimal fermentation time in the tea industry. Similarly visual inspection, sniffing, tasting by tongue, chemical analysis (TF-TR) etc. are commonly used methods by the sensory panel for the final quality judgment during the grading process. The human sensory panel (tea tasters) of trained personnel scores the product on the basis of a number of flavor terminologies (flavor descriptors). These human experts, since tea industry was established, have been traditionally determining tea aroma and color during fermentation process by human olfaction and visual approximation respectively. In the other method (colorimetric approach) a colorimeter is used. This technique measures the optical density of the ethyl acetate extract produced from the fermented tea sample to determine the color intensity. This ties the TF concentration measurement to the fermentation process. As a result, the maximum concentration of TF content in tea is used to determine the ideal fermentation duration. During sorting process, the tea granules are separated from each other using some different sized severs. Then they are categorized using the visual approximation. The tea tasters finally create their own convention for characterizing the tea liquor and infused leaf in order to judge quality. Backey, Body, Bright, Brisk, Burnt, Color, Cream, Dry, Dull, Full, Pungent, Strength/Strong, Thin, Coppery, Green, and Even are some of the main words used by tea tasters to describe the quality of the tea liquid. These adjectives are positively correlated with the accuracy of several tea manufacturing processing phases. In addition to this, the TF-TR analysis, a spectro photometric technique, is employed to assess tea quality.

1.6. Machine Vision in Food and Agriculture

Color, texture, flavor etc. are the important process parameters for the subjective determination of the overall product quality of food processing industries and agricultural

products (Giese, 2000). In a broader context, color perception can be used to estimate degree of ripeness (e.g., tomatoes, bananas), extent of cooking (meat, cereal products), and even anticipated flavor strength (tea, fruit juices) etc. (Wright, 1969; Nielsen, 1998; Hutchings, 1999). In the case of processed food like sugar, juice, jam, jelly, chips, chocolates, tea etc. these color, size and shape are important process parameters determining the overall quality of the products. These facts reveal the importance of analysis and classification of color and flavor of food products as the major techniques of quality determination. Similarly, texture analysis and classification can be implemented in sorting processed food into different grades in terms of size, shape etc. (Brosnan, et. al., 2002; Quevedo, et. al., 2002). Therefore, these process parameters can

be implemented as the quality control and automation tool in food processing industries, which can be achieved by continuous process monitoring. It has been elsewhere reported that the quality monitoring has been performed by manual methods or by using different physiochemical methods such as GC (Dewulf, et. al., 2002); colorimeter (Melendez, et. al., 2003) etc. in different food processing industries. More recently, the developments of machine vision techniques supported by for example the ISE techniques, such as artificial neural networks (ANN), fuzzy logic etc., in food processing industries, have grown extensively. Novel system prototypes, such as computer vision (He, et. al., 1998; Sun, 2000), electronic sensors (For example: EN (Bartlett, et. al., 1997; Huang, et. al., 2001)) for food quality measurement, analysis, and prediction have come into the limelight. The Electronic Tongue (ET) application is also being broadly implemented in flavor profile analysis mainly in milk, fruit juice and wine processing industries (Natale, et. al., 2000; Bleibaum, et. al., 2002; Buratti, et. al., 2004)

etc. These advanced process control techniques have proven to be efficient for their various advantages over the conventional quality evaluation methods. Objectives of using of such systems can be summarized as the improvement of food quality, saving energy and increasing productivity etc. Moreover, such advanced techniques explore the possibilities of automation in food quality evaluation, monitoring and process control by analysis, classification of complex parameters. In fact, such advanced techniques and their application to the foods processing industries as a quality-monitoring tool have become a very active field of research. The goal has been, and still is, to have a machine performing the same functions that a human brain does but with higher accuracy; speed for accurate judgment of process parameters; and efficient repeatability.

1.7. Machine Vision in Tea

It has been observed that the various attributes such as color, aroma, taste, astringent etc., both natural and controlled, are the important quality parameters in production of black tea. Some of the important constituents in black tea are given in Table. 1 [Akuli, et.al. 2016], along with their color contribution.

Table 1. Chemical constituents contributing towards color formation in tea

Compounds	Colors
Pheophytin	Blackish
Flavanol glycosides	Light yellow
Thearubigins	Reddish brown
Theaflavin	Yellowish brown
Phephorbide	Brownish

The human organoleptic methods, which are used for the measurement these parameters, are not standardized and vary (refer to section 1.1) due to various reasons. Therefore, some possibilities are being proposed to assess some of the attributes like color and aroma of tea with the aid of instruments (physiochemical methods) (Hazarika, et. al., 1984; Takeo, et. al., 1983; Liang, et. al., 2003). The use of colorimeter in fermenting tea color is one of such physiochemical method (Ullah, et. al., 1979). In terms of the perception of tea quality, such initiatives are fairly successful and based on accepted methods, but they take time, effort, and money. Machine vision techniques can be considered an effective alternative technique to assist the traditional procedures employed by the human sensory panel for tea quality assessment in order to address such inaccuracy and inefficiency. The rapid growth of electronic systems, such as the development of charge coupled device (CCD), EN, ET, etc. and the possibilities in using them in repeated manner enrich such prospects. Moreover, the implication of ISE based techniques, which can learn from and adapt to the input data, to analyze the data gathered by this system enhances the prospect in more transparent way. This study is with regards mainly to color matching of fermenting tea images with standard database images. One of the key phenomena of this research is the efficient color feature extraction from the images. For that the selection of the actual color model is also important aspect. Color features are extracted by dissimilarity measurements of the color histograms generated from the images. The HSI (hue, saturation and intensity) color model is considered as one of the most useful models for such color feature extraction. The color indexing, a technique developed by Swain and Ballard (Swain, et. al., 1991), and then modified for object recognition on the basis of color content (Funt, et. al., 1995) is considered in this research. The enhanced version of the technique has been faithfully implemented for the color comparisons and analysis for fermenting tea images. Such approach is needed for optimum fermentation time judgment in tea industries.

1.8. Machine Vision in tea Granule Size Estimation

The tea granule size estimation has been addressed by texture analysis of the images for image surface roughness. As the images consist of stochastic natural textures, efficient texture feature extraction methods also play an important role in this research. Wavelets transform (WT) based sub-band images are considered for the statistical texture feature extraction from the images. The Daubechies' wavelets are implemented using the fast wavelet transform (FWT), which has been shown to be successful technique in the texture analysis of natural images (Wang, et. al., 1997). Such a tea granule size estimation technique would be useful as a monitoring tool for effective tea grading.

1.9. Development of Color

However, fermentation is an enzymatic oxidation process of catechins found in Tea leaves that have been bent and damaged into two groups of colorful Theaflavins (TF) and Therubigins are chemical molecules (TR). Golden yellow, the TF is The color of TR is reddish-brown. The combination of these two chemicals gives the tea beverage its color preferences and personal preferences (Roberts, 1962; Wickremasinghe, 1978; Yamanishi, 1981). Additional chemical modifications occur in addition to the synthesis of TF and TR. during the fermentation process in the leaf tissues. Examples include the degradation of proteins, the Pheophytins are created from chlorophyll, and certain volatile chemicals are also produced. due to changes in several fragrance precursors found in tea leaves.

Additionally, these modifications affect the flavor and color of prepared tea (Mahanta, et. al., 1985; Hazarika, et al., 1983; Takeo, et. al., 1987). When determining quality, these ingredients have a significant impact on the tea liquor's brightness, briskness, strength, body, and color. Based on the season, growth, and other factors, the color provides a gauge of the tea's depth.. Brightness, briskness, and strength are the three qualities that stand out among these; while body and color are connected to TR contents, these are determined by TF contents. The market value of black tea and the TF content are positively correlated. The concentration of TF rises as fermentation progresses, peaks, and then begins to fall if the fermentation is continued. In this step, TF is downgraded to TR. On the other hand, when the fermentation time increases and the liquor body thickens, the concentration of TR continues to rise. The overly fermented tea has "body," but it lacks other qualities that make a good cup of tea enjoyable. In general, the production of tea aims to have the maximum TF content feasible.

The TF and TR contents rise as the procedure goes on, and the tea's color gradually changes as well. The color of the processed tea turns coppery red at its best when the ratio of TF to TR approaches 1:10 or 1:9. If the process keeps going, TF eventually degrades to TR, the color deepens, and the Quality of the tea decreases. Thus, it can be seen that the molecules giving liquor its color and flavor gradually alter, and after reaching their peak, they drastically decline. Due to this constantly altering lipid production event, it might be challenging to determine the quality of the tea during fermentation using chemical analysis of the tea. As a result, in a controlled environment, tea fermentation requires exact subjective monitoring and prompt process management after the desired quality is reached. In such circumstances, the industries typically need to apply tests for fermentation process optimization. Therefore, for the sake of quality in the tea processing business, reliable detection of the process' ideal state is the most crucial requirement of the fermentation process. After ensuring the best possible fermentation, the tea leaves are delivered to the drier to remove any remaining moisture and halt fermentation. These electronic sensing platforms measure the aroma, the taste, and color profiles, respectively.

1.10. Human Sensory Panel Judgment Based on Color of Tea

This visual assessment is used to consistently identify the deep coppery red color achieved during fermentation. If precise color recognition is achievable, this method is thought to offer a trustworthy and accurate subjective judgement of quality. However, it has been noted that due to variations in the sensory panel's eye approximation and other circumstances, it is not always possible to match the color of tea exactly at the optimal fermentation setting. Individual variability, adaptation (becoming less sensitive due to prolonged exposure), weariness, infection, mental state, and other factors may all be contributing factors in these variances. Also, now-a-days, the human sensory panel has been partially replaced by 'Electronic Tongue' and 'Electronic nose'.

1.11. Use of Colorimeter

Chemical tests are carried out in various tea facilities like Tocklai Tea Research Instirue, Jorhat, Assam, etc. to determine the proper fermentation of Orthodox and CTC tea. This method makes use of an instrument named as 'colorimeter'. Since the completion of the test is verified by measuring the optical density of the ethyl acetate extract made from the fermented tea sample, this method is purely subjective. This links the measurement of theaflavin (TF) concentration during tea leaf fermentation and determines the best period for fermentation based on the highest TF concentration. The steps in this experiment include collecting an ethyl acetate extract of the fermenting tea sample in a test tube and comparing the color intensity by looking through the liquid's column against a white background. Then, using water as a blank, a colorimeter is used to measure optical density, or the intensity of color, at 460 nm. When the crucial period of fermentation is approaching, this test is performed every 10 minutes throughout fermentation and every 5 minutes after that. The greatest color intensity of the ethyl acetate extract at a specific moment indicates optimal fermentation for CTC production. On the other hand, in the case of Orthodox manufacturing, the procedure is continued for an additional 10 minutes after the ethyl acetate extract's color has reached its maximum depth (Ullah, et al., 1979). Because the chemical tests take a long time to complete, this procedure is difficult and the results are not always reliable. Additionally, the concerned person must have a lot of patience and focus for a satisfactory outcome.

1.12. Machine Vision for Color Classification

Computer vision based recognition of color in image of food products has been reported as an efficient method of quality inspection (Gunasekaran, 1996; Brosnan, et. al., 2002). Color provides an important feature depending on which an image can be analyzed and classified. Since visual color inspection is affected by wide variety of factors such as lighting condition, angle of observation and individual variability of color perception etc. machine vision color analysis method provides a subjective and consistent method of color analysis. In this research image of tea under fermentation process is analyzed in terms of color variation. In other words, machine vision technique has been carried out to evaluate the optimum fermenting condition by detecting the color in a non-intrusive way as an alternative subjective method over the human organoleptic methods.

1.13. Image Color Processing

The purpose of the color image capture of fermenting tea during ongoing fermentation process is to detect the images with particular color. This section provides a idea of color processing in the images, which is the necessary background of computer vision work in the color analysis purpose. It is elsewhere mentioned that the main uses of color in image processing and computer vision (Gonzalez, et. al., 1992; Castleman, 1996) are usually for image enhancement and image analysis respectively. Therefore, one of the key aspects of computer vision for such color analysis of the images is to get image data into the right color model for the specific application. For assessing the color of objects in an image, models that separate light's chromatic qualities from its intensity are typically more helpful. Basically, any of the approaches can be used to process color photos, but caution is advised because each color pixel contains at least three pieces of information: red ®, green (G), and blue (B). (Funt, 1995). In the YIQ model, the I and Q sections remain the same while the Y part changes in response to variations in lighting across an orange patch. Therefore, selection of a suitable color model for the specific purpose of computer vision is important. This section discusses some of the relevant color models relevant to the research purpose and explores their advantages in such color based system. A color model is a mathematical representation of a set of color. Several color models are available in image processing for various purposes. A color model is used to make it easier to specify color in a uniform, widely recognized manner. RGB model, CMY model, CIE L*a*b color space, HSI model are some common examples of such color model used in machine vision or computer vision.

1.13.1. RGB color model

In actual systems, ultraviolet and infrared light represent the electromagnetic spectrum's lower and higher limits, respectively, and visible light is found there between 380 and 780 nm. White light strikes an object, reflects back, and the color of the thing is determined by this reflected light. The amount of light, the type of light source, and the observer's viewpoint all affect how this color appears. For instance, when considering a black and white image, the image can be quantified into a single area made up of a number of pixels, with the ability to assign an integer value to each pixel. The brightness of the image at the specified point is represented by the value of the integers. If an eight-bit register is used to hold values, the integer can have a maximum value of 255 and a minimum value of 0, or a total of 256 different values. White and black are represented by the values 255 and 0, respectively. Red ®, green (G), and blue (B) are the three additive primaries that make up the RGB color space, as was already explained (B). The final color is created by the additive combination of the spectral elements of various hues. The RGB model is represented by a three-dimensional cube, with the colors red, green, and blue located at the corners of each axis. Black is the starting point. On the other side of the cube is white. The diagonal of the cube, which runs from black to white, is followed by the grey scale. Red is represented by 8 bits in a 24-bit color graphics system (255, 0, 0). The normalized values of R, G, and B to [0,1] are shown as (1, 0, 0) on the color cube. To explain color, a variety of color models or schemes are employed. For instance, HIS (hue, saturation, and intensity), CMY (cyan, yellow, and magenta), and other color models are frequently used for color picture processing. All color models are made up of these three-color planes, R, G, and B; some of them will be briefly discussed in the final section of this chapter.

1.13.2. CIE color model

The international authority on light, illumination, color, and color spaces is the International Commission on Illumination, also known as the CIE after its French names "Commission Internationale de l'Eclairage" (CIE publication, 1987). It has established a system for categorizing color in accordance with the human visual system. A color space is a geometric representation of colors in space, typically in three dimensions, according to CIE. The CIE created a system for describing color spaces that has had a significant impact and is mostly utilized in the food sector for color measurement. Device independence is a feature of the CIE spaces, which include CIE XYZ, CIE L*a*b*, and CIE L*u*v*. A color is identified by two co-ordinates, x, and y, in this color space. L* is based on a perceptual measure of brightness, and others two (a*, b* or u*, v*) are chromatic coordinates. The palette of colors available in these color spaces is not constrained by the rendering power of a given device or the eye-hand coordination of a particular observer. Additionally, in this color space, color variations in any direction are roughly equivalent. Therefore, the relative distance between two colors can be calculated using a distance measurement such as the Euclidean distance. As close as any color space is expected to sensibly approach to being linear with visual perception is the CIE L*a*b and CIE L*u*v. Both the L*a*b* and L*u*v* color spaces were recommended by the CIE in 1976 as perceptually approximate uniform color spaces (McLaren, 1976; Agoston, 1979). Since the illuminant, lighting, and viewing geometry will all affect the measurements, the three tri-stimulus values X, Y, and Z each separately indicate a color.

1.13.3. HSI color model

Both the RGB and CIE color models are not independent of lighting. However, they provide a helpful color model for computer vision under constant lighting conditions. On the other hand, the HIS color model is made to be used more naturally when modifying color and to simulate how people see and recognize color. To distinguish between the three color components, hue (H), saturation (S), and intensity (I) are defined. This color model's H and S components are unaffected by surface orientation, viewing angle, illumination source, and intensity. The letter "H" stands in for the color's real wavelength (pure color), which is represented by the color's name, such as red, green, etc. The letter "S" represents a measurement of color purity. This illustrates how the pure color is altered to become white light. The letter "I" stands for how pale the color is.

1.14. Data Processing

This section describes the methods that are adopted for the process of the data in all the three aspects, namely color of image (fermenting tea image), texture of image (image of tea grades) and EN signals (aroma of tea grades). The data processing is being carried out in three serial steps.

- Data visualization
- Data clustering
- Data classification

1.14.1. Data Visualization

The principal component analysis (PCA) is a useful unsupervised pattern analysis technique, which is widely used for such pattern visualization (D. Das, et al.). A mathematical process is used in the method to reduce the number of correlated variables to a greater number of uncorrelated variables. The primary components of each data vector in the data set are these uncorrelated variables. Most of the variability in the data is accounted for by the first main component. The remaining data variability is then accounted for to the same extent by each following component. Three principal components have been used in the case of data visualization of the data in this research.

1.14.2. Data Clustering

Data clustering is a common technique for data processing and analysis. It consists of partitioning a dataset into subsets (clusters), so that the data in each cluster share some common properties (often similarity). Two different algorithms, namely k-mean and SOM based data clustering are mainly used for data clustering. Both the technique obeys the unsupervised clustering technique algorithm.

1.14.3. K-mean Clustering

By iteratively adding and changing cluster centers, this algorithm provides a method for estimating the number of heterogeneous clusters existing in a given dataset. It divides the input set of N observations (x_i) in 'd'-dimensional spaces into k clusters with centroids (y_1–y_k), where k is specified by the user [Bezdek, 1981]. The partition that minimizes a cost function Is taken into account: w_i is the scalar weight assigned to each observation in this case. In other words, the centroid nearest to each observation is chosen to minimize the sum of the squared Euclidean distances that is weighted inside the cluster. In contrast to other clustering algorithms that create hierarchical discrimination, K means produces a flat or non-hierarchical grouping.

1.14.4. SOM Data Clustering

Unsupervised neural networks (ISE method) can map high-dimensional input data onto a lower-dimensional output space using the self-organizing map (SOM) (Kohonen, 1982). In data mining, the SOM has shown to be an invaluable tool. It is also successfully used in a variety of technical applications, including process monitoring, image analysis, and pattern recognition. Mapping the data patterns onto an N-dimensional grid of neurons or units is the fundamental concept behind a SOM. SOM is a network made up of N neurons that are arranged as the nodes of a planner grid, with each neuron having four close neighbors. In response to the input space, where the data patterns were present, the grid creates what is referred to as the output space. In order to preserve topological relationships, this mapping maps near patterns in the input space to close units in the output space and vice versa. The output space is typically 2 dimensional to facilitate easy viewing (2-D). The SOM relies on unsupervised learning and uses the competitive learning approach.

1.14.5. Data Classification

A system that can be considered to learn from and adapt to incoming stimuli is known as an intelligent system engineering (ISE), even though there is no formal definition for the specific problem to be handled [Bishop, 1995]. They can specifically utilize their knowledge from prior experiences to tackle the issues. Artificial Neural Networks (ANNs), Fuzzy Logic (FL), Genetic Algorithms (Gas), and hybrid techniques like the Neural-Fuzzy System (NFS), among others, are only a few of the diverse ISE that are covered here. Some of the fundamental tenets of ANNs are discussed in this section. The constructive probabilistic neural network (CPNN), multilayer perceptron (MLP), and radial basis function network architectures are described in particular. A network of artificial neurons that are coupled together makes up an ANN. These neurons typically receive input data, which they then transform mathematically as a group, producing output data as a result of these calculations. McCulloch and Pitts initially proposed the artificial neuron in 1943. (McCulloch, et. Al., 1943).

1.14.6. Multi-layer Perceptron Network

The multi-layer perceptron (MLP) network is a widely reported and used neural network in a number of practical problems. It consists of an input layer of neurons, one or more hidden layers of neurons, and an output layers. The layers of neurons are inter-connected in such a way that the outputs of one layer are propagated to the subsequent layer. In the MLP, each neuron performs a weighted sum of inputs and transforms it using a non-linear activation function (For example, the most commonly used non-linear function is sigmoid function). The MLP uses the supervised training phase as it is presented with training vectors together with the associated targets. A MLP network learns from the input data by adjusting the weights in the network using some specific techniques. Many different training algorithms exist that can be applied to the MLP, but the most commonly used algorithm is error back-propagation (Rumelhart, et. al., 1986). The purpose of this algorithm is to minimize the difference between the generated network output and the desired output, termed as error.

1.14.7. Radial Basis Function Network

Similar to the MLP, the radial basis function (Moody, et. al., 1989) network also consists of artificial neurons and number of layers is three. It has a similar architecture as like MLP, exhibiting fully inter-connected layers. The only difference with three layers MLP is that the hidden layer employs a different type of neuron, called the Radial Basis (RB) neurons.

Although radial basis function network does not provide the error estimates, it has an intrinsic ability to indicate when it is extrapolating. This is because the activation function of the receptive fields is directly related to the proximity of the last pattern to the training data. One advantage of the radial basis function network over MLP, with error back propagation, is that it can be trained with relatively lower computational overheads than would be required by the later method.

1.14.8. Constructive Probabilistic Neural Network

The constructive probabilistic neural network (CPNN) (Berthold, et. al., 1998) is essentially a PNN (Probabilistic Neural Network) that is grown by sequential addition of the neurons in the hidden and output layers. The neurons are added in response to patterns presented in the training dataset. Prior to adding neurons, an assessment is firstly made as to whether existing neurons can perform the same function. If they can, then they are adjusted to encompass the new training pattern, otherwise a new neuron is added. This method of constructing CPNN often results in a highly compact and efficient architectures requiring low computational overheads. Barthold, et. al., found that CPNN algorithm can offer substantial advantages in terms of network size and generalization capability. So, in comparison, whilst the MLP demands higher computational overhead, the CPNN is inherently computationally lightweight ANN.

1.15. CONCLUSION

The dominant parameters that are considered for decision making about completion of fermentation process are 'color' and 'aroma' of fermenting tea. This chapter discusses about the method of computer vision that adopted to analyze the color of fermenting tea. The methods recognize the required color by the using color constraints of the captured images during ongoing fermentation process. It was mentioned that though the color formation in fermenting tea during fermentation process is not the only quality determining factors, it is one of the most desired phenomenon. Because, this phenomenon (color of fermenting tea becomes deep coppery red) is useful for judging the completion of fermentation process in tea industries. Such importance of color attribute can be observed in some other food processing and allied industries such as sugar, paper etc. also, where color is one of their important parameters for maintaining product quality. Various color analysis techniques are adopted in such industries as a quality monitoring tool. Some traditional methods include the optical absorption and optical reflection methods. These electronic sensing platforms measure the aroma, the taste, and color profiles, respectively. The data obtained from the corresponding platforms are fed as input to the mathematical classification algorithms. To evaluate the quality of food products, nowadays, precise and non-contact color measuring equipment is required and hence the machine vision technique comes into picture for standardization of tea quality evaluation.

REFERENCES

Agoston, G. A. (1979). *Color Theory and Its Application in Art and Design.* DOI: 10.1007/978-3-662-15801-2

Akuli, 2016Akuli, A., Pal, A., Bej, G., Dey, T., Ghosh, A., Tudu, B., Bhattacharyya, N., & Bandyopadhyay, R. (2016). A Machine Vision System For Estimation Of Theaflavins And Thearubigins In Orthodox Black Tea. *International Journal on Smart Sensing and Intelligent Systems,* 9(2), 709–731. DOI: 10.21307/ijssis-2017-891

Bartlett, et al., 1997Bartlett, P. N., Elliot, J. M., & Gardner, J. W. (1997). Electronic noses and their application in the food Industry. *Food Technology,* 51, 44–48.

Berthold, et al., 1998Berthold, M. J., & Diamond, J. (1998). Contructive training of probabilistic neural networks. *Neurocomputing,* 19(1-3), 167–183. DOI: 10.1016/S0925-2312(97)00063-5

Bezdek, J. C. (1981). *Pattern Recognition with Fuzzy Objective Function Algorithms.* Plenum Press. DOI: 10.1007/978-1-4757-0450-1

Bishop, C. M. (1995). *Neural Network for pattern recognition.* Oxford University Press. DOI: 10.1093/oso/9780198538493.001.0001

Bleibaum, et al., 2002Bleibaum, R. N., Stone, H., Tan, T., Labreche, S., Saint-Martin, E., & Isz, S. (2002). Comparison of sensory and consumer results with electronic nose and tongue sensors for apple juices. *Food Quality and Preference,* 13(6), 409–422. DOI: 10.1016/S0950-3293(02)00017-4

Borah, et al., 2003 Borah, S., & Bhuyan, M. 2003, "Quality indexing by machine vision during fermentation in black tea manufacturing", Sixth International Conference on Quality Control by Artificial Vision, Fabrice Meriaudeau Gatlinburg, United States, SPIE, Volume 5132, PP. 468-475 (2003)

Borah, et al, 2005Borah, S., & Bhuyan, M. (2005). A computer based system for matching colours during the monitoring of tea fermentation. *International Journal of Food Science & Technology,* 40(6), 675–682. DOI: 10.1111/j.1365-2621.2005.00981.x

Brosnan, et al., 2002Brosnan, T., & Sun, D. (2002). Inspection and grading of agricultural and food productsby computer vision systems – a review. *Computers and Electronics in Agriculture,* 36(2-3), 193–313. DOI: 10.1016/S0168-1699(02)00101-1

Buratti, et al., 2004Buratti, S., Benedetti, S., Scampicchio, M., & Pangerod, E. C. (2004). Characterization and classification of Italian Barbera wines by using an electronic nose and an amperometric electronic tongue. *Analytica Chimica Acta*, 525(1), 133–139. DOI: 10.1016/j.aca.2004.07.062

Castleman, K. R. (1996). *Digital Image Processing*. Prentice Hall.

D. Das, et al.Das, D., Nandy Chatterjee, T., Banerjee Roy, R., Tudu, B., Hazarika, A. K., Sabhapondit, S., & Bandyopadhyay, R. (2020). Titanium oxide nanocubes embedded molecularly imprinted polymer-based electrode for selective detection of caffeine in green tea. *IEEE Sensors Journal*, 20(12), 6240–6247. DOI: 10.1109/JSEN.2020.2972773

Dewulf, et al., 2002Dewulf, J., & Langenhove, H. V. (2002). Analysis of volatile organic compounds using gas chromatography. *Trends in Analytical Chemistry*, 21(9-10), 637–646. DOI: 10.1016/S0165-9936(02)00804-X

[Funt, 1995] Funt., Funt, B. V., Finlayson, G. D., 1995. Color Constant Color Indexing, IEEE Transaction on PAMI, Vol. 17, No. 5, pages 522-529.

Giese, 2000Giese, J. (2000). Colour measurement in foods as a quality parameter. *Food Technology*, 54(2), 62–65.

Gonzalez, R. C., & Woods, R. E. (1992). *Digital Image Processing, Addison-Wesley Publishing Company; Graves, M., Batchelor, B., Machine Vision for the Inspection of Natural Products*. Springer London., DOI: 10.1007/b97526

Gunasekaran, 1996Gunasekaran, S. (1996). Computer vision technology for food quality assurance. *Trends in Food Science & Technology*, 7(8), 245–256. DOI: 10.1016/0924-2244(96)10028-5

Hazarika, et al., 1984Hazarika, M., Mahanta, P. K., & Takeo, T. (1984). Studies on some volatile flavor constituents in orthodox black tea of various clones and flushes in N. E. India. *Journal of the Science of Food and Agriculture*, 35(11), 1201–1207. DOI: 10.1002/jsfa.2740351111

[He, et. al., 1998] He, D. J., Yang, Q., Xue, S. P., Geng, N., 1998. Computer vision for colour sorting of fresh fruits, Trans. Scinese society of Agricultural Engineering, Vol. 14, No. 3, pages 202-205.

Huang, Y., Whittaker, A. D., & Lacey, R. E. (2001). *Automation for food engineering: Food Quality Quantization and Process Control*. Culinary and Hospitality Industry. DOI: 10.1201/9781420039023

Hutchings, J. B. (1999). *Food colour & appearance* (2nd ed.). DOI: 10.1007/978-1-4615-2373-4

[Kamisoyama, et. al., 2012] Kamisoyama, Y., and Tadahiro Kitahashi, "Image feature extraction from tea leaves for measuring the degree of the steaming", Sept.2012 IEEE DOI , pages 295-296DOI: 10.1109/IIAI-AAI.2012.vol.83

Kohonen, 1982Kohonen, T. (1982). Self-organised formation of topologically correct feature maps. *Biological Cybernetics*, 43(1), 59–69. DOI: 10.1007/BF00337288

Liang, et al., 2003Liang, Y., Lu, J., Zhang, L., Wu, S., & Wu, Y. (2003). Estimation of black tea quality by analysis of chemical composition and colour difference of tea infusions. *Food Chemistry*, 80(2), 283–290. DOI: 10.1016/S0308-8146(02)00415-6

Mahanta, 1988Mahanta, P. K. [Mahanta, 1988]Mahanta, P. K., 1988. Biochemical basis of colour and flavour of black tea, *Proc. of 30th Tocklai Conference*, Assam, India, pages 124-134.

Mahanta, et al., 1985Mahanta, P. K., & Hazarika, M. (1985). Chlorophyll and degradation products in orthodox and CTC black teas and their influence on shade of colour and sensory quality in relation to thearubigins. *Journal of the Science of Food and Agriculture*, 36(11), 1133–1139. DOI: 10.1002/jsfa.2740361117

[Mark, et. al., 2014]. Mark, D. F., 1998. Color Appearance Models, Addison-Wesley, Reading, MA

McCulloch, et Al., 1943 McCulloch, W. S., & Pitts, W. 1943. A logical calculus of ideas immanent in nervous activity, Bulletin of Mathematical Biophysics, Vol. 5, pages.115-133.

McLaren, 1976McLaren, K. (1976). The development of the CIE 1976 (L*a*b*) uniform colour-spaceand colour-difference formula. *Journal of the Society of Dyers and Colourists*, 92(9), 338–341. DOI: 10.1111/j.1478-4408.1976.tb03301.x

Melendez, et al., 2003Melendez-Martinez, A. J., Vicario, I. M., & Heredia, F. J. (2003). Application of tristimulus colorimetry to estimate the carotenoids content in ultrafrozen orange juices. *Journal of Agricultural and Food Chemistry*, 51(25), 7266–7270. DOI: 10.1021/jf034873z PMID: 14640568

Moody, et al., 1989 Moody, J., & Darken, C. 1989. Fast learning in networks of locally-tuned processing units, Neural Computing, Vol. 1, No. 2, pages.281-294.

Natale, et al., 2000Natale, C. D., Paolesse, R., Macagnano, A., & Mantini, A. (2000). Electronic nose and electronic tongue integration for improved classification of clinical and food samples. *Sensors and Actuators. B, Chemical*, 64(1-3), 15–21. DOI: 10.1016/S0925-4005(99)00477-3

Nielsen, S. S. (1998). *Food Analysis* (2nd ed.). Kluwer Academic/Plenum Publishers.

Okinda, et al., 1998Okinda, P., & Obanda, M. (1998). The changes in black tea quality due to variations of plucking standard and fermentation time. *Food Chemistry*, 61(4), 435–441. DOI: 10.1016/S0308-8146(97)00092-7

[Patil, et.al., 2021]. Patil, A.B.; Bachute, M.; Kotecha, K. Artificial Perception of the Beverages: An in Depth Review of the Tea Sample. IEEE Access 2021, 7, pages 82761–82785.

Publications Services, Texas

Quevedo, et al., 2002Quevedo, R., Carlos, L. G., Aguilera, J. M., & Cadoche, L. (2002). Description of food surfaces and microstructural changes using fractal image texture analysis. *Journal of Food Engineering*, 53(4), 361–371. DOI: 10.1016/S0260-8774(01)00177-7

Roberts, E. A. H. (1962). *Tea fermentation, Economic Importance of Flavonoid Substance*. Pergamon Press.

Rumelhart, et al., 1986Rumelhart, D. E., Hinton, G. E., & Williams, R. J. (1986). Learning representations by backpropagating errors. *Nature*, 323(6088), 533–536. DOI: 10.1038/323533a0

[Sharma, et.al., 2013] Sharma, M., Edathiparambil Vareed Thomas, "Electronic vision study of tea grains color during Infusion", International Journal of Engineering Science Invention ISSN Vol. 2 Issue 4th April. 2013, pages.52-58

Sun, 2000Sun, D. W. (2000). Inspecting pizza topping percentage and distribution by a computer visison method. *Journal of Food Engineering*, 44(4), 245–249. DOI: 10.1016/S0260-8774(00)00024-8

Swain, et al., 1991Swain, M. J., & Ballard, D. H. (1991). Color Indexing. *International Journal of Computer Vision*, 7(1), 11–32. DOI: 10.1007/BF00130487

Takeo, et al., 1983Takeo, T., & Mahanta, P. K. (1983). Comparison of black tea aromas of orthodox and CTC tea and black teas made from different varieties. *Journal of the Science of Food and Agriculture*, 34(3), 307–310. DOI: 10.1002/jsfa.2740340315

Ullah, M. R., Bajaj, K. L., & Chakraborty, S. (1979). *Fermentation test by Dr. M R Ullah, Factory Floor Fermentation Test*. Tocklai Experimental Station.

Wang, et al., 1997 Wang, J. [Wang, et al., 1997] Wang, J. Ze., Wiederhold, G., Firschein, O., 1997. Wavelet based image indexing techniques with partial sketch retrieval capability, In Proc. 4th Forum on Research and Technology Advances in Digital Libraries, pages 13-24.

Wright, W. D. (1969). *The measurement of colour* (4th ed.). Hilger & Watts.

Xu, et al., 2018 Xu, M., & Wang, J. [Xu, et al., 2018] Xu, M.; Wang, J. The Qualitative and Quantitative Assessment of Tea Quality Based on E-nose, E-tongue and E-eye Signals Combining With Chemometrics Methods. In Proceedings of the 2018 ASABE Annual International Meeting, Detroit, MI, USA, 29 July–1 August 2018; pages 1800610.

Xu, et al., 2019Xu, M., Wang, J., & Zhu, L. (2019). The Qualitative and Quantitative Assessment of Tea Quality Based on E-nose, E-tongue and E-eye Combined with Chemometrics. *Food Chemistry*, 289, 482–489. DOI: 10.1016/j.foodchem.2019.03.080 PMID: 30955639

Zhu, et al., 2019Zhu, H., Liu, F., Ye, Y., Chen, L., Liu, J., Gui, A., Zhang, J., & Dong, C. (2019). Jingyuan Liu, Anhui Gui, Jianqiang Zhang, Chunwang Dong Application of machine learning algorithms in quality assurance of fermentation process of black tea– based on electrical properties. *Journal of Food Engineering*, 263, 165–172. DOI: 10.1016/j.jfoodeng.2019.06.009

Compilation of References

Abass, T., Itua, E. O., Bature, T., & Eruaga, M. A. (2024). Concept paper: Innovative approaches to food quality control: AI and machine learning for predictive analysis. *World Journal of Advanced Research and Reviews*, 21(3), 823–828. DOI: 10.30574/wjarr.2024.21.3.0719

Abdellah, A. M., & Ishag, K. E. A. (2012). Effect of storage packaging on sunflower oil oxidative stability. *American Journal of Food Technology*, 7(11), 700–707. DOI: 10.3923/ajft.2012.700.707

Abdulmalek, S., Nasir, A., Jabbar, W. A., Almuhaya, M. A. M., Bairagi, A. K., Khan, M. A.-M., & Kee, S.-H. (2022, October 11). IoT-Based Healthcare-Monitoring System towards Improving Quality of Life: A Review. *Healthcare (Basel)*, 10(10), 1993. DOI: 10.3390/healthcare10101993 PMID: 36292441

Abid, H. M. R., Khan, N., Hussain, A., Anis, Z. B., Nadeem, M., & Khalid, N. (2024). Quantitative and qualitative approach for accessing and predicting food safety using various web-based tools. *Food Control*, 162, 110471. DOI: 10.1016/j.foodcont.2024.110471

Abimbola, O. F., & Okpara, M. O. (2023). Artificial Intelligence in the Food Packaging Industry. In *Sensing and Artificial Intelligence Solutions for Food Manufacturing* (pp. 165–171). CRC Press. DOI: 10.1201/9781003207955-12

Adnan, M., Oh, K. K., Cho, D. H., & Alle, M. (2021). Nutritional, pharmaceutical, and industrial potential of forest-based plant gum. Non-Timber Forest Products: Food, Healthcare and Industrial Applications, 105-128.

Agbo, F. J., Oyelere, S. S., Suhonen, J., & Tukiainen, M. (2021). Scientific production and thematic breakthroughs in smart learning environments: A bibliometric analysis. *Smart Learning Environments*, 8(1), 1–25. DOI: 10.1186/s40561-020-00145-4

Agoston, G. A. (1979). *Color Theory and Its Application in Art and Design*. DOI: 10.1007/978-3-662-15801-2

Agrawal, A. V., Shashibhushan, G., Pradeep, S., Padhi, S., Sugumar, D., & Boopathi, S. (2023). Synergizing Artificial Intelligence, 5G, and Cloud Computing for Efficient Energy Conversion Using Agricultural Waste. In *Sustainable Science and Intelligent Technologies for Societal Development* (pp. 475–497). IGI Global.

Agrawal, A. V., Bakkiyaraj, M., Das, S., Reddy, C. M. S., Kiran, P. B. N., & Boopathi, S. (2024). Digital Strategies for Modern Workplaces and Business Through Artificial Intelligence Techniques. In *Multidisciplinary Applications of Extended Reality for Human Experience* (pp. 231–258). IGI Global., DOI: 10.4018/979-8-3693-2432-5.ch011

Agrawal, A. V., Magulur, L. P., Priya, S. G., Kaur, A., Singh, G., & Boopathi, S. (2023). Smart Precision Agriculture Using IoT and WSN. In *Handbook of Research on Data Science and Cybersecurity Innovations in Industry 4.0 Technologies* (pp. 524–541). IGI Global. DOI: 10.4018/978-1-6684-8145-5.ch026

Akuli, 2016Akuli, A., Pal, A., Bej, G., Dey, T., Ghosh, A., Tudu, B., Bhattacharyya, N., & Bandyopadhyay, R. (2016). A Machine Vision System For Estimation Of Theaflavins And Thearubigins In Orthodox Black Tea. *International Journal on Smart Sensing and Intelligent Systems*, 9(2), 709–731. DOI: 10.21307/ijssis-2017-891

Alam, A. U., Rathi, P., Beshai, H., Sarabha, G. K., & Deen, M. J. (2021). Fruit quality monitoring with smart packaging. *Sensors (Basel)*, 21(4), 1509. DOI: 10.3390/s21041509 PMID: 33671571

Alam, F., Ashfaq Ahmed, M., Jalal, A., Siddiquee, I., Adury, R., Hossain, G., & Pala, N. (2024, March 30). Recent Progress and Challenges of Implantable Biodegradable Biosensors. *Micromachines*, 15(4), 475. DOI: 10.3390/mi15040475 PMID: 38675286

Albahri, A., Khaleel, Y. L., Habeeb, M. A., Ismael, R. D., Hameed, Q. A., Deveci, M., Homod, R. Z., Albahri, O., Alamoodi, A., & Alzubaidi, L. (2024). A systematic review of trustworthy artificial intelligence applications in natural disasters. *Computers & Electrical Engineering*, 118, 109409. DOI: 10.1016/j.compeleceng.2024.109409

Aliac, C. J. G., & Maravillas, E. (2018). IOT Hydroponics Management System. *2018 IEEE 10th International Conference on Humanoid, Nanotechnology, Information Technology,Communication and Control, Environment and Management (HNICEM)*, 1–5. DOI: 10.1109/HNICEM.2018.8666372

Ali, M. N., Senthil, T., Ilakkiya, T., Hasan, D. S., Ganapathy, N. B. S., & Boopathi, S. (2024). IoT's Role in Smart Manufacturing Transformation for Enhanced Household Product Quality. In *Advanced Applications in Osmotic Computing* (pp. 252–289). IGI Global. DOI: 10.4018/979-8-3693-1694-8.ch014

Alimonti, G., Brambilla, R., Pileci, R. E., Romano, R., Rosa, F., & Spinicci, L. (2017). Edible energy: Balancing inputs and waste in the food supply chain and biofuels from algae. *The European Physical Journal Plus*, 132(1), 132. DOI: 10.1140/epjp/i2017-11301-8

Aliqab, K., Nadeem, I., & Khan, S. R. (2023, July 21). A Comprehensive Review of In-Body Biomedical Antennas: Design, Challenges and Applications. *Micromachines*, 14(7), 1472. DOI: 10.3390/mi14071472 PMID: 37512782

Almeida, C., Vaz, S., & Ziegler, F. (2015). Environmental life cycle assessment of a canned sardine product from Portugal. *Journal of Industrial Ecology*, 19(4), 607–617. DOI: 10.1111/jiec.12219

Al-Turjman, F., Salama, R., & Altrjman, C. (2023). Overview of IoT solutions for sustainable transportation systems. NEU Journal for Artificial Intelligence and Internet of Things, 2(3).

AlZubi, A. A., & Galyna, K. (2023). Artificial intelligence and internet of things for sustainable farming and smart agriculture. *IEEE Access : Practical Innovations, Open Solutions*, 11, 78686–78692. DOI: 10.1109/ACCESS.2023.3298215

Ambrożkiewicz, B., & Rounak, A. (2022). ENERGY HARVESTING – NEW GREEN ENERGY. Journal of Technology and Exploitation in Mechanical Engineering.

Amin, T., Mobbs, R. J., Mostafa, N., Sy, L. W., & Choy, W. J. (2021, July 20). Wearable devices for patient monitoring in the early postoperative period: A literature review. *mHealth*, 7, 50. DOI: 10.21037/mhealth-20-131 PMID: 34345627

Anand, R., Pandey, D., Gupta, D. N., Dharani, M. K., Sindhwani, N., & Ramesh, J. V. N. (2024). Wireless Sensor-based IoT System with Distributed Optimization for Healthcare. In Anand, R., Juneja, A., Pandey, D., Juneja, S., & Sindhwani, N. (Eds.), *Meta Heuristic Algorithms for Advanced Distributed Systems.*, DOI: 10.1002/9781394188093.ch16

Annese, V. F., Coco, G., Galli, V., Cataldi, P., & Caironi, M. (2023). Edible Electronics and Robofood: A Move Towards Sensors for Edible Robots and Robotic Food. *2023 IEEE International Conference on Flexible and Printable Sensors and Systems (FLEPS)*, 1–4. DOI: 10.1109/FLEPS57599.2023.10220412

Anto, S., Mukherjee, S. S., Muthappa, R., Mathimani, T., Deviram, G., Kumar, S. S., Verma, T. N., & Pugazhendhi, A. (2020). Algae as green energy reserve: Technological outlook on biofuel production. *Chemosphere*, 242, 125079. DOI: 10.1016/j.chemosphere.2019.125079 PMID: 31678847

Arshak, K., Moore, E., Lyons, G. M., Harris, J., & Clifford, S. (2007). A review of gas sensors employed in electronic nose applications. *Sensor Review*, 27(1), 7–20.

Atitallah, S. B., Driss, M., Boulila, W., & Ghézala, H. B. (2020). Leveraging Deep Learning and IoT big data analytics to support the smart cities development: Review and future directions. *Computer Science Review*, 38, 100303. DOI: 10.1016/j.cosrev.2020.100303

Avancini, D. B.. (2021). A new IoT-based smart energy meter for smart grids, Volume45, Issue1, Special Issue:Smart. *Energy Technology (Weinheim)*, •••, 189–202.

Baek, S., Jeon, E., Park, K. S., Yeo, K.-H., & Lee, J. (2018). Monitoring of Water Transportation in Plant Stem With Microneedle Sap Flow Sensor. *Journal of Microelectromechanical Systems*, 27(3), 440–447. DOI: 10.1109/JMEMS.2018.2823380

Bai, L., Liu, M., & Sun, Y. (2023). Overview of Food Preservation and Traceability Technology in the Smart Cold Chain System. *Foods*, 2023(12), 2881. DOI: 10.3390/foods12152881 PMID: 37569150

Bakhtar, N., Chhabria, V., Chougle, I., Vidhrani, H., & Hande, R. (2018). IoT based Hydroponic Farm. *2018 International Conference on Smart Systems and Inventive Technology (ICSSIT)*, 205–209. DOI: 10.1109/ICSSIT.2018.8748447

Balamurugan, S., Ayyasamy, A., & Joseph, K. S. (2022). IoT-Blockchain driven traceability techniques for improved safety measures in food supply chain. *International Journal of Information Technology : an Official Journal of Bharati Vidyapeeth's Institute of Computer Applications and Management*, 14(2), 1087–1098. DOI: 10.1007/s41870-020-00581-y

Bandal, A., & Thirugnanam, M. (2016). Quality measurements of fruits and vegetables using sensor network. *Proceedings of the 3rd International Symposium on Big Data and Cloud Computing Challenges (ISBCC–16)*, 121–130. DOI: 10.1007/978-3-319-30348-2_11

Bande, S., & Shete, V. V. (2017). Smart flood disaster prediction system using IoT & neural networks. *2017 International Conference On Smart Technologies For Smart Nation (SmartTechCon)*, 189–194. DOI: 10.1109/SmartTechCon.2017.8358367

Baranov, D. G., Zuev, D. A., Lepeshov, S. I., Kotov, O. V., Krasnok, A. E., Evlyukhin, A. B., & Chichkov, B. N. (2017). All-dielectric nanophotonics: The quest for better materials and fabrication techniques. *Optica*, 4(7), 814–825. DOI: 10.1364/OPTICA.4.000814

Bartlett, et al., 1997Bartlett, P. N., Elliot, J. M., & Gardner, J. W. (1997). Electronic noses and their application in the food Industry. *Food Technology*, 51, 44–48.

Batt, C. A., & Tortorello, M. L. (2014). *Encyclopedia of food microbiology*. Elsevier.

Beardslee, L. A., Banis, G. E., Chu, S., Liu, S., Chapin, A. A., Stine, J. M., Pasricha, P. J., & Ghodssi, R. (2020). Ingestible Sensors and Sensing Systems for Minimally Invasive Diagnosis and Monitoring: The Next Frontier in Minimally Invasive Screening. *ACS Sensors*, 5(4), 891–910. DOI: 10.1021/acssensors.9b02263 PMID: 32157868

Belesova, K., Gasparrini, A., Sié, A., Sauerborn, R., & Wilkinson, P. (2017). Household cereal crop harvest and children's nutritional status in rural Burkina Faso. *Environmental Health*, 16(1), 1–11. DOI: 10.1186/s12940-017-0258-9 PMID: 28633653

Benkeblia, N. (2012). Post-harvest diseases and disorders of potato tuber Solanum tuberosum L. *Potatoes Prod. Consum. Health Benefits*, 7, 99–114.

Berean, K. J., Ha, N., Ou, J., Chrimes, A. F., Grando, D., Yao, C. K., Muir, J. G., Ward, S. A., Burgell, R. E., Gibson, P. R., & Kalantar-zadeh, K. (2018). The safety and sensitivity of a telemetric capsule to monitor gastrointestinal hydrogen production in vivo in healthy subjects: A pilot trial comparison to concurrent breath analysis. *Alimentary Pharmacology & Therapeutics*, 48(6), 646–654. DOI: 10.1111/apt.14923 PMID: 30067289

Berthold, et al., 1998Berthold, M. J., & Diamond, J. (1998). Contructive training of probabilistic neural networks. *Neurocomputing*, 19(1-3), 167–183. DOI: 10.1016/S0925-2312(97)00063-5

Betlej, K., Rybak, M., Nowacka, A., Antczak, S., Borysiak, B., Krochmal-Marczak, K., Lipska, P., & Boruszewski, P. (2022). Pomace from Oil Plants as a New Type of Raw Material for the Production of Environmentally Friendly Biocomposites. *Coatings*, 13(10), 1722. Advance online publication. DOI: 10.3390/coatings13101722

Bettinger, C. J. (2019). Edible hybrid microbial-electronic sensors for bleeding detection and beyond. *Hepatobiliary Surgery and Nutrition*, 8(2), 157–160. Advance online publication. DOI: 10.21037/hbsn.2018.11.14 PMID: 31098367

Beuchat, L. R. (2006). Control of foodborne pathogens. In Doyle, M. P., & Beuchat, L. R. (Eds.), *Springer*.

Bezdek, J. C. (1981). *Pattern Recognition with Fuzzy Objective Function Algorithms*. Plenum Press. DOI: 10.1007/978-1-4757-0450-1

Bhardwaj, A., Dagar, V., Khan, M. O., Aggarwal, A., Alvarado, R., Kumar, M., Irfan, M., & Proshad, R. (2022). Smart IoT and machine learning-based framework for water quality assessment and device component monitoring. *Environmental Science and Pollution Research International*, 29(30), 46018–46036. DOI: 10.1007/s11356-022-19014-3 PMID: 35165843

Bharti, N. K., Dongargaonkar, M. D., Kudkar, I. B., Das, S., & Kenia, M. (2019). Hydroponics System for Soilless Farming Integrated with Android Application by Internet of Things and MQTT Broker. *2019 IEEE Pune Section International Conference (PuneCon)*, 1–5. DOI: 10.1109/PuneCon46936.2019.9105847

Bharti, S., Jaiswal, S., & Sharma, V. (2023). Perspective and challenges: Intelligent to smart packaging for future generations. In *Green sustainable process for chemical and environmental engineering and science* (pp. 171–183). Elsevier. DOI: 10.1016/B978-0-323-95644-4.00015-2

Bikash Chandra Saha, M. S., Deepa, R., Akila, A., & Sai Thrinath, B. V. (2022). Iot Based Smart Energy Meter For Smart Grid.

Bishop, C. M. (1995). *Neural Network for pattern recognition*. Oxford University Press. DOI: 10.1093/oso/9780198538493.001.0001

Bishop, R. R., Kubiak-Martens, L., Warren, G. M., & Church, M. J. (2023). Getting to the root of the problem: New evidence for the use of plant root foods in Mesolithic hunter-gatherer subsistence in Europe. *Vegetation History and Archaeobotany*, 32(1), 65–83. DOI: 10.1007/s00334-022-00882-1

Blagg, C.R. (2011). Preface. Hemodialysis International, 15.

Bleibaum, et al., 2002Bleibaum, R. N., Stone, H., Tan, T., Labreche, S., Saint-Martin, E., & Isz, S. (2002). Comparison of sensory and consumer results with electronic nose and tongue sensors for apple juices. *Food Quality and Preference*, 13(6), 409–422. DOI: 10.1016/S0950-3293(02)00017-4

Boopathi, S. (2023). Deep Learning Techniques Applied for Automatic Sentence Generation. In *Promoting Diversity, Equity, and Inclusion in Language Learning Environments* (pp. 255–273). IGI Global. DOI: 10.4018/978-1-6684-3632-5.ch016

Boopathi, S. (2024). Minimization of Manufacturing Industry Wastes Through the Green Lean Sigma Principle. *Sustainable Machining and Green Manufacturing*, 249–270.

Boopathi, S. (2024a). Balancing Innovation and Security in the Cloud: Navigating the Risks and Rewards of the Digital Age. In *Improving Security, Privacy, and Trust in Cloud Computing* (pp. 164–193). IGI Global.

Boopathi, S. (2024b). Balancing Innovation and Security in the Cloud: Navigating the Risks and Rewards of the Digital Age. In *Improving Security, Privacy, and Trust in Cloud Computing* (pp. 164–193). IGI Global.

Boopathi, S. (2024b). Enabling Machine Control Through Virtual Methods: Harnessing the Power of 5G. In *Advances in Logistics, Operations, and Management Science* (pp. 50–74). IGI Global. DOI: 10.4018/979-8-3693-1862-1.ch004

Boopathi, S. (2024c). Sustainable Development Using IoT and AI Techniques for Water Utilization in Agriculture. In *Sustainable Development in AI, Blockchain, and E-Governance Applications* (pp. 204–228). IGI Global. DOI: 10.4018/979-8-3693-1722-8.ch012

Boopathi, S., Karthikeyan, K. R., Jaiswal, C., Dabi, R., Sunagar, P., & Malik, S. (2024). *IoT based Automatic Cooling Tower*.

Boopathi, S., Kumar, P. K. S., Meena, R. S., Sudhakar, M., & Associates. (2023). Sustainable Developments of Modern Soil-Less Agro-Cultivation Systems: Aquaponic Culture. In *Human Agro-Energy Optimization for Business and Industry* (pp. 69–87). IGI Global.

Boopathi, S. (2023). Internet of Things-Integrated Remote Patient Monitoring System: Healthcare Application. In *Dynamics of Swarm Intelligence Health Analysis for the Next Generation* (pp. 137–161). IGI Global. DOI: 10.4018/978-1-6684-6894-4.ch008

Boopathi, S. (2024a). Advancements in Machine Learning and AI for Intelligent Systems in Drone Applications for Smart City Developments. In *Futuristic e-Governance Security With Deep Learning Applications* (pp. 15–45). IGI Global. DOI: 10.4018/978-1-6684-9596-4.ch002

Borah, et al, 2005Borah, S., & Bhuyan, M. (2005). A computer based system for matching colours during the monitoring of tea fermentation. *International Journal of Food Science & Technology*, 40(6), 675–682. DOI: 10.1111/j.1365-2621.2005.00981.x

Borah, et al., 2003 Borah, S., & Bhuyan, M. 2003, "Quality indexing by machine vision during fermentation in black tea manufacturing", Sixth International Conference on Quality Control by Artificial Vision, Fabrice Meriaudeau Gatlinburg, United States, SPIE, Volume 5132, PP. 468-475 (2003)

Brasil, B. S. A. F., Silva, F. C. P., & Siqueira, F. G. (2017). Microalgae biorefineries: The Brazilian scenario in perspective. *New Biotechnology*, 39, 90–98. DOI: 10.1016/j.nbt.2016.04.007 PMID: 27343427

Brosnan, et al., 2002Brosnan, T., & Sun, D. (2002). Inspection and grading of agricultural and food productsby computer vision systems – a review. *Computers and Electronics in Agriculture*, 36(2-3), 193–313. DOI: 10.1016/S0168-1699(02)00101-1

Buratti, et al., 2004Buratti, S., Benedetti, S., Scampicchio, M., & Pangerod, E. C. (2004). Characterization and classification of Italian Barbera wines by using an electronic nose and an amperometric electronic tongue. *Analytica Chimica Acta*, 525(1), 133–139. DOI: 10.1016/j.aca.2004.07.062

Butpheng, C., Yeh, K.-H., & Xiong, H. (2020). Security and Privacy in IoT-Cloud-Based e-Health Systems—A Comprehensive Review. *Symmetry*, 2020(12), 1191. DOI: 10.3390/sym12071191

Calabrese, A., Battistoni, P., Ceylan, S., Zeni, L., Capo, A., Varriale, A., D'Auria, S., & Staiano, M. (2023). An impedimetric biosensor for detection of volatile organic compounds in food. *Biosensors (Basel)*, 13(3), 341. DOI: 10.3390/bios13030341 PMID: 36979553

Campbell-Platt, G. (2009). *Food science and technology*. Wiley-Blackwell.

Castleman, K. R. (1996). *Digital Image Processing*. Prentice Hall.

Chai, P. R., Castillo-Mancilla, J. R., Buffkin, E., Darling, C. E., Rosen, R. K., Horvath, K. J., Boudreaux, E. D., Robbins, G. K., Hibberd, P. L., & Boyer, E. W. (2015). Utilizing an Ingestible Biosensor to Assess Real-Time Medication Adherence. *Journal of Medical Toxicology; Official Journal of the American College of Medical Toxicology*, 11(4), 439–444. DOI: 10.1007/s13181-015-0494-8 PMID: 26245878

Chai, P. R., Vaz, C., Goodman, G. R., Albrechta, H., Huang, H., Rosen, R. K., Boyer, E. W., Mayer, K. H., & O'Cleirigh, C. (2022, February 28). Ingestible electronic sensors to measure instantaneous medication adherence: A narrative review. *Digital Health*, 8, 20552076221083119. DOI: 10.1177/20552076221083119 PMID: 35251683

Chen, C.-J., Huang, Y.-Y., Li, Y.-S., Chang, C.-Y., & Huang, Y.-M. (2020). An AIoT based smart agricultural system for pests detection. *IEEE Access : Practical Innovations, Open Solutions*, 8, 180750–180761. DOI: 10.1109/ACCESS.2020.3024891

Chen, C., Zhang, H., Dong, C., Ji, H., Zhang, X., Li, L., Ban, Z., Zhang, N., & Xue, W. (2019). Effect of ozone treatment on the phenylpropanoid biosynthesis of post-harvest strawberries. *RSC Advances*, 9(44), 25429–25438. DOI: 10.1039/C9RA03988K PMID: 35530059

Chen, J., Chen, H., Hu, D., Pan, J. Z., & Zhou, Y. (2015). Smog disaster forecasting using social web data and physical sensor data. *2015 IEEE International Conference on Big Data (Big Data)*, 991–998. DOI: 10.1109/BigData.2015.7363850

Chhetri, K. B. (2024). Applications of Artificial Intelligence and Machine Learning in Food Quality Control and Safety Assessment. *Food Engineering Reviews*, 16(1), 1–21. DOI: 10.1007/s12393-023-09363-1

Chisenga, S., Tolesa, G., & Workneh, T. (2020). Biodegradable food packaging materials and prospects of the fourth industrial revolution for tomato fruit and product handling. *International Journal of Food Sciences*, 2020(1), 8879101. DOI: 10.1155/2020/8879101 PMID: 33299850

Chong, K. P., & Woo, B. K. P. (2021, March 28). Emerging wearable technology applications in gastroenterology: A review of the literature. *World Journal of Gastroenterology*, 27(12), 1149–1160. DOI: 10.3748/wjg.v27.i12.1149 PMID: 33828391

Corigliano, O., Florio, G., & Fragiacomo, P. (2016). Energy Valorization of Edible Organic Matter for Electrical, Thermal and Cooling Energy Generation: Part One. *Energy Procedia*, 101, 89–96. DOI: 10.1016/j.egypro.2016.11.012

Cruz, F., Batista-Santos, P., Monteiro, S., Neves-Martins, J., & Ferreira, R. B. (2021). Maximizing Blad-containing oligomer fungicidal activity in sweet cultivars of Lupinus albus seeds. *Industrial Crops and Products*, 162, 113242. DOI: 10.1016/j.indcrop.2021.113242

D. Das, et al.Das, D., Nandy Chatterjee, T., Banerjee Roy, R., Tudu, B., Hazarika, A. K., Sabhapondit, S., & Bandyopadhyay, R. (2020). Titanium oxide nanocubes embedded molecularly imprinted polymer-based electrode for selective detection of caffeine in green tea. *IEEE Sensors Journal*, 20(12), 6240–6247. DOI: 10.1109/JSEN.2020.2972773

Danchuk, A. I., Komova, N. S., Mobarez, S. N., Doronin, S. Y., Burmistrova, N. A., Markin, A. V., & Duerkop, A. (2020). Optical sensors for determination of biogenic amines in food. *Analytical and Bioanalytical Chemistry*, 412(17), 4023–4036. DOI: 10.1007/s00216-020-02675-9 PMID: 32382967

Dewulf, et al., 2002Dewulf, J., & Langenhove, H. V. (2002). Analysis of volatile organic compounds using gas chromatography. *Trends in Analytical Chemistry*, 21(9-10), 637–646. DOI: 10.1016/S0165-9936(02)00804-X

Dey, S. (2023). Design & Development of new age IOT based smart garbage monitoring & controlling system for smart city. *Journal of Emerging Technologies and Innovative Research*, 10(6), 694–700.

Dey, S. (2024). Phenomenon of Excess of Artificial Intelligence: Quantifying the Native AI, Its Leverages in 5G/6G and beyond. In *Radar and RF Front End System Designs for Wireless Systems* (pp. 245–274). IGI Global. DOI: 10.4018/979-8-3693-0916-2.ch010

Dhanya, D., Kumar, S. S., Thilagavathy, A., Prasad, D., & Boopathi, S. (2023). Data Analytics and Artificial Intelligence in the Circular Economy: Case Studies. In *Intelligent Engineering Applications and Applied Sciences for Sustainability* (pp. 40–58). IGI Global.

Dias, R. M., Marques, G., & Bhoi, A. K. (2020). Internet of things for enhanced food safety and quality assurance: A literature review. *International Conference on Emerging Trends and Advances in Electrical Engineering and Renewable Energy*, 653–663.

Doyle, M. P., & Buchanan, R. L. (2013). *Food microbiology: Fundamentals and frontiers*. ASM Press.

Eni, L. N., Groenewald, E. S., Hamidi, I. A., & Garg, A. (2024). Optimizing Supply Chain Processes through Deep learning Algorithms: A Managerial Approach. *Journal of Informatics Education and Research*, 4(1). Advance online publication. DOI: 10.52783/jier.v4i1.567

Fitzgerald, J. E., Wang, C., Jamieson, R. P., & Coughlan, A. P. (2001). Temperature and humidity sensors for use in food storage. *Sensors and Actuators. B, Chemical*, 81(1), 115–119.

Floreano, D., Kwak, B., Pankhurst, M., Shintake, J., Caironi, M., Annese, V. F., Qi, Q., Rossiter, J., & Boom, R. M. (2024). Towards edible robots and robotic food. *Nature Reviews. Materials*, •••, 1–11.

Food and Agriculture Organization (FAO). (2011). Global food losses and food waste – Extent, causes and prevention. Rome.

Frias, J., & Vidal-Valverde, C. (2008). *Advances in food and nutrition research*. Elsevier.

Fuangthong, M., & Pramokchon, P. (2018). Automatic control of electrical conductivity and PH using fuzzy logic for hydroponics system. *2018 International Conference on Digital Arts, Media and Technology (ICDAMT)*, 65–70. DOI: 10.1109/ICDAMT.2018.8376497

Ganesan, V., Manoj, C., Ramaswamy, V., Tejaswi, T., Akilan, T., & Kumar, G. A. (2022, December). Food safety checking measures by artificial intelligent IOT. In 2022 4th International Conference on Advances in Computing, Communication Control and Networking (ICAC3N) (pp. 1385-1389). IEEE. DOI: 10.1109/ICAC3N56670.2022.10074055

Garcia-Breijo, E., Garrigues, J., Sanchez, L. G., & Laguarda-Miro, N. (2013). An embedded simplified fuzzy ARTMAP implemented on a microcontroller for food classification. *Sensors (Basel)*, 13(8), 10418–10429. DOI: 10.3390/s130810418 PMID: 23945736

Gbashi, S., & Njobeh, P. B. (2024). Enhancing food integrity through artificial intelligence and machine learning: A comprehensive review. *Applied Sciences (Basel, Switzerland)*, 14(8), 3421. DOI: 10.3390/app14083421

Giese, 2000Giese, J. (2000). Colour measurement in foods as a quality parameter. *Food Technology*, 54(2), 62–65.

Glady, J. B. P., D'Souza, S. M., Priya, A. P., Amuthachenthiru, K., Vikram, G., & Boopathi, S. (2024). A Study on AI-ML-Driven Optimizing Energy Distribution and Sustainable Agriculture for Environmental Conservation. In *Harnessing High-Performance Computing and AI for Environmental Sustainability* (pp. 1–27). IGI Global., DOI: 10.4018/979-8-3693-1794-5.ch001

Gnanaprakasam, C., Vankara, J., Sastry, A. S., Prajval, V., Gireesh, N., & Boopathi, S. (2023). Long-Range and Low-Power Automated Soil Irrigation System Using Internet of Things: An Experimental Study. In *Contemporary Developments in Agricultural Cyber-Physical Systems* (pp. 87–104). IGI Global.

Gonzalez, R. C., & Woods, R. E. (1992). *Digital Image Processing, Addison-Wesley Publishing Company; Graves, M., Batchelor, B., Machine Vision for the Inspection of Natural Products*. Springer London., DOI: 10.1007/b97526

Gopal, M., Lurdhumary, J., Bathrinath, S., Priya, A. P., Sarojwal, A., & Boopathi, S. (2024). Energy Harvesting and Smart Highways for Sustainable Transportation Infrastructure: Revolutionizing Roads Using Nanotechnology. In *Principles and Applications in Speed Sensing and Energy Harvesting for Smart Roads* (pp. 136–165). IGI Global. DOI: 10.4018/978-1-6684-9214-7.ch005

Gram, L., Ravn, L., Rasch, M., Bruhn, J. B., Christensen, A. B., & Givskov, M. (2002). Food spoilage—Interactions between food spoilage bacteria. *International Journal of Food Microbiology*, 78(1-2), 79–97. DOI: 10.1016/S0168-1605(02)00233-7 PMID: 12222639

Grazioli, C., Faura, G., Dossi, N., Toniolo, R., Abate, M., Terzi, F., & Bontempelli, G. (2020). 3D printed portable instruments based on affordable electronics, smartphones and open-source microcontrollers suitable for monitoring food quality. *Microchemical Journal*, 159, 105584. DOI: 10.1016/j.microc.2020.105584

Grimmer, J., Roberts, M. E., & Stewart, B. M. (2021). Machine learning for social science: An agnostic approach. *Annual Review of Political Science*, 24(1), 395–419. DOI: 10.1146/annurev-polisci-053119-015921

Gunasekaran, 1996Gunasekaran, S. (1996). Computer vision technology for food quality assurance. *Trends in Food Science & Technology*, 7(8), 245–256. DOI: 10.1016/0924-2244(96)10028-5

Hafezi, H., Robertson, T. L., Moon, G. D., Au-Yeung, K. Y., Zdeblick, M. J., & Savage, G. M. (2015). An Ingestible Sensor for Measuring Medication Adherence. *IEEE Transactions on Biomedical Engineering*, 62(1), 99–109. DOI: 10.1109/TBME.2014.2341272 PMID: 25069107

Hälterlein, J. (2021). Epistemologies of predictive policing: Mathematical social science, social physics and machine learning. *Big Data & Society*, 8(1), 20539517211003118. DOI: 10.1177/20539517211003118

Hanumanthakari, S., Gift, M. M., Kanimozhi, K., Bhavani, M. D., Bamane, K. D., & Boopathi, S. (2023). Biomining Method to Extract Metal Components Using Computer-Printed Circuit Board E-Waste. In *Handbook of Research on Safe Disposal Methods of Municipal Solid Wastes for a Sustainable Environment* (pp. 123–141). IGI Global. DOI: 10.4018/978-1-6684-8117-2.ch010

Harnkarnsujarit, Nathdanai. "Glass-Transition and Non-Equilibrium States of Edible Films and Barriers." Elsevier eBooks, January 1, 2017. .DOI: 10.1016/B978-0-08-100309-1.00019-5

Hassan, S. I., Alam, M. M., Illahi, U., Al Ghamdi, M. A., Almotiri, S. H., & Su'ud, M. M. (2021). A systematic review on monitoring and advanced control strategies in smart agriculture. *IEEE Access : Practical Innovations, Open Solutions*, 9, 32517–32548. DOI: 10.1109/ACCESS.2021.3057865

Hassoun, A., Boukid, F., Ozogul, F., Aït-Kaddour, A., Soriano, J. M., Lorenzo, J. M., Perestrelo, R., Galanakis, C. M., Bono, G., Bouyahya, A., Bhat, Z., Smaoui, S., Jambrak, A. R., & Câmara, J. S. (2023). Creating new opportunities for sustainable food packaging through dimensions of industry 4.0: New insights into the food waste perspective. *Trends in Food Science & Technology*, 142, 104238. DOI: 10.1016/j.tifs.2023.104238

Hazarika, et al., 1984Hazarika, M., Mahanta, P. K., & Takeo, T. (1984). Studies on some volatile flavor constituents in orthodox black tea of various clones and flushes in N. E. India. *Journal of the Science of Food and Agriculture*, 35(11), 1201–1207. DOI: 10.1002/jsfa.2740351111

Hema, N., Krishnamoorthy, N., Chavan, S. M., Kumar, N., Sabarimuthu, M., & Boopathi, S. (2023). A Study on an Internet of Things (IoT)-Enabled Smart Solar Grid System. In *Handbook of Research on Deep Learning Techniques for Cloud-Based Industrial IoT* (pp. 290–308). IGI Global. DOI: 10.4018/978-1-6684-8098-4.ch017

Hoyos, M. C., Morales, R. S., & Akhavan-Tabatabaei, R. (2015). OR models with stochastic components in disaster operations management: A literature survey. *Computers & Industrial Engineering*, 82, 183–197. DOI: 10.1016/j.cie.2014.11.025

Htwe, T., Chotikarn, P., Duangpan, S., Onthong, J., Buapet, P., & Sinutok, S. (2022). Integrated biomarker responses of rice associated with grain yield in copper-contaminated soil. *Environmental Science and Pollution Research International*, 29(6), 1–10. DOI: 10.1007/s11356-021-16314-y PMID: 34498193

Huang, W. D., Deb, S., Seo, Y. S., Rao, S., Chiao, M., & Chiao, J. C. (2011). A passive radio-frequency pH-sensing tag for wireless food-quality monitoring. *IEEE Sensors Journal*, 12(3), 487–495. DOI: 10.1109/JSEN.2011.2107738

Huang, Y., Whittaker, A. D., & Lacey, R. E. (2001). *Automation for food engineering: Food Quality Quantization and Process Control*. Culinary and Hospitality Industry. DOI: 10.1201/9781420039023

Hutchings, J. B. (1999). *Food colour & appearance* (2nd ed.). DOI: 10.1007/978-1-4615-2373-4

Hwa, L. S., & Te Chuan, L. (2024). A brief review of artificial intelligence robotic in food industry. *Procedia Computer Science*, 232, 1694–1700. DOI: 10.1016/j.procs.2024.01.167

Idumah, C., Zurina, M., Ogbu, J., Ndem, J., & Igba, E. (2020). A review on innovations in polymeric nanocomposite packaging materials and electrical sensors for food and agriculture. *Composite Interfaces*, 27(1), 1–72. DOI: 10.1080/09276440.2019.1600972

Ilic, I.K., Galli, V., Lamanna, L., Cataldi, P., Pasquale, L., Annese, V.F., Athanassiou, A. & Caironi, M., (2023) An Edible Rechargeable Battery. *Adv Mater*. 2023 May;35(20): e2211400. doi: . Epub. PMID: 36919977.DOI: 1002/adma.202211400

Istif, E., Mirzajani, H., Dağ, Ç., Mirlou, F., Ozuaciksoz, E. Y., Cakır, C., Koydemir, H. C., Yilgor, I., Yilgor, E., & Beker, L. (2023). Miniaturized wireless sensor enables real-time monitoring of food spoilage. *Nature Food*, 4(5), 427–436. DOI: 10.1038/s43016-023-00750-9 PMID: 37202486

Ivan, K. (2023, May). Ilic., Valerio, Galli., Leonardo, Lamanna., Pietro, Cataldi., Lea, Pasquale., V., F., Annese., Athanassia, Athanassiou., Mario, Caironi. (2023). An Edible Rechargeable Battery. *Advanced Materials*, 35(20), 2211400. Advance online publication. DOI: 10.1002/adma.202211400

Jabbar, W. A., Kian, T. K., Ramli, R. M., Zubir, S. N., Zamrizaman, N. S. M., Balfaqih, M., Shepelev, V., & Alharbi, S. (2019). Design and Fabrication of Smart Home With Internet of Things Enabled Automation System. *IEEE Access : Practical Innovations, Open Solutions*, 7, 144059–144074. DOI: 10.1109/ACCESS.2019.2942846

Jagtap, S., Bader, F., Garcia-Garcia, G., Trollman, H., Fadiji, T., & Salonitis, K. (2020). Food logistics 4.0: Opportunities and challenges. *Logistics (Basel)*, 5(1), 2. DOI: 10.3390/logistics5010002

Jain, A. K. (2012). Emerging Dimensions in the Energy Harvesting. *IOSR Journal of Electrical and Electronics Engineering*, 3(1), 70–80. DOI: 10.9790/1676-0317080

Jay, J. M. (2000). *Modern food microbiology*. Springer. DOI: 10.1007/978-1-4615-4427-2

Jedermann, R., Pötsch, T., & Lloyd, C. (2014). Communication techniques and challenges for wireless food quality monitoring. *Philosophical Transactions of the Royal Society A: Mathematical, Physical and Engineering Sciences, 372*(2017), 20130304.

Jeevanantham, Y. A., Saravanan, A., Vanitha, V., Boopathi, S., & Kumar, D. P. (2022). Implementation of Internet-of Things (IoT) in Soil Irrigation System. *IEEE Explore*, 1–5.

Jha, S. K., & Beevi, S. J. P., H., Babitha, M. N., Chinnusamy, S., & Boopathi, S. (2024). Artificial Intelligence-Infused Urban Connectivity for Smart Cities and the Evolution of IoT Communication Networks. In *Blockchain-Based Solutions for Accessibility in Smart Cities* (pp. 113–146). IGI Global. DOI: 10.4018/979-8-3693-3402-7.ch005

Johnson, M., & Brown, T. (2020). Ingestible Sensors for Chronic Disease Management. *Journal of Medical Devices*.

Joppi R et al, (2019) Food and Drug Administration vs European Medicines Agency: Review times and clinical evidence on novel drugs at the time of approval. Br J Clin Pharmacol. 2020 Jan;86(1):170-174. . Epub 2019 Dec 16. PMID: 31657044; PMCID: PMC6983504.DOI: 10.1111/bcp.14130

Joshi, N., Pransu, G., & Conte-Junior, A. (2023). Critical review and recent advances of 2D materials-based gas sensors for food spoilage detection. *Critical Reviews in Food Science and Nutrition*, 63(30), 10536–10559. DOI: 10.1080/10408398.2022.2078950 PMID: 35647714

Joshitha, C., Kanakaraja, P., Kumar, K. S., Akanksha, P., & Satish, G. (2021). An eye on hydroponics: The IoT initiative. *2021 7th International Conference on Electrical Energy Systems (ICEES)*, 553–557. DOI: 10.1109/ICEES51510.2021.9383694

Jun, X., Huiyun, W., Yaoyao, G., & Liping, Z. (2019). Research progress in forecasting methods of rainstorm and flood disaster in China. *Torrential Rain and Disasters*, 38(5), 416–421.

K M A et al. (2024, March 14). Internet of Things enabled open source assisted real-time blood glucose monitoring framework. *Scientific Reports*, 14(1), 6151. DOI: 10.1038/s41598-024-56677-z PMID: 38486038

Kaewwiset, T., & Yooyativong, T. (2017). Electrical conductivity and pH adjusting system for hydroponics by using linear regression. *2017 14th International Conference on Electrical Engineering/Electronics, Computer, Telecommunications and Information Technology (ECTI-CON)*, 761–764. DOI: 10.1109/ECTICon.2017.8096350

Kalantar-Zadeh, K., Berean, K. J., Ha, N., Chrimes, A. F., Xu, K., Grando, D., Ou, J. Z., Pillai, N., Campbell, J. L., Brkljača, R., Taylor, K. M., Burgell, R. E., Yao, C. K., Ward, S. A., McSweeney, C. S., Muir, J. G., & Gibson, P. R. (2018). A human pilot trial of ingestible electronic capsules capable of sensing different gases in the gut. *Nature Electronics*, 1(1), 79–87. DOI: 10.1038/s41928-017-0004-x

Kamal, T., & Rahman, S. M. A. (2024). *Productivity optimization in the electronics industry using simulation-based modeling approach.* International Journal of Research in Industrial Engineering., DOI: 10.22105/riej.2024.445760.1423

Kapoor, A., Ramamoorthy, S., Sundaramurthy, A., Vaishampayan, V., Sridhar, A., Balasubramanian, S., & Ponnuchamy, M. (2024). Paper-based lab-on-a-chip devices for detection of agri-food contamination. *Trends in Food Science & Technology*, 147, 104476. DOI: 10.1016/j.tifs.2024.104476

Kaushik, S., & Singh, C. (2013). Monitoring and controlling in food storage system using wireless sensor networks based on zigbee & bluetooth modules. *International Journal of Multidisciplinary in Cryptology and Information Security*, 2(3).

Kav, R. P., Pandraju, T. K. S., Boopathi, S., Saravanan, P., Rathan, S. K., & Sathish, T. (2023). Hybrid Deep Learning Technique for Optimal Wind Mill Speed Estimation. *2023 7th International Conference on Electronics, Communication and Aerospace Technology (ICECA)*, 181–186.

Kavitha, C., Varalatchoumy, M., Mithuna, H., Bharathi, K., Geethalakshmi, N., & Boopathi, S. (2023). Energy Monitoring and Control in the Smart Grid: Integrated Intelligent IoT and ANFIS. In *Applications of Synthetic Biology in Health, Energy, and Environment* (pp. 290–316). IGI Global.

Kelleher, J. D., Mac Namee, B., & D'arcy, A. (2020). *Fundamentals of machine learning for predictive data analytics: Algorithms, worked examples, and case studies*. MIT press.

Kettunen, M., & D'Amato, D. (2013). PROVISIONING SERVICES AND RELATED GOODS. In Social and Economic Benefits of Protected Areas: An assessment guide. DOI: 10.4324/9780203095348-10

Khalid, M. S. B., & Shafiai, S. B. (2015). Flood disaster management in Malaysia: An evaluation of the effectiveness flood delivery system. *International Journal of Social Science and Humanity*, 5(4), 398–402. DOI: 10.7763/IJSSH.2015.V5.488

Khandelwal, G., Joseph Raj, N. P. M., Alluri, N. R., & Kim, S. J. (2021). Enhancing hydrophobicity of starch for biodegradable material-based triboelectric nanogenerators. *ACS Sustainable Chemistry & Engineering*, 9(27), 9011–9017. DOI: 10.1021/acssuschemeng.1c01853

Khanh, P. T., Ngoc, T. T. H., & Pramanik, S. (2023). Future of smart agriculture techniques and applications. In *Handbook of Research on AI-Equipped IoT Applications in High-Tech Agriculture* (pp. 365–378). IGI Global. DOI: 10.4018/978-1-6684-9231-4.ch021

Khan, P. W., Byun, Y.-C., & Park, N. (2020). IoT-blockchain enabled optimized provenance system for food industry 4.0 using advanced deep learning. *Sensors (Basel)*, 20(10), 2990. DOI: 10.3390/s20102990 PMID: 32466209

Khansili, N., Rattu, G., & Krishna, P. M. (2018). Label-free optical biosensors for food and biological sensor applications. *Sensors and Actuators. B, Chemical*, 265, 35–49. DOI: 10.1016/j.snb.2018.03.004

Kheirabadi, N. R., Karimzadeh, F., Enayati, M. H., & Kalali, E. N. (2023). Green flexible triboelectric nanogenerators based on edible proteins for electrophoretic deposition. *Advanced Electronic Materials*, 9(2), 2200839. DOI: 10.1002/aelm.202200839

Khovrat, A., Kobziev, V., Nazarov, A., & Yakovlev, S. (2022). Parallelization of the VAR Algorithm Family to Increase the Efficiency of Forecasting Market Indicators During Social Disaster. *IT&I*, 222–233.

Kim, J., Jeerapan, I., Ciui, B., Hartel, M. C., Martin, A., & Wang, J. (2017). Edible Electrochemistry: Food Materials Based Electrochemical Sensors. *Advanced Healthcare Materials*, 6(22), 1700770. Advance online publication. DOI: 10.1002/adhm.201700770 PMID: 28783874

Kohonen, 1982Kohonen, T. (1982). Self-organised formation of topologically correct feature maps. *Biological Cybernetics*, 43(1), 59–69. DOI: 10.1007/BF00337288

Koshariya, A. K., Kalaiyarasi, D., Jovith, A. A., Sivakami, T., Hasan, D. S., & Boopathi, S. (2023). AI-Enabled IoT and WSN-Integrated Smart Agriculture System. In *Artificial Intelligence Tools and Technologies for Smart Farming and Agriculture Practices* (pp. 200–218). IGI Global. DOI: 10.4018/978-1-6684-8516-3.ch011

Koshariya, A. K., Khatoon, S., Marathe, A. M., Suba, G. M., Baral, D., & Boopathi, S. (2023). Agricultural Waste Management Systems Using Artificial Intelligence Techniques. In *AI-Enabled Social Robotics in Human Care Services* (pp. 236–258). IGI Global. DOI: 10.4018/978-1-6684-8171-4.ch009

Kouhi, M., Prabhakaran, M. P., & Ramakrishna, S. (2020). Edible polymers: An insight into its application in food, biomedicine and cosmetics. *Trends in Food Science & Technology*, 103, 248–263. DOI: 10.1016/j.tifs.2020.05.025

Kulkarni, V., Mrad, R.B., El-Diraby, T.E., & Prasad, E. (2010). Energy Harvesting Using Piezoceramics.

Kumar, P. R., Meenakshi, S., Shalini, S., Devi, S. R., & Boopathi, S. (2023). Soil Quality Prediction in Context Learning Approaches Using Deep Learning and Blockchain for Smart Agriculture. In *Effective AI, Blockchain, and E-Governance Applications for Knowledge Discovery and Management* (pp. 1–26). IGI Global. DOI: 10.4018/978-1-6684-9151-5.ch001

Kumar, R. R., & Cho, J. Y. (2014). Reuse of hydroponic waste solution. *Environmental Science and Pollution Research International*, 21(16), 9569–9577. DOI: 10.1007/s11356-014-3024-3 PMID: 24838258

Kumar, S., Sharma, D., Rao, S., Lim, W. M., & Mangla, S. K. (2022). Past, present, and future of sustainable finance: Insights from big data analytics through machine learning of scholarly research. *Annals of Operations Research*, •••, 1–44. DOI: 10.1007/s10479-021-04410-8 PMID: 35002001

Kuswandi, B., Wicaksono, Y., Abdullah, A., Heng, L. Y., & Ahmad, M. (2011). Smart packaging: Sensors for monitoring of food quality and safety. *Sensing and Instrumentation for Food Quality and Safety*, 5(3-4), 137–146. DOI: 10.1007/s11694-011-9120-x

Labuza, T. P., & Sinskey, A. J. (2006). *Chemical deterioration and physical instability of food and beverages*. Academic Press.

Landi, G., Pagano, S., Granata, V., Avallone, G., La Notte, L., Palma, A. L., Sdringola, P., Puglisi, G., & Barone, C. (2024). Regeneration and Long-Term Stability of a Low-Power Eco-Friendly Temperature Sensor Based on a Hydrogel Nanocomposite. *Nanomaterials (Basel, Switzerland)*, 14(3), 283. DOI: 10.3390/nano14030283 PMID: 38334553

Law, E. P., Arnow, E., & Diemont, S. A. (2020). Ecosystem services from old-fields: Effects of site preparation and harvesting on restoration and productivity of traditional food plants. *Ecological Engineering*, 158, 105999. DOI: 10.1016/j.ecoleng.2020.105999

Liang, et al., 2003Liang, Y., Lu, J., Zhang, L., Wu, S., & Wu, Y. (2003). Estimation of black tea quality by analysis of chemical composition and colour difference of tea infusions. *Food Chemistry*, 80(2), 283–290. DOI: 10.1016/S0308-8146(02)00415-6

Licardo, J. T., Domjan, M., & Orehovački, T. (2024). Intelligent robotics—A systematic review of emerging technologies and trends. *Electronics (Basel)*, 13(3), 542. DOI: 10.3390/electronics13030542

Linardos, V., Drakaki, M., Tzionas, P., & Karnavas, Y. L. (2022). Machine learning in disaster management: Recent developments in methods and applications. *Machine Learning and Knowledge Extraction*, 4(2).

Liu, B., Gurr, P. A., & Qiao, G. G. (2020). Irreversible spoilage sensors for protein-based food. *ACS Sensors*, 5(9), 2903–2908. DOI: 10.1021/acssensors.0c01211 PMID: 32869625

Liu, X., Lu, D., Zhang, A., Liu, Q., & Jiang, G. (2022). Data-driven machine learning in environmental pollution: Gains and problems. *Environmental Science & Technology*, 56(4), 2124–2133. DOI: 10.1021/acs.est.1c06157 PMID: 35084840

Li, Z., Wang, N., Raghavan, G., & Cheng, W. (2006). A microcontroller-based, feedback power control system for microwave drying processes. *Applied Engineering in Agriculture*, 22(2), 309–314. DOI: 10.13031/2013.20277

Lu, P., Guo, X., Liao, X., Liu, Y., Cai, C., Meng, X., Wei, Z., Du, G., Shao, Y., Nie, S., & Wang, Z. (2024). Advanced application of triboelectric nanogenerators in gas sensing. Nano Energy, 126, 109672. ISSN 2211-2855. https://doi.org/DOI: 10.1016/j.nanoen.2024.109672

Luk, Iversen Jun Lam, Kobun Rovina, Joseph Merillyn Vonnie, Patricia Matanjun, Kana Husna Erna, Nur'Aqilah Nasir Md, Felicia Xia Ling Wen, and Andree Alexander Funk. "The emergence of edible and food-application coatings for food packaging: a review." (2022): 20220435220.

Luning, P. A., & Marcelis, W. J. (2006). A techno-managerial approach in food quality management research. *Trends in Food Science & Technology*, 17(7), 378–385. DOI: 10.1016/j.tifs.2006.01.012

Luo, J., Zhu, Z., Lv, W., Wu, J., Yang, J., Zeng, M., Hu, N., Su, Y., Liu, R., & Yang, Z. (2023). E-nose system based on Fourier series for gases identification and concentration estimation from food spoilage. *IEEE Sensors Journal*, 23(4), 3342–3351. DOI: 10.1109/JSEN.2023.3234194

Maguluri, L. P., Ananth, J., Hariram, S., Geetha, C., Bhaskar, A., & Boopathi, S. (2023). Smart Vehicle-Emissions Monitoring System Using Internet of Things (IoT). In *Handbook of Research on Safe Disposal Methods of Municipal Solid Wastes for a Sustainable Environment* (pp. 191–211). IGI Global.

Mahalik, N. P. (2009a). Processing and packaging automation systems: A review. *Sensing and Instrumentation for Food Quality and Safety*, 3(1), 12–25. DOI: 10.1007/s11694-009-9076-2

Mahanta, 1988Mahanta, P. K. [Mahanta, 1988]Mahanta, P. K., 1988. Biochemical basis of colour and flavour of black tea, *Proc. of 30th Tocklai Conference*, Assam, India, pages 124-134.

Mahanta, et al., 1985Mahanta, P. K., & Hazarika, M. (1985). Chlorophyll and degradation products in orthodox and CTC black teas and their influence on shade of colour and sensory quality in relation to thearubigins. *Journal of the Science of Food and Agriculture*, 36(11), 1133–1139. DOI: 10.1002/jsfa.2740361117

Maheswari, B. U., Imambi, S. S., Hasan, D., Meenakshi, S., Pratheep, V., & Boopathi, S. (2023). Internet of things and machine learning-integrated smart robotics. In *Global Perspectives on Robotics and Autonomous Systems: Development and Applications* (pp. 240–258). IGI Global. DOI: 10.4018/978-1-6684-7791-5.ch010

Majer-Baranyi, K., Székács, A., & Adányi, N. (2023). Application of electrochemical biosensors for determination of food spoilage. *Biosensors (Basel)*, 13(4), 456. DOI: 10.3390/bios13040456 PMID: 37185531

Malathi, J., Kusha, K., Isaac, S., Ramesh, A., Rajendiran, M., & Boopathi, S. (2024). IoT-Enabled Remote Patient Monitoring for Chronic Disease Management and Cost Savings: Transforming Healthcare. In *Advances in Explainable AI Applications for Smart Cities* (pp. 371–388). IGI Global.

Manikandan, R., Ranganathan, G., & Bindhu, V. (2023). Deep learning based IoT module for smart farming in different environmental conditions. *Wireless Personal Communications*, 128(3), 1715–1732. DOI: 10.1007/s11277-022-10016-5

Manisha, N., & Jagadeeshwar, M. (2023). BC driven IoT-based food quality traceability system for dairy product using deep learning model. *High-Confidence Computing*, 3(3), 100121. DOI: 10.1016/j.hcc.2023.100121

Matindoust, S., Farzi, A., Baghaei Nejad, M., Shahrokh Abadi, M. H., Zou, Z., & Zheng, L. R. (2017). Ammonia gas sensor based on flexible polyaniline films for rapid detection of spoilage in protein-rich foods. *Journal of Materials Science Materials in Electronics*, 28(11), 7760–7768. DOI: 10.1007/s10854-017-6471-z

Matindoust, S., Farzi, G., Nejad, M. B., & Shahrokhabadi, M. H. (2021). Polymer-based gas sensors to detect meat spoilage: A review. *Reactive & Functional Polymers*, 165, 104962. DOI: 10.1016/j.reactfunctpolym.2021.104962

Ma, Z., Chen, P., Cheng, W., Yan, K., Pan, L., Shi, Y., & Yu, G. (2018). Highly sensitive, printable nanostructured conductive polymer wireless sensor for food spoilage detection. *Nano Letters*, 18(7), 4570–4575. DOI: 10.1021/acs.nanolett.8b01825 PMID: 29947228

McCulloch, et Al., 1943 McCulloch, W. S., & Pitts, W. 1943. A logical calculus of ideas immanent in nervous activity, Bulletin of Mathematical Biophysics, Vol. 5, pages.115-133.

McEvoy, A. K., Von Bueltzingsloewen, C., McDonagh, C. M., MacCraith, B. D., Klimant, I., & Wolfbeis, O. S. (2003). Optical sensors for application in intelligent food-packaging technology. In Opto-Ireland 2002: Optics and Photonics Technologies and Applications (Vol. 4876, pp. 806-815). SPIE. DOI: 10.1117/12.464210

McGranahan, G., & Leslie, C. (1991). Walnuts (Juglans). *Genetic Resources of Temperate Fruit and Nut Crops*, 290, 907–974.

McLaren, 1976McLaren, K. (1976). The development of the CIE 1976 (L*a*b*) uniform colour-spaceand colour-difference formula. *Journal of the Society of Dyers and Colourists*, 92(9), 338–341. DOI: 10.1111/j.1478-4408.1976.tb03301.x

Melendez, et al., 2003Melendez-Martinez, A. J., Vicario, I. M., & Heredia, F. J. (2003). Application of tristimulus colorimetry to estimate the carotenoids content in ultrafrozen orange juices. *Journal of Agricultural and Food Chemistry*, 51(25), 7266–7270. DOI: 10.1021/jf034873z PMID: 14640568

Merz, B., Kuhlicke, C., Kunz, M., Pittore, M., Babeyko, A., Bresch, D. N., Domeisen, D. I., Feser, F., Koszalka, I., Kreibich, H., & others. (2020). Impact forecasting to support emergency management of natural hazards. *Reviews of Geophysics, 58*(4), e2020RG000704.

Mete, B., Durukan, D., Keskin, Y., Dinçer, O., Ogeday, M., Cicek, B., Yildiz, S., Aygün, B., Ercan, H., & Unalan, E. (2023). An Edible Supercapacitor Based on Zwitterionic Soy Sauce-Based Gel Electrolyte. *Advanced Functional Materials*. Advance online publication. DOI: 10.1002/adfm.202307051

Miller, R. (2014). *Green Technologies in Electronics*. Sustainable Technology Journal.

Mohamed, E. S., Belal, A., Abd-Elmabod, S. K., El-Shirbeny, M. A., Gad, A., & Zahran, M. B. (2021). Smart farming for improving agricultural management. *The Egyptian Journal of Remote Sensing and Space Sciences*, 24(3), 971–981. DOI: 10.1016/j.ejrs.2021.08.007

Mohanty, A., Venkateswaran, N., Ranjit, P., Tripathi, M. A., & Boopathi, S. (2023). Innovative Strategy for Profitable Automobile Industries: Working Capital Management. In *Handbook of Research on Designing Sustainable Supply Chains to Achieve a Circular Economy* (pp. 412–428). IGI Global.

Molakatala, N., Kumar, D. A., Patil, U., Mhatre, P. J., Sambathkumar, M., & Boopathi, S. (2024). Integrating 5G and IoT Technologies in Developing Smart City Communication Networks. In *Blockchain-Based Solutions for Accessibility in Smart Cities* (pp. 147–170). IGI Global. DOI: 10.4018/979-8-3693-3402-7.ch006

Moody, et al., 1989 Moody, J., & Darken, C. 1989. Fast learning in networks of locally-tuned processing units, Neural Computing, Vol. 1, No. 2, pages.281-294.

Mu, B., Cao, G., Zhang, L., Zou, Y., & Xiao, X. (2021). Flexible wireless pH sensor system for fish monitoring. *Sensing and Bio-Sensing Research*, 34, 100465. DOI: 10.1016/j.sbsr.2021.100465

Mu, B., Dong, Y., Qian, J., Wang, M., Yang, Y., Nikitina, M. A., Zhang, L., & Xiao, X. (2022). Hydrogel coating flexible pH sensor system for fish spoilage monitoring. *Materials Today. Chemistry*, 26, 101183. DOI: 10.1016/j.mtchem.2022.101183

Musundire, R., Ngonyama, D., Chemura, A., Ngadze, R. T., Jackson, J., Matanda, M. J., Tarakini, T., Langton, M., & Chiwona-Karltun, L. (2021). Stewardship of wild and farmed edible insects as food and feed in Sub-Saharan Africa: A perspective. *Frontiers in Veterinary Science*, 8, 601386. DOI: 10.3389/fvets.2021.601386 PMID: 33681322

Musundire, R., Zvidzai, C. J., Chidewe, C., Samende, B. K., & Manditsera, F. A. (2014). Nutrient and anti-nutrient composition of Henicus whellani (Orthoptera: Stenopelmatidae), an edible ground cricket, in south-eastern Zimbabwe. *International Journal of Tropical Insect Science*, 34(4), 223–231. DOI: 10.1017/S1742758414000484

Nanda, A. K., Sharma, A., Augustine, P. J., Cyril, B. R., Kiran, V., & Sampath, B. (2024). Securing Cloud Infrastructure in IaaS and PaaS Environments. In *Improving Security, Privacy, and Trust in Cloud Computing* (pp. 1–33). IGI Global. DOI: 10.4018/979-8-3693-1431-9.ch001

Narsaiah, K., Jha, S. N., Bhardwaj, R., Sharma, R., & Kumar, R. (2012). Optical biosensors for food quality and safety assurance—A review. *Journal of Food Science and Technology*, 49(4), 383–406. DOI: 10.1007/s13197-011-0437-6 PMID: 23904648

Natale, et al., 2000Natale, C. D., Paolesse, R., Macagnano, A., & Mantini, A. (2000). Electronic nose and electronic tongue integration for improved classification of clinical and food samples. *Sensors and Actuators. B, Chemical*, 64(1-3), 15–21. DOI: 10.1016/S0925-4005(99)00477-3

Nath, S. (2024). *Advancements in food quality monitoring: Integrating biosensors for precision detection*. Sustainable Food Technology.

Nayak, J., Vakula, K., Dinesh, P., Naik, B., & Pelusi, D. (2020). Intelligent food processing: Journey from artificial neural network to deep learning. *Computer Science Review*, 38, 100297. DOI: 10.1016/j.cosrev.2020.100297

Nayik, G. A., Muzaffar, K., & Gull, A. (2015). Robotics and food technology: A mini review. *Journal of Nutrition & Food Sciences*, 5(4), 1–11.

Nerkar, P. M., Shinde, S. S., Liyakat, K. K. S., Desai, S., & Kazi, S. S. L. (2023). Monitoring fresh fruit and food using Iot and machine learning to improve food safety and quality. *Tuijin Jishu/Journal of Propulsion Technology*, 44(3), 2927–2931.

Nguyen, L. H., Naficy, S., McConchie, R., Dehghani, F., & Chandrawati, R. (2019). Polydiacetylene-based sensors to detect food spoilage at low temperatures. *Journal of Materials Chemistry. C, Materials for Optical and Electronic Devices*, 7(7), 1919–1926. DOI: 10.1039/C8TC05534C

Nielsen, S. S. (1998). *Food Analysis* (2nd ed.). Kluwer Academic/ Plenum Publishers.

Nurcholis, M., Ahlasunnah, W., Utami, A., Krismawan, H., & Wibawa, T. (2021, November). Management of degraded land for developing biomass energy industry. In AIP Conference Proceedings (Vol. 2363, No. 1). AIP Publishing. DOI: 10.1063/5.0061179

Oberoi, H. S., & Dinesh, M. R. (2019). Trends and innovations in value chain management of tropical fruits. *Journal of Horticultural Sciences*, 14(2), 87–97. DOI: 10.24154/jhs.v14i2.773

Ojianwuna, C. C., Enwemiwe, V. N., Esiwo, E., Orji, G. O., & Nkeze, A. J. (2023). Food and Feed Additive of Insects: Economic and Environmental Impacts. *International Journal of Child Health and Nutrition*, 12(3), 107–119. DOI: 10.6000/1929-4247.2023.12.03.5

Okinda, et al., 1998Okinda, P., & Obanda, M. (1998). The changes in black tea quality due to variations of plucking standard and fermentation time. *Food Chemistry*, 61(4), 435–441. DOI: 10.1016/S0308-8146(97)00092-7

Oliveira, I. S., da Silva, A. G.Junior, de Andrade, C. A., & Oliveira, M. D. (2019). Biosensors for early detection of fungi spoilage and toxigenic and mycotoxins in food. *Current Opinion in Food Science*, 29, 64–79. DOI: 10.1016/j.cofs.2019.08.004

Pachiappan, K., Anitha, K., Pitchai, R., Sangeetha, S., Satyanarayana, T., & Boopathi, S. (2024). Intelligent Machines, IoT, and AI in Revolutionizing Agriculture for Water Processing. In *Handbook of Research on AI and ML for Intelligent Machines and Systems* (pp. 374–399). IGI Global.

Pacquit, A., Lau, K. T., McLaughlin, H., Frisby, J., Quilty, B., & Diamond, D. (2007). Development of a smart packaging for the monitoring of fish spoilage. *Food Chemistry*, 102(2), 466–470. DOI: 10.1016/j.foodchem.2006.05.052

Parasuraman, K., Anandan, U., & Anbarasan, A. (2021). IoT based smart agriculture automation in artificial intelligence. *2021 Third International Conference on Intelligent Communication Technologies and Virtual Mobile Networks (ICICV)*, 420–427. DOI: 10.1109/ICICV50876.2021.9388578

Patel, T. K., Vasundhara, S., Rajesha, S., Priyalakshmi, B., Chinnusamy, S., & Boopathi, S. (2024). Leveraging Drone and GPS Technologies For Precision Agriculture: Pest Management Perspective. In *Revolutionizing Pest Management for Sustainable Agriculture* (pp. 285–308). IGI Global. DOI: 10.4018/979-8-3693-3061-6.ch012

Patel, K., & Sharma, V. (2013). Edible sensors in personalized medicine. *Journal of Personalized Medicine*.

Patil, D. D., Singh, A. K., Shrivastava, A., & Bairagi, D. (2022). IOT Sensor-Based Smart Agriculture Using Agro-robot. In *IoT Based Smart Applications* (pp. 345–361). Springer.

Pecunia, V., Silva, S. R. P., Phillips, J. D., Artegiani, E., Romeo, A., Shim, H., Park, J., Kim, J. H., Yun, J. S., Welch, G. C., Larson, B. W., Creran, M., Laventure, A., Sasitharan, K., Flores-Diaz, N., Freitag, M., Xu, J., Brown, T. M., Li, B., & Joshi, A. P. (2023). Roadmap on energy harvesting materials. *JPhys Materials*, 6(4), 042501. DOI: 10.1088/2515-7639/acc550

Pereira, J. A., Saraiva, J. A., Casal, S., & Ramalhosa, E. (2018). The effect of different post-harvest treatments on the quality of borage (Borago officinalis) petals. *Acta Scientiarum Polonorum. Technologia Alimentaria*, 17(1), 5–10. PMID: 29514420

Peuchpanngarm, C., Srinitiworawong, P., Samerjai, W., & Sunetnanta, T. (2016). DIY sensor-based automatic control mobile application for hydroponics. *2016 Fifth ICT International Student Project Conference (ICT-ISPC)*, 57–60. DOI: 10.1109/ICT-ISPC.2016.7519235

Peyroteo, M., Ferreira, I. A., Elvas, L. B., Ferreira, J. C., & Lapão, L. V. (2021, December 21). Remote Monitoring Systems for Patients With Chronic Diseases in Primary Health Care: Systematic Review. *JMIR mHealth and uHealth*, 9(12), e28285. DOI: 10.2196/28285 PMID: 34932000

Piñeros-Hernandez, D., Medina-Jaramillo, C., López-Córdoba, A., & Goyanes, S. (2017, February 1). Edible Cassava Starch Films Carrying Rosemary Antioxidant Extracts for Potential Use as Active Food Packaging. *Food Hydrocolloids*, 63, 488–495. Advance online publication. DOI: 10.1016/j.foodhyd.2016.09.034

Poltronieri, P., Mezzolla, V., Primiceri, E., & Maruccio, G. (2014). Biosensors for the detection of food pathogens. *Foods*, 3(3), 511–526. DOI: 10.3390/foods3030511 PMID: 28234334

Popa, A., Hnatiuc, M., Paun, M., Geman, O., Hemanth, D. J., Dorcea, D., Son, L. H., & Ghita, S. (2019). An intelligent IoT-based food quality monitoring approach using low-cost sensors. *Symmetry*, 11(3), 374. DOI: 10.3390/sym11030374

Prajwal, A., Vaishali, P., & Sumit, D. (2020). Food quality detection and monitoring system. *2020 IEEE International Students' Conference on Electrical, Electronics and Computer Science (SCEECS)*, 1–4.

Pramila, P., Amudha, S., Saravanan, T., Sankar, S. R., Poongothai, E., & Boopathi, S. (2023). Design and Development of Robots for Medical Assistance: An Architectural Approach. In *Contemporary Applications of Data Fusion for Advanced Healthcare Informatics* (pp. 260–282). IGI Global.

Prasad, S. (2017). Application of robotics in dairy and food industries: A review. *International Journal of Science, Environment and Technology*, 6(3), 1856–1864.

Prasanna, A. P. S., Vivekananthan, V., Khandelwal, G., Alluri, N. R., Maria Joseph Raj, N. P., Anithkumar, M., & Kim, S. J. (2022). Green energy from edible materials: Triboelectrification-enabled sustainable self-powered human joint movement monitoring. *ACS Sustainable Chemistry & Engineering*, 10(20), 6549–6558. DOI: 10.1021/acssuschemeng.1c08030

Preethichandra, D. M., Gholami, M. D., Izake, E. L., O'Mullane, A. P., & Sonar, P. (2023). Conducting polymer based ammonia and hydrogen sulfide chemical sensors and their suitability for detecting food spoilage. *Advanced Materials Technologies*, 8(4), 2200841. DOI: 10.1002/admt.202200841

Publications Services, Texas

Puttinaovarat, S., & Horkaew, P. (2020). Flood forecasting system based on integrated big and crowdsource data by using machine learning techniques. *IEEE Access : Practical Innovations, Open Solutions*, 8, 5885–5905. DOI: 10.1109/ACCESS.2019.2963819

Qazi, S., Khawaja, B. A., & Farooq, Q. U. (2022). IoT-equipped and AI-enabled next generation smart agriculture: A critical review, current challenges and future trends. *IEEE Access : Practical Innovations, Open Solutions*, 10, 21219–21235. DOI: 10.1109/ACCESS.2022.3152544

Quevedo, et al., 2002Quevedo, R., Carlos, L. G., Aguilera, J. M., & Cadoche, L. (2002). Description of food surfaces and microstructural changes using fractal image texture analysis. *Journal of Food Engineering*, 53(4), 361–371. DOI: 10.1016/S0260-8774(01)00177-7

Raak, N., Symmank, C., Zahn, S., Aschemann-Witzel, J., & Rohm, H. (2017). Processing- and product-related causes for food waste and implications for the food supply chain. *Waste Management (New York, N.Y.)*, 61, 461–472. DOI: 10.1016/j.wasman.2016.12.027 PMID: 28038904

Radovanović, M., ... (2023), "Edible electronics components for biomedical applications," 2023 22nd International Symposium INFOTEH-JAHORINA (INFOTEH), East Sarajevo, Bosnia and Herzegovina, 2023, pp. 1-4, DOI: 10.1109/INFOTEH57020.2023.10094196

Radovanović, Milan R., Goran M. Stojanović, Mitar Simić, Dragana Suvara, Lazar Milić, Sanja Kojić, and Biljana D. Škrbić. "Edible Electronic Components Made from Recycled Food Waste." Advanced Electronic Materials: 2300905.

Radovanović, M. R., Stojanović, G. M., Simić, M., Suvara, D., Milić, L., Kojic, S. P., & Škrbić, B. D. (2023). Edible Electronic Components Made from Recycled Food Waste. *Advanced Electronic Materials*.

Ragaveena, S., Shirly Edward, A., & Surendran, U. (2021). Smart controlled environment agriculture methods: A holistic review. *Reviews in Environmental Science and Biotechnology*, 20(4), 887–913. DOI: 10.1007/s11157-021-09591-z

Ragavi, B., Pavithra, L., Sandhiyadevi, P., Mohanapriya, G., & Harikirubha, S. (2020). Smart agriculture with AI sensor by using Agrobot. *2020 Fourth International Conference on Computing Methodologies and Communication (ICCMC)*, 1–4. DOI: 10.1109/ICCMC48092.2020.ICCMC-00078

Ramírez, I. J., & Briones, F. (2017). Understanding the El Niño costero of 2017: The definition problem and challenges of climate forecasting and disaster responses. *International Journal of Disaster Risk Science*, 8(4), 489–492. DOI: 10.1007/s13753-017-0151-8

Ramudu, K., Mohan, V. M., Jyothirmai, D., Prasad, D., Agrawal, R., & Boopathi, S. (2023). Machine Learning and Artificial Intelligence in Disease Prediction: Applications, Challenges, Limitations, Case Studies, and Future Directions. In *Contemporary Applications of Data Fusion for Advanced Healthcare Informatics* (pp. 297–318). IGI Global.

Ramy Ghanim et al,(2023),Communication protocols integrating wearables, ingestibles, and implantables for closed-loop therapies,Device,1, 3,100092,ISSN 2666-9986,DOI: 10.1016/j.device.2023.100092

Ray, B., & Bhunia, A. (2013). *Fundamental food microbiology*. CRC Press. DOI: 10.1201/b16078

Revathi, S., Babu, M., Rajkumar, N., Meti, V. K. V., Kandavalli, S. R., & Boopathi, S. (2024). Unleashing the Future Potential of 4D Printing: Exploring Applications in Wearable Technology, Robotics, Energy, Transportation, and Fashion. In *Human-Centered Approaches in Industry 5.0: Human-Machine Interaction, Virtual Reality Training, and Customer Sentiment Analysis* (pp. 131–153). IGI Global.

Rivas-Sánchez, Y. A., Moreno-Pérez, M. F., & Roldán-Cañas, J. (2019). Environment control with low-cost microcontrollers and microprocessors: Application for green walls. *Sustainability (Basel)*, 11(3), 782. DOI: 10.3390/su11030782

Roberts, E. A. H. (1962). *Tea fermentation, Economic Importance of Flavonoid Substance*. Pergamon Press.

Rout, S., Tambe, S., Deshmukh, R. K., Mali, S., Cruz, J., Srivastav, P. P., Amin, P. D., Gaikwad, K. K., Andrade, E. H. de A., & de Oliveira, M. S. (2022). Recent trends in the application of essential oils: The next generation of food preservation and food packaging. *Trends in Food Science & Technology*, 129, 421–439. DOI: 10.1016/j.tifs.2022.10.012

Rumelhart, et al., 1986Rumelhart, D. E., Hinton, G. E., & Williams, R. J. (1986). Learning representations by backpropagating errors. *Nature*, 323(6088), 533–536. DOI: 10.1038/323533a0

S., B. (2024). Advancements in Optimizing Smart Energy Systems Through Smart Grid Integration, Machine Learning, and IoT. In *Advances in Environmental Engineering and Green Technologies* (pp. 33–61). IGI Global. DOI: 10.4018/979-8-3693-0492-1.ch002

Saaid, M. F., Yahya, N. A. M., Noor, M. Z. H., & Ali, M. S. A. M. (2013). A development of an automatic microcontroller system for Deep Water Culture (DWC). *2013 IEEE 9th International Colloquium on Signal Processing and Its Applications*, 328–332. DOI: 10.1109/CSPA.2013.6530066

Sadeghi, K., Kim, J., & Seo, J. (2022). Packaging 4.0: The threshold of an intelligent approach. *Comprehensive Reviews in Food Science and Food Safety*, 21(3), 2615–2638. DOI: 10.1111/1541-4337.12932 PMID: 35279943

Saibi, W., Brini, F., Hanin, M., & Masmoudi, K. (2013). Development of energy plants and their potential to withstand various extreme environments. *Recent Patents on DNA & Gene Sequences*, 7(1), 13–24. DOI: 10.2174/1872215611307010004 PMID: 22779438

Sakai, K., Hassan, M. A., Vairappan, C. S., & Shirai, Y. (2022). Promotion of a green economy with the palm oil industry for biodiversity conservation: A touchstone toward a sustainable bioindustry. *Journal of Bioscience and Bioengineering*, 133(5), 414–424. DOI: 10.1016/j.jbiosc.2022.01.001 PMID: 35151536

Salama, R., & Al-Turjman, F. (2024). An Overview of Blockchain Applications and Benefits in Cloud Computing. AIoT and Smart Sensing Technologies for Smart Devices, 66-76.

Salama, R., & Al-Turjman, F. (2024). Security And Privacy in Mobile Cloud Computing and the Internet of Things. NEU Journal for Artificial Intelligence and Internet of Things, 3(1).

Salama, R., Al-Turjman, F., Bhatla, S., & Mishra, D. (2023, April). Mobile edge fog, Blockchain Networking and Computing-A survey. In 2023 International Conference on Computational Intelligence, Communication Technology and Networking (CICTN) (pp. 808-811). IEEE.

Salama, R., Cacciagrano, D., & Al-Turjman, F. (2024, April). Blockchain and Financial Services a Study of the Applications of Distributed Ledger Technology (DLT) in Financial Services. In International Conference on Advanced Information Networking and Applications (pp. 124-135). Cham: Springer Nature Switzerland.

Salama, R., & Al-Turjman, F. (2023). Cyber-security countermeasures and vulnerabilities to prevent social-engineering attacks. In *Artificial Intelligence of Health-Enabled Spaces* (pp. 133–144). CRC Press. DOI: 10.1201/9781003322887-7

Salama, R., & Al-Turjman, F. (2024). A Description of How AI and Blockchain Technology Are Used in Business. In *AIoT and Smart Sensing Technologies for Smart Devices* (pp. 1–15). IGI Global. DOI: 10.4018/979-8-3693-0786-1.ch001

Salama, R., & Al-Turjman, F. (2024). Overview of the global value chain and the effectiveness of artificial intelligence (AI) techniques in reducing cyber-security risks. In *Smart Global Value Chain* (pp. 137–149). CRC Press. DOI: 10.1201/9781003461432-8

Salama, R., & Al-Turjman, F. (2024). Using artificial intelligence in education applications. In *Computational Intelligence and Blockchain in Complex Systems* (pp. 77–84). Morgan Kaufmann. DOI: 10.1016/B978-0-443-13268-1.00012-1

Salama, R., Al-Turjman, F., Aeri, M., & Yadav, S. P. (2023, April). Intelligent hardware solutions for covid-19 and alike diagnosis-a survey. In *2023 International Conference on Computational Intelligence, Communication Technology and Networking (CICTN)* (pp. 796-800). IEEE. DOI: 10.1109/CICTN57981.2023.10140850

Salama, R., Al-Turjman, F., Aeri, M., & Yadav, S. P. (2023, April). Internet of intelligent things (IoT)–An overview. In *2023 International Conference on Computational Intelligence, Communication Technology and Networking (CICTN)* (pp. 801-805). IEEE. DOI: 10.1109/CICTN57981.2023.10141157

Salama, R., Al-Turjman, F., Altrjman, C., & Bordoloi, D. (2023, April). The ways in which Artificial Intelligence improves several facets of Cyber Security-A survey. In *2023 International Conference on Computational Intelligence, Communication Technology and Networking (CICTN)* (pp. 825-829). IEEE. DOI: 10.1109/CICTN57981.2023.10141376

Salama, R., Al-Turjman, F., Altrjman, C., & Gupta, R. (2023, April). Machine learning in sustainable development–an overview. In *2023 International Conference on Computational Intelligence, Communication Technology and Networking (CICTN)* (pp. 806-807). IEEE.

Salama, R., Al-Turjman, F., Altrjman, C., Kumar, S., & Chaudhary, P. (2023, April). A comprehensive survey of blockchain-powered cybersecurity-a survey. In *2023 International Conference on Computational Intelligence, Communication Technology and Networking (CICTN)* (pp. 774-777). IEEE. DOI: 10.1109/CICTN57981.2023.10141282

Salama, R., Al-Turjman, F., Bhatla, S., & Gautam, D. (2023, April). Network security, trust & privacy in a wiredwireless Environments–An Overview. In *2023 International Conference on Computational Intelligence, Communication Technology and Networking (CICTN)* (pp. 812-816). IEEE. DOI: 10.1109/CICTN57981.2023.10141309

Salama, R., Al-Turjman, F., Bhatla, S., & Yadav, S. P. (2023, April). Social engineering attack types and prevention techniques-A survey. In *2023 International Conference on Computational Intelligence, Communication Technology and Networking (CICTN)* (pp. 817-820). IEEE. DOI: 10.1109/CICTN57981.2023.10140957

Salama, R., Al-Turjman, F., Bordoloi, D., & Yadav, S. P. (2023, April). Wireless sensor networks and green networking for 6G communication-an overview. In *2023 International Conference on Computational Intelligence, Communication Technology and Networking (CICTN)* (pp. 830-834). IEEE. DOI: 10.1109/CICTN57981.2023.10141262

Salama, R., Al-Turjman, F., Chaudhary, P., & Banda, L. (2023, April). Future communication technology using huge millimeter waves—an overview. In *2023 International Conference on Computational Intelligence, Communication Technology and Networking (CICTN)* (pp. 785-790). IEEE. DOI: 10.1109/CICTN57981.2023.10140666

Salama, R., Al-Turjman, F., Chaudhary, P., & Yadav, S. P. (2023, April). (Benefits of Internet of Things (IoT) Applications in Health care-An Overview). In *2023 International Conference on Computational Intelligence, Communication Technology and Networking (CICTN)* (pp. 778-784). IEEE. DOI: 10.1109/CICTN57981.2023.10141452

Salama, R., Al-Turjman, F., & Culmone, R. (2023, March). AI-powered drone to address smart city security issues. In *International Conference on Advanced Information Networking and Applications* (pp. 292-300). Cham: Springer International Publishing. DOI: 10.1007/978-3-031-28694-0_27

Sampath, B., Pandian, M., Deepa, D., & Subbiah, R. (2022). Operating parameters prediction of liquefied petroleum gas refrigerator using simulated annealing algorithm. *AIP Conference Proceedings*, 2460(1), 070003. DOI: 10.1063/5.0095601

Sangeetha, M., Kannan, S. R., Boopathi, S., Ramya, J., Ishrat, M., & Sabarinathan, G. (2023). Prediction of Fruit Texture Features Using Deep Learning Techniques. *2023 4th International Conference on Smart Electronics and Communication (ICOSEC)*, 762–768.

Sankar, K. M., Booba, B., & Boopathi, S. (2023). Smart Agriculture Irrigation Monitoring System Using Internet of Things. In *Contemporary Developments in Agricultural Cyber-Physical Systems* (pp. 105–121). IGI Global. DOI: 10.4018/978-1-6684-7879-0.ch006

Saraswathi, D., Manibharathy, P., Gokulnath, R., Sureshkumar, E., & Karthikeyan, K. (2018). Automation of Hydroponics Green House Farming using IOT. *2018 IEEE International Conference on System, Computation, Automation and Networking (ICSCA)*, 1–4. DOI: 10.1109/ICSCAN.2018.8541251

Saravanan, S., Khare, R., Umamaheswari, K., Khare, S., Krishne Gowda, B. S., & Boopathi, S. (2024). AI and ML Adaptive Smart-Grid Energy Management Systems: Exploring Advanced Innovations. In *Principles and Applications in Speed Sensing and Energy Harvesting for Smart Roads* (pp. 166–196). IGI Global. DOI: 10.4018/978-1-6684-9214-7.ch006

Saravanavel, G., John, S., Wyatt-Moon, G., Flewitt, A., & Sambandan, S. (2022). Edible resonators. *arXiv preprint arXiv*:2202.13782.

Sarker, M. N. I., Peng, Y., Yiran, C., & Shouse, R. C. (2020). Disaster resilience through big data: Way to environmental sustainability. *International Journal of Disaster Risk Reduction*, 51, 101769. DOI: 10.1016/j.ijdrr.2020.101769

Schaertel, B. J., & Firstenberg-Eden, R. (1988). Biosensors in the food industry: Present and future. *Journal of Food Protection*, 51(10), 811–820. DOI: 10.4315/0362-028X-51.10.811 PMID: 28398862

Sekhar, K. Ch., Ingle, R. B., Banu, E. A., Rinawa, M. L., Prasad, M. M., & Boopathi, S. (2024). Integrating VR and AR for Enhanced Production Systems: Immersive Technologies in Smart Manufacturing. In *Advances in Computational Intelligence and Robotics* (pp. 90–112). IGI Global. DOI: 10.4018/979-8-3693-6806-0.ch005

Sekiya, N., Abe, J., Shiotsu, F., & Morita, S. (2015). Effects of partial harvesting on napier grass: Reduced seasonal variability in feedstock supply and increased biomass yield. *Plant Production Science*, 18(1), 99–103. DOI: 10.1626/pps.18.99

Semeano, A. T., Maffei, D. F., Palma, S., Li, R. W., Franco, B. D., Roque, A. C., & Gruber, J. (2018). Tilapia fish microbial spoilage monitored by a single optical gas sensor. *Food Control*, 89, 72–76. DOI: 10.1016/j.foodcont.2018.01.025 PMID: 29503510

Shaikh, F. K., Memon, M. A., Mahoto, N. A., Zeadally, S., & Nebhen, J. (2021). Artificial intelligence best practices in smart agriculture. *IEEE Micro*, 42(1), 17–24. DOI: 10.1109/MM.2021.3121279

Shaikh, T. A., Rasool, T., & Lone, F. R. (2022). Towards leveraging the role of machine learning and artificial intelligence in precision agriculture and smart farming. *Computers and Electronics in Agriculture*, 198, 107119. DOI: 10.1016/j.compag.2022.107119

Sharova, Alina S., et al. "Edible electronics: The vision and the challenge." *Advanced Materials Technologies* 6.2 (2021): 2000757.

Sharova, A. S., Melloni, F., Lanzani, G., Bettinger, C. J., & Caironi, M. (2021). Edible electronics: The vision and the challenge. *Advanced Materials Technologies*, 6(2), 2000757. DOI: 10.1002/admt.202000757

Sharova, A., & Caironi, M. (2021). Sweet Electronics: Honey-Gated Complementary Organic Transistors and Circuits Operating in Air. *Advanced Materials*, 33(40), 2103183. Advance online publication. DOI: 10.1002/adma.202103183 PMID: 34418204

Sharova, A., Modena, F., Luzio, A., Melloni, F., Cataldi, P., Viola, F., Lamanna, L., Zorn, N., Sassi, M., Ronchi, C., Zaumseil, J., Beverina, L., Antognazza, M., & Caironi, M. (2023). Chitosan gated organic transistors printed on ethyl cellulose as a versatile platform for edible electronics and bioelectronics. *Nanoscale*, 15(25), 10808–10819. Advance online publication. DOI: 10.1039/D3NR01051A PMID: 37334549

Shen, S. C., Khare, E., Lee, N. A., Saad, M. K., Kaplan, D. L., & Buehler, M. J. (2023). Computational Design and Manufacturing of Sustainable Materials through First-Principles and Materiomics. *Chemical Reviews*, 123(5), 2242–2275. DOI: 10.1021/acs.chemrev.2c00479 PMID: 36603542

Shi, D., Zhao, B., Zhang, P., Li, P., Wei, X., & Song, K. (2024). Edible composite films: Enhancing the post-harvest preservation of blueberry. *Horticulture, Environment and Biotechnology*, 65(3), 1–19. DOI: 10.1007/s13580-023-00581-4

Shuit, S. H., Tee, S. F., Sim, L. C., & Lim, S. (2021). Biofuels Production from Microalgae: Processes and Conversion Technologies. In Biofuel Production from Microalgae, Macroalgae and Larvae: Processes and Conversion Technologies.

Siddiqui, S. A., Tettey, E., Yunusa, B. M., Ngah, N., Debrah, S. K., Yang, X., Fernando, I., Povetkin, S. N., & Shah, M. A. (2023). Legal situation and consumer acceptance of insects being eaten as human food in different nations across the world–A comprehensive review. *Comprehensive Reviews in Food Science and Food Safety*, 22(6), 4786–4830. DOI: 10.1111/1541-4337.13243 PMID: 37823805

Singh, A., Kuila, A., Adak, S., Bishai, M., & Banerjee, R. (2012). Utilization of Vegetable Wastes for Bioenergy Generation. *Agricultural Research*, 1(3), 213–222. DOI: 10.1007/s40003-012-0030-x

Sitharthan, R., Rajesh, M., Vimal, S., Kumar, S., Yuvaraj, S., Kumar, A., Raglend, J., & Vengatesan, K. (2023). A novel autonomous irrigation system for smart agriculture using AI and 6G enabled IoT network. *Microprocessors and Microsystems*, 101, 104905. DOI: 10.1016/j.micpro.2023.104905

Sit, M., Demiray, B. Z., Xiang, Z., Ewing, G. J., Sermet, Y., & Demir, I. (2020). A comprehensive review of deep learning applications in hydrology and water resources. *Water Science and Technology*, 82(12), 2635–2670. DOI: 10.2166/wst.2020.369 PMID: 33341760

Solaymani, S. (2023). Biodiesel and its potential to mitigate transport-related CO_2 emissions. *Carbon Research*, 2(1), 38. DOI: 10.1007/s44246-023-00067-z

Solinas, S., Fazio, S., Seddaiu, G., Roggero, P. P., Deligios, P. A., Doro, L., & Ledda, L. (2015). Environmental consequences of the conversion from traditional to energy cropping systems in a Mediterranean area. *European Journal of Agronomy*, 70, 124–135. DOI: 10.1016/j.eja.2015.07.008

Sonia, R., Gupta, N., Manikandan, K., Hemalatha, R., Kumar, M. J., & Boopathi, S. (2024). Strengthening Security, Privacy, and Trust in Artificial Intelligence Drones for Smart Cities. In *Analyzing and Mitigating Security Risks in Cloud Computing* (pp. 214–242). IGI Global. DOI: 10.4018/979-8-3693-3249-8.ch011

Sowmya, Y. (2017). A short review on milk spoilage. *Journal of Food Dairy Technology*, 5, 1–5.

Sperber, W. H., & Doyle, M. P. (2009). *Compendium of the microbiological spoilage of foods and beverages*. Springer. DOI: 10.1007/978-1-4419-0826-1

Sreedhar, P. S. S., Sujay, V., Rani, M. R., Melita, L., Reshma, S., & Boopathi, S. (2024). Impacts of 5G Machine Learning Techniques on Telemedicine and Social Media Professional Connection in Healthcare. In *Advances in Medical Technologies and Clinical Practice* (pp. 209–234). IGI Global. DOI: 10.4018/979-8-3693-1934-5.ch012

Stegelmeier, A. A., Rose, D. M., Joris, B. R., & Glick, B. R. (2022). The Use of PGPB to Promote Plant Hydroponic Growth. *Plants*, 11(20), 2783. DOI: 10.3390/plants11202783 PMID: 36297807

Subha, S., Inbamalar, T., Komala, C., Suresh, L. R., Boopathi, S., & Alaskar, K. (2023). A Remote Health Care Monitoring system using internet of medical things (IoMT). *IEEE Explore*, 1–6.

Sumathi, P., Subramanian, R., Karthikeyan, V., & Karthik, S. (2021). Retracted: Soil monitoring and evaluation system using EDL-ASQE: Enhanced deep learning model for IoT smart agriculture network. *International Journal of Communication Systems*, 34(11), e4859. DOI: 10.1002/dac.4859

Sun, 2000Sun, D. W. (2000). Inspecting pizza topping percentage and distribution by a computer visison method. *Journal of Food Engineering*, 44(4), 245–249. DOI: 10.1016/S0260-8774(00)00024-8

Sun, H., Cao, Y., Kim, D., & Marelli, B. (2022). Biomaterials technology for Agro-Food resilience. *Advanced Functional Materials*, 32(30), 2270173. Advance online publication. DOI: 10.1002/adfm.202270173

Sun, R., Li, Y., Du, T., & Qi, Y. (2023). Recent advances in integrated dual-mode optical sensors for food safety detection. *Trends in Food Science & Technology*, 135, 14–31. DOI: 10.1016/j.tifs.2023.03.013

Susanti, N. D., Sagita, D., Apriyanto, I. F., Anggara, C. E. W., Darmajana, D. A., & Rahayuningtyas, A. (2022). Design and implementation of water quality monitoring system (temperature, ph, tds) in aquaculture using iot at low cost. *6th International Conference of Food, Agriculture, and Natural Resource (IC-FANRES 2021)*, 7–11. DOI: 10.2991/absr.k.220101.002

Swain, et al., 1991Swain, M. J., & Ballard, D. H. (1991). Color Indexing. *International Journal of Computer Vision*, 7(1), 11–32. DOI: 10.1007/BF00130487

Tagesse, T., Arulkumar, S., Mahalingam, S., Subbarao Tadepalli, N. V. R., Munjal, N., & Boopathi, S. (2024). Analyzing Fuel Cell Vehicles in India via the PESTLE Framework and Intelligent Battery Management Systems (BMS): Blockchain in E-Mobility. In *Advances in Mechatronics and Mechanical Engineering* (pp. 107–129). IGI Global. DOI: 10.4018/979-8-3693-5247-2.ch007

Tahmasebi, P., Kamrava, S., Bai, T., & Sahimi, M. (2020). Machine learning in geo-and environmental sciences: From small to large scale. *Advances in Water Resources*, 142, 103619. DOI: 10.1016/j.advwatres.2020.103619

Takemoto, M., Yunoki, A., Miao, S., & Dowaki, K. (2023). Environmental Impact Analysis of Food Considering Upcycling. *IOP Conference Series. Earth and Environmental Science*, 1187(1), 1187. DOI: 10.1088/1755-1315/1187/1/012032

Takeo, et al., 1983Takeo, T., & Mahanta, P. K. (1983). Comparison of black tea aromas of orthodox and CTC tea and black teas made from different varieties. *Journal of the Science of Food and Agriculture*, 34(3), 307–310. DOI: 10.1002/jsfa.2740340315

Takeuchi, Y. (2019). 3D Printable Hydroponics: A Digital Fabrication Pipeline for Soilless Plant Cultivation. *IEEE Access : Practical Innovations, Open Solutions*, 7, 35863–35873. DOI: 10.1109/ACCESS.2019.2905233

Tao, F., Qi, Q., Liu, A., & Kusiak, A. (2018). Data-driven smart manufacturing. *Journal of Manufacturing Systems*, 48, 157–169. DOI: 10.1016/j.jmsy.2018.01.006

Teixeira, S. C., Gomes, N. O., Oliveira, T. V., Fortes-Da-Silva, P., Soares, N. D., & Raymundo-Pereira, P. A. (2023). Review and Perspectives of sustainable, biodegradable, eco-friendly and flexible electronic devices and (Bio)sensors. *Biosensors and Bioelectronics: X.*, Volume 14,2023,100371, ISSN 2590-1370, https://doi.org/ DOI: 10.1016/j.biosx.100371

Teja, N. B., Kannagi, V., Chandrashekhar, A., Senthilnathan, T., Pal, T. K., & Boopathi, S. (2024). Impacts of Nano-Materials and Nano Fluids on the Robot Industry and Environments. In *Advances in Computational Intelligence and Robotics* (pp. 171–194). IGI Global. DOI: 10.4018/979-8-3693-5767-5.ch012

Thwaites, P. A., Yao, C. K., Halmos, E. P., Muir, J. G., Burgell, R. E., Berean, K. J., Kalantar-zadeh, K., & Gibson, P. R. (2024, February). Review article: Current status and future directions of ingestible electronic devices in gastroenterology. *Alimentary Pharmacology & Therapeutics*, 59(4), 459–474. DOI: 10.1111/apt.17844 PMID: 38168738

Tirlangi, S., Teotia, S., Padmapriya, G., Senthil Kumar, S., Dhotre, S., & Boopathi, S. (2024). Cloud Computing and Machine Learning in the Green Power Sector: Data Management and Analysis for Sustainable Energy. In *Developments Towards Next Generation Intelligent Systems for Sustainable Development* (pp. 148–179). IGI Global. DOI: 10.4018/979-8-3693-5643-2.ch006

Tomar, P., & Kaur, G. (2021). *Artificial Intelligence and IoT-based Technologies for Sustainable Farming and Smart Agriculture*. IGI global. DOI: 10.4018/978-1-7998-1722-2

Trajkovska Petkoska, A., Daniloski, D., D'Cunha, N. M., Naumovski, N., & Broach, A. T. (2021). Edible packaging: Sustainable solutions and novel trends in food packaging. *Food Research International*, 140, 109981. DOI: 10.1016/j.foodres.2020.109981 PMID: 33648216

Tripathi, S., Hakim, L., Gupta, D., Deshmukh, R. K., & Gaikwad, K. K. (2024). Recent Trends in Films and Gases for Modified Atmosphere Packaging of Fresh Produce. In Novel Packaging Systems for Fruits and Vegetables (1st ed., pp. 27). Apple Academic Press. eBook ISBN: 9781003415701 DOI: 10.1201/9781003415701-3

Turasan, H., & Kokini, J. (2021). Novel nondestructive biosensors for the food industry. *Annual Review of Food Science and Technology*, 12(1), 539–566. DOI: 10.1146/annurev-food-062520-082307 PMID: 33770468

Tutul, M. J. I., Alam, M., & Wadud, M. A. H. (2023). Smart food monitoring system based on iot and machine learning. *2023 International Conference on Next-Generation Computing, IoT and Machine Learning (NCIM)*, 1–6. DOI: 10.1109/NCIM59001.2023.10212608

Ugandar, R., Rahamathunnisa, U., Sajithra, S., Christiana, M. B. V., Palai, B. K., & Boopathi, S. (2023). Hospital Waste Management Using Internet of Things and Deep Learning: Enhanced Efficiency and Sustainability. In *Applications of Synthetic Biology in Health, Energy, and Environment* (pp. 317–343). IGI Global.

Ulfa, M., Setyonugroho, W., Lestari, T., Widiasih, E., & Nguyen Quoc, A. (2022). Nutrition-Related Mobile Application for Daily Dietary Self-Monitoring. *Journal of Nutrition and Metabolism*, 30, 2476367. DOI: 10.1155/2022/2476367 PMID: 36082357

Ullah, M. R., Bajaj, K. L., & Chakraborty, S. (1979). *Fermentation test by Dr. M R Ullah, Factory Floor Fermentation Test*. Tocklai Experimental Station.

Umor, N. A., Ismail, S., Abdullah, S., Huzaifah, M. H. R., Huzir, N. M., Mahmood, N. A. N., & Zahrim, A. Y. (2021). Zero waste management of spent mushroom compost. *Journal of Material Cycles and Waste Management*, 23(5), 1726–1736. DOI: 10.1007/s10163-021-01250-3

Upadhyaya, A. N., Saqib, A., Devi, J. V., Rallapalli, S., Sudha, S., & Boopathi, S. (2024). Implementation of the Internet of Things (IoT) in Remote Healthcare. In *Advances in Medical Technologies and Clinical Practice* (pp. 104–124). IGI Global. DOI: 10.4018/979-8-3693-1934-5.ch006

Usigbe, M. J., Asem-Hiablie, S., Uyeh, D. D., Iyiola, O., Park, T., & Mallipeddi, R. (2024). Enhancing resilience in agricultural production systems with AI-based technologies. *Environment, Development and Sustainability*, 26(9), 21955–21983. DOI: 10.1007/s10668-023-03588-0

Vangeri, A. K., Bathrinath, S., Anand, M. C. J., Shanmugathai, M., Meenatchi, N., & Boopathi, S. (2024). Green Supply Chain Management in Eco-Friendly Sustainable Manufacturing Industries. In *Environmental Applications of Carbon-Based Materials* (pp. 253–287). IGI Global., DOI: 10.4018/979-8-3693-3625-0.ch010

Vashishth, R., Pandey, A. K., Kaur, P., & Semwal, A. D. (2022). Smart technologies in food manufacturing. In *Smart and Sustainable Food Technologies* (pp. 125–155). Springer. DOI: 10.1007/978-981-19-1746-2_5

Velazquez, L. A., Hernandez, M. A., Leon, M., Dominguez, R. B., & Gutierrez, J. M. (2013). First advances on the development of a hydroponic system for cherry tomato culture. *2013 10th International Conference on Electrical Engineering, Computing Science and Automatic Control (CCE)*, 155–159. DOI: 10.1109/ICEEE.2013.6676029

Velusamy, V., Arshak, K., Korostynska, O., Oliwa, K., & Adley, C. (2010). An overview of foodborne pathogen detection: In the perspective of biosensors. *Biotechnology Advances*, 28(2), 232–254. DOI: 10.1016/j.biotechadv.2009.12.004 PMID: 20006978

Venkatesh, A., Saravanakumar, T., Vairamsrinivasan, S., Vigneshwar, A., & Kumar, M. S. (2017). A food monitoring system based on bluetooth low energy and Internet of Things. *International Journal of Engineering Research and Applications*, 7(3), 30–34. DOI: 10.9790/9622-0703063034

Venkateswaran, N., Kunduru, K. R., Ashwin, N., Sundar Ganesh, C. S., Hema, N., & Boopathi, S. (2024). Navigating the Future of Ultra-Smart Computing Cyberspace: Beyond Boundaries. In *Advances in Computational Intelligence and Robotics* (pp. 170–199). IGI Global. DOI: 10.4018/979-8-3693-2399-1.ch007

Venkateswaran, N., Kiran Kumar, K., Maheswari, K., Kumar Reddy, R. V., & Boopathi, S. (2024). Optimizing IoT Data Aggregation: Hybrid Firefly-Artificial Bee Colony Algorithm for Enhanced Efficiency in Agriculture. *AGRIS On-Line Papers in Economics and Informatics*, 16(1), 117–130. DOI: 10.7160/aol.2024.160110

Venkateswaran, N., Kumar, S. S., Diwakar, G., Gnanasangeetha, D., & Boopathi, S. (2023). Synthetic Biology for Waste Water to Energy Conversion: IoT and AI Approaches. *Applications of Synthetic Biology in Health. Energy & Environment*, •••, 360–384.

Vergara, A., Llobet, E., Ramírez, J., Ivanov, P., Fonseca, L., Zampolli, S., Scorzoni, A., Becker, T., Marco, S., & Wöllenstein, J. (2007). An RFID reader with onboard sensing capability for monitoring fruit quality. *Sensors and Actuators. B, Chemical*, 127(1), 143–149. DOI: 10.1016/j.snb.2007.07.107

Verma, R., Christiana, M. B. V., Maheswari, M., Srinivasan, V., Patro, P., Dari, S. S., & Boopathi, S. (2024). Intelligent Physarum Solver for Profit Maximization in Oligopolistic Supply Chain Networks. In *AI and Machine Learning Impacts in Intelligent Supply Chain* (pp. 156–179). IGI Global. DOI: 10.4018/979-8-3693-1347-3.ch011

Vidhya, R., & Valarmathi, K. (2018). Survey on Automatic Monitoring of Hydroponics Farms Using IoT. *2018 3rd International Conference on Communication and Electronics Systems (ICCES)*, 125–128. DOI: 10.1109/CESYS.2018.8724103

Vieira, F., Santana, H. E., Jesus, M., Santos, J., Pires, P., Vaz-Velho, M., Silva, D. P., & Ruzene, D. S. (2024). Coconut Waste: Discovering Sustainable Approaches to Advance a Circular Economy. *Sustainability (Basel)*, 16(7), 3066. DOI: 10.3390/su16073066

Vlahova-Vangelova, D., Balev, D., Kolev, N., Dragoev, S., Petkov, E., & Popova, T. (2024). Comparison of the Effect of Drying Treatments on the Physicochemical Parameters, Oxidative Stability, and Microbiological Status of Yellow Mealworm (Tenebrio molitor L.) Flours as an Alternative Protein Source. *Agriculture*, 14(3), 436. DOI: 10.3390/agriculture14030436

Vushe, A. (2021). Proposed research, science, technology, and innovation to address current and future challenges of climate change and water resource management in Africa. Climate Change and Water Resources in Africa: Perspectives and Solutions Towards an Imminent Water Crisis, 489-518.

Waheed, H., Hassan, S.-U., Aljohani, N. R., Hardman, J., Alelyani, S., & Nawaz, R. (2020). Predicting academic performance of students from VLE big data using deep learning models. *Computers in Human Behavior*, 104, 106189. DOI: 10.1016/j.chb.2019.106189

Waimin, J., Gopalakrishnan, S., Heredia-Rivera, U., Kerr, N. A., Nejati, S., Gallina, N. L., Bhunia, A. K., & Rahimi, R. (2022). Low-cost nonreversible electronic-free wireless pH sensor for spoilage detection in packaged meat products. *ACS Applied Materials & Interfaces*, 14(40), 45752–45764. DOI: 10.1021/acsami.2c09265 PMID: 36173396

Wakchaure, M., Patle, B., & Mahindrakar, A. (2023). Application of AI techniques and robotics in agriculture: A review. *Artificial Intelligence in the Life Sciences*, 3, 100057. DOI: 10.1016/j.ailsci.2023.100057

Wang, et al., 1997 Wang, J. [Wang, et al., 1997] Wang, J. Ze., Wiederhold, G., Firschein, O., 1997. Wavelet based image indexing techniques with partial sketch retrieval capability, In Proc. 4th Forum on Research and Technology Advances in Digital Libraries, pages 13-24.

Wang, S., Cao, J., & Philip, S. Y. (2020). Deep learning for spatio-temporal data mining: A survey. *IEEE Transactions on Knowledge and Data Engineering*, 34(8), 3681–3700. DOI: 10.1109/TKDE.2020.3025580

Wang, W., Wang, Y., Gong, Z., Yang, S., & Jia, F. (2021). Comparison of the nutritional properties and transcriptome profiling between the two different harvesting periods of Auricularia polytricha. *Frontiers in Nutrition*, 8, 771757. DOI: 10.3389/fnut.2021.771757 PMID: 34765633

Wang, X., Fu, D., Fruk, G., Chen, E., & Zhang, X. (2018). Improving quality control and transparency in honey peach export chain by a multi-sensors-managed traceability system. *Food Control*, 88, 169–180. DOI: 10.1016/j.foodcont.2018.01.008

Wei, Y., Li, W., An, D., Li, D., Jiao, Y., & Wei, Q. (2019). Equipment and Intelligent Control System in Aquaponics: A Review. *IEEE Access : Practical Innovations, Open Solutions*, 7, 169306–169326. DOI: 10.1109/ACCESS.2019.2953491

Woittiez, L. S., Rufino, M. C., Giller, K. E., & Mapfumo, P. (2013). The use of woodland products to cope with climate variability in communal areas in Zimbabwe. *Ecology and Society*, 18(4), art24. DOI: 10.5751/ES-05705-180424

Wright, W. D. (1969). *The measurement of colour* (4th ed.). Hilger & Watts.

Wu JY et al, (2023) IoT-based wearable health monitoring device and its validation for potential critical and emergency applications. Front Public Health. 16;11:1188304. .DOI: 10.3389/fpubh.2023.1188304

Wu, W., Emerton, R., Duan, Q., Wood, A. W., Wetterhall, F., & Robertson, D. E. (2020). Ensemble flood forecasting: Current status and future opportunities. *WIREs. Water*, 7(3), e1432. DOI: 10.1002/wat2.1432

Xiang, W., Wang, H. W., Tian, Y., & Sun, D. W. (2021). Effects of salicylic acid combined with gas atmospheric control on post-harvest quality and storage stability of wolfberries: Quality attributes and interaction evaluation. *Journal of Food Process Engineering*, 44(8), e13764. DOI: 10.1111/jfpe.13764

Xu, et al., 2018 Xu, M., & Wang, J. [Xu, et al., 2018] Xu, M.; Wang, J. The Qualitative and Quantitative Assessment of Tea Quality Based on E-nose, E-tongue and E-eye Signals Combining With Chemometrics Methods. In Proceedings of the 2018 ASABE Annual International Meeting, Detroit, MI, USA, 29 July–1 August 2018; pages 1800610.

Xu, et al., 2019Xu, M., Wang, J., & Zhu, L. (2019). The Qualitative and Quantitative Assessment of Tea Quality Based on E-nose, E-tongue and E-eye Combined with Chemometrics. *Food Chemistry*, 289, 482–489. DOI: 10.1016/j.foodchem.2019.03.080 PMID: 30955639

Xu, W., Yang, H., Zeng, W., Houghton, T., Wang, X., Murthy, R., Kim, H., Lin, Y., Mignolet, M., Duan, H., Yu, H., Slepian, M., & Jiang, H. (2017). Food-Based Edible and Nutritive Electronics. *Advanced Materials Technologies*, 2(11), 1700181. DOI: 10.1002/admt.201700181

Ya'acob, N., Dzulkefli, N., Yusof, A., Kassim, M., Naim, N., & Aris, S. (2021). Water quality monitoring system for fisheries using internet of things (iot). *IOP Conference Series. Materials Science and Engineering*, 1176(1), 012016. DOI: 10.1088/1757-899X/1176/1/012016

Yang, C., Huang, Y., & Zheng, W. (n.d.). Research of hydroponics nutrient solution control technology. *Fifth World Congress on Intelligent Control and Automation (IEEE Cat. No.04EX788), 1*, 642–644. DOI: 10.1109/WCICA.2004.1340657

Yang, K., Bai, C., Liu, B., Liu, Z., & Cui, X. (2023). Self-Powered, Non-Toxic, Recyclable Thermogalvanic Hydrogel Sensor for Temperature Monitoring of Edibles. *Micromachines*, 14(7), 1327. DOI: 10.3390/mi14071327 PMID: 37512638

Yang, S. Y., Sencadas, V., You, S. S., Jia, N. Z.-X., Srinivasan, S. S., Huang, H.-W., Ahmed, A. E., Liang, J. Y., & Traverso, G. (2021, October 26). Powering Implantable and Ingestible Electronics. *Advanced Functional Materials*, 31(44), 2009289. DOI: 10.1002/adfm.202009289 PMID: 34720792

Yan, H., Moradkhani, H., & Zarekarizi, M. (2017). A probabilistic drought forecasting framework: A combined dynamical and statistical approach. *Journal of Hydrology (Amsterdam)*, 548, 291–304. DOI: 10.1016/j.jhydrol.2017.03.004

Yeatman, E.M. (2009). Energy harvesting: small scale energy production from ambient sources. Smart Structures and Materials + Nondestructive Evaluation and Health Monitoring.

Yin, H.-Y., Fang, T. J., Li, Y.-T., Fung, Y.-F., Tsai, W.-C., Dai, H.-Y., & Wen, H.-W. (2019). Rapidly detecting major peanut allergen-Ara h2 in edible oils using a new immunomagnetic nanoparticle-based lateral flow assay. *Food Chemistry*, 271, 505–515. DOI: 10.1016/j.foodchem.2018.07.064 PMID: 30236709

Yin, S., Bao, J., Zhang, Y., & Huang, X. (2017). M2M security technology of CPS based on Blockchains. *Symmetry*, 9(9), 193. DOI: 10.3390/sym9090193

Youn, S.B., Yeo, J., Joung, H., & Yang, Y. (2015). Energy harvesting from food waste by inoculation of vermicomposted organic matter into Microbial Fuel Cell (MFC). 2015 IEEE SENSORS, 1-4.

Yuan, Q., Shen, H., Li, T., Li, Z., Li, S., Jiang, Y., Xu, H., Tan, W., Yang, Q., Wang, J., Gao, J., & Zhang, L. (2020). Deep learning in environmental remote sensing: Achievements and challenges. *Remote Sensing of Environment*, 241, 111716. DOI: 10.1016/j.rse.2020.111716

Zeng, Z., Piao, S., & Li, L. (2017). *Climate mitigation from vegetation biophysical feedbacks during the past three decades.* Nature Clim Change., DOI: 10.1038/nclimate3299

Zhang, Z., Ao, Y., Su, N., Chen, Y., Wang, K., & Ou, L. (2022). Dynamic changes in morphology and composition during seed development in Xanthoceras sorbifolium Bunge. *Industrial Crops and Products*, 190, 115899. DOI: 10.1016/j.indcrop.2022.115899

Zhong, S., Zhang, K., Bagheri, M., Burken, J. G., Gu, A., Li, B., Ma, X., Marrone, B. L., Ren, Z. J., Schrier, J., Shi, W., Tan, H., Wang, T., Wang, X., Wong, B. M., Xiao, X., Yu, X., Zhu, J.-J., & Zhang, H. (2021a). Machine learning: New ideas and tools in environmental science and engineering. *Environmental Science & Technology*, 55(19), 12741–12754. DOI: 10.1021/acs.est.1c01339 PMID: 34403250

Zhu, et al., 2019Zhu, H., Liu, F., Ye, Y., Chen, L., Liu, J., Gui, A., Zhang, J., & Dong, C. (2019). Jingyuan Liu, Anhui Gui, Jianqiang Zhang, Chunwang Dong Application of machine learning algorithms in quality assurance of fermentation process of black tea– based on electrical properties. *Journal of Food Engineering*, 263, 165–172. DOI: 10.1016/j.jfoodeng.2019.06.009

Zou, Q., Lin, G., Jiang, X., Liu, X., & Zeng, W. (2019). Smart technologies for food quality assurance and safety: Status and perspectives. *Trends in Food Science & Technology*, 91, 18–28.

Zou, Y., Ren, Z., Xiang, Y., Guo, H., Yang, G.-Z., & Tao, G. (2024). Flexible fiberbotic laser scalpels: Material and fabrication challenges. *Matter*, 7(3), 758–771. DOI: 10.1016/j.matt.2024.01.007

Zvezdin, A., Mauro, E. D., Rho, D., Santato, C., & Khalil, M. S. (2020). En route toward sustainable organic electronics. *MRS Energy & Sustainability : a Review Journal*, 7(1), 1–8. DOI: 10.1557/mre.2020.16

About the Contributors

Shilpa Mehta has completed her PhD degree from Auckland University of Technology, New Zealand. She has five years of teaching experience from different reputed colleges and universities of India. Currently, she is working as a Teaching Assistant at Auckland University of Technology (AUT), Auckland, New Zealand. She was awarded Summer Doctoral Research Scholarship for her PhD research. She has worked on various interdisciplinary research projects and her research interests include Radio Frequency Integrated Circuits, RF front ends, Optimization, Internet of Things, Wireless Communication, Artificial Intelligence, Healthcare, Radars, and Software-defined Radios, Smart Cities. She is working as a Series Editor for one of the Bentham book series. As an editor, she has edited four books with IGI Global and one with Nova Science Publishers. Additionally, she is working on 11 edited books to be published by IGI Global, Apple Academic Press, Scrivener-Wiley, and Springer Nature, respectively.

Fadi Al-Turjman received his Ph.D. in computer science from Queen's University, Canada, in 2011. He is the associate dean for research and the founding director of the International Research Center for AI and IoT at Near East University, Nicosia, Cyprus. Prof. Al-Turjman is the head of Artificial Intelligence Engineering Dept., and a leading authority in the areas of smart/intelligent IoT systems, wireless, and mobile networks' architectures, protocols, deployments, and performance evaluation in Artificial Intelligence of Things (AIoT). His publication history spans over 650 SCI/E publications, in addition to numerous keynotes and plenary talks at flagship venues. He has authored and edited more than 70 books about cognition, security, and wireless sensor networks' deployments in smart IoT environments, which have been published by well-reputed publishers such as Taylor and Francis, Elsevier, IET, and Springer. He has received several recognitions and best papers' awards at top international conferences. He also received the prestigious Best Research Paper Award from Elsevier Computer Communications Journal for the period 2015-2018, in addition to the Top Researcher Award for 2018 at Antalya Bilim University, Turkey and the Lifetime Golden-award of Dr. Suat Gunsel from Near East University, Cyprus in 2022. Prof. Al-Turjman has led a number of international symposia and workshops in flagship communication society conferences. Currently, he serves as book series editor and the lead guest/associate editor for several top tier journals, including the IEEE Communications Surveys and Tutorials (IF 23.9) and the Elsevier Sustainable Cities and Society (IF 10.8), in addition to organizing international conferences and symposiums on the most up to date research topics in AI and IoT.

Sampath Boopathi is an accomplished individual with a strong academic background and extensive research experience. He completed his undergraduate studies in Mechanical Engineering and pursued his postgraduate studies in the field of Computer-Aided Design. Dr. Boopathi obtained his Ph.D. from Anna University, focusing his research on Manufacturing and optimization. Throughout his career, Dr. Boopathi has made significant contributions to the field of engineering. He has authored and published over 300 research articles in internationally peer-reviewed journals, highlighting his expertise and dedication to advancing knowledge in his area of specialization. His research output demonstrates his commitment to conducting rigorous and impactful research. In addition to his research publications, Dr. Boopathi has also been granted one patent and has three published patents to his name. This indicates his innovative thinking and ability to develop practical solutions to real-world engineering challenges. With 17 years of academic and research experience, Dr. Boopathi has enriched the engineering community through his teaching and mentorship roles.

Sahil Manoj Chavan is a faculty member in the Department of Electrical Power System at Sandip University, located in Nashik, Maharashtra 422213, India

E. Kannan, Registrar & Professor in School of Computing at Vel Tech University, Avadi, Chennai. He obtained his Ph.D from NIT, Trichy in the year 2006. His research interest spans across computer networking and parallel computing. Much of his work has been on improvising the understanding, design and the performance of parallel and networked computer systems, mainly through the application of data mining, statistics and performance evaluation. He has published more than 100 research publications in Scopus and SCI indexed international journals and conferences. He began his career in 1991 and has held various responsible positions in reputed institutions. He is also a member in many professional's society and a senior member in IEEE.

Vishal Eswaran is an accomplished Senior Big Data Engineer with an impressive career spanning over 6 years. His fervor for constructing robust data pipelines, unearthing insights from intricate datasets, identifying trends, and predicting future trajectories has fueled his journey. Throughout his tenure, Vishal has lent his expertise to empower numerous prominent US healthcare clients, including CVS Health, Aetna, and Blue Cross and Blue Shield of North Carolina, with informed business decisions drawn from expansive datasets. Vishal's ability to distill intricate data into comprehensive documents and reports stands as a testament to his proficiency in managing multifaceted internal and external data analysis responsibilities. His aptitude for synthesizing complex information ensures that insights are both accessible and impactful for strategic decision-making. Moreover, Vishal's distinction extends to his role as a co-author of the book "Internet of Things - Future Connected Devices." This book not only underscores his prowess in the field but also showcases his visionary leadership in the realm of Internet of Things (IoT). His insights resonate with a forward-looking perspective, emphasizing the convergence of technology and human life. As the author of "Secure Connections: Safeguarding the Internet of Things (IoT) with Cybersecurity," Vishal Eswaran's reputation as a thought leader is further solidified. His work is a manifestation of his commitment to ensuring the security of interconnected devices within the IoT landscape, a vital consideration in our digitally driven world. Vishal's dedication to enhancing the safety and integrity of IoT ecosystems shines through in his work.

Vivek eswaran With 8 years of experience as a Senior Software Engineer specializing in front-end development, Vivek Eswaran brings a vital perspective to securing the Internet of Things (IoT). At Medallia, Vivek played an instrumental role in crafting engaging user interfaces and optimized digital experiences. This profound expertise in front-end engineering equips them to illuminate the crucial synergy between usability and cybersecurity as IoT adoption accelerates. In the new book "Secure Connections: Safeguarding the Internet of Things with Cybersecurity," Vivek combines their real-world experiences building intuitive and secure software systems with cutting-edge insights into strengthening IoT ecosystems. Drawing parallels between front-end best practices and security imperatives, they offer readers an invaluable guide for fortifying IoT without compromising usability. As businesses and consumers continue rapidly connecting people, processes, and devices, Vivek's contribution provides timely insights. Blending user empathy with security proficiency, Vivek empowers audiences to realize the potential of IoT through resilient and human-centered systems designed for safety without friction.

Ushaa Eswaran is an esteemed author, distinguished researcher, and seasoned educator with a remarkable journey spanning over 34 years, dedicated to advancing academia and nurturing the potential of young minds. Currently serving as a Principal and Professor in Andhra Pradesh, India, her vision extends beyond imparting cutting-edge technical expertise to encompass the nurturing of universal human values. With a foundation in Electronics Engineering, Dr. Eswaran delved into the realm of biosensors, carving a pioneering path in nanosensor models, a remarkable achievement that earned her a well-deserved Doctorate. Her insights have been encapsulated in her acclaimed book, "Internet of Things: Future Connected Devices," offering profound insights into the evolving IoT landscape. Her expertise also finds its place in upcoming publications centered around computer vision and IoT technologies. Dr. Eswaran's commitment to literature is rooted in her unwavering passion to equip the younger generation with the latest knowledge fortified by ethical principles. Her book stands as a beacon of practical wisdom, providing a roadmap through the intricate IoT terrain while shedding light on its future societal impacts. Her forthcoming contributions unveil her interdisciplinary perspective, seamlessly integrating electronics, nanotechnology, and computing. Bolstering her scholarly contributions is her ORCID identifier, 0000-0002-5116-3403, a testament to her prolific research journey that encompasses over a hundred published papers. Dr. Eswaran thrives in merging her profound academic insights with her dedication to nurturing holistic student growth. Her tireless exploration of the dynamic interface between technology and human values continues to shape her works. As the author of "Secure Connections: Safeguarding the Internet of Things (IoT) with Cybersecurity," Dr. Ushaa Eswaran's voice emerges as a beacon of wisdom in the realm of IoT. Her work encapsulates her dedication to enhancing the interconnected world while ensuring its resilience against cyber threats. Dr. Ushaa Eswaran is an esteemed author, distinguished researcher, and seasoned educator with a remarkable journey spanning over 34 years, dedicated to advancing academia and nurturing the potential of young minds. Currently serving as a Principal and Professor in Andhra Pradesh, India, her vision extends beyond imparting cutting-edge technical expertise to encompass the nurturing of universal human values. With a foundation in Electronics Engineering, Dr. Eswaran delved into the realm of biosensors, carving a pioneering path in nanosensor models, a remarkable achievement that earned her a well-deserved Doctorate. Her insights have been encapsulated in her acclaimed book, "Internet of Things: Future Connected Devices," offering profound insights into the evolving IoT landscape. Her expertise also finds its place in upcoming publications centered around computer vision and IoT technologies. Dr. Eswaran's commitment to literature is rooted in her unwavering passion to equip the younger generation with the latest knowledge fortified by ethical principles. Her book stands as a beacon of practical wisdom, providing a roadmap through the intricate IoT terrain while shedding light on its future societal impacts. Her forthcoming contributions unveil her interdisciplinary perspective, seamlessly integrating electronics, nanotechnology, and computing. Bolstering her scholarly contributions is her ORCID identifier, 0000-0002-5116-3403, a testament to her prolific research journey that encompasses over a hundred published papers. Dr. Eswaran thrives in merging her profound academic insights with her dedication to nurturing holistic student growth. Her tireless exploration of the dynamic interface between technology and human values continues to shape her works. As the author of "Secure Connections: Safeguarding the Internet of Things (IoT) with Cybersecurity," Dr. Ushaa Eswaran's voice emerges as a beacon of wisdom in the realm of IoT. Her work encapsulates her dedication to enhancing the interconnected world while ensuring its resilience against cyber threats.

Babitha Hemanth is an Electrical and Electronics Engineer with a Master's degree in Industrial Automation and Robotics. With a blend of 4 years of teaching experience and 6.4 years in industry, she brings a unique perspective to her role as Assistant Professor at Sahyadri College of Engineering and Management. Babitha is also a published author, contributing her expertise to the field.

J. Mangaiyarkkarasi is currently working as an Associate Professor of Physics in NMSS Vellaichamy Nadar College, Nagamalai, Madurai- 625 019, Tamil Nadu, India. She has 23 years of teaching experience. Her major research interest includes synthesis of ceramic materials, powder X-ray analysis and electron density studies. She has published 11 research papers in the refereed International journals and five book chapters. She has prepared study materials (UG& PG Physics) for Madurai Kamaraj University Distance Education Course.

S. N Kumar is currently an Associate. Professor, Dept. of Electrical and Electronics Engineering., Amal Jyothi College of Engineering, Kanjirappally, Kerala. He was awarded PhD in Electronics and Communication Engineering from Sathyabama Institute of Science and Technology, Chennai. He has 15 years of teaching experience, and his areas of interest include Medical Image Processing and Embedded System Applications in Telemedicine. He was the Co- Principal Investigator of the DST IDP-funded project and research fellow of the RCA scheme, NTU Singapore. He has edited 3 books, authored 32 book chapters and 30 publications in peer reviewed journals. He had authored 12 text books in engineering discipline.

Keerthana Murali With over 5 years of experience as a Site Reliability Engineer at Dell, Keerthna Murali has honed an intricate expertise in maintaining and optimizing robust digital infrastructures. On the frontlines of ensuring seamless online experiences, Keerthna specialized in troubleshooting complex issues and proactively enhancing system performance and availability. These capabilities uniquely position them to tackle the critical challenge of security for rapidly emerging IoT ecosystems. In their new book "Secure Connections," Keerthna channels their real-world experiences maintaining enterprise-scale platforms into a compelling vision for building security into the foundation of IoT systems. Blending software engineering best practices with cybersecurity insights, they offer a prescient guide for developers, IT leaders, and security experts seeking to realize IoT's potential while mitigating its risks.

Index

A

Access Control 321, 333, 334, 335, 343, 344
Agriculture 75, 76, 105, 106, 107, 108, 109, 119, 128, 133, 146, 162, 193, 195, 238, 240, 244, 260, 265, 266, 277, 325, 351, 352, 353, 354, 355, 357, 358, 359, 360, 361, 362, 363, 366, 367, 368, 370, 371, 372, 373, 374, 375, 376, 377, 378, 379, 380, 381, 382, 409, 446, 447, 460, 472, 473, 474, 475
AI-controlled robotics 351, 352, 353, 354, 355, 358, 359
anomaly detection 27, 31, 37, 65, 66, 85, 89, 90, 92, 93, 94, 95, 103, 104, 307, 320, 324, 335, 340, 393
Artificial Intelligence 24, 25, 27, 28, 31, 37, 43, 44, 46, 49, 72, 101, 105, 106, 107, 110, 136, 152, 167, 178, 192, 193, 194, 212, 213, 220, 235, 244, 260, 296, 299, 300, 302, 303, 328, 329, 331, 332, 346, 347, 349, 350, 352, 358, 359, 363, 369, 374, 379, 380, 381, 382, 386, 390, 393, 400, 409, 410, 411, 412, 442, 443
Authentication 6, 168, 169, 174, 259, 282, 295, 307, 316, 317, 320, 322, 324, 327, 337, 338, 340, 341, 343, 344
Automated Alerts 73, 86, 87, 384, 386
automation 51, 52, 53, 55, 61, 62, 63, 175, 176, 179, 183, 186, 187, 193, 239, 263, 280, 281, 282, 283, 287, 288, 289, 290, 291, 292, 293, 294, 295, 297, 298, 307, 309, 310, 311, 337, 354, 363, 366, 367, 373, 374, 380, 460, 461, 473

B

biosensors 7, 57, 58, 72, 76, 125, 149, 151, 152, 154, 156, 162, 163, 164, 165, 166, 200, 207, 214, 231, 262

C

climate modeling 423, 425, 435, 437, 439, 440
Consumer empowerment 18, 290, 310
Cybersecurity 71, 105, 107, 282, 287, 294, 295, 296, 298, 299, 300, 302, 308, 310, 314, 318, 322, 324, 327, 328, 329, 331, 346, 347, 349, 354, 374, 376, 409

D

data analytics 52, 60, 64, 66, 67, 73, 95, 104, 105, 106, 230, 307, 308, 323, 352, 360, 371, 372, 377, 384, 387, 405, 409, 441
Data Encryption 333, 335
Data Security 20, 28, 41, 42, 44, 138, 259, 299, 305, 306, 307, 309, 324, 325, 326, 327, 329, 347, 354, 371, 374, 378, 404
deep learning 27, 31, 37, 46, 79, 80, 81, 82, 83, 88, 89, 90, 91, 92, 93, 94, 95, 96, 98, 99, 100, 101, 102, 103, 104, 105, 106, 107, 108, 109, 265, 277, 360, 375, 380, 391, 405, 412, 415, 416, 417, 418, 419, 421, 422, 423, 424, 425, 426, 427, 428, 429, 430, 431, 432, 433, 434, 435, 436, 437, 438, 439, 440, 441, 442
Disaster Forecasting 383, 384, 385, 386, 387, 388, 389, 390, 392, 393, 394, 395, 397, 402, 403, 404, 405, 406, 407, 408, 409
Disaster Management 384, 386, 387, 389, 391, 394, 403, 405, 406, 407, 408, 410, 440
dynamic packaging 168, 171, 175, 177, 179

E

Edible electronics 1, 2, 3, 4, 5, 6, 7, 8, 9, 10, 11, 12, 13, 14, 16, 17, 18, 19, 20, 21, 23, 24, 25, 26, 27, 28, 31, 32, 33, 35, 36, 37, 38, 39, 40, 41, 42, 43, 44, 45, 47, 127, 134, 137, 149, 153, 154, 159, 160, 161, 167, 168, 169, 170, 171, 172, 174, 175, 189, 190, 191,

192, 194, 197, 198, 199, 200, 201,
202, 203, 205, 206, 207, 208, 209,
211, 212, 214, 217, 218, 219, 220,
221, 222, 225, 226, 227, 228, 229,
231, 233, 236, 237, 238, 239, 240,
241, 242, 243, 244, 245, 246, 247,
248, 249, 250, 253, 254, 255, 256,
257, 258, 259, 260, 261, 263
Edible Energy Harvesting 111, 112, 113,
114, 115, 116, 117, 130, 136, 137,
139, 140, 141
Edible Sensors 3, 4, 8, 9, 12, 13, 14, 15, 16,
17, 18, 26, 27, 34, 35, 45, 102, 149,
153, 154, 160, 165, 168, 171, 172,
174, 220, 226, 231, 242, 248
electronic tongue 445, 448, 460, 464,
473, 475
Energy 5, 8, 29, 33, 35, 41, 42, 47, 63, 64,
73, 75, 77, 102, 105, 106, 107, 108,
109, 111, 112, 113, 114, 115, 116,
117, 118, 119, 120, 121, 122, 123,
124, 125, 126, 127, 128, 129, 130,
131, 132, 133, 134, 135, 136, 137,
138, 139, 140, 141, 142, 143, 144,
145, 146, 161, 180, 194, 200, 205,
207, 210, 214, 215, 236, 238, 239,
242, 243, 246, 248, 249, 250, 251,
252, 253, 254, 255, 256, 260, 262,
279, 281, 282, 283, 284, 285, 286,
287, 288, 290, 291, 292, 293, 294,
296, 298, 306, 308, 310, 311, 313,
314, 317, 321, 324, 325, 327, 363,
379, 381, 411, 461
environmental sciences 415, 416, 423,
424, 440, 441
Environment Monitoring 197, 206
Extended shelf life 82, 97, 175, 182, 366

F

Food Packaging 1, 2, 4, 5, 6, 7, 8, 9, 10,
11, 12, 13, 14, 15, 17, 18, 19, 20, 21,
22, 23, 24, 25, 26, 33, 34, 35, 39, 44,
47, 160, 167, 168, 169, 170, 171, 172,
174, 175, 176, 177, 179, 180, 181, 182,
187, 189, 190, 191, 192, 193, 206, 236
Food Quality Monitoring 23, 24, 25, 26, 27,
28, 31, 32, 35, 36, 37, 40, 41, 42, 44,
45, 49, 50, 51, 52, 53, 55, 56, 58, 59,
61, 64, 67, 68, 69, 70, 71, 72, 73, 74,
75, 76, 79, 80, 81, 82, 83, 85, 87, 88,
89, 90, 94, 96, 98, 101, 103, 104, 168
Food Safety 2, 4, 5, 7, 8, 9, 10, 12, 13, 14,
15, 18, 19, 23, 24, 28, 32, 34, 38, 39,
42, 43, 44, 45, 46, 51, 52, 58, 63, 64,
65, 68, 74, 75, 82, 85, 88, 90, 97, 101,
103, 104, 107, 127, 132, 138, 145,
150, 151, 153, 154, 159, 160, 163,
166, 168, 170, 171, 173, 174, 175,
176, 178, 179, 184, 188, 189, 191,
194, 231, 244, 260, 376
food supply chain 2, 8, 21, 39, 41, 43, 47,
58, 72, 74, 80, 81, 82, 86, 87, 91, 94,
112, 128, 142, 151, 153, 162, 183,
184, 186, 187, 191, 198
Food waste 2, 3, 4, 6, 8, 14, 15, 17, 18,
22, 34, 35, 47, 112, 146, 162, 168,
178, 182, 186, 187, 188, 191, 193,
197, 198, 214
freshness 2, 4, 5, 8, 9, 11, 12, 13, 14, 15,
16, 17, 18, 21, 23, 24, 26, 28, 30, 31,
34, 35, 37, 38, 44, 73, 79, 80, 82, 83,
85, 87, 88, 90, 94, 95, 97, 101, 102,
103, 104, 126, 152, 153, 168, 171, 173,
174, 178, 182, 185, 186, 188, 189, 190
freshness indicators 34, 168, 182
fusion 2, 3, 92, 102, 103, 389, 406, 411,
435, 436, 445, 448

G

Green Technology 198, 221

H

Hydroponics 265, 266, 267, 268, 273,
277, 279, 280

I

Identity Management 333, 334, 335, 341,
343, 344
Ingestible 2, 19, 111, 113, 114, 125, 127,
135, 138, 163, 170, 175, 197, 198,

199, 200, 203, 205, 207, 213, 218, 221, 225, 226, 228, 230, 231, 232, 234, 235, 237, 238, 239, 242, 243, 244, 245, 247, 248, 249, 250, 251, 252, 253, 254, 255, 256, 259, 260, 261, 262, 263, 264

Internet of Things 23, 25, 30, 35, 44, 59, 64, 75, 76, 77, 80, 85, 86, 106, 109, 114, 140, 152, 169, 183, 192, 194, 219, 237, 238, 239, 263, 279, 281, 282, 283, 287, 289, 291, 293, 294, 298, 299, 301, 302, 303, 305, 306, 307, 309, 310, 313, 314, 315, 316, 318, 320, 324, 326, 328, 330, 331, 332, 333, 334, 335, 336, 338, 339, 340, 342, 343, 344, 345, 346, 348, 349, 350, 379, 409, 410

IoT 23, 25, 27, 28, 30, 35, 36, 44, 46, 54, 59, 60, 64, 71, 73, 75, 76, 77, 79, 80, 81, 82, 83, 85, 86, 87, 88, 90, 91, 92, 93, 94, 95, 98, 99, 100, 101, 102, 103, 104, 105, 106, 107, 108, 109, 111, 114, 140, 149, 151, 152, 162, 163, 168, 169, 170, 175, 179, 182, 183, 184, 185, 186, 187, 188, 190, 191, 194, 195, 213, 220, 237, 238, 239, 240, 244, 245, 246, 247, 249, 250, 253, 254, 255, 256, 257, 258, 259, 260, 261, 262, 264, 265, 266, 269, 270, 279, 280, 281, 282, 286, 287, 288, 289, 290, 291, 292, 294, 295, 296, 297, 298, 299, 300, 301, 302, 305, 306, 307, 308, 309, 310, 311, 312, 313, 314, 315, 316, 317, 318, 319, 320, 321, 322, 323, 324, 325, 326, 327, 328, 329, 330, 331, 333, 334, 335, 336, 337, 338, 340, 341, 342, 343, 344, 345, 346, 347, 348, 349, 375, 379, 380, 381, 382, 384, 385, 386, 387, 388, 389, 394, 395, 396, 397, 399, 400, 402, 403, 406, 407, 409, 410, 436, 441

IoT-enabled Supply Chain 306

IoT Integration 28, 36, 91, 246, 258, 295

L

LSTM 65, 110, 265, 266, 270, 271, 272, 273, 275, 276, 277, 422, 438, 442

M

machine learning 27, 28, 31, 37, 46, 65, 72, 80, 88, 95, 96, 105, 106, 107, 108, 109, 136, 151, 152, 155, 178, 194, 195, 212, 220, 229, 296, 300, 302, 308, 329, 331, 340, 342, 344, 347, 349, 352, 356, 358, 361, 365, 366, 369, 375, 381, 384, 386, 388, 389, 390, 391, 392, 393, 396, 398, 399, 400, 403, 405, 407, 408, 410, 411, 412, 413, 415, 416, 418, 419, 434, 441, 442, 457, 476

Machine vision 445, 446, 447, 448, 449, 450, 456, 460, 461, 462, 464, 465, 471, 472, 473

micro-controllers 51, 52, 53, 60, 64, 68

N

nutritional monitoring 174

P

pH levels 5, 14, 15, 17, 18, 60, 127, 152, 153, 173, 226, 242, 245, 256, 257

precision farming 354, 356, 357, 359, 360, 370

predictive analytics 65, 66, 72, 73, 82, 88, 90, 93, 101, 104, 294, 307, 326, 352, 369, 370, 390, 391, 392, 393, 407, 415, 416, 417, 418, 439, 440

Predictive Modeling 27, 28, 31, 37, 44, 65, 384, 386, 392, 398, 400, 401

product authentication 168, 174

productivity 46, 129, 143, 265, 267, 277, 298, 351, 352, 353, 354, 355, 358, 359, 363, 364, 365, 367, 369, 370, 371, 372, 374, 375, 376, 378, 461

Q

quality control 43, 51, 57, 63, 72, 77, 81, 82, 87, 105, 106, 132, 135, 187, 319, 367, 402, 404, 459, 460, 472

R

real-time data 2, 24, 33, 44, 51, 52, 56, 58, 60, 61, 62, 72, 80, 82, 87, 90, 93, 94, 95, 97, 101, 104, 127, 173, 183, 184, 185, 186, 190, 238, 242, 244, 245, 246, 247, 254, 256, 257, 258, 282, 291, 292, 294, 307, 325, 326, 359, 365, 369, 384, 385, 387, 388, 389, 391, 392, 394, 396, 401, 407
Real-time monitoring 2, 4, 5, 8, 12, 13, 14, 17, 18, 23, 24, 30, 33, 35, 44, 50, 59, 61, 71, 73, 81, 82, 86, 95, 97, 102, 104, 151, 153, 154, 163, 168, 170, 171, 173, 174, 175, 176, 177, 190, 239, 242, 245, 247, 258, 305, 308, 326, 352, 365, 367, 396
resource optimization 399
RGB model 465, 466
Risk Assessment 96, 132, 371, 384, 386, 397, 400, 401
robotics 59, 76, 107, 167, 168, 169, 170, 171, 175, 176, 177, 178, 179, 189, 190, 191, 193, 194, 226, 351, 352, 353, 354, 355, 358, 359, 360, 361, 362, 363, 364, 365, 366, 367, 371, 372, 373, 374, 375, 376, 377, 378, 380, 382, 410, 412

S

safety 2, 4, 5, 6, 7, 8, 9, 10, 11, 12, 13, 14, 15, 16, 17, 18, 19, 20, 21, 23, 24, 26, 27, 28, 30, 32, 33, 34, 35, 37, 38, 39, 40, 41, 42, 43, 44, 45, 46, 50, 51, 52, 53, 56, 57, 58, 61, 63, 64, 65, 67, 68, 70, 72, 74, 75, 79, 80, 82, 83, 85, 87, 88, 90, 94, 95, 97, 101, 103, 104, 106, 107, 111, 112, 115, 116, 121, 122, 127, 128, 131, 132, 134, 135, 138, 139, 140, 141, 145, 150, 151, 152, 153, 154, 159, 160, 162, 163, 165, 166, 168, 169, 170, 171, 172, 173, 174, 175, 176, 177, 178, 179, 185, 187, 188, 189, 190, 191, 193, 194, 209, 213, 218, 228, 231, 233, 238, 239, 243, 244, 249, 250, 259, 260, 284, 290, 295, 376, 377, 401
safety standards 8, 50, 53, 80, 122, 132, 135, 377
security 20, 28, 32, 41, 42, 43, 44, 71, 72, 75, 82, 83, 98, 99, 100, 101, 102, 103, 104, 106, 107, 129, 138, 151, 153, 166, 259, 262, 281, 282, 283, 286, 287, 288, 289, 290, 291, 292, 293, 294, 295, 296, 297, 298, 299, 300, 301, 302, 305, 306, 307, 308, 309, 310, 311, 312, 313, 314, 315, 316, 317, 318, 319, 320, 321, 322, 323, 324, 325, 326, 327, 328, 329, 330, 331, 333, 334, 335, 336, 337, 338, 339, 340, 341, 342, 343, 344, 345, 346, 347, 348, 349, 352, 354, 359, 371, 374, 378, 385, 402, 403, 404, 405, 406, 407, 408, 409, 411, 417, 435
sensors 1, 2, 3, 4, 5, 6, 8, 9, 12, 13, 14, 15, 16, 17, 18, 19, 20, 23, 25, 26, 27, 28, 29, 30, 31, 32, 33, 34, 35, 36, 37, 39, 40, 43, 44, 45, 50, 51, 52, 53, 54, 55, 56, 57, 58, 59, 60, 61, 62, 63, 64, 67, 68, 69, 71, 72, 73, 75, 76, 77, 80, 81, 82, 83, 85, 86, 87, 88, 89, 90, 91, 92, 93, 95, 96, 98, 99, 100, 101, 102, 103, 104, 106, 112, 124, 127, 128, 136, 146, 149, 151, 152, 153, 154, 155, 156, 157, 159, 160, 162, 163, 164, 165, 166, 168, 169, 170, 171, 172, 173, 174, 176, 177, 178, 182, 185, 186, 187, 188, 189, 190, 192, 193, 199, 200, 204, 205, 206, 209, 212, 213, 214, 217, 218, 220, 221, 226, 227, 228, 229, 231, 233, 235, 236, 237, 238, 239, 242, 245, 248, 250, 253, 256, 258, 262, 263, 265, 266, 267, 269, 270, 281, 282, 291, 292, 294, 315, 323, 324, 325, 334, 352, 355, 356, 357, 358, 359, 364, 365, 368, 372, 375, 376, 384, 385,

386, 387, 388, 389, 394, 395, 396, 397, 398, 399, 400, 402, 403, 406, 407, 408, 416, 425, 426, 436, 446, 460, 472, 473, 475
shelf life 2, 4, 7, 8, 11, 26, 33, 34, 35, 44, 51, 57, 80, 83, 97, 104, 119, 120, 126, 127, 129, 136, 149, 150, 168, 169, 175, 176, 177, 178, 180, 181, 182, 183, 188, 191, 366
smart agriculture 76, 107, 108, 351, 352, 375, 378, 379, 380, 381, 382
Smart Food 14, 15, 17, 79, 109, 111, 137, 167, 168, 169, 170, 171, 172, 174, 179, 187, 189, 190, 191
Smart Food Packaging 14, 15, 17, 167, 168, 169, 170, 171, 172, 174, 179, 187, 189, 190, 191
smart grid 75, 105, 194, 281, 282, 283, 284, 285, 286, 287, 288, 289, 291, 292, 293, 294, 295, 296, 297, 298, 305, 306, 307, 308, 309, 310, 311, 312, 313, 314, 315, 316, 317, 318, 320, 321, 322, 323, 324, 326, 327
Smart Grid Environment 305, 306, 307, 308, 309, 310, 313, 317, 324, 326
smart packaging 4, 16, 17, 18, 19, 73, 74, 75, 87, 94, 102, 124, 125, 126, 138, 163, 165, 169, 170, 175, 177, 178, 179, 182, 183, 184, 185, 186, 187, 188, 189, 190, 191, 192
smart sensors 80, 218, 229, 294
Smart Technology 1, 383, 387, 403
spoilage detection 12, 15, 26, 52, 149, 151, 152, 153, 154, 155, 156, 157, 159, 163, 164, 166, 170
Supply Chain Ecosystem Transformation 305, 307, 309, 324, 326
sustainability 1, 2, 3, 6, 7, 9, 12, 14, 16, 17, 18, 19, 21, 28, 41, 42, 43, 44, 76, 82, 103, 104, 106, 109, 112, 114, 116, 117, 131, 133, 138, 140, 141, 146, 168, 169, 170, 171, 177, 178, 179, 182, 186, 187, 189, 191, 194, 206, 207, 208, 212, 215, 222, 227, 231, 232, 233, 236, 265, 282, 290, 291, 292, 298, 307, 309, 310, 311, 314, 351, 353, 354, 355, 359, 361, 363, 367, 370, 371, 374, 378, 379, 382, 407, 409, 411, 446, 447
Sustainable packaging 5, 14, 18, 176

T

Technology 1, 2, 3, 6, 7, 8, 11, 12, 13, 14, 15, 16, 17, 18, 19, 20, 21, 22, 23, 28, 35, 38, 39, 40, 41, 42, 43, 44, 45, 46, 47, 49, 50, 52, 53, 55, 61, 64, 71, 72, 74, 76, 79, 82, 85, 100, 101, 106, 107, 111, 112, 113, 114, 116, 122, 126, 127, 128, 132, 135, 137, 138, 139, 140, 142, 146, 151, 153, 154, 162, 164, 165, 166, 167, 170, 171, 175, 176, 178, 181, 184, 185, 188, 190, 191, 193, 194, 198, 208, 211, 212, 217, 218, 220, 221, 222, 231, 237, 238, 239, 244, 253, 257, 260, 261, 262, 263, 265, 279, 280, 282, 284, 285, 286, 287, 288, 291, 296, 299, 300, 301, 302, 303, 310, 325, 327, 328, 329, 330, 331, 332, 333, 334, 335, 337, 341, 344, 346, 347, 348, 349, 350, 351, 352, 354, 357, 358, 361, 366, 372, 373, 375, 376, 377, 383, 386, 387, 388, 389, 393, 395, 397, 402, 403, 404, 405, 406, 407, 435, 441, 442, 445, 447, 472, 473, 476
temperature control 56, 63